Informatik-Fachberichte 269

Herausgeber: W. Brauer
im Auftrag der Gesellschaft für Informatik (GI)

Informatik-Fachberichte 269

Herausgeber: W. Brauer
im Auftrag der Gesellschaft für Informatik (GI)

G. Hommel (Hrsg.)

Prozeßrechensysteme '91

Automatisierungs- und Leitsysteme
in den neunziger Jahren
Berlin, 25.-27. Februar 1991

Proceedings

Springer-Verlag

Berlin Heidelberg New York London
Paris Tokyo Hong Kong Barcelona

Herausgeber

G. Hommel
TU Berlin, Institut für Technische Informatik
Franklinstr. 28/29, W-1000 Berlin 10

CR Subject Classification (1987): C.3, D.2, J.7

ISBN-13: 978-3-540-53808-0   e-ISBN-13: 978-3-642-76501-8
DOI: 10.1007/978-3-642-76501-8

2133/3140-543210 – Gedruckt auf säurefreiem Papier

Vorsitzender des Programmkomitees:

Günter Hommel

Vorsitzender des Organisationskomitees:

Alois Knoll

Tagungsleitung:

Günter Hommel

Programmkomitee:

G. Duelen, Berlin
G. Färber, München
W. Halang, Groningen
R.G. Herrtwich, Berlin
G. Hirschberg, Erlangen
F. Hofmann, Erlangen
R. Isermann, Darmstadt
A. Küchle, Friedrichshafen
E. Kühn, München
R. Lauber, Stuttgart
W. Merker, Berlin
H. Meyerhoff, Bremen
M. Polke, Leverkusen
W. Pollmann, Sindelfingen
U. Rembold, Karlsruhe
H. Steusloff, Karlsruhe
E. Schmitter, München
G. Schweizer, Karlsruhe
H. Trauboth, Karlsruhe

Organisationskomitee:

G. Pourshirazi
K. Bartscht
U.W. Brandenburg
R.G. Herrtwich

Vorwort

Die Fachtagung „Prozeßrechensysteme '91" steht in der Tradition der Prozeßrechner-tagungen, die seit 1974 mit großem Erfolg in der Bundesrepublik Deutschland veranstaltet wurden. Diese Fachtagung wurde unter dem neuen Titel „Prozeßrechensysteme" zum ersten Mal 1988 in Stuttgart durchgeführt. Diese Titeländerung war bedingt durch den technischen Fortschritt und charakterisiert auch heute noch in zutreffender Weise, daß klassische zentrale Prozeßrechner durch immer komplexer werdende Prozeßrechensysteme abgelöst wurden. Diese bestehen aus verteilten Rechensystemen, welche durch Nachrichtenaustausch miteinander kommunizieren. Bei der Programmierung dieser Systeme hat man es also mit nebenläufigen, verteilten und echtzeitabhängigen Programmsystemen zu tun, die je nach Anwendung zusätzliche Eigenschaften, wie z. B. Fehlertoleranz und Sicherheit, aufweisen müssen.

Die Anwendung solcher Systeme bei der Automatisierung von Prozessen der Produktions-, Energie-, Verfahrens-, Umwelt-, Verkehrs- und Raumfahrttechnik steht heutzutage im Mittelpunkt des Interesses, wobei die Innovationsgeschwindigkeit zunehmend von der Einführung neuer Softwaretechnologien bestimmt wird. Objektorientierte und wissensbasierte Programmiertechniken mit den zusätzlichen oben genannten Randbedingungen der Verteiltheit, Nebenläufigkeit und Echtzeitabhängigkeit spielen hierbei eine besondere Rolle.

Andererseits hat die rationelle Erstellung solcher Systeme eine erhebliche Bedeutung. Sie wird durch eine weitgehende Normung und Standardisierung der Komponenten unterstützt. Im Kommunikationsbereich ist diese Standardisierung schon weit fortgeschritten und wird sich auch noch in Zukunft fortsetzen. Es zeichnet sich aber auch bereits ab, daß im Betriebssystembereich, ausgehend von de facto Standards für UNIX, in absehbarer Zeit ein Standard für echtzeitfähige UNIX-Betriebssysteme verfügbar sein wird. Solche Entwicklungen haben wieder zum verstärkten Nachdenken über den geeigneten Aufbau von Echtzeitbetriebssystemen geführt.

In einigen Anwendungsbereichen ist die Standardisierung so weit fortgeschritten, daß die eigentliche Programmierung vollständig in den Hintergrund tritt und Anwendungssysteme aus vorhandenen Bausteinen konfiguriert werden. Es ist absehbar, daß dieser Trend weiter anhält und auf andere Anwendungsgebiete ausgedehnt wird.

Die hier angesprochenen Aussagen für gegenwärtige und zukünftige Entwicklungen im Bereich der Prozeßrechensysteme führten dazu, daß die Tagung „Prozeßrechensysteme '91" unter das Motto „Automatisierungs- und Leitsysteme in den neunziger Jahren" gestellt wurde und folgende vier Schwerpunktthemen erhielt:

- Anforderungsdefinition, Projektierung und Konfigurierung

- Architektur und Kommunikation

- Softwaretechnologien

- Anwendungen

Neben rein technischen Aspekten werden auf dieser Tagung zusätzlich Fragen der zukünftigen Ausbildung von Informatikern und des Urheberrechts für Software angesprochen. Gerade in einem Fachgebiet, das raschem technischem Wandel unterliegt, ist es besonders wichtig, von Zeit zu Zeit Ausbildungsziele und -inhalte kritisch zu beleuchten. Die juristische Frage nach dem Urheberrecht für Software spielt schon derzeit eine wesentliche Rolle für alle softwareproduzierenden Unternehmen, wird sich aber bei der Harmonisierung des europäischen Marktes noch verschärfen.

Insgesamt soll die Tagung ein Forum schaffen zum Austausch von Forschungs- und Entwicklungsergebnissen, zur Einschätzung zukünftiger Trends und Erfordernisse sowie für Erfahrungsberichte aus der Sicht der Anwender und Planer. Sie wendet sich an Betriebs- und Automatisierungsingenieure ebenso wie an Informatiker, Entwicklungsingenieure und Kommunikationsfachleute.

Am ersten Tag der Fachtagung werden Tutorien zu den Themen

- Dialogtechniken

- Principles of Hard Real-Time Systems

- Technische Expertensysteme

- Fehlertolerante Systeme

angeboten, die einen aktuellen Überblick über den Stand der Technik vermitteln und einen besseren Zugang zu den auf der Fachtagung angesprochenen Themen geben sollen.

Das Programmkomitee stand vor der Aufgabe, aus zahlreichen guten Beiträgen eine Auswahl zu treffen, die auf hohem fachlichem Niveau stehen, ansprechend dargestellt sind und die oben genannten Schwerpunktthemen abdecken. Diese Aufgabe war insofern nicht allzu schwierig, als aufgrund des vorgegebenen Zeitrahmens nur 40% aller Beiträge akzeptiert werden konnten und damit eine gute Auswahlmöglichkeit bestand. Es muß allerdings auch gesagt werden, daß damit natürlich eine erhebliche Anzahl von Beiträgen zurückgewiesen wurde, die von ihrer Qualität her durchaus in die Tagung gepaßt hätten.

Ganz besonders freue ich mich, daß wir erstmals in der Geschichte dieser Tagungsreihe einen erheblichen Anteil von Papieren aus dem Gebiet der ehemaligen DDR erhalten haben. Die Entscheidung, die Tagung zum ersten Mal in Berlin durchzuführen, wurde bereits vor dem historischen 9. November 1989 getroffen und stellt sich jetzt als besonders glücklich heraus. Die zentrale Lage Berlins für die fünf neuen Bundesländer hat bestimmt schon bei der Einreichung im Frühjahr 1990 die Zahl der eingereichten Papiere positiv beeinflußt und wird auch allen Interessierten aus diesen Ländern die Teilnahme an der Tagung erleichtern.

An dieser Stelle danke ich allen Autorinnen und Autoren für ihre Mühe bei der Erstellung ihrer Manuskripte sehr herzlich. Wir hatten im Gegensatz zu früheren Tagungen keine Kurzfassungen, sondern nur Langfassungen der Beiträge zur Beurteilung akzeptiert, was sowohl auf der Seite der Schreibenden als auch der Lesenden mit erheblich mehr Arbeit verbunden war. Wir denken, daß sich dies insgesamt positiv auf die Qualität der Tagung ausgewirkt hat.

Allen Mitgliedern des Programmkomitees sei an dieser Stelle für ihre Mühe bei der Beurteilung der eingereichten Papiere und ihre zahlreichen konstruktiven Hinweise zur inhaltlichen und organisatorischen Gestaltung der Tagung sehr herzlich gedankt.

Weiterhin bedanke ich mich bei allen, die mich bei der Vorbereitung und Durchführung der Fachtagung unterstützt haben. Dies waren vor allem

- das Organisationskomitee unter Leitung von Herrn Dr.-Ing. Knoll und Mitarbeit von Frau Pourshirazi, Frau Bartscht, Herrn Dipl.-Inform. Brandenburg und Herrn Dr.-Ing. Herrtwich,

- der Präsident der Technischen Universität Berlin, Herr Prof. Dr.-Ing. Fricke, der die Tagung großzügig unterstützt hat,

- der Senat von Berlin, der den Empfang ausrichtete,

- der Direktor des Pergamon-Museums, der es uns ermöglichte, den Empfang in den wundervollen Museumsräumen durchzuführen,

- die industriellen Förderer der Fachtagung und nicht zuletzt

- der Springer-Verlag, der in bewährter Weise für die rechtzeitige Erstellung des Tagungsbands sorgte.

Allen Teilnehmerinnen und Teilnehmern der Tagung wünsche ich anregende Vorträge, Diskussionen und fachliche Gespräche, aber auch schöne Tage in Berlin und erfreuliche, über das Fachliche hinausgehende Kontakte.

Berlin, im November 1990   Günter Hommel

Inhaltsverzeichnis

Softwaretechnologien II

W. Mala, C. Grein; ESG München
Objektorientierter Entwurf und Programmierung von Echtzeitsystemen   113

T. Tempelmeier; Fachhochschule Rosenheim
Eine kritische Bewertung der Software-Entwurfsmethode HOOD   122

G. Schmiedel; Siemens München
Neuartige Entwicklung wissensbasierter Systeme mit DIWA   132

Softwaretechnologien III

K. Fischer; TU München
Ein Agentensystem für eine flexible Fertigungssteuerung   140

J. Lunze; Zentralinstitut für Kernforschung Rossendorf
Prozeßdiagnose auf der Grundlage einer prädikatenlogischen   150
Prozeßbeschreibung

K.-P. Aupperle, R. Rößler; ESG München
Wissensbasierte Unterstützung der Musterzulassung   160

Softwaretechnologien IV

M. Bayer, K. Gresser; TU München
Betriebssystem-Funktionen in einem fehlertoleranten Echtzeit-   169
system

R. Weigel; Siemens Nürnberg
Die Realisierung eines Echtzeit-UNIX für den Einsatz in der   179
Automatisierungstechnik

F. Saglietti; Ges. für Reaktorsicherheit Garching
The Impact of Forced Diversity on the Failure Behaviour of Multi-   193
Version Software

Softwaretechnologien V

P. Fritz; Siemens Karlsruhe
Bausteintechnik - eine durchgängige Softwaretechnologie für   206
Prozeßrechner- und Leitsysteme

B. Rehwaldt; Humboldt-Univ. Berlin
Ein Werkzeug zur funktionsbausteinorientierten Programm-   216
entwicklung für die Geräte- und Kleinautomatisierung

A. Knoll, A. Schweikard, M. Freericks; TU Berlin
Eine datenflußorientierte, funktionale Programmiersprache für   226
die Echtzeitdatenverarbeitung

Architektur und Kommunikation I

Architektur und Kommunikation II

Architektur und Kommunikation III

Anwendungen I

Anwendungen II

Anwendungen III

Eingeladene Vorträge

On the Reflective Nature of the Spring Kernel (Invited Paper)*

John A. Stankovic
Dept. of Computer and Information Science
University of Massachusetts
Amherst, Mass. 01003

1 Introduction

The Spring Kernel is a research oriented kernel designed to form the basis of a flexible, real–time operating system for complex, next generation, real-time applications. The Spring Kernel is being implemented in stages on a network of 68020 and 68030 based multiprocessors called SpringNet. Version 1 of the Kernel is now operational. While much has already been written on the Spring Kernel [10,11,12,4], the purpose of this *invited paper* is to combine ideas found in separate papers and presents them with a different emphasis. In particular, in Section 2.1 we categorize real-time systems to clearly indicate the difficulty in building such systems. Since most real-time systems aspire to being *predictable*, we discuss exactly what predictability means for such systems. We show that it is important to fully understand the implications of predictability and to not over estimate its value. In Section 2.2 we discuss how the notion of predictability can likely be achieved by an integrated approach to designing and building next generation hard real-time systems. Much of Section 2 is taken from [16]. In Section 3 we review the main ideas found in the Spring Kernel. Much of the material found in Section 3 is from [12]. In Section 4 we discuss the reflective nature of the Kernel providing the emphasis that has not appeared in previous papers. By reflection is meant the ability of the Kernel to maintain and act on information concerning the application, the environment, and the Kernel itself. This includes identifying what information is to be used, how to monitor this information, and how to dynamically adapt the system. Section 5 makes summary remarks.

---

*This work was supported by ONR under contract NOOO14-85-K-0389 and NSF under grant DCR-8500332.

2   Predictability for Real-Time Systems

2.1   Categorizing Real-Time Systems

Real-Time Systems are those systems in which the correctness of the system depends not only on the logical results of computations, but also on the time at which the results are produced. However, the full meaning of this definition takes on various subtleties depending on, at least, five dimensions. They are:

- what is the granularity of deadlines and what are the laxities for tasks,

- how strict are the deadlines,

- how reliable must the system be,

- what is the size of the system and what is the degree of interaction (coordination) among components, and

- what are the characteristics of the environment in which the system operates.

The characteristics of the environment, in turn, seem to give rise to how static or dynamic the system has to be. As can be imagined, depending on the answers to these questions many different system designs occur. However, one common denominator seems to be that all designers want their real-time system to be *predictable*. But what does predictability mean? It means that it should be possible to show, demonstrate, or prove that requirements are met subject to any assumptions made, e.g., concerning failures and workloads. In other words, predictability is always subject to the underlying assumptions being made. Let us now consider each of these 5 dimensions, in turn.

Granularity of the Deadline and Laxity of the Tasks: In a real-time system some of the tasks have deadlines and/or periodic timing constraints. If the time between when a task is activated (required to be executed) and when it must complete execution is short then the deadline is tight (i.e., the granularity of the deadline is small, or alternatively said, the deadline is close). This implies that the operating system reaction time has to be short, and the scheduling algorithm to be executed must be fast and very simple. Tight time constraints may also arise when the deadline granularity is large (i.e., from the time of activation), but the amount of computation required is also great. In other words even large granularity deadlines can be tight when the laxity (deadline minus computation time) is small. In many real-time systems tight timing constraints predominate and consequently designers focus on developing very fast and simple techniques to react to this type of task activation. In large, complex, real-time systems we find both loose and tight deadlines, and short and long laxities.

Strictness of Deadline: The strictness of the deadline refers to the value of executing a task after its deadline. For a *hard real-time task* there is no value to executing the task after the deadline has passed. A *soft real-time task* retains some diminished value after its deadline so it should still be executed. Very different techniques are usually used for hard and soft real-time tasks. In many cases hard real-time tasks are preallocated and prescheduled resulting in 100% of them making their deadlines. Soft real-time tasks are often scheduled either with non-real-time scheduling algorithms, or with algorithms that explicitly address the timing constraints, but aim only at good average case performance, or with algorithms that combine importance and timing requirements (e.g., cyclic scheduling). In complex, real-time systems both soft and hard real-time tasks exist simultaneously.

Reliability: Many real-time systems operate under severe reliability requirements. That is, if certain tasks, called critical tasks, miss their deadline then a catastrophe may occur. These tasks are usually guaranteed to make their deadlines by an off-line analysis and by schemes that reserve resources for these tasks even if it means that those resources are idle most of the time. In other words, the requirement for critical tasks should be that all of them always make their deadline (a 100% guarantee), subject to certain failure and workload assumptions. However, it is our opinion that too many systems treat all the tasks that have hard timing constraints as critical tasks (when, in fact, only some of those tasks are truly critical). This can result in erroneous requirements and an overdesigned and inflexible system. It is also common to see hard real-time tasks defined as those with both strict deadlines and of critical importance. We prefer to keep a clear separation between these notions because they are not always related.

Size of System and Degree of Coordination: Real-time systems vary considerably in size and complexity. In most current real-time systems the entire system is loaded into memory, or if there are well defined phases, each phase is loaded just prior to the beginning of the phase. In many applications, subsystems are highly independent of each other and there is limited cooperation among tasks. The ability to load entire systems into memory and to limit task interactions simplifies many aspects of building and analyzing real-time systems. However, for next generation large, complex, real-time systems, having completely resident code and highly independent tasks will not always be practical. Consequently, increased size and coordination give rise to many new problems that must be addressed and complicates the notion of predictability.

Environment: The environment in which a real-time system is to operate plays an important role in the design of the system. Many environments are very well defined (such as a lab experiment, an automobile engine, or an assembly line). Designers think of these as deterministic environments (even though they may not be intrinsically deterministic, they are forced to be). These environments give rise to small, static real-time systems where all deadlines can be guaranteed *a priori*. Even in these simple environments we need to place restrictions on the inputs. For example, the assembly line can only cope with five items per minute; given more than that, the system fails. Taking this approach enables an off-line analysis where a quantitative analysis of the timing properties can be made. Since we know exactly what to expect given the assumptions about the well defined environment we can consider these systems to be predictable.

The problem is that the approaches taken in relatively small, static systems do not scale to other environments which are larger, much more complicated, and less controllable. Consider a next generation real-time system such as a team of cooperating mobile robots on Mars. This next generation real–time system will be large, complex, distributed, adaptive, contain many types of timing constraints, need to operate in a highly non-deterministic environment, and evolves over a long system lifetime. It is much more difficult to force this environment to look deterministic - in fact, that is exactly what you do not want to do because the system would be too inflexible and would not be able to react to unexpected events or combinations of events. We consider this type of real-time system to be a dynamic real-time system operating in a non-deterministic environment. Such systems are required in many applications. It is much more difficult to define predictability for these systems and the typical semantics (all tasks make their deadlines 100% of the time) associated with the term for small static real-time systems is not sufficient. Many advances are required to address predictability of these next generation systems in a scientific manner. For example, one of the most difficult aspects will be in demonstrating that these systems meet both their overall performance requirements (which are generally average case statistics but with respect to meeting deadlines and maximizing value of executed tasks), as well as specific deadline and periodicity timing requirements of individual tasks or groups of tasks, or instances thereof. If both types of timing requirements can be demonstrated, then we can refer to the system as being predictable.

2.2 Achieving Predictability

While there may be many ways to achieve predictability in complex real-time systems, here we consider one that we call the *layer-by-layer* approach. This approach is advocated by the Spring Kernel to be discussed in the next section of this paper.

Before we discuss the layer-by-layer approach we have a few preparatory remarks. A real-time system can be considered to be composed of entities at various hardware and software layers. Broadly speaking these levels are: semiconductor components, the hardware/architecture layer, the operating system layer, and the application layer. The layer-by-layer method assumes that a higher layer is predictable, if and only if, the lower layer is predictable.

In the layer-by-layer approach, in order to obtain a predictable system, it is necessary to have a tight interaction between all aspects of the system starting from the design rules and constraints used, to the programming language, to the compiler, to the operating system, and to the hardware [15]. Then, based on a careful software and hardware design it should be possible to achieve both microscopic and macroscopic predictability. In the microscopic view, we can compute the worst case execution time of any task. This is not as simple as it first may seem. First, we require a simplified architecture so that instructions times are well defined. Second, we must be able to account for resource requirements and calls to system primitives made on behalf of this task. This can be accomplished via various techniques including a *planning* scheduler such as found in the Spring system [7,11,2,18]. In this way, the execution time of a particular invocation of a task with its resource needs can be accurately computed. In many other approaches predictability breaks down here because they have no good method for dealing with delays for resources.

Further, the layer-by-layer approach enables a macroscopic view of predictability. That is, first, we require the *macroscopic* view that *all* critical tasks will *always* make their deadlines (subject to the assumptions of the analysis). In other words, for critical tasks the requirement is a 100% guarantee. Some (small) systems force all their tasks to be critical. This amounts to overdesign, has a number of disadvantages, and will not scale to next generation, large, and dynamic systems. Second, by on-line planning and through microscopic predictability, at any point in time we know *exactly* which non-critical but hard real-time tasks in the entire system will make their deadlines given the current load. In other words we have a dynamic and macroscopic picture of the capabilities of the current state of the system with respect to timing requirements. This has several advantages with respect to fault tolerance and graceful degradation. Third, it is also possible to develop an overall quantitative, but probabilistic assessment of the performance of non-critical hard real-time tasks given expected normal and overload workloads. For example, via simulation one might compute the average percentage of non-critical tasks that make their deadlines or the expected value of tasks that make their deadline. We then would need to show that on the average these tasks meet the system requirements or add resources until this is true. Fourth, we require the macroscopic view of the capabilities of the I/O front ends. For example, it may be possible to state that the tasks on the I/O processor, scheduled according to the rate monotonic algorithm, will always make all their deadlines, because the load is less than 69% and because there are no resource conflicts.

In some circles this four pronged macroscopic view may seem unsatisfying because *everything* is not 100% guaranteed. However, we believe that this is necessary and unavoidable given that we are operating in a complex, non-deterministic environment. In these environments it seems necessary to *carefully* develop the requirements as actually needed, and then to employ different means to meet the different types of requirements. It is also important to not over estimate what a 100% guarantee for a set of tasks means. This guarantee is a logical analysis based on assumptions which may become false due to overloads, failures, or errors (such as an incorrectly specified worst case time for a task). Consequently, even with 100% guarantees there is a need for error handlers and other reliability techniques.

3 The Spring Kernel

In this Section we present the main ideas supported by the Spring Kernel. See [1,3,8,9,17] for descriptions of other interesting real-time kernels.

3.1 Types of Tasks

Our approach categorizes the types of tasks found in real–time applications depending on their interaction with and impact on the environment. This gives rise to two main criteria on the basis of which to classify tasks: importance and timing requirements. Our Kernel then treats the different classes of tasks differently thereby reducing the overall complexity.

Based on importance and timing requirements we define three types of tasks: critical tasks, essential tasks, and non-essential tasks. Tasks' timing requirements may range over a wide spectrum including hard deadlines, soft deadlines, periodic execution requirements, while other tasks may have no explicit timing requirements. *Critical* tasks are those tasks which must make their deadline, otherwise a catastrophic result might occur (missing their deadlines will contribute a minus infinity value to the system). It must be shown *a priori* that these tasks will always meet their deadlines subject to some specified number of failures. Resources will be reserved for such tasks. That is, a worst case analysis must be done for these tasks to guarantee that their deadlines are met. Using current OS paradigms such a worst case analysis, even for a small number of tasks is complex. Our new, more predictable Kernel facilitates this worst case analysis. Note that the number of truly critical tasks (even in very large systems) will be small in comparison to the total number of tasks in the system. *Essential* tasks are tasks that are necessary to the operation of the system, have specific timing constraints, and will degrade the performance of the system if their timing constraints are not met. However, essential tasks will not cause a catastrophe if they are not finished on time. There are a large number of such tasks. It is necessary to treat such tasks in a dynamic manner as it is impossible to reserve enough resources for all contingencies with respect to these tasks. Our approach applies an on-line, dynamic guarantee algorithm (see [7]) to this collection of tasks. Importance levels of essential tasks may differ. Also, the importance level of a given task may change with time. *Non-essential* tasks, whether they have deadlines or not, execute when they do not impact critical or essential tasks. Many background tasks, long range planning tasks, maintenance functions, etc. fall into this category.

3.2   The New Paradigm

In light of the complexities of real–time systems, the key to next generation real–time operating systems will be finding the correct approach to make the systems predictable yet flexible in such a way as to be able to assess the performance of the system with respect to requirements, especially timing requirements. In particular, the Spring Kernel stresses the real–time and flexibility requirements, and also contains several features to support fault tolerance. Our approach combines the following ideas resulting, we believe, in a new paradigm. They ideas are:

- resource segmentation/partitioning,

- functional partitioning,

- selective preallocation,

- *a priori* guarantee for critical tasks,

- an on-line guarantee for essential tasks,

- integrated CPU scheduling and resource allocation,

- use of the scheduler in a planning mode,

- the separation of importance and timing constraints, e.g., a deadline,

- end-to-end scheduling, and

- the utilization of significant information about tasks at *run time* including timing, task importance, fault tolerance requirements, etc. and the ability to dynamically alter this information.

The first three ideas are not new, but are quite useful and, consequently, we make use of them. We now briefly indicate how the Spring Kernel incorporates the above ideas, thereby supporting predictability and flexibility.

Resource Segmentation: All resources in the system are partitioned into well defined entities. The Kernel supports the resource abstractions of tasks and task groups, and various resource segments such as code, stacks, task control blocks (TCBs), task descriptors (TDs), local data, global data, ports, virtual disks, and non segmented memory. It is important to note that tasks and task groups (which includes the operating system primitives) are *time and resource segmented and bounded* meaning that they are composed of well defined segments and that both the worst case execution times and the worst case resource requirements for these tasks are known. Kernel primitives are also time and resource segmented and bounded. There exists a prologue (as part of an Invoke primitive) that uses formulas for worst case needs to compute the timing and resource requirements for the current invocation. Resource segmentation thereby provides the scheduling algorithm with a clear picture of all the individual resources that must be allocated and scheduled. This contributes to the *microscopic* predictability, i.e., each task upon being activated is bounded in time and resource requirements. Microscopic predictability is necessary, but not sufficient condition for overall system predictability.

Functional Partitioning: Each node in SpringNet is a multiprocessor. Structuring a Spring node as a multiprocessor with specialized components is a prerequisite for functional partitioning. There is a system processor, a communications processor, one or more application processors, and one or more front end I/O processors. Application processors execute previously guaranteed and relatively high level application tasks. System processors offload the scheduling algorithm and other OS overhead from the application tasks both for speed, and so **that external interrupts and OS overhead do not cause uncertainty in executing guaranteed tasks.** Upon failure of the system processor, one of the application processors can become the systems processor. Functional partitioning provides many benefits including dividing a large problem into more manageable pieces, allowing us to treat critical, essential and non-essential tasks differently, allowing different solutions for different levels of granularity of timing constraints, and enabling the isolation of tasks that run on the application processors from unpredictable interrupts generated by the non-deterministic environment. This latter point is extremely important and together with our *guarantee algorithm* allows us to construct a more macroscopic view of predictable performance since the collection of tasks currently guaranteed to execute by their deadline are not subject to unknown, environment-driven interrupts. The unexpected interrupts can occur, but they affect the current tasks in a very predictable manner due to our on-line guarantee approach.

The I/O subsystem is partitioned away from the Spring Kernel and it handles non-critical I/O, slow I/O devices, and fast sensors. This shields the application processors from external processors. Many real–time constraints arise due to I/O devices including sensors and actuators. The set of I/O devices that exist for a given application will be relatively static in most systems. Even if the I/O devices change, since they can be partitioned from the application processors and changes to them are isolated, these changes have minimal impact on the Kernel. Special independent driver processes must be designed to handle the special timing needs of these devices. In Spring we separate slow and fast I/O devices. Slow I/O devices are multiplexed through a front end dedicated I/O processor. System support for this is predetermined and not part of the dynamic on-line guarantee. For example, the I/O processor might be running a cyclic scheduler or a rate monotonic scheduler, etc. However, the slow I/O devices might invoke a task which does have a deadline and which is subject to the guarantee. Fast I/O devices such as sensors are handled with a dedicated processor, or have dedicated cycles on a given processor or bus. The processors might be front-end I/O processors or one or more of the application processors. The fast I/O devices are critical since they interact more closely with the real–time application and have tight time constraints. They might invoke subsequent higher level real–time tasks. However, it is precisely because of the tight timing constraints and the relatively static nature of the collection of sensors that we preallocate resources for the fast I/O sensors. In summary, our strategy suggests that some of the tasks which have real–time constraints can be dealt with statically, and others by a dynamic scheduling algorithm in the front-end. This leaves a smaller number of tasks which typically have higher levels of functionality and can tolerate a greater latency, for the dynamic, on-line guarantee routine.

Selective Preallocation: Critical tasks and tasks with very fast I/O requirements are preallocated. Further, the Spring Kernel contains task management primitives that utilize the notion of preallocation where possible to improve speed and to eliminate unpredictable delays. For example, all essential tasks are memory resident, or are made memory resident before they can be invoked. In addition, a system initialization program loads code, and sets up stacks, TCBs, TDs, local data, global data, ports, virtual disks and non segmented memory using the Kernel primitives. Multiple instances of a task or task group may be created at initialization time and multiple free TCBs, TDs, ports and virtual disks may also be created at initialization time. Subsequently, dynamic operation of the system only needs to free and allocate (the first item on a list) these segments rather than creating them. While facilities also exist for dynamically creating new segments of any type, such facilities should not be used under hard real–time constraints. Using this approach, the system can be fast and predictable, yet still be flexible enough to accommodate major changes in non hard real–time mode.

A Priori Guarantee for Critical Tasks: The notion of guaranteeing timing constraints is central to our approach. However, because we are dealing with large, complex systems in non-deterministic environments, the guarantee is separated into two main parts: an *a priori* guarantee for critical tasks and an on-line guarantee for essential tasks. All critical tasks are guaranteed *a priori* and resources are reserved for them either in dedicated processors, or as a dedicated collection of resource slices on the application processors (this is part of the selective preallocation policy used in Spring). Resources are provided under specified failure assumptions. For example, if t Byzantine processor failures should be accommodated, resources are provided for $2t + 1$ replicates of a task. Hence, critical tasks are guaranteed for the entire lifetime of the system. While *a priori* dedicating resources to critical tasks is, of course, not flexible, due to the importance of these tasks, we have no other choice! On the positive side, typically, the ratio of critical tasks to essential tasks is very small.

On-line Guarantee for Essential Tasks: Due to the large numbers of essential tasks and to the extremely large number of their possible invocation orders, preallocation of resources to essential tasks is not possible due to cost, nor desirable due to its inflexibility. Hence, this class of tasks is guaranteed on-line via the algorithm presented in [7]. This allows for many task invocation scenarios to be handled dynamically (partially supporting the flexibility requirement). However, the notion of on-line guarantee has a very specific meaning as described in the first itemized point below. The basic notion and properties of guarantee for essential tasks have been developed elsewhere [7] and have the following characteristics:

- it allows the unique abstraction that at any point in time the operating system knows exactly which tasks have been guaranteed to make their deadlines[1], what, where and when spare resources exist or will exist, a complete schedule for the guaranteed tasks, and which tasks are running under non-guaranteed assumptions. However, because of the non-deterministic environment the capabilities of the system may change over time, so the on-line guarantee for essential tasks is an *instantaneous* guarantee that refers to the current state. Consequently, at any point in time we have the *macroscopic* view that *all* critical tasks will make their deadlines and we know *exactly* which essential tasks will make their deadlines given the current load[2],

- conflicts over resources are *avoided* thereby eliminating the random nature of waiting for resources found in timesharing operating systems (this same feature also tends to minimize context switches since tasks are not being context switched to wait for resources). Basically, resource conflicts are solved by scheduling tasks at different times if they contend for a given resource,

- there is a separation of dispatching and guarantee allowing these system functions to run in parallel; the dispatcher is always working with a set of tasks which have been previously guaranteed to make their deadlines and the guarantee routine operates on the current set of guaranteed tasks plus any newly invoked tasks,

- provides early notification; by performing the guarantee calculation when a task arrives there may be time to reallocate the task to another host of the system via the distributed scheduling module of the scheduling approach; early notification also has *fault tolerance* implications in that it is now possible to run alternative error handling tasks early, before a deadline is missed,

- within this approach there is the notion of still "possibly" meeting the deadline even if the task is not guaranteed, that is, if a task is not guaranteed it could receive idle cycles at this node, and, in parallel, there can be an attempt to get the task guaranteed on another host of the system subject to location dependent constraints, or based on the fault tolerance semantics of the task, various alternatives could be invoked,

- the guarantee routine supports the co-existence of real–time and non real–time tasks, and note that this is non-trivial when non real–time tasks might use some of the same resources as real–time tasks,

- the guarantee can be subject to computation time requirements, deadline or periodic time constraints, resource requirements where resources are segmented, im-

---

[1]In contrast, current real–time scheduling algorithms, such as earliest deadline, have no global knowledge of the task set nor of the system's ability to meet deadlines; they only know which task to run next.

[2]It is also possible to develop an overall quantitative, but probabilistic assessment of the performance of essential tasks. For example, given expected normal and overload workloads, we can compute the average percentage of essential tasks that are guaranteed, i.e., make their deadlines.

portance levels for tasks, precedence constraints, I/O requirements, etc. depending on the specific guarantee algorithm being used in a given system.

Integrated CPU Scheduling and Resource Allocation: Current real–time scheduling algorithms schedule the CPU independently of other resources. For example, consider a typical real–time scheduling algorithm, earliest deadline first. Scheduling a task which has the earliest deadline does no good if it subsequently blocks because a resource it requires is unavailable. Our approach integrates CPU scheduling and resource allocation so that this blocking never occurs. Scheduling is an integral part of the Kernel and the abstraction provided is one of a guaranteed task set.

By integrating CPU scheduling and resource allocation at run time, we are able to understand (at each point in time), the current resource contention and completely control it so that task performance with respect to deadlines is predictable, rather than letting resource contention occur in a random pattern resulting in an unpredictable system.

Use of Scheduler in Planning Mode: Another important feature of our scheduling approach is how and when we use the scheduler, i.e., we use it in a *planning* mode when a new task is invoked. When a new task is invoked, the scheduler attempts to plan a schedule for it and some number of other tasks so that all tasks can make their deadlines. This enables our system to understand the total load of the system and to make intelligent decisions when a guarantee cannot be made, e.g. see the next point below. This is at odds with other real–time scheduling algorithms which, as mentioned earlier, have a myopic view of the set of tasks. That is, these algorithms only know *which task to run next* and have no understanding of the total load or current capabilities of the system. This planning is done on the system processor in parallel with the previously guaranteed tasks so it must account for those tasks which may be completed before it itself completes.

Separation of Importance and Deadline: A major advantage of our approach is that we can separate deadlines from importance. This is necessary since importance and deadline are orthogonal task characteristics. Again, all critical tasks are of the utmost importance and are *a priori* scheduled. Essential tasks are not critical, but each is assigned a level of importance which may vary as system conditions change. To maximize the value of executed tasks, *all* critical tasks should make their deadlines and as many essential tasks as possible should also make their deadlines. Ideally, if any essential tasks cannot make their deadlines, then those tasks which do not execute should be the least important ones. In the first phase of the guarantee algorithm, scheduling is done ignoring importance. If all tasks are guaranteed then the importance value plays no part. On the other hand, when a newly invoked essential task is not guaranteed, then the guarantee routine will remove the least important tasks from the system task table if those preemptions contribute to the subsequent guarantee of the new task. The low importance eliminated tasks, or the original task, if none, are then subject to distributed scheduling. Various algorithms for this combination of deadlines and importance have been developed and analyzed [2]. It is important to point out that our approach is much more flexible at handling the combination of timing and importance than a static priority scheduling mechanism typically found in real–time systems. For example, using static priority scheduling a designer may have a task with a short deadline and low importance, and another task with a long deadline and high importance. For average loads it is usually acceptable to assign the short deadline task the higher priority, and under these loads all tasks probably make their deadlines. However, if there is overload, it will be the high importance task which ends up missing its deadline. This condition would not occur with our scheme.

End-to-End Scheduling: Most *application* level functions (such as stop the robot before it hits the wall) which must be accomplished under a timing constraint are actually composed of a set of smaller dispatchable tasks. Previous real–time kernels do not provide support for a collection of tasks with a single deadline. The Spring Kernel supports tasks and task groups and is currently developing support for dependent task groups. A task group is a collection of simple tasks that have precedence constraints among themselves, but have a single deadline. Each task acquires resources before it begins and can release the resources upon its completion. For task groups, it is assumed that when the task group is invoked the worst case computation time and resource requirements of each task can be determined. A dependent task group is the same as a task group except that computation time and resource requirements of only those tasks with no precedence constraints are known at invocation time. Needs of the remaining tasks of the dependent group can only be known when all preceding tasks are completed. The dependent task group requires some special handling with respect to guarantees which we have not done at this time. Precedence constraints are used to model end-to-end timing constraints both for a single node and across nodes and the scheduling heuristic we use can account for precedence constraints.

Dynamic Utilization of Task Information: Information about tasks and task groups is retained at run time and includes formulas describing worst case execution time, deadlines or other timing requirements, importance level, precedence constraints, resource requirements, fault tolerance requirements, task group information, etc. The Kernel then dynamically utilizes this information to guarantee timing and other requirements of the system. In other words, our approach retains significant amounts of semantic information about a task or task group which can be utilized at run time. Kernel primitives exist to inquire about this information and to dynamically alter the information. This enhances the flexibility of the system.

4 The Reflective Nature of the Spring Kernel

Computational reflection is normally defined as an activity performed by a computational system when doing computation about (and by that, possibly affecting) its own computation. In our context reflection means the ability of the Kernel to maintain and act on information concerning the application, the environment, and the Kernel itself.

This includes identifying what information is to be used, how to monitor this information, and how to dynamically adapt the system. Examples of reflection include keeping and using performance statistics, keeping information for debugging or on-line decision making, performing computation to decide what computation to pursue next (or for the next interval), self-optimization, and self-modification. Features covering all these examples appear or are planned for the Spring Kernel [14,6]. Much of the reflective capabilities of the Kernel arise from the task management and scheduling features of the Kernel. We restrict our discussion to these areas.

4.1   Reflection in Task Management and Scheduling

Tasks arise when real-time programs - specified in the form of communicating processes - are decomposed into schedulable entities, namely tasks, with precedence relationships, resource requirements, fault tolerance requirements, importance levels, and timing constraints. The task management primitives support executable and guaranteeable entities called tasks and task groups. A task consists of reentrant code, local data, global data, a stack, a TD, and a TCB. Multiple instances of a task may be invoked. In this case the (reentrant) code and task descriptor are shared. A task group is a collection of simple tasks that have precedence constraints among themselves, but have a single group deadline. Each task acquires resources before it begins and releases the resources upon its completion. For task groups, it is assumed that when the task group is invoked, all tasks in the group can be sized (this means that the worst case computation time and resource requirements of each task can be determined at invocation time). More flexible types of task groups are currently being investigated.

We require that designers follow strict rules and guidelines in programming. The purpose is to facilitate subsequent analysis of timing requirements. In order to support on-line analysis we use reflection. We require that tasks be characterized by:

- C (a worst case execution time - may be a formula that depends on various input data and/or state information pertaining to a specific task invocation),

- D (Deadline) or period or other real–time constraint

- preemptive or non-preemptive property

- maximum number and type of resources needed (this includes memory segments, ports, etc.)

- type: critical, essential, or non-essential

- importance level for essential and non-essential tasks (this is an indication of the value imparted to the system by the execution of the task)

- incremental task or not (incremental tasks compute an initial answer quickly and then continue to refine the answer for the rest of its requested computation time)

- location of task copies indicating the various nodes in the distributed system and on which processor of each node where the task resides,

- Group ID, if any (tasks may be part of a task group)

- precedence graph (describes the required precedence among tasks in a task group or a dependent task group)

- communication graph (list of tasks with which a task communicates), and type of communication (asynchronous or synchronous)

- a fault model (described below).

All the above information concerning a task is maintained in the task descriptor (TD) and used as part of the on-line decision making. In other words, the system maintains information about itself and subsequently uses that information to make more intelligent decisions. Much of the above information is also maintained in the task control block (TCB) with the difference being that the information in the task control block is specific to a particular instance of the task. For example, a task descriptor might indicate that the worst case execution time for TASK A is $5z$ milliseconds where z is the number of input data items at the time the task is invoked. At invocation time a short procedure is executed to compute the actual worst case time for this module and this value is then inserted into the TCB. The guarantee is then performed against this specific task instance. All the other fields dealing with time, computation, resources or importance are handled in a similar way. Further, all these fields can be modified (via the Modify primitive), e.g., the importance of a task may vary depending on the overall state of the system or the environment.

Our scheduling approach separates policy from mechanism and is composed of 4 levels. The 3 highest levels exhibit many of the features of reflection, while the lowest level does not. At the lowest level multiple dispatchers exist; one type of dispatcher runs on each of the application processors, and another type executes on the system processor. The *application dispatchers* simply remove the next (ready) task from a system task table (STT) that contains previously guaranteed tasks arranged in the proper order for each application processor. The *system dispatcher* provides for the periodic execution of systems tasks, and asynchronous invocation when it can determine that allowing these extra invocations will not adversely affect guaranteed tasks, nor the minimum guaranteed periodic rate of other system tasks. Asynchronous invocation of system tasks are ordered by importance, e.g., the local scheduler is of higher importance than the meta level controller (see below).

The three higher level scheduling modules are executed on the system processor. The second level is a *local scheduler*. The local scheduler is responsible for locally *guaranteeing* that a new task or task group can make its deadline, and for ordering the tasks properly in the STT using information about the environment, the system state, and the semantics of the application task. Another reflective property is that we are planning the computation of the system (far) out into the future. The local scheduler, when invoked, attempts to guarantee any new tasks or task groups that arrived since its last activation. It guarantees the new task if the task can be scheduled to complete before its deadline and if the previously guaranteed tasks are not jeopardized by the execution of the new task. If it cannot make the guarantee then it invokes the fault tolerance model for this task which indicates one of the following alternatives:

- try to guarantee a shorter, error handler,

- remove low importance tasks until guaranteed,

- perform distributed scheduling,

- try to execute without a guarantee, or

- abort.

The third scheduling level is the *distributed scheduler* which attempts to find a node for execution for any task or for components of a task group that have to execute on different nodes [5], because they cannot be locally guaranteed. The fourth level is a *Meta Level Controller* (MLC) which has the responsibility of adapting various parameters or switching scheduling algorithms for both local and distributed scheduling by noticing significant changes in the environment. The MLC is a decentralized controller based on heuristics with the purpose of controlling the scheduling itself. The capabilities of the MLC support some of the adaptability and flexibility needs of next generation real–time systems. The MLC provides a user interface allowing dynamic changes to meta-level control policies as well as providing a means for the user to provide an even higher level of control. See [6] for more details. The MLC exhibits the self-optimization and self-modification features of reflection.

In the future, we plan to have the scheduler dynamically monitor the success ratio of essential tasks. If the success ratio drops below a certain level, immediate and long term corrective actions can be planned, e.g., performing distributed reallocation and scheduling, or simply announcing degraded service available from the system, or notifying the system managers that additional processing power must be added to the system.

To these basic scheduling modules we plan to add one or more Time Planners which have access to the task descriptors, and a condition monitor facility [14]. The Time Planners could be considered versions of the local scheduler being invoked for different purposes, i.e., to suggest tradeoffs, to perform planning of future schedules for tasks, or to suggest causes of the problems in the current plan. Multiple Time Planners may be invoked simultaneously, each assessing different alternatives. An example of a tradeoff that might be identified is either Tasks 1,2,3 and 4 can complete, or Tasks 1,2, and 3 can complete together with short error handlers for Tasks 4 and 5. The condition monitor facility permits various (situations,action) pairs to be defined and monitored. The set of situations to monitor and the actions to perform are dynamically modifiable and can relate to the state of the environment or the system. Various levels of importance can be attached to the different situations being modified.

5   Summary

One goal of our research is to show that a real-time system meets its timing requirements. Achieving this goal is non-trivial and requires research breakthroughs in many aspects of system design and implementation. For example, good design rules and constraints must be used to guide real–time system developers so that subsequent implementation and *analysis* can be facilitated. Programming language features must be tailored to these rules and constraints, must limit its features to enhance predictability, and must provide the ability to specify timing, fault tolerance and other information for subsequent use at run time. Execution time of each primitive of the Kernel must be bounded and predictable, and the operating system should provide explicit support for all the requirements including the real–time requirements. Further, the operating system should support flexibility, adaptability and long-lived systems. We believe that reflection is an important property to achieve these goals. The hardware must also adhere to the rules and constraints and be simple enough so that predictable timing information can be obtained, e.g., caching, memory refresh and wait states, pipelining, and some complex instructions all contribute to timing analysis difficulties. An insidious aspect of critical real–time systems, especially with respect to the real–time requirements, is that the weakest link in the entire system can undermine careful design and analysis at other levels. Our research is attempting to address these issues in an integrated fashion.

References

[1] Alger, L. and J. Lala, "Real–Time Operating System For A Nuclear Power Plant Computer," *Proc. 1986 Real–Time Systems Symposium*, Dec. 1986.

[2] Biyabani, S., J. Stankovic, and K. Ramamritham, "The Integration of Criticalness and Deadline In Scheduling Hard Real–Time Tasks," *Real–Time Systems Symposium*, Dec. 1988

[3] Holmes, V. P., D. Harris, K. Piorkowski, and G. Davidson, "Hawk: An Operating System Kernel for a Real–Time Embedded Multiprocessor," Sandia National Labs Report, 1987.

[4] Molesky, L., K. Ramamritham, C. Shen, J. Stankovic, and G. Zlokapa, "Implementing a Predictable Real-Time Multiprocessor Kernel - The Spring Kernel," extended abstract, *IEEE Workshop on Real-Time Operating Systems and Software*, Jan. 1990.

[5] Ramamritham, K., J. Stankovic, and W. Zhao, "Distributed Scheduling of Tasks With Deadlines and Resource Requirements," *IEEE Transactions on Computers*, Vol. 38, No. 8, August 1989, pp. 1110-1123.

[6] Ramamritham, K., J. Stankovic, and W. Zhao, "Meta Level Control in Distributed Real-Time Systems," *Proc. 7th Int. Conf. on Distributed Computing Systems*, Sept. 1987.

[7] Ramamritham, K., J. Stankovic, and P. Shiah, "Efficient Scheduling Algorithms for Real–Time Multiprocessor Systems," *IEEE Transactions on Parallel and Distributed Systems*, Vol. 1, No. 2, April 1990, pp. 184-194.

[8] Ready, J., "VRTX: A Real–Time Operating System for Embedded Microprocessor Applications," *IEEE Micro*, pp. 8-17, Aug. 1986.

[9] Schwan, K., W. Bo and P. Gopinath, "A High Performance, Object-Based Operating System for Real–Time, Robotics Application," *Proc. 1986 Real–Time Systems Symposium*, Dec. 1986.

[10] Stankovic, J. and K. Ramamritham, "The Design of the Spring Kernel," *Proc. 1987 Real–Time Systems Symposium*, Dec. 1987.

[11] Stankovic, J. and K. Ramamritham, "The Spring Kernel: A New Paradigm for Real–Time Operating Systems," *ACM Operating Systems Review*, Vol. 23, No. 3, July, 1989, pp. 54-71.

[12] Stankovic, J. and K. Ramamritham, "The Spring Kernel: Support for Next Generation Real-Time Systems," submitted to *IEEE Computer*, August 1990.

[13] Stankovic, J., "Misconceptions About Real–Time Computing," *IEEE Computer*, Vol. 21, No. 10, Oct. 1988.

[14] Stankovic, J., K. Ramamritham, and D. Niehaus, "On Using the Spring Kernel to Support Real-Time AI Applications," *Proc. Euromicro Workshop on Real-Time Systems*, June 1989.

[15] Stankovic, J., "The Spring Architecture," *Proc. Euromicro Workshop on Real-Time Systems*, June 1990.

[16] Stankovic, J. and K. Ramamritham, "What is Predictability for Real-Time Systems," *Real-Time Systems*, Dec. 1990.

[17] Tokuda, H., and C. Mercer, "ARTS: A Distributed Real–Time Kernel," *ACM Operating Systems Review*, Vol. 23, No. 3, July, 1989.

[18] Zhao, W., K. Ramamritham, and J. Stankovic, "Scheduling Tasks with Resource Requirements," *IEEE Trans. on Software Engineering*, May 1987.

<u>GESETZLICHER SCHUTZ VON PROGRAMMEN</u>
Hans Goldrian

Mein Thema ist der gesetzliche Schutz von Programmen, allgemeiner
gesprochen von Software. Das ist ein so weites Feld, daß man daraus
leicht eine eigene Studienrichtung machen könnte. Ich fange also
zunächst damit an, das Thema einzugrenzen.

Es gibt im Immaterialgüterrecht kaum ein anderes Rechtsgut, das
so verletzlich ist wie Software. Sei sie mit noch so hohem Aufwand
und kreativen Leistungen der Programmierer erstellt, das Ergebnis
ist leicht zu kopieren: Selbst den Vorgang des Kopierens nimmt
einem noch eine Maschine ab. Das verlockt zu mißbräuchlicher Verwendung der Kopie.

Betrachtet man die wirtschaftliche Seite des Schutzes, was ich in
diesem Referat tun will, so geht es darum, das Ergebnis der eigenen
Entwicklungsarbeit vor unberechtigtem Gebrauch durch einen Dritten,
insbesondere vor Verwertung in seiner geschäftlichen Tätigkeit zu
schützen. Diesem Wunsch kommt eine Reihe von Gesetzen entgegen,
die verschiedene Aspekte des Entwicklungsergebnisses gegen Nachahmung und Ausbeutung schützen; es gibt z.B. Patente für die zugrunde liegende Erfindung, Urheberrecht für das Programm, Waren-
zeichen für den Programmnamen.

Da ein dringender Bedarf nach einem wirksamen Schutz für Programme
und Software besteht, hat man schon jeden erdenklichen Weg für den
Rechtsschutz untersucht und teilweise auch in der Praxis erprobt,
ganz abgesehen von mechanischen Schutzmaßnahmen, die viele Hersteller in ihre Programme einbauen. Es würde zu weit führen, hier auch
nur andeutungsweise alle gesetzlichen Schutzmöglichkeiten zu behandeln. Wer daran Interesse hat, muß auf die umfangreiche Literatur
dieses Sachgebiets verwiesen werden, z.B. [1].

Besonders wertvolle Software wird dem Benutzer in der Regel durch
Überlassungsverträge lizenziert, in denen vertragsrechtliche Schutz-
klauseln enthalten sind. Solche Verträge haben allerdings den Nach-

teil, daß sie nur zwischen den Parteien wirksam sind. Nur gegenüber
bestimmten natürlichen oder juristischen Personen wirkt auch der
Schutz aus dem Wettbewerbsrecht, sei es, daß den Handlungen im
jeweiligen Fall ein spezifisches Element der Unlauterkeit anhaftet,
z.B. eine Täuschung des Marktes, sei es, daß es um Geheimnisverrat
eines Arbeitnehmers geht.

Es ist also sehr verständlich, daß man gern einen absoluten Schutz
hätte, der gegen jedermann wirkt, ohne daß persönliche Umstände
oder Eigenschaften eine Rolle spielen. Einen solchen Schutz bieten
u.a. das Urheberrecht und das Patentrecht. Mit diesen beiden Möglichkeiten, die für die Industrie von besonderer Bedeutung sind,
möchte ich mich im folgenden näher beschäftigen.

Zuvor einige warnende Worte:
Erwarten Sie bitte keine wissenschaftliche Abhandlung des Themas.
Ich fühle mich nicht dazu berufen, die gelehrte Auseinandersetzung
weiterzuführen, ob Urheberrecht bzw. Patentrecht rechtsdogmatisch
überhaupt als Schutzinstrumente geeignet sind. Ich spreche hier als
Vertreter der Industrie, genauer der Systemanbieter, und wir haben
sehr pragmatische Ansichten darüber, was wir brauchen und ob es aus
den Schutzgesetzen zu gewinnen ist. Was diese nicht ausdrücklich
ausschließen, wollen wir zur Befriedigung des Schutzbedürfnisses
benutzen. Dies führt gelegentlich dazu, daß wir zum Unbehagen der
Rechtsgelehrten in Neuland der Reichweite mancher Gesetze hinein-
argumentieren, und daß wir Handhabungen von Behörden und Urteile
von Gerichten in sehr zielgerichteter Weise kritisieren. Legen Sie
diesen Maßstab bitte auch an die folgenden Aussagen an; sie sind
interessenorientiert, wenn Sie wollen, unakademisch.

I. Patentrecht

In Kreisen der Softwarefachleute ist die Überzeugung weit verbreitet:
      Programme sind nicht patentierbar.
Diese Überzeugung ist irreführend.

Erlauben Sie mir zur Erläuterung der Situation einen kleinen Abste-
cher in das Patentrecht: Patente werden für Erfindungen erteilt,
die neu sind, auf einer erfinderischen Tätigkeit beruhen und ge-

werblich anwendbar sind. So beginnt § 1 des deutschen Patentgesetzes. Die jahrzehntelange Rechtspraxis hat eine weitere Bedingung
hinzuformuliert: Die Erfindung muß eine technische Erfindung sein,
keine ästhetische, philosophische, sprachliche oder dergl. [2]

Nicht alle Erfindungen jedoch, die die vier Bedingungen Neuheit,
erfinderische Tätigkeit, gewerbliche Anwendbarkeit und technischer
Charakter erfüllen, sind dem Patentschutz zugänglich. Im Patentge-
setz stehen Ausnahmen von der Patentierbarkeit, z.B. mathematische
Methoden, Pläne, Regeln und Verfahren für gedankliche Tätigkeit,
und ausdrücklich auch Programme für Datenverarbeitungsanlagen. Das
Gesetz definiert zwar den Begriff "Programm" nicht, aber im § 1
steht nun die Schutzausnahme für jeden lesbar, und damit scheint
der Patentschutz erledigt.

Nun hat aber § 1 noch einen Absatz 3, wonach die Schutzausnahme nur
insoweit gilt, als für die ausgenommenen Gegenstände oder Tätigkeiten als solche Schutz begehrt wird.

Nicht patentierbar sind also Programme als solche. Ein großer Teil
der Literatur auf diesem Gebiet und eine Reihe von Gerichtsent-
scheidungen bis hinauf zum Bundesgerichtshof betreffen die Frage,
was ein "Programm als solches" sein mag. Wüßte man dies genauer, so
wäre auch deutlich, was aus dem Bereich der Software patentiert
werden kann. Obwohl diese Fassung des Patentgesetzes seit über zehn
Jahren geltendes Recht ist, und obwohl nach übereinstimmender Meinung die ausdrückliche Erwähnung von Programmen bei den Schutzausnahmen nur eine Klarstellung der auch vorher geltenden Rechtslage
ist, steht eine allgemein anerkannte Definition des "Programms als
solches" bis heute noch aus.

Die Haltung der Patentbehörden gegenüber einem Programm als Schutz-
objekt leitete sich aus der langjährigen Rechtsprechung zur sogenannten bloßen Anweisung an den menschlichen Geist ab. Damit ist
folgendes gemeint:

Wenn sich eine Information ausschließlich an die Sinne bzw. das Be-
griffsvermögen des Menschen richtet, fehlt der technische Bezug und
damit die Patentierbarkeit. Da nun zumindest gedanklich ein Programm
auch von einem Menschen mit Papier und Bleistift vollzogen werden
könnte, hat man Programme in den Bereich der bloßen Anweisung an

den menschlichen Geist einzuordnen versucht. Der Computer ist dann
nichts weiter als ein Rechenhilfsmittel, wie z.B. ein Rechenschie-
ber.

Wenn Sie das Gefühl haben, daß diese Betrachtungsweise mit der Bedeutung und dem Einfluß der Informatik auf die moderne Technik
nicht vereinbar ist, so kann ich Ihnen nur recht geben. Es ist kein
Wunder, daß das, was man heute unter Software zu verstehen hat, die
Fachleute auf dem Patentgebiet in erhebliche Probleme der Beurteilung des technischen Charakters stürzt.

Woran soll sich der Praktiker halten? Er braucht eine Arbeitshypo-
these. Das Programm darf kein "Programm als solches" sein, und es
muß technischen Charakter haben. Faßt man diese beiden Bedingungen
zusammen, dann lautet die Hypothese:
> Wenn Software einen technischen Effekt hervorbringt,
> liegt patentierbare Materie vor.

Diese Behauptung blieb nicht unwidersprochen, beginnt sich aber immer mehr durchzusetzen. Im Zeitalter der Informationstechnologie ist
man immer mehr bereit, Informationen mit technischen Effekten als
Elemente der Technik zu akzeptieren. [3] Das geht inzwischen so weit,
daß selbst bei unveränderter Computerhardware ein Verfahren, das
unter der Steuerung eines neuen Programms einen neuen technischen
Effekt erzielt, als patentierbar angesehen werden kann. Dies ist
für viele Patentfachleute bis vor kurzem noch undenkbar gewesen, da
bei unveränderter Computerhardware doch nur ein "Programm als solches" vorliegen könne. Diese Anschauung, die die modernen Verwen-
dungsmöglichkeiten des Computers außer acht läßt, ist der Hintergrund der sogenannten "Kerntheorie".

Die Kerntheorie ist aus der Grundsatzentscheidung "Dispositionsprogramm" [4] des Bundesgerichtshofs abgeleitet worden, ohne daß ich
hier die Rechtsmeinung des Bundesgerichtshofs mit dem gleichsetzen
will, was bei diesen Ableitungen herausgekommen ist. Die Kerntheorie fordert, grob vereinfacht, daß der Kern der Erfindung technisch
sein müsse. Häufig ist diese Bedingung so ausgelegt worden, daß der
Kern der Erfindung gegenständlich sein, also im Hardwarebereich liegen müsse. Eine solche Auslegung ermöglichte Entscheidungen, die den
Ablauf eines Programms auf einem Computer zum bestimmungsgemäßen
Gebrauch dieses Computers erklärt haben: damit war man wieder beim

"Programm als solchem" und dem Ausschluß von Patentschutz.

Die auf dieser Basis gefällten Entscheidungen von Senaten des
Bundespatentgerichts und von Prüfungsstellen des Deutschen Patentamts haben in der Branche lebhafte Opposition erzeugt, Opposition
gegen die als antiquiert, formalistisch und neben dem Bedarf lie-
gend empfundene Beurteilung der Patentierbarkeit von Softwareerfindungen.

In den Entwicklungslaboratorien wurde immer wieder die unbefriedi-
gende Situation beklagt. Man argumentierte: solange wir technische
Aufgaben mit Hardware lösten, stand die Patentierbarkeit des Er-
gebnisses außer Frage. Werden gleichartige Aufgaben nun mit einem
Computer und geeigneter Software gelöst, ist Aufgabe und Lösung so
technisch wie zuvor und nur der Weg ein anderer. Soll eine solche
Lösung auf einmal nicht mehr patentierbar sein? Müßte man damit
rechnen, so wäre in ein paar Jahrzehnten ein Großteil der technischen Entwicklung schutzlos. Es wurde daher nachdrücklich die Forderung gestellt, daß Hardware- und Softwarelösung einer technischen
Aufgabe patentrechtlich gleichwertig behandelt werden müßten.

Diese Forderung erfüllt die Ganzheitsbetrachtung, die nicht nach
einem Kern der Erfindung fragt, sei er technisch oder gegenständlich oder nicht, sondern danach, ob der Erfindungsgegenstand in
seiner Gesamtheit eine technische Aufgabe löst, eine technische
Wirkung hervorbringt. Ein positives Indiz dafür ist z.B. auch, ob
der programmgesteuerte Computer etwas in der physikalischen Umwelt
verändert, also z.B. einen Produktionsprozeß steuert. Es genügt
z.B. nicht, daß der Computer Rechenergebnisse darstellt oder aus-
druckt. Das könnte als bloße Information an den menschlichen Geist
ausgelegt werden. Der Computer müßte auf Grund der Rechenergebnisse
in den Produktionsprozeß eingreifen. Ähnliches gilt für Computer-
einsätze in der Regeltechnik.

Unter Berücksichtigung solcher Kriterien kann man nun versuchen,
die Situation durch ein Schlagwort zu charakterisieren, das auch in
der Rechtsprechung des Bundesgerichtshofs schon angesprochen ist.
Ich verkürze also die oben aufgestellte Hypothese und sage:
          Ein technisches Programm ist patentierbar.

Dieses Schlagwort ist als Korrektur der weit verbreiteten Meinung

von der Unpatentierbarkeit von Programmen zu verstehen und gilt
natürlich nur, wenn man den Begriff "technisches Programm" auf-
faßt als "Programm, das mittels eines Computers einen technischen
Effekt hervorruft".

Ich habe schon erwähnt, daß die Rechtsprechung auf diesem Gebiet im
Fluß ist. Es existieren Prüfungsrichtlinien des Deutschen [5] und
des Europäischen [6] Patentamts, die auf dem Boden der Ganzheitsbetrachtung stehen und weitgehend akzeptierbar sind. Die Beschwerde-
kammern des Europäischen [7] Patentamts und mehrere Beschwerdesenate
des Bundespatentgerichts [8] haben Entscheidungen gefällt, die die
Prüfungsrichtlinien stützen. Andere Senate des Bundespatentgerichts [9] sind noch nicht überzeugt und berufen sich auf die aus
den Entscheidungen des Bundesgerichtshofs abgeleiteten Rechtsmei-
nungen. Derzeit laufen zwei Rechtsbeschwerden beim Bundesgerichtshof, so daß dieser Gelegenheit hätte, seine frühere Rechtsprechung
und vor allem die Auslegung dieser Rechtsprechung zu überprüfen. Da
er auch sonst die Harmonisierung rechtlicher Bestimmungen im Rahmen
des Gemeinsamen Marktes befürwortet, besteht berechtigte Hoffnung,
daß er auch auf diesem Gebiet eine Angleichung der deutschen
Rechtspraxis an die europäische einleiten wird.

Bis dahin kann nur dringend empfohlen werden, schutzwürdige Erfindungen zum Patent anzumelden, bei denen mit Hilfe von Programmen
bzw. Software eine technische Aufgabe gelöst bzw. ein technischer
Effekt erzielt wird. Vor allem den Gerichtsinstanzen muß deutlich
gemacht werden, daß ein Bedarf für den Schutz solcher Erfindungen
besteht, ja, daß sogar für weite Gebiete der Technik ein Beharren
auf überholten Anschauungen des Technikbegriffs zur Schutzverwei-
gerung führen müßte. Vielleicht erreichen wir bald den Zustand, daß
man sich auf ein erteiltes Patent auf diesem Gebiet verlassen kann
- heute ist das Risiko noch groß, daß die Gerichte den von der Er-
teilungsbehörde festgestellten technischen Charakter nicht anerkennen und das Patent für nichtig erklären.

Das Patent ist vor allem deswegen wichtig, weil es auf dem vorliegenden Rechtsgebiet bei weitem den stärksten und wirkungsvollsten
Schutz bietet. Es verbietet jedermann, die Erfindung ohne Zustim-
mung des Patentinhabers zu benutzen. Es schützt alle vom Patentan-
spruch überdeckten Ausführungsmöglichkeiten, mit anderen Worten also die Idee, die der Erfindung zugrunde liegt, gegen Nachahmung.

Dies ist beim Urheberrecht nicht der Fall. Das Patent schützt allerdings nicht das bei der Realisierung der Erfindung verwendete Programm als solches, z.B. gegen Kopieren.

Bei der Betrachtung des Patentschutzes für Software habe ich mich im wesentlichen nur mit dem Erfordernis des technischen Charakters befaßt. Es ist ganz selbstverständlich, daß bei technischen Programmen Neuheit, erfinderische Tätigkeit und gewerbliche Anwendbarkeit vorliegen müssen, wie bei jeder anderen Erfindung auch. Ich erwähne dies hier nur deswegen, weil bei den Patentbehörden gelegentlich die Überprüfung auf technischen Charakter mit der auf erfinderische Tätigkeit in einer unzulässigen Weise verknüpft wurden. Die Berechtigung für eine solche Handlungsweise wurde auch aus der Kerntheorie abgeleitet: Der Kern der Erfindung, also das, was Erfindungshöhe hat, soll technisch sein, am besten gegenständlich, also ganz simpel: Hardware. Hier zeigt sich ganz deutlich, wie wichtig die Ganzheitsbetrachtung ist. Sie entscheidet zunächst über den technischen Charakter der Erfindung, ob sie sich nun in Hardware oder in Software ausprägt. Ob sie auch Erfindungshöhe hat und die übrigen Erteilungsvorausssetzungen erfüllt, ist erst nachher zu untersuchen.

Zum Abschluß des Kapitels Patentschutz noch ein paar Worte zu USA und Japan. Das trilaterale Harmonisierungskommittee aus Beamten des US, des japanischen und des europäischen Patentamts hat festgestellt, daß die Haltung der drei Ämter gegenüber Softwareerfindungen so vergleichbar wäre, daß kein Harmonisierungsbedarf bestünde. Wenn die Anmelderschaft dieser Auffassung auch nicht uneingeschränkt zustimmen kann, so kann man doch mit der Patentierbarkeit solcher Erfindungen in den USA und Japan rechnen. Das US Patentamt scheint besonders großzügig, wenn man beispielsweise die US-Patente 4 722 055 oder 4 752 877 auf "financial products" betrachtet; solche Erfindungen hätten in Europa wohl keine Erteilungschance.

II. Urheberrecht

Während sich das Patent dazu eignet, eine erfinderische Idee zu schützen, ist das Urheberrecht auf den Schutz eines Werkes gerichtet, also des Ausdrucks einer Idee. Zunächst gilt dies für die wörtliche Fassung und gewisse Abwandlungen. Ob darüber hinaus auch die Struktur eines Programms und damit Aspekte der zugrunde liegen-

den Idee geschützt sein können, ist ein heiß diskutiertes Thema;
ich möchte mich hier auf den gesicherten Bereich beschränken. Das
Urheberrecht ist demnach kein Ersatz für das Patentrecht. Es kann
aber bei technischen Programmen eine wertvolle Ergänzung des Patentschutzes bieten. Bei nichttechnischen Programmen, wo der Patentschutz ausfällt, gewinnt das Urheberrecht unter den vorhandenen
Schutzmöglichkeiten besonders an Bedeutung.

In Deutschland sind seit Juni 1985 Programme für die Datenverarbeitung im Werkekatalog des Urheberrechtsgesetzes ausdrücklich aufgeführt, und zwar unter den Sprachwerken, insbesondere Schriftwerken.
Nach Auffassung des Bundesjustizministeriums handelte es sich dabei nur um eine Klarstellung der ohnehin gegebenen Rechtslage.

Leider ist diese Klarstellung aber mißlungen. Während das Gesetz
feststellt, daß Programme zu den urheberrechtsschutzfähigen Werken gehören, stellt die Rechtsprechung so strenge Bedingungen auf,
daß nach verbreiteter Ansicht nur ein kleiner Teil aller Programme
wirksamen Schutz genießt.

Diese Rechtsprechung ist abgeleitet aus der Entscheidung "Inkassoprogramm" des Bundesgerichtshofs [10]. Darin wird zunächst die Urheberrechtsschutzfähigkeit bejaht, unabhängig davon in welcher Phase
der Erstellung eines Programms (Problemanalyse, Datenflußplan, Codierung) die urheberrechtlich relevanten Formgestaltungen wahrnehm-
bar werden. Dies ist erfreulich.

Was jedoch die erforderliche persönliche geistige Schöpfung angeht,
errichtet der Bundesgerichtshof eine hohe Hürde. Er kann einen hinreichend schöpferischen Eigentümlichkeitsgrad nur dann anerkennen,
wenn ein deutliches Überragen der Gestaltungstätigkeit in Auswahl,
Sammlung, Anordnung und Einteilung der Informationen und Anweisungen gegenüber dem allgemeinen Durchschnittskönnen gegeben ist. Er
stellt also einen Vergleich zum Schaffen eines Durchschnittsprogram-
mierers an. Einen Vergleich zum Können des Durchschnittsfachmanns
kennen wir aus dem Patentrecht sehr gut. Für das Urheberrecht ist
ein solches Vorgehen allerdings ungewöhnlich. Bemerkt sei noch, daß
der Aufwand an Zeit und Geld und der Umfang des erstellten Programms
nach Meinung des Bundesgerichtshofs keine Rolle spielen, in deutli-
chem Gegensatz zu der Idee des Leistungsschutzes aus dem Wettbewerbsrecht.

Die durch eine solche Rechtsprechung geschaffene Situation ist für
die Industrie aus mehreren Gründen unbefriedigend. Zunächst fällt
es schon schwer, einen Nachweis für eine überdurchschnittliche
schöpferische Leistung zu führen, ähnlich wie sich auch im Patentrecht die Erfindungshöhe kaum beweisen läßt. Man kann dafür immer
nur Anhaltspunkte geben.

Noch bedenklicher ist aber, daß bei einer so restriktiven Rechtsprechung der durch das Gesetz in Aussicht gestellte Schutz im
Effekt nicht existiert; zumindest aber ist die Frage, ob ein Programm urheberrechtlich geschützt ist oder nicht, mit ganz erheb-
licher Rechtsunsicherheit behaftet. Im Urheberrecht kann man sonst
von der Vermutung ausgehen, daß ein geschaffenes Werk geschützt
ist. Hier trifft eher die umgekehrte Vermutung zu. Dies ist ganz
besonders deswegen unhaltbar, weil der Urheberrechtsschutz für
Programme in den maßgeblichen Ländern tatsächlich zur Verfügung
steht und Deutschland in dieser Hinsicht eine deutliche Ausnahme
darstellt. Leider akzeptiert der einschlägige Bericht der Bundes-
regierung [11] diese Situation und verweist als Ausweg auf ein
spezifisches Leistungsschutzrecht für Computerprogramme, was die
Sonderstellung Deutschlands zum Nachteil der an internationalem
Schutz interessierten Industrie aufrechterhalten würde. Wir wün-
schen uns kein neues Gesetz, sondern eine harmonisierte Rechtsprechung. Ist also - zumindest in Deutschland - der urheberrechtliche
Schutz von Programmen an sich schon nicht verläßlich, so werden von
interessierten Kreisen noch besondere Schutzausnahmen gefordert.

Dazu gehört beispielsweise die Zulässigkeit der Rückwärtsanalyse
(Dekompilation) von Programmen. Dies wäre eine Ausnahme vom aus-
schließlichen Vervielfältigungsrecht, das der Urheberrechtsinhaber
genießt, da die zur Rückwärtsanalyse nötigen Maßnahmen als Verviel-
fältigung zu gelten haben. Es werden vor allem zwei Aspekte ins
Treffen geführt:
- Es müsse zur Förderung des wissenschaftlichen Fortschritts gestattet sein, durch Rückwärtsanalyse bis zum Quellencode die darin
  enthaltenen Ideen (die ja nicht urheberrechtlich geschützt sind)
  kennenzulernen und zur Vorwärtsprogrammierung zu verwenden (dies
  wird auch als "reverse engineering" bezeichnet).
- Um zur Förderung des Wettbewerbs die Erstellung interoperabler
  Programme zu ermöglichen, müsse es gestattet sein, zumindest die
  Umgebung der in Betracht kommenden Schnittstellen zu analysieren.

Wenn die Industrie diesen beiden Forderungen kritisch gegenüber-
steht, so vor allem deswegen, weil bei allen derartigen Schutzein-
schränkungen die Anreize und Möglichkeiten zur mißbräuchlichen Verwendung der auf dieser Basis gewonnenen Informationen rasch zuneh-
men. Die Rückwärtsanalyse eines Programms mag wissenschaftliche Erkenntnisse bringen. Diese Erkenntnisse verlocken aber zur Herstellung eines Substituts für das Originalprogramm derart, daß sowohl
in die schöpferische Leistung als auch in die Verwertungsrechte des
Programmerstellers eingegriffen wird. Dies ist auch der Fall bei
der sogenannten "clean room-Technik", wo Rückwärtsanalyse und Erstellen des Substitutprogramms von verschiedenen Arbeitsgruppen
durchgeführt werden, die angeblich keinen Kontakt miteinander haben. Wenn sie beide dem gleichen Unternehmen angehören, ist eine
wirksame Trennung wenig wahrscheinlich.

Vergleichbare Überlegungen sind auch bei der Interoperabilität von
Programmen anzustellen. Es ist durchaus akzeptabel, daß zu vorhandenen Programmen, die an eine geeignete Betriebssystemschnittstelle
angeschlossen sind, Konkurrenzprogramme geschrieben werden, die natürlich Informationen über die Schnittstelle benötigen. Auch hier jedoch dürfen die Informationen nicht dazu benutzt werden, um ein Sub-
stitut für das Originalprogramm, also z.B. ein Betriebssystem, zu
schreiben.

Die Industrie plädiert nachdrücklich dafür, daß alle Schutzein-
schränkungen die Berner Verbandsübereinkunft zum Schutz von Werken
von Literatur und Kunst berücksichtigen müssen, einem internationalen Abkommen, in dem sich die Mitgliedsstaaten zur Einhaltung bestimmter Vorschriften auf dem Gebiet des Urheberrechts verpflichtet
haben. Der Artikel 9.2 schreibt vor, daß Ausnahmen vom ausschließ-
lichen Vervielfältigungsrecht die legitimen Interessen des Rechts-
inhabers nicht beeinträchtigen dürfen. Wir sind daher sehr daran
interessiert, daß die Berner Übereinkunft auch für den urheber-
rechtlichen Schutz von Programmen als maßgebliche internationale
Richtlinie akzeptiert wird. Mit Befriedigung können wir feststellen,
daß die meisten Staaten, die den Urheberrechtsschutz für Programme
anerkennen, auch bereit sind, ihn den Bedingungen der Berner Übereinkunft für Werke der Literatur zu unterwerfen und keine Vorschriften zu erlassen, die mit diesen Bedingungen im Widerspruch stehen,
sei es für den Umfang des Schutzes oder für seine Dauer. Es wird,
wie auch im sonstigen Urheberrecht, keine Anmeldung oder Registrie-

rung verlangt; der Rechtsschutz beginnt mit der Schaffung des Werkes. Urheberrechtsschutz ist also - im Gegensatz zum Patentschutz - sehr einfach, ohne Kosten und Mühen zu erwerben - nur die spätere Durchsetzung kann teuer werden.

Auch hier noch ein paar Worte zu USA und Japan. In beiden Staaten ist der Urheberrechtsschutz anerkannt und durchsetzbar. Während die japanische Industrie im internationalen Rahmen versucht, Sonderregelungen durchzusetzen, die dem spezifischen Charakter von Programmen als industrielle Produkte Rechnung tragen sollen, haben die Gerichte in Japan bisher nach "normalem" Urheberrecht geurteilt.

III. Direktive der EG

Wenn auch die Staaten der Europäischen Gemeinschaft die Urheberrechtsschutzfähigkeit von Programmen bejahen, bestehen doch noch Unterschiede in den Auffassungen darüber, wie die Einzelheiten zu regeln sind. Die Kommission hat daher im Januar 1989 [12] den Entwurf einer Direktive für den Urheberrechtsschutz von Programmen vorgelegt, der seitdem von den interessierten Kreisen lebhaft diskutiert wird. In der Direktive, die bis zur Bildung des Binnenmarktes in nationales Recht der EG-Staaten umgesetzt werden muß, sind einige Grundregeln mit erfreulicher Klarheit ausgesprochen, z.B. daß Programme als Werke der Literatur geschützt sein sollen und nicht "als ob" sie Werke der Literatur wären oder diesen gleichgestellt. Dies zielt genau in die gewünschte Richtung, nämlich Programme so wie andere Werke den Vorschriften der Berner Übereinkunft zu unterwerfen. Es dürfen auch keine qualitativen Anforderungen an das Programm gestellt werden - vielleicht ein weiterer Anlaß für den Bundesgerichtshof, seine Haltung zu überdenken.

Neben diesen und anderen erfreulichen Bestimmungen muß sich die Direktive auch mit den oben schon erwähnten Problemen der Schutzausnahmen auseinandersetzen. Auf diesem Gebiet ist offenbar die Einigung besonders schwierig. Die verschiedensten Vorschläge aus Kreisen der Industrie, der Softwarehäuser und der Anwender, aber auch aus dem europäischen Parlament und dem Wirtschafts- und Sozialausschuß der Gemeinschaft, sind bisher noch nicht zu einem Kompromiß zu verbinden gewesen. Neben reverse engineering und Interoperabilität von Software sind noch weitere Probleme ins Gespräch gebracht worden, z.B. daß auch für Zwecke der Wartung von Programmen die Dekompilation gestattet werden sollte, oder daß die Herstellung nicht nur interoperabler Software, sondern auch interoperabler Hardware durch Rückwärtsanalyse von Originalprogrammen ermöglicht werden müsse.

Der Stand im Oktober 1990 ist, daß die Kommission unter Berücksichtigung eines Votums des Europäischen Parlaments über einen von diesem für richtig erachteten Text einen Vorschlag an den Ministerrat der Europäischen Gemeinschaft zu erstellen hat. Beim Ministerrat liegt die endgültige Entscheidung darüber, was im Rahmen der Direktive dem Gesetzgeber in den einzelnen Mitgliedsstaaten vorgeschrieben werden soll.

IV. Zusammenfassung

Technische Programme, d.h. technische Erfindungen, die zur Realisierung mit Software arbeiten, sind dem Patentschutz grundsätzlich zugänglich.

Alle Programme "als solche" sind grundsätzlich dem Urheberrechtsschutz zugänglich.

Im Gegensatz zu maßgeblichen anderen Ländern werden in Deutschland dem Rechtsuchenden spezielle Hindernisse in den Weg gelegt, zu deren Begründung auf die Rechtsprechung des Bundesgerichtshofs verwiesen wird:
- im Patentrecht das Problem, daß der technische Charakter bei "bestimmungsgemäßem Gebrauch" des Computers nicht anerkannt wird,
- im Urheberrecht das Erfordernis der überdurchschnittlichen Gestaltungshöhe, die nur von wenigen Programmen erfüllt wird.

Da die beiden Gesetze eine so restriktive Rechtsprechung keineswegs vorschreiben, ist zu hoffen, daß der Bundesgerichtshof im Zuge der Harmonisierung der Rechtssysteme in der EG den Weg zu einer liberaleren Haltung der deutschen Gerichte weisen wird. Die Direktive der EG-Kommission zum Urheberrecht von Computerprogrammen mag hier mithelfen; mindestens ebenso wichtig aber ist es, geeignete Streitfälle durch die Instanzen bis zum Bundesgerichtshof zu führen, um ihn zu Äußerungen über seine Rechtsmeinung zu veranlassen. Der

Rechtsschutz in Deutschland muß den engen Rahmen, der aus den Entscheidungen "Dispositionsprogramm" und "Inkassoprogramm" gezimmert worden ist, in Richtung auf eine angemessene Berücksichtigung der Elemente der Informationstechnologie so bald wie möglich verlassen.

1) Rechtsschutz und Verwertung von Computerprogrammen; M. Lehmann (Hrsg.), Otto Schmidt Verlag, Köln 1988
2) Kraßer, in 1) S. 109
3) Beyer, GRUR 1990, S. 399
4) GRUR 1977, S. 96
5) Blatt PMZ 1987, S. 1
6) Prüfungsrichtlinien des Europäischen Patentamts, März 1985, Teil C, Kapitel IV, S. 37
7) z.B. "VICOM", Amtsblatt EPA 1987, S. 14
8) z.B. "Elektronisches Stellwerk", GRUR 1987, S. 799 "Rolladensteuerung", Blatt PMZ 1989, S. 116
9) z.B. "Arbeitsspeichersystem" CuR 1988, S. 652
10) GRUR 1985, S. 141
11) Bundestagsdrucksache 11/4929, S. 41
12) Official Journal of the EC, No C 91/4 vom 12.4.1989

Objektorientierte Softwaretechnik

H.J. Schneider [1]

Zusammenfassung:

Dieser Aufsatz will einen allgemein gehaltenen Überblick über die Ideen geben, die die objektorientierte Programmierung von anderen Programmierstilen unterscheidet. Wir halten das Konzept der Klassifikation für den entscheidenden Unterschied. Der erste Abschnitt charakterisiert die verschiedenen Entwicklungsstadien des Programmierens. Dann wird der Begriff des Objektes diskutiert, wie ihn die objektorientierte Programmierung verwendet. Im dritten Abschnitt führen wir die Klassenhierarchie ein und skizzieren ihre Beziehung zur Wiederverwendbarkeit der Software. Schließlich betrachten wir den Einfluß, den die objektorientierte Programmierung auf den Prozeß der Software-Entwicklung ausübt.

Abstract:

The aim of this paper is to give a general survey of the ideas that distinguish object-oriented programming from other programming styles. We think that the main difference is the concept of classification. The first section characterizes the different stages in the development of programming. Then the notion of an object is discussed as it is used in object-oriented programming. In the third section, we introduce the class hierarchy and sketch its relationship to software reusability. Finally, we consider the influence object-oriented programming exerts on the design process.

1. Einleitung

Betrachtet man die eigentlich doch noch recht junge Geschichte der Software-Entwicklung, so kann man verschiedene Perioden unterscheiden, in denen jeweils neue Problemgrößenordnungen in den Vordergrund traten und dementsprechend neue Methoden zu ihrer Beherrschung. Die treibende Kraft dieser Entwicklung war der rasante Fortschritt der Hardware, der es uns ermöglichte, immer größere Softwaresysteme zu konstruieren. Heute unterscheidet man drei Phasen, die auf verschiedene Weise charakterisiert werden können: Durch den Detaillierungsgrad der Sicht auf das Programm, durch die Methode der Problembeherrschung, aber auch durch den Umfang des Problems.

---

[1] Prof. Dr. Hans J. Schneider, Lehrstuhl für Programmiersprachen der Universität Erlangen-Nürnberg, Martensstraße 3, 8520 Erlangen

1.1. Programmieren im Kleinen

In der ersten Phase, die wir den sechziger Jahren zuordnen können, betrachtete man ein Programm als Folge von Anweisungen, die auf einem Zustandsraum operieren. Jede Anweisung und jede Folge von Anweisungen beschreiben einen mehr oder weniger komplizierten Zustandsübergang. Die typische Vorgehensweise ist das Prinzip der *schrittweisen Verfeinerung*. Das Problem definiert einen Zustandsübergang, der immer feiner in eine Sequenz von Zustandsübergängen unterteilt wird, bis jeder einzelne Schritt dieser Sequenz genau einer Anweisung der Programmiersprache entspricht. Die Sprachen dieser Phase beginnen nach FORTRAN mit ALGOL und enden etwa mit PASCAL. Man spricht auch vom Programmieren im Kleinen.

1.2. Programmieren im Großen

Als die Programme Ende der sechziger Jahre immer größer wurden, ergab sich eine Vielzahl von Querverbindungen, die beherrscht werden mußten. Es ging nunmehr darum, auf den ersten Blick von den Abläufen im einzelnen zu abstrahieren und sich zunächst auf die Querverbindungen zu konzentrieren. Um diese zu überblicken, betrachtete man das Programm auf einer höheren Abstraktionsstufe als eine Menge von Komponenten, die miteinander nur über wohldefinierte Nahtstellen verbunden sind. Der entscheidende Schritt beim Entwurf einer Problemlösung ist die Zerlegung des Problems in ein System von Moduln und deren Spezifikation, die die präzise Festlegung der nach außen sichtbaren Information und der Modulfunktionen umfaßt. Dabei geht es aber nur darum, *was* der Modul tut, und nicht um das *Wie*. Da insbesondere offen bleibt, welche Datenstrukturen die Implementierung schließlich verwendet, sprechen wir von dem Prinzip der *Datenabstraktion*. Dieses Prinzip, das die Verwendung der Modulfunktionen von ihrer Implementierung trennt, erleichtert sowohl die Wiederverwendung bereits vorhandener Moduln als auch das Austauschen einer Implementierung gegen eine andere. Programmiersprachen, die dieses Prinzip unterstützen, sind beispielsweise CLU, MODULA und ADA. (Selbstverständlich beachten diese Sprachen auch das Prinzip der schrittweisen Verfeinerung.)

1.3. Programmieren im ganz Großen

Sowohl die schrittweise Verfeinerung als auch die Datenabstraktion gehen davon aus, daß die Funktion des Gesamtsystems a priori genau bekannt ist. Diese Prämisse trifft aber auf viele der heute zu entwickelnden Systeme nicht mehr zu. Dies kann verschiedene Ursachen haben. So stoßen wir immer mehr in Anwendungsbereiche vor, in denen formalisierte Aufgabenstellungen selten sind. Man denke z.B. an die benutzergerechte Gestaltung von Dialogkomponenten. Ein anderes Beispiel ist die Entwicklung und Pflege sehr großer und sehr langlebiger Systeme. In diesen Fällen werden im Laufe der Zeit Änderungen und Ergänzungen nötig, die zu Beginn der Entwicklung nicht vorhergesehen werden konnten. Ein besonderer, früher nicht beachteter Aspekt ist dabei, daß Modulvarianten entstehen können, die im System nebeneinander existieren müssen. Wir müssen nun nicht nur die gegenseitigen Zugriffe der Moduln kontrollieren, sondern auch den Grad der Modulverwandtschaft. Sind Moduln identisch? Unterscheiden sie sich nur durch die Implementierung, verhalten sie sich also gleich? Oder verhalten sie sich nur in bezug auf einige Eigenschaften gleich, bezüglich anderer verschieden? Um diese Fragen zu beherrschen, benötigen wir das Prinzip der *Klassifikation*, das in vielen Wissensgebieten erfolgreich dazu dient, umfangreiches Material übersichtlich zu organisieren.

Man charakterisiert die einzelnen Objekte, indem man auf der einen Seite ihre Gemeinsamkeiten, auf der anderen ihre Unterschiede hervorhebt. Objekte, die sich bezüglich einer bestimmten Betrachtungsweise gleich verhalten, werden in eine gemeinsame Klasse eingeordnet. Sie unterscheiden sich in wenigstens einem relevanten Punkt von den Objekten anderer Klassen. Dabei dürfen die Klassen nicht unverbunden nebeneinander stehen; zur Beherrschung von Komplexität durch Klassenbildung gehört auch, daß die Beziehungen der Klassen untereinander deutlich werden. Beispielsweise kann eine Klasse A Abstraktion zweier Klassen B_1 und B_2 sein; B_1 und B_2 sind also unterschiedliche Spezialisierungen von A. Die Objekte von B_1 und B_2 weisen dann alle Eigenschaften der Objekte von A auf; darüber hinaus interessieren jedoch, wenn wir von der Klasse B_1 sprechen, weitere Eigenschaften, die die Objekte von B_2 nicht aufweisen.

1.4. Objektorientierte Programmierung

Die objektorientierte Software-Entwicklung bildet dieses Prinzip nach. Das Programm besteht aus simultan existierenden Objekten, die sich zu jedem Zeitpunkt in einem bestimmten Zustand befinden, gewisse Operationen ausführen und untereinander kommunizieren können. Die Objekte mit gleichem Verhalten werden in einer Klasse zusammengefaßt. Das Lösen eines Problems besteht dann „nur noch" im Definieren problemspezifischer Objektklassen, Erzeugen von Objekten dieser Klassen und dem Festlegen der Kommunikation zwischen den erzeugten Objekten. Programmiersprachen, die das Prinzip der Klassifikation unterstützen, sind SIMULA, SMALLTALK, C++, TRELLIS/OWL, EIFFEL, verschiedene LISP-Dialekte u.a.

Charakteristisch für den objektorientierten Programmierstil sind also nicht irgendwelche Äußerlichkeiten wie etwa die Verwendung einer graphischen Bedienoberfläche oder einer ganz bestimmten Programmiersprache. *D.A. Moon* faßt das Wesentliche an diesem Stil folgendermaßen zusammen [7]:

> An object-oriented program consists of a set of objects and a set of operations on these objects. These entities are not defined in a monolithic way. Instead, the definitions of the operations are distributed among the various objects that they can operate upon. At the same time, the definitions of the objects are distributed among the various facets of their behavior. An object-oriented programming system is an organizational framework for combining these distributed definitions and managing the interactions among them.

Auf der gleichen Linie liegt *P. Wegner*, der die Gleichung aufstellt [19]:

$$\text{objektorientierte Programmierung} = \text{Objekte} + \text{Objektklassen} + \text{Vererbung}$$

Je nach Interessengebiet der Autoren finden sich weitere Forderungen. So legen beispielsweise viele Autoren aus dem Gebiet der artifiziellen Intelligenz Wert auf das dynamische Binden zur Laufzeit, während Datenbankfachleute eher an der Persistenz der Objekte interessiert sind und bei der Programmierung von Prozeßsteuerungen in erster Linie Objekte interessieren, die Nebenläufigkeit realisieren. Selbstverständlich schließt die Wegnersche Definition keinen dieser Gesichtspunkte aus, aber genuin ist der Gedanke der Klassifikation durch Herausarbeiten der Gemeinsamkeiten und der Unterschiede.

2.   Objekt und Klasse

Der zentrale Begriff der objektorientierten Software-Entwicklung ist das Objekt. Es handelt sich dabei um eine programmiersprachliche Einheit, die über einen inneren Zustand verfügt und die in Abhängigkeit von diesem Zustand Berechnungen ausführt oder ausführen kann. Die Objekte einer Klasse sind dadurch charakterisiert, daß die Mengen ihrer potentiell möglichen Zustände und Berechnungen gleich sind. Diese werden in der Klassendefinition festgelegt. In Fig. 1 sind zwei Beispiele für Klassendefinitionen angegeben[2]. Um das Verhalten einer ganzen Klasse von Objekten zu ändern, genügt also eine Änderung an einer einzigen Stelle, nämlich in der Klassendefinition.

2.1.   Zustand

Wenn wir sagen, daß sich die Objekte einer Klasse gleich verhalten, so bedeutet dies, daß die Menge der möglichen Zustände und Berechnungen übereinstimmt. Aktuell werden sie sich aber in unterschiedlichen Zuständen befinden. Der Zustand eines Objektes reflektiert seine Vorgeschichte, hängt also davon ab, welche Operationen bisher ausgeführt wurden. Er wird durch einen Satz *lokaler Variablen* beschrieben, die selbstverständlich selbst wieder Objekte sind. Im ersten Beispiel von Fig. 1 ist dies die aktuelle Stellung der Weiche. Diese Variable wird für jedes Objekt separat festgelegt und verändert. In der Terminologie der objektorientierten Programmierung handelt es sich um eine *Instanzenvariable*. Im zweiten Beispiel von Fig. 1 ist die Stationsbezeichnung eine Instanzenvariable.

Dieses Beispiel enthält darüber hinaus eine weitere lokale Variable, nämlich die Festlegung der maximalen Bezeichnerlänge. Bei ihr handelt es sich um eine sog. *Klassenvariable*: Ihr

---

[2]Die Beispiele sind nicht in einer konkreten Programmiersprache geschrieben. Sie stammen aus der Bearbeitung einer Paketverteilanlage; eine Beschreibung dieser Aufgabe findet sich beispielsweise bei F. Hofmann [3, Kap.1].

```
Klassenname               weiche
Oberklasse(n)             object
Klassenvariable(n)
Instanzenvariable(n)      aktuelle_stellung: richtung
Unsichtbare Methode(n)    initialisiere()
                          protokolliere()
Sichtbare Methode(n)      stelle_weiche(r: richtung): boolean
                          stellung(): richtung

Klassenname               station
Oberklasse(n)             task
Klassenvariable(n)        bez_lng: const int = 27
Instanzenvariable(n)      bezeichnung: array [bez_lng] of char
Unsichtbare Methode(n)    initialisiere()
                          terminiere()
                          protokolliere()
Sichtbare Methode(n)      melde_an(p: paket)
```

Fig. 1: Beispiel zweier Klassendefinitionen

Wert ist für alle Objekte einer Klasse identisch und kann deshalb bereits in der Klassendefinition festgelegt werden.

Der Begriff der lokalen Variablen darf in diesem Zusammenhang nicht unbedingt so interpretiert werden, daß ihre Werte während der Lebensdauer des Objektes verändert werden. Im Fall der Stationsbezeichnung ist dies beispielsweise unerwünscht. Ob und wie einzelne Programmiersprachen erlauben, dies in der Definition festzulegen, soll uns hier nicht interessieren.

2.2. Operationen

Zur Sicherung der Konsistenz des Objektzustandes befolgt auch die objektorientierte Programmierung das Prinzip der Datenabstraktion: Ein anderes Objekt darf auf die lokalen Variablen unmittelbar weder lesend noch schreibend zugreifen. Vielmehr muß es dem Zielobjekt eine Nachricht zukommen lassen, welche Operation dieses ausführen soll. Dadurch werden Art und Zeitpunkt der Reaktion ausschließlich von der Implementierung des angesprochenen Objektes bestimmt und hängen nicht vom Zusammenspiel mit logisch oder textuell weit entfernten Objekten ab. (Bei Systemen mit nebenläufigen Aktivitäten betrifft dies beispielsweise die Sequentialisierung überlappender Zugriffe.)

Welche Operationen - die objektorientierte Szene spricht von *Methoden* - ein Objekt ausführen kann, ergibt sich aus der Klassendefinition. Im einfachsten Fall enthält die Klassendefinition eine vollständige Aufstellung aller Methoden. Diese umfaßt jeweils den Methodenbezeichner, ggf. eine Liste formaler Parameter und schließlich eine Implementierung der Methode. Letztere ist in jedem Fall vertraulich und in unseren Abbildungen weggelassen. Der

Methodenkopf (Bezeichner und Parameterliste) kann dagegen öffentlich sein. Die Entscheidung hierüber trifft, wer die Klasse definiert. Die objektorientierten Programmiersprachen stellen dann über ihre Compiler sicher, daß die Implementatoren anderer Systemkomponenten nur die sichtbaren Methoden benutzen.

Im ersten Beispiel von Fig. 1 stehen nach außen zwei Methoden zur Verfügung: die den Zustand des Objektes verändernde Operation *stelle_weiche* und die den Zustand nicht verändernde Abfrage der aktuellen Stellung. Die Implementierung dieser Methoden kann auf die nach außen nicht sichtbare Protokollroutine zurückgreifen.

2.3. Verwendung von Objekten

Ist einmal eine Klasse definiert, so können beliebig viele Objekte dieser Klasse generiert und mit Bezeichnern belegt werden:

```
a, b, c: weiche;
weichenfolge: array[10] of weiche;
```

Jede solche Deklaration veranlaßt die Schaffung neuer Objekte, in diesem Fall drei Objekte, die mit *a*, *b* und *c* bezeichnet sind und eine aus zehn Objekten bestehende Reihung. Über die Bezeichner sind die Objekte dann ansprechbar:

```
a.stelle_weiche(links)
b.stelle_weiche(rechts)
weichenfolge[4].stelle_weiche(rechts)
weichenfolge[6].stellung()
```

Der letzte Aufruf liefert als Funktionswert die aktuelle Stellung der bezeichneten Weiche, die übrigen eine Aussage darüber, ob der Auftrag erfolgreich ausgeführt werden konnte.

2.4. Konstruktoren und Destruktoren

Bei der Deklaration eines Objektes wird der Speicherplatz für die lokalen Variablen bereitgestellt. Ferner wird dem Objekt ein Verweis auf seine Klassenzugehörigkeit mitgegeben, damit beim Aufruf einer Methode die richtige Routine ausgeführt wird. Schließlich wird eine Initialisierungsroutine aufgerufen, sofern die Klassendefinition eine solche enthält. Im zweiten Beispiel von Fig. 1 könnte sie beispielsweise die Eingabe der Stationsbezeichnung veranlassen. In unseren Beispielen haben wir der Konstruktorroutine die Bezeichnung *initialisiere* gegeben. Dies ist eine willkürliche Festlegung; in einigen objektorientierten Programmiersprachen ist festgelegt, daß sie die gleiche Bezeichnung wie die Klasse selbst trägt. Aber dies sind programmiersprachliche Feinheiten, die für das zugrundeliegende Prinzip unwesentlich sind.

Symmetrisch zur Konstruktormethode kann auch das Löschen eines Objektes durch eine eigene Methode (Destruktor) festgelegt werden. Bei der Modellierung von Prozeßrechneranwendungen zeigt sich deutlicher als bei den üblichen Standardbeispielen der Informatiker, wie wichtig das wohldefinierte Löschen von Objekten ist. So kann es beispielsweise Aufgabe des Destruktors sein, den von ihm gesteuerten Anlagenteil geordnet herunterzufahren.

3. Beziehungen zwischen Klassen

Der Gedanke der Wiederverwendung von Softwarebausteinen ist nicht neu; selbstverständlich darf nicht bei jeder neuen Software-Entwicklung das Rad neu erfunden werden. Im Idealfall entsteht ein neues System dadurch, daß man aus einer Bibliothek die geeigneten Moduln entnimmt und nur noch richtig zusammensetzt. Diese Betrachtungsweise versteht Wiederverwendbarkeit in einem strengen Sinn: Der Softwarebaustein wird so eingesetzt, wie er ist; lediglich formale Parameter werden durch tatsächliche ersetzt. Eine solche Modifikation hat der Entwickler vorhergesehen und deshalb formale Parameter statt fester Werte eingesetzt. Diese Technik gibt es seit der Einführung der Unterprogramme[3].

3.1. Unvorhergesehene Varianten

Bei der Pflege sehr großer und langlebiger Softwaresysteme bekommt der Begriff der Wiederverwendung einen etwas anderen Anstrich. Häufig tritt hierbei die Situation auf, daß ein Modul zwar geringfügig, aber in einer ursprünglich nicht vorhergesehenen Weise geändert werden muß. Dabei denken wir nicht an Fehlerbeseitigung oder lokale Effizienzverbesserung; das Austauschen von Implementierungen war schließlich einer der Anstöße für die Datenabstraktion. Wir meinen hier vielmehr eine Änderung des funktionalen Verhaltens, die für neue Komponenten erforderlich ist, ohne daß andere Komponenten davon etwas merken sollen. Ein typisches Beispiel sind Ergänzungen zur Bearbeitung bestimmter Sonderfälle, die bisher keine Rolle spielten.

J. Micaleff diskutiert in ihrer Übersicht über objektorientierte Programmiersprachen drei Möglichkeiten, die den Softwarepflegern in einer solchen Situation zur Verfügung stehen [6]:
- Modifikation des Moduls,
- Modifikation einer Kopie des Moduls,
- Definition einer verwandten Klasse.

Die erste Alternative führt zu Bausteinen immer höherer Komplexität, weil schließlich die bisherigen Leistungen unverändert erbracht werden sollen. In der Regel entstehen umfangreiche Folgen von Fallunterscheidungen, die sicherstellen, daß jede Systemkomponente über die von ihr gewünschte Version verfügen kann. Am Beispiel einer Ressourcenverwaltung

---

[3]Im Prinzip sind auch die Modulschablonen von ADA nichts anderes, obwohl dort z.B. Datentypen als Parameter auftreten dürfen.

```
Klassenname                verteilstation
Oberklasse(n)              station
Klassenvariable(n)
Instanzenvariable(n)       linker_nachfolger:  station
                           rechter_nachfolger: station
                           stationsweiche:     weiche
Unsichtbare Methode(n)     initialisiere()
                           protokolliere()
Sichtbare Methode(n)       paketankunft()
                           paketauslauf(r: richtung)
```

Fig. 2: Unterklassenbildung durch Erweiterung

habe ich auf einer früheren Prozeßrechnertagung deutlich gemacht, wie aufwendig solche Modifikationen beispielsweise in ADA werden [15].

Die zweite Lösung fügt einen neuen Baustein in das System ein, der nur die veränderte Leistung erbringt. Er wird durch eine neue Bezeichnung von der alten Version unterschieden; die Systemkomponenten, die weiterhin nur die bisherige Version benötigen, werden durch die Änderung überhaupt nicht berührt. Aber nun tritt das Aktualisierungsproblem auf. Da die verschiedenen Versionen durch Modifikation auseinander hervorgegangen sind, enthalten sie gemeinsame Anteile, die mehrfach im System enthalten sind. Wird an einer solchen Stelle eine Aktualisierung notwendig, so muß man durch organisatorische Maßnahmen sicherstellen, daß diese Änderungen in allen verwandten Moduln nachgetragen werden.

3.2. Modifikation durch Unterklassenbildung

Bei der vorangegangenen Abwägung der Modifikationsmöglichkeiten habe ich ganz bewußt den Begriff der Klasse vermieden, weil das Prinzip der Klassifikation durch Herausarbeiten der Gemeinsamkeiten und der Unterschiede einen gangbaren Lösungsweg eröffnet. Jeder Baustein gehört als Objekt einer bestimmten Klasse an; das Verhalten ist durch die Klassendefinition bestimmt. Um das modifizierte Objekt einzuführen, wird eine neue Klasse definiert. Dabei wird jedoch im Idealfall ausschließlich der Unterschied zur bereits existierenden Klasse spezifiziert; die Gemeinsamkeiten beider Klassen werden nicht wiederholt, sondern automatisch von der vorhandenen an die neue Klasse weitergegeben. Dies wird dadurch erreicht, daß die neue Klasse als Unterklasse der alten firmiert. Wenn man dieses *Vererbungsprinzip* konsequent einsetzt, ist jede Information nur einmal im System vorhanden.

3.3. Unterklassenbildung durch Erweiterung

Die einfachste Form der Modifikation ist die *Erweiterung.* Die Liste der lokalen Variablen und der Methoden, die die Objekte der Oberklasse charakterisieren, wird durch weitere lokale Variablen und Methoden ergänzt. Diese sind nur für die Objekte der Unterklasse definiert. Um dies an einem Beispiel zu erläutern, gehen wir von der bereits eingeführten Klasse *station* aus und führen als Spezialfall Verteilstationen ein, die eine Weiche steuern (Fig. 2). Der Zustand eines Objektes der neuen Klasse wird also durch die lokalen Variablen in der Oberklassendefinition und zusätzlich durch die in der neuen Definition beschrieben. Konkret bedeutet dies, daß auch für die Objekte der Klasse *verteilstation* eine Bezeichnung definiert ist. Dagegen kennen die Objekte der Oberklasse keinen Nachfolger[4]. Analog sind die Methoden der Oberklasse automatisch für Objekte der neuen Klasse definiert:

```
x: verteilstation;
x.paketankunft();
x.melde_an(p);
x.delay;
```

Die Methode *paketankunft* kann von dem Objekt x ausgeführt werden, weil sie in dessen Klassendefinition selbst enthalten ist. Die Methode *melde_an* ist dort zwar nicht definiert, aber in der Klasse *station*, und die Klasse *verteilstation* ist eine Unterklasse von dieser. Nehmen wir an, daß die Methode *delay* für Objekte der Klasse *task* definiert ist, so ist auch die dritte Anweisung zulässig; die Methode wird über zwei Klassenebenen hinweg vererbt.

In die Unterklassendefinition haben wir eine neue Konstruktorroutine *initialisiere* aufgenommen. Diese kann die Konstruktorroutine der Oberklasse ergänzen. Soll ein Objekt der Klasse *verteilstation* kreiert werden, so wird zunächst die Konstruktorroutine der Klasse *station* aufgerufen, dann die der Unterklasse[5]. Verwendet man diese Routinen zur Erfassung der Anfangsdaten, so werden also zunächst die für alle Stationen gültigen Daten (z.B. die Bezeichnung) erfragt, danach die speziellen Daten für Objekte der Unterklasse. Existieren verschiedene Unterklassen, so verlangt die Initialisierung automatisch nur die für die gewünschte Klasse relevanten Daten.

3.4. Überschreiben von Methoden

Eine andere Möglichkeit der Unterklassenbildung ist, die Implementierung von einer oder mehreren Methoden zu ändern (Fig.3). Wenn wir die bereits erwähnten Zielstationen definieren wollen, stellen wir fest, daß das Anmelden eines Paketes anders implementiert werden kann. Bei Verteilstationen führt der Aufruf dieser Methode zu einer Eintragung in eine Warteschlange. Bei den Zielstationen ist dies nicht mehr nötig; hier genügt ein Test, ob das Paket an der richtigen Station angekommen ist. Die Objekte der Unterklasse verhalten sich

---

[4]Man beachte, daß wir von den Nachfolgerstationen lediglich verlangen, daß es sich um Stationen handelt; dies können weitere Verteilstationen, aber auch Zielstationen sein.

[5]Falls die verwendete Programmiersprache dies nicht von sich aus macht, erlaubt sie vielleicht, es entsprechend zu programmieren.

```
Klassenname              zielstation
Oberklasse(n)            station
Klassenvariable(n)
Instanzenvariable(n)
Unsichtbare Methode(n)   initialisiere()
                         fehlermeldung(z: paketziel)
Sichtbare Methode(n)     melde_an(p: paket)
```

Fig. 3: Überschreiben von Methoden

also bei Erhalt der gleichen Nachricht anders als die der Oberklasse. Erhält ein Objekt eine Nachricht, so muß zunächst in der Definition der Klasse dieses Objektes nach einer passenden Methodendefinition gesucht werden. Wird keine solche Methode gefunden, so wird in der Oberklasse gesucht, aus der die Klasse abgeleitet wurde, dann in deren Oberklasse usw., bis die passende Definition gefunden wird. Bleibt die Suche erfolglos, so führt dies zu einer Fehlermeldung. Existieren andererseits in den Klassendefinitionen mehrere Methoden mit gleicher Bezeichnung und will der Implementator eine ganz bestimmte ansprechen, so muß er dies explizit angeben[6].

Einige der Programmiersprachen erlauben die Mehrfachvererbung. Darunter versteht man die Möglichkeit, daß eine Klasse mehrere gleichberechtigte Oberklassen besitzt: Auf ein Textfenster beispielsweise kann man sowohl die Fensteroperationen (*öffnen, verschieben, ...*) als auch die Textoperationen (*einfügen, Schreibmarke bewegen, ...*) anwenden. Es liegt auf der Hand, daß die Benennung und die Identifikation der Methoden in diesem Fall besondere Sorgfalt erfordern.

4. Klassenbildung und Entwurf

Welche Auswirkungen hat die konsequente Anwendung der objektorientierten Programmierung auf den Prozeß der Software-Entwicklung? Programmieren mit Blick auf die Wiederverwendbarkeit ist schwieriger als für den einmaligen Programmentwurf, denn von den gefällten Entscheidungen hängen die Kosten zukünftiger Entwicklungen ab. Je größer bei einer Entwurfsentscheidung der Spielraum ist, desto gewissenhafter muß von den Möglichkeiten Gebrauch gemacht werden, um spätere Entwicklungen nicht zu erschweren. E. Sandewall hat einmal von dem Prinzip des *strukturierten Wachstums* gesprochen [12]. Wildwuchs ist nicht angesagt. Das bedeutet aber, daß es grundverkehrt wäre, aus der leichten Modifizierbarkeit objektorientierter Programme zu schließen, man könne bei deren Entwicklung und Weiterentwicklung auf einen sauberen Entwurf verzichten. Aufgabe dieser Tätigkeit ist es,
- die Objekte und ihre relevanten Attribute zu identifizieren,
- die Kommunikation zwischen den Objekten herauszuarbeiten und daraus den Methodenvorrat abzuleiten,

---

[6]Nicht alle Programmiersprachen erlauben dies.

– die Objekte auf gemeinsame Methoden und Attribute hin zu untersuchen und in vorhandene oder neu zu definierende Klassen einzuordnen.

Insbesondere die Wiederverwendung vorhandener Klassen erfordert einen sehr guten Überblick über die Klassenhierarchie.

4.1. Identifizieren der Objekte

Die zentrale Frage beim objektorientierten Software-Entwurf ist die nach den Objekten. Im Bericht über eine Arbeitstagung lesen wir [9]:

> Where do objects come from? This is the central question posed at the Workshop on Specification and Design. The consensus is that it is difficult to find the right objects for the design of a program, and mapping specifications to objects usually doesn't work the first time around.

In der Tat findet man in der Literatur nur vage Hinweise darauf, was ein Objekt ist. Beispielsweise charakterisieren H. Stoyan und G. Görz die Objekte vom Ergebnis her [16]:

> Jeder Gegenstand in einem objektorientierten System ist ein Objekt [...] Objekte sind individuelle Artefakte, die sowohl aktive als auch passive Rollen spielen.

I. Jacobson macht deutlich, daß nicht nur materielle Gegenstände zu beachten sind [4]:

> Entities are things or events in reality which are of interest to the system that is being developed. They are things close to reality about which you want to make assertions and about which you want information.

Verläßt man die in der Literatur weit verbreiteten Spielbeispiele, so muß man sehr bald feststellen, daß viele Objekte nicht auf der Hand liegen und erst nach sehr genauer Analyse der Anforderungsdefinition erkannt werden[7].

Auf der bereits erwähnten Arbeitstagung wurden mehrere Vorschläge gemacht, wie bekannte Ansätze zur Lösung dieses Problems herangezogen werden können. E. Seidewitz schlug in seiner Diskussionsbemerkung vor [9, p.11], die Objekte und ihr Verhältnis zueinander mit Entity-Relationship-Diagrammen zu erfassen. Diese Vorgehensweise erscheint mir bei Aufgabenstellungen im CAD-Bereich erfolgversprechend. (Man vergleiche hierzu auch das Beispiel von H. Wedekind [18].) Für Aufgabenstellungen aus dem Bereich der Steuerung technischer Prozesse empfahl R. Hilliard [9, p.15], zunächst die Kontrollabläufe zu identifizieren und diese als aktive Objekte aufzufassen. (Dieser Weg liegt auch unserem Beispiel zugrunde [14].)

---

[7]Dies bestätigt auch die psychologische Studie von M.B. Rosson und E. Gold [11], die erfahrene Programmierer bei der Arbeit beobachtet haben.

4.2. Identifizieren des Methodenvorrats

Nur durch den Aufruf von Methoden kommunizieren die Objekte miteinander. Der Methodenvorrat, der ein Objekt charakterisiert, muß sich also aus dessen Kommunikation mit anderen Objekten ergeben. Eine genaue Analyse der Kommunikationsstruktur kann beispielsweise durch Datenflußdiagramme unterstützt werden[8]. Bei der Frage nach der Zuordnung von Methoden zu Objekten sollte man an einen Werkvertrag denken: Der Auftraggeber erwartet, daß der Auftragnehmer eine bestimmte Leistung erbringt, überläßt diesem aber die Einzelheiten der Ausführung. Der Software-Entwickler hat sich an dieser Stelle zwei Fragen zu stellen:
- Für welche Aktionen ist das Objekt verantwortlich?
- Zu welchen Informationen muß es deshalb Zugang haben?

Die Antwort auf die erste Frage beinhaltet selbstverständlich die genaue Festlegung der Nachrichtenformate. Man beachte jedoch, daß die Antwort auf die zweite Frage keine Aussage darüber macht, wo diese Informationen gespeichert werden. Denn es liegt in der Verantwortung eines Auftragnehmers, ob er Unterauftragnehmer einschaltet oder alles selbst erledigt. In unserem Beispiel benötigt etwa die Verteilstation eine Aussage über die Stellung der von ihr verwalteten Weiche; diese Information kann durch eine weitere Instanzenvariable in der Klasse *verteilstation* gespeichert oder durch Abfrage des Weichenobjektes beschafft werden. Dieser Ansatz, den R. Wirfs-Brock und B. Wilkerson *„responsibility-driven"* nennen [20], legt also nur das Objektverhalten, noch nicht die Objektstruktur fest.

4.3. Herausarbeiten der Klassenhierarchie

Sind alle Objekte durch die notwendigen Attribute und Methoden charakterisiert, so kann man daran gehen, sie zu klassifizieren. Hierzu ist es erforderlich, gemeinsame Attribute und Methoden zu erkennen. Dabei ist der Fall, daß mehrere Objekte über die gleichen Attribute und Methoden verfügen, am einfachsten: man wird sie (zumindest anfangs) in eine Klasse einordnen. (Falls sich später das Erfordernis unterschiedlicher Implementierung herausstellt, ergeben sich doch noch Unterklassen.) Wesentlich häufiger dürfte aber der Fall sein, daß keine völlige Übereinstimmung vorliegt. In diesem Fall muß man Gruppen gemeinsamer Attribute und Methoden herausarbeiten. Diese definieren dann eine Oberklasse, während die vollständigen Verhaltensbeschreibungen unterschiedliche Unterklassen einführen. Je feiner man bei der Charakterisierung der Objekte die Methoden und Attribute unterteilt hat, desto größer ist die Chance, Gemeinsamkeiten zu finden. J. Waite zitiert ein Projekt mit 100474 Programmzeilen (ohne Kommentare), das 312 Klassen und 6044 Methoden umfaßt [10, p.106], also pro Methode etwa 16 Zeilen[9].

Selbstverständlich gilt dieses Granularitätsprinzip auch für die Klassenbildung: je kleiner

---

[8]Erfahrungsgemäß tauchen bei dieser Analyse neue Objekte auf, so daß der Entwurfsprozeß iterativ wird.

[9]Das Projekt ist in OBJECTIVE-C realisiert.

die Unterschiede zwischen Klassen sind, desto eher läßt sich eine spätere Änderung auf eine oder zwei Klassen beschränken. Bei einer hinreichend fein gestalteten Hierarchie gibt es nämlich mit größerer Wahrscheinlichkeit eine gemeinsame Oberklasse, wo dann die Änderung vorgenommen werden kann[10].

4.4. Werkzeuge

In realistischen Projekten steigt die Zahl der Klassen rasant an; objektorientierte Programmiersysteme verfügen häufig bereits standardmäßig über mehrere Hundert Klassen. Die Gefahr ist groß, daß die Software-Entwickler den Überblick verlieren und unproduktive Doppelarbeit geleistet wird. Von ganz besonderer Bedeutung für den wirtschaftlichen Einsatz der objektorientierten Software-Entwicklung sind daher Werkzeuge, die den Entwickler auf dem laufenden halten und ihm die systematische Analyse der verfügbaren Klassen und ihrer Beziehungen untereinander ermöglichen:

- Der Entwickler muß in der Klassenhierarchie navigieren können. Er muß beim Betrachten einer Klasse wissen, welche Ober- und Unterklassen, aber u.U. auch welche Nachbarklassen existieren.
- Er benötigt eine Liste aller Methoden, die für die betrachtete Klasse definiert sind. Diese sollte auch die ererbten Methoden umfassen, jedoch mit Quellenangabe.
- Er muß sich Klassen- und Methodendefinitionen anschauen können, wobei in vielen Fällen der Kommentar wichtiger ist als die Implementierung.

Diese Informationen benötigt er selbstverständlich gleichzeitig, d.h. wir setzen ein geeignetes Fenstersystem voraus. Bei großen Projekten ist außerdem erforderlich, daß auch die von einem Entwickler nur angedachten Klassendefinitionen bereits anderen Entwicklern bekannt gemacht werden, da sonst Doppelarbeit möglich ist. Die Anwendung des objektorientierten Entwurfs ohne integrierte Projektbibliothek gerät leicht zu einem Torso.

Man kann aus diesen wenigen Anmerkungen schon erkennen, daß der systematische Einsatz der objektorientierten Software-Entwicklung mehr ist als eine Programmiermethode oder das Umsteigen auf eine bestimmte Programmiersprache. Er erfordert mächtige Werkzeuge und umfangreiche Investitionen. Außerdem gibt es erst sehr wenige Arbeiten zum theoretischen Hintergrund. Diese sind aber erforderlich, wenn man zu objektorientierten Spezifikationstechniken mit formalisierten Schnittstellen zur Implementierung kommen will. Die kurze Geschichte der praktischen Informatik lehrt bereits, daß Modeströmungen kommen und gehen; langfristige Bedeutung erlangten nur diejenigen Techniken, die man theoretisch beherrscht hat.

---

[10]Wer das Paketverteilbeispiel mit der früheren Fassung [14] vergleicht, wird feststellen, daß die Einteilung feiner geworden ist.

5. Schlußbemerkung

Als Ursprung der objektorientierten Programmierung kann die Einführung des Klassenbegriffes in der Programmiersprache SIMULA angesehen werden [1]. Eine Sprache, die konsequent auf dem Prinzip der Objektorientierung aufbaut, ist aber erst SMALLTALK [2]. Die LISP-Familie bot sich auf Grund ihrer Flexibiltät an, verschiedene Varianten zu erproben; aber auch ganz andere Programmiersprachen, wie beispielsweise C [17], wurden durch Konzepte des neuen Programmierstils erfolgreich erweitert. Eine Übersicht aus jüngerer Zeit über die Programmiersprachenlandschaft findet der interessierte Leser bei J.H. Saunders [13]. Diese Arbeit enthält auch ein sehr ausführliches Literaturverzeichnis[11].

Die Lehrbücher über objektorientierte Programmierung stellen regelmäßig eine bestimmte Programmiersprache in den Mittelpunkt. B. Meyer [5] geht von EIFFEL aus, M. Mullin [8] setzt auf C++, während R. Wirfs-Brock et al. [21] SMALLTALK verwenden.

Meinen Kollegen H. Wedekind und F. Hofmann, sowie meinen Mitarbeitern K. Barthelmann, D. Kips, G. Schied, C. Schiedermeier und M. Zanzinger danke ich für die kritische Durchsicht des Manuskriptes und die wertvollen Hinweise.

Literatur

[1] O.J. Dahl/K. Nygaard: *Class and subclass declarations*, Proc. IFIP Working Conf. Simulation Languages (Oslo, May 1967; Ed.: J.W. Buxton), p.158-174, Amsterdam: North-Holland, 1968

[2] A. Goldberg/D. Robson: *Smalltalk 80 - The Language and Its Implementation*, Reading, Mass.: Addison-Wesley, 1983

[3] F. Hofmann: *Betriebssysteme - Grundkonzepte und Modellvorstellungen*, Stuttgart: Teubner, 1984

[4] I. Jacobson: *Object oriented development in an industrial environment*, ACM SIGPLAN Not. 22, 12 (1987), p.183-191

[5] B. Meyer: *Object-Oriented Software Construction*, Englewood Cliffs, N.J.: Prentice-Hall, 1988

[6] J. Micaleff: *Encapsulation, reusability and extensibility in object-oriented programming languages*, J. Object-Oriented Programming 1,1 (1988), p.12-39

[7] D.A. Moon: *Object-oriented programming with* FLAVORS, ACM SIGPLAN Not. 21, 11 (1986), p.1-8

[8] M. Mullin: *Object-Oriented Program Design With Examples in C++*, Reading, Mass.: Addison-Wesley, 1989

[9] L. Power: *Addendum to the Proceedings OOPSLA'87 - Specification and design of objects*, ACM SIGPLAN Not. 23, 5 (1988), p.7-50

[10] L. Power: *Addendum to the Proceedings OOPSLA'87 - Engineering object-oriented products*, ACM SIGPLAN Not. 23, 5 (1988), p.89-128

---

[11]Nach Abschluß des Manuskriptes erreichte mich das September-Heft 1990 der *Communications of ACM*, das fast ausschließlich der objektorientierten Programmierung gewidmet ist und einige interessante Aufsätze enthält.

[11] M.B. Rosson/E. Gold: *Problem-solution mapping in object-oriented design*, ACM SIGPLAN Not. 24, 10 (1989), p.7-10

[12] E. Sandewall: *Programming in an interactive environment - the „LISP" experience*, ACM Computing Surveys 10, 1 (1978), p.35-71

[13] J.H. Saunders: *A survey of object-oriented programming languages*, J. Object-Oriented Programming 1,6 (1989), p.5-11

[14] H.J. Schneider: *Programming language concept for structuring control software in correspondence to the technical process*, Computing 85 (Eds.: G. Bucci/G. Valle), p.261-268, Amsterdam: North-Holland, 1985

[15] H.J. Schneider: *Objektorientierte Strukturierung verteilter Software und statische Typprüfung*, Informatik-Fachberichte 167 (1988), p.546-555

[16] H. Stoyan/G. Görz: *Was ist objektorientierte Programmierung?*, Objektorientierte Software- und Hardware-Architekturen (Hrsg.: H. Stoyan/H. Wedekind), p.9-31, Stuttgart: Teubner, 1983

[17] B. Stroustrup: *The C++ Programming Language*, Reading, Mass.: Addison-Wesley, 1986

[18] H. Wedekind: *Objektorientierung und Vererbung*, Informationstechnik it 32, 2 (1990), p.79-86

[19] P. Wegner: *The object-oriented classification paradigm*, Research Directions in Object-Oriented Programming (Eds.: B. Shriver/P. Wegner), p. 479-560, Cambridge, Mass.: MIP Press, 1987

[20] R. Wirfs-Brock/B. Wilkerson: *Object-oriented design - A responsibility-driven approach*, ACM SIGPLAN Not. 24, 10 (1989), p.71-75

[21] R. Wirfs-Brock et al.: *Designing Object-Oriented Software*, Englewood Cliffs, N.J.: Prentice-Hall, 1990

Anwendungsdefinition, Projektierung und Konfigurierung

Rapid-Prototyping und CASE-
Gegensatz oder Ergänzung?

Dr. Peter Baur, GPP mbH, Oberhaching

Kurzfassung:

Rapid-Prototyping hat das Ziel, möglichst frühzeitig bzw. entwicklungsbegleitend Fehler und Unzulänglichkeiten eines zu entwickelnden Software-Systems zu erkennen. Ansätze die dazu notwendigen rechnergestützten Hilfsmittel bereitzustellen, waren jedoch bislang methodisch und werkzeugmäßig von den industriell eingesetzten Softwareentwicklungsumgebungen isoliert. Prototyping bedeutete hohen zusätzlichen Aufwand. Deshalb wurde ein neues Prototypingkonzept erarbeitet und implementiert, das die Projektdatenbank einer CASE-Umgebung inhaltlich als Prototypmodell betrachtet und darauf basierend unterschiedlichste Analyse- und Simulationswerkzeuge bereitstellt. Damit kann mit geringstem Aufwand jederzeit ein Entwicklungsstand eingefroren und validiert werden. [1]

1.   Ausgangsproblematik

Entwicklungsprojekte in der Automatisierungstechnik sind zunehmend gekennzeichnet durch eine hohe Komplexität. Es sind Echtzeitanforderungen zu berücksichtigen. Die Realisierung erfolgt in Form verteilter Systeme. Hohe Zuverlässigkeitsanforderungen als auch Sicherheitskriterien sind zu beachten. Die Gefahr des Auftretens von Entwicklungsfehlern ist damit in solchen Projekten sehr hoch und ohne geeignete Werkzeuge nicht mehr beherrschbar.

[1]   Die hier beschriebenen Arbeiten zur Realisierung einer Prototyping-Umgebung wurden mit Mitteln des BMFT unter dem Förderkennzeichen 01RS8907 gefördert

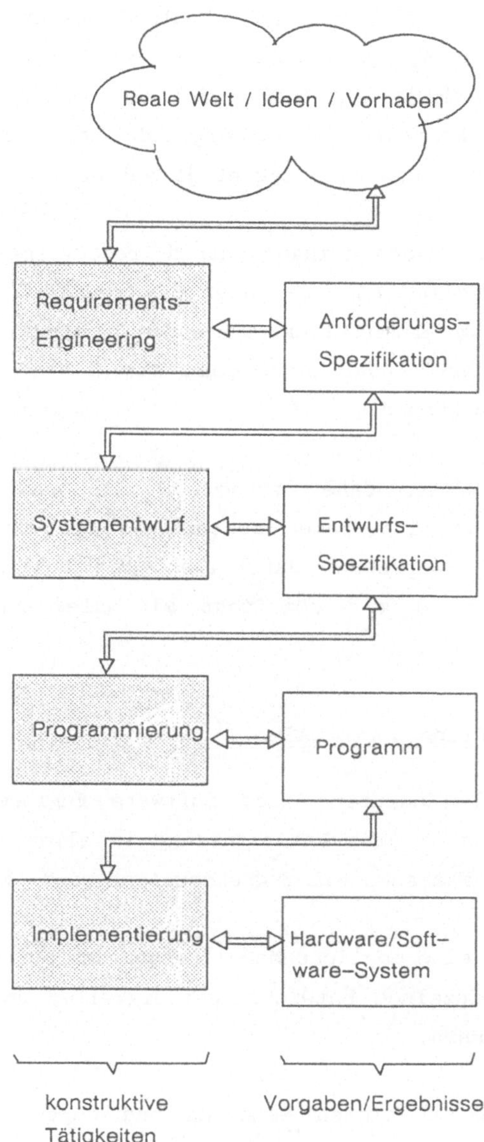

Bild 1:   Softwareentwicklungsprozeß: Aktivitäten, Ergebnisse

Im Gegensatz dazu können aber inhaltliche Denkfehler z.B. im Lösungsalgorithmus, Mißverständnisse über die Funktionsweise des zu entwickelnden Programms, zu lange Antwortzeiten, usw. von CASE-Tools nicht aufgedeckt werden. Diese Art von Fehler läßt sich bisher erst nach der Implementierung feststellen, d.h. beim Testen des Programms bzw. bei der Abnahme durch den Auftraggeber.

Zudem wird sehr häufig technisches Neuland beschritten bzw. technisch sehr anspruchsvolle Aufgaben bearbeitet, so daß gerade in den frühen Phasen einer Entwicklung oft die Frage der Machbarkeit bzw. der Wahl geeigneter Lösungskonzepte im Vordergrund steht. Fehlerbehebungen und Konzeptkorrekturen werden umso teurer je später sie durchgeführt werden.

Aus diesen Gründen werden dringend Rapid-Prototyping-Werkzeuge benötigt, um so früh wie möglich Aussagen über das Verhalten und die Leistungsfähigkeit eines in Entwicklung stehenden Systems zu erhalten. Diese sollten mit fortschreitender Entwicklung aktualisiert bzw. präzisiert und detailliert werden können.

Diese Werkzeuge müssen daher methodisch als auch hinsichtlich ihres Datenbestandes einen integralen Bestandteil der existierenden Entwicklungsmethoden und -werkzeuge darstellen. Die Realisierung und Analyse eines Prototypen muß mit möglichst minimalem zusätzlichem Aufwand durchführbar sein.

2. Nutzen und Grenzen von CASE

Durchgängige CASE (Computer Aided Software Engineering) - Umgebungen bieten dem Entwickler eine Unterstützung in allen Projektphasen, wobei das traditionelle Phasenmodell zugrundegelegt wird /1/.

Bild 1 zeigt dazu eine mögliche Modellierung des Softwareentwicklungsprozesses mit den einzelnen Entwicklungstätigkeiten und den daraus resultierenden Ergebnissen.

Diese CASE-Umgebungen bieten Methoden und darauf aufbauende Spezifikationsmodelle und -sprachen mit entsprechend abgestimmten Auswertewerkzeugen (Analyse, Dokumentation, Konfigurationsmanagement). Sie sind bereits so mächtig, daß aus den Spezifikationen eine automatische Programmgenerierung ermöglicht wird.

Durch umfangreiche Analysefunktionen kann der Entwickler diese Spezifikationen jederzeit durch verschiedene Prüfungen auf formale Eigenschaften wie Vollständigkeit, Widerspruchsfreiheit usw. untersuchen lassen. Dies trägt bedeutend dazu bei, die Qualität eines Softwareproduktes zu verbessern.

3. Mängel existierender Prototyping-Konzepte

Prototyping dagegen hat insbesondere das Ziel, inhaltliche und konzeptionelle Fehler frühzeitig zu offenbaren. Es werden im allgemeinen zwei Zielrichtungen unterschieden /2/, /3/, /4/:

1. "Horizontales Prototyping": Das Verhalten des zu entwickelnden Systems soll dem späteren Benutzer demonstriert werden, um gemeinsam mit diesem den Funktionsumfang und die Benutzeroberfläche festzulegen.

2. "Vertikales Prototyping": Bestimmte Teilaspekte des zu entwickelnden Systems sollen genauer untersucht werden, z.B. Untersuchung von verschiedenen Realisierungsmöglichkeiten, funktionelle Aspekte oder das Zeitverhalten.

Bei diesen Rapid-Prototyping-Konzepten wird meistens weder ein sequentielles Phasenmodell noch eine formale Spezifikation zugrunde gelegt. Die Realisierung eines Prototypen wird als eine von der eigentlichen Systementwicklung isolierte Aktivität betrachtet, wobei das Ergebnis der Prototypenanalyse nur manuell in die weitere Systementwicklung eingebracht werden kann.

Dabei besteht die latente Gefahr, diesen Prototyp zu einem Endprodukt mit vollem Funktionsumfang zu vervollständigen. Auf diese Weise entstehen im allgemeinen schwer wartbare und fehleranfällige Softwaresysteme, weil ausgehend von einer ad-hoc-Vorversion eines Software-Systems keine strukturierte Weiterentwicklung möglich ist.

Wird der Prototyp jedoch "weggeworfen", und ein Redesign unter Zuhilfenahme von Spezifikationsmethoden- und werkzeugen vorgenommen, ergibt sich ein erhöhter Aufwand, der, zumindest aus der Sicht des Auftraggebers, nicht zu rechtfertigen ist.

Dem steht wiederum entgegen, daß man bei einer strikten phasenorientierten Entwicklung erst auf der Code-Ebene, nach Abschluß der Spezifikation, Aufschluß über die Korrektheit der Lösungskonzeption und die Erfüllung der Wünsche der Auftraggeber erhält.

4. Integriertes Prototyping-Konzept

Aufgrund der genannten Probleme ist das Rapid-Prototyping als eine Ergänzung zu Softwareentwicklungsumgebungen zu betrachten. Es muß in einem laufenden Entwicklungsprozeß jederzeit ermöglicht werden, den jeweils aktuellen Entwicklungsstand zu validieren, um damit Fehler und Fehlentwicklung frühzeitig zu erkennen und zu korrigieren.

Gerade der Einsatz einer rechnergestützten Softwareentwicklungsumgebung gewährleistet, daß die verschiedenen Phasenergebnisse (und Zwischenergebnisse) in einer rechnergestützt auswertbaren Form in einer Datenbank frühzeitig verfügbar sind und damit für das Rapid-Prototyping ausgewertet werden können. Damit liegt es nahe, eine solche Projektdatenbank inhaltlich zu jedem Entwicklungszeitpunkt als eine prototypische Beschreibung des zu entwickelnden Zielsystems zu betrachten.

Somit werden Rapid-Prototyping-Werkzeuge benötigt, die die im Rahmen der Entwicklung entstehenden Beschreibungen (Spezifikationen) ablauffähig machen bzw. simulieren und analysieren (siehe Bild 2).

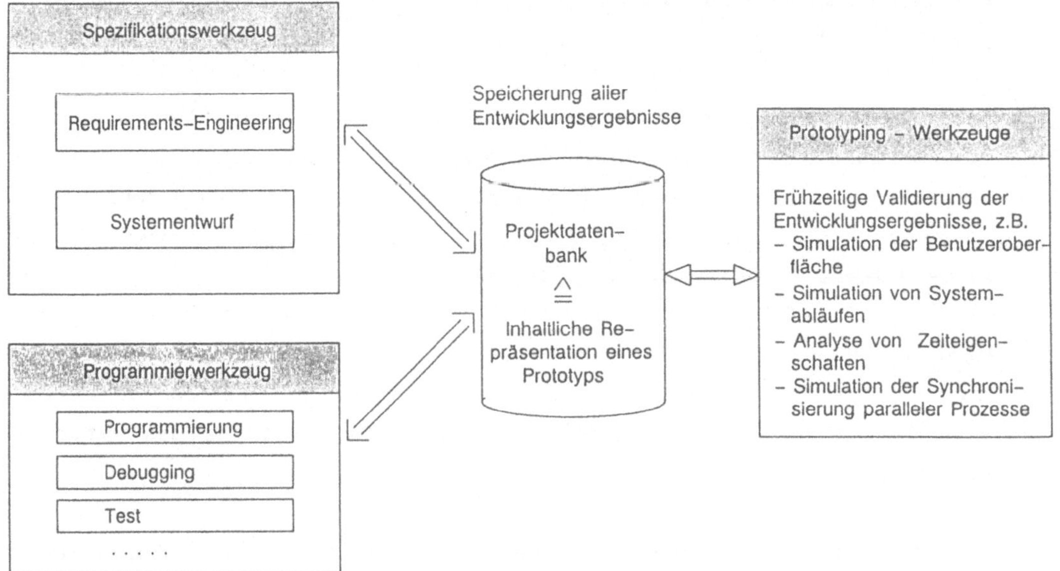

Bild 2:   Gesamtkonzept

5. Prototyping im Rahmen der CASE-Umgebung EPOS

Das EPOS-System ist eine in vielen (Groß-)Projekten, unter anderem auch bei der Entwicklung von verteilten Realzeitsystemen, bewährte und breit eingesetzte Softwareproduktionsumgebung. Sie unterstützt methodisch und werkzeugmäßig den gesamten Softwareentwicklungsprozeß, beginnend bei der Anforderungsdefinition bis hin zu den Wartungs- und Pflegeaufgaben. Alle anfallenden Entwicklungsergebnisse werden dabei laufend und aktuell in einer Projektdatenbank gehalten und zur Verfügung gestellt.

Im Hinblick auf die Entwicklung hochzuverlässiger bzw. sicherer Software ist zudem von Bedeutung, daß EPOS bereits eine Vielzahl von rechnergestützten Spezifikationsanalysen bereitstellt, um Fehler möglichst frühzeitig im Entwicklungsprozeß zu erkennen /8/, /9/.

EPOS bietet damit eine geeignete Ausgangsbasis für die Entwicklung spezieller Prototyping-Werkzeuge. Ziel der Entwicklung war es, die existierende Projektdatenbank als Prototyp zu betrachten und verschiedene Analyse und Simulationsfunktionen bereitzustellen, um diesen Prototyp "validieren" zu können. Beispiele für geforderte Validierungsfunktionen sind:

- Simulation der Benutzeroberfläche
- Simulation von Systemabläufen
- Analyse von Zeiteigenschaften
- Simulation der Synchronisierung paralleler Prozesse
- usw.

Dabei wurden zwei Realisierungskonzepte verfolgt:
a) Transformation der in Form von EPOS-Spezifikationen vorliegenden Information, so daß diese unter Zuhilfenahme existierender Analyse- und Simulationswerkzeuge validiert werden kann (siehe 5.1).

b) Realisierung eines Spezifikationsinterpreters, der es jederzeit ermöglicht, eine Spezifikation auszuführen und interaktiv zu evaluieren (siehe 5.2).

Bild 3 zeigt in einer Übersicht die damit realisierte Prototyping-Umgebung und die Schnittstellen zu bestehenden Werkzeugen.

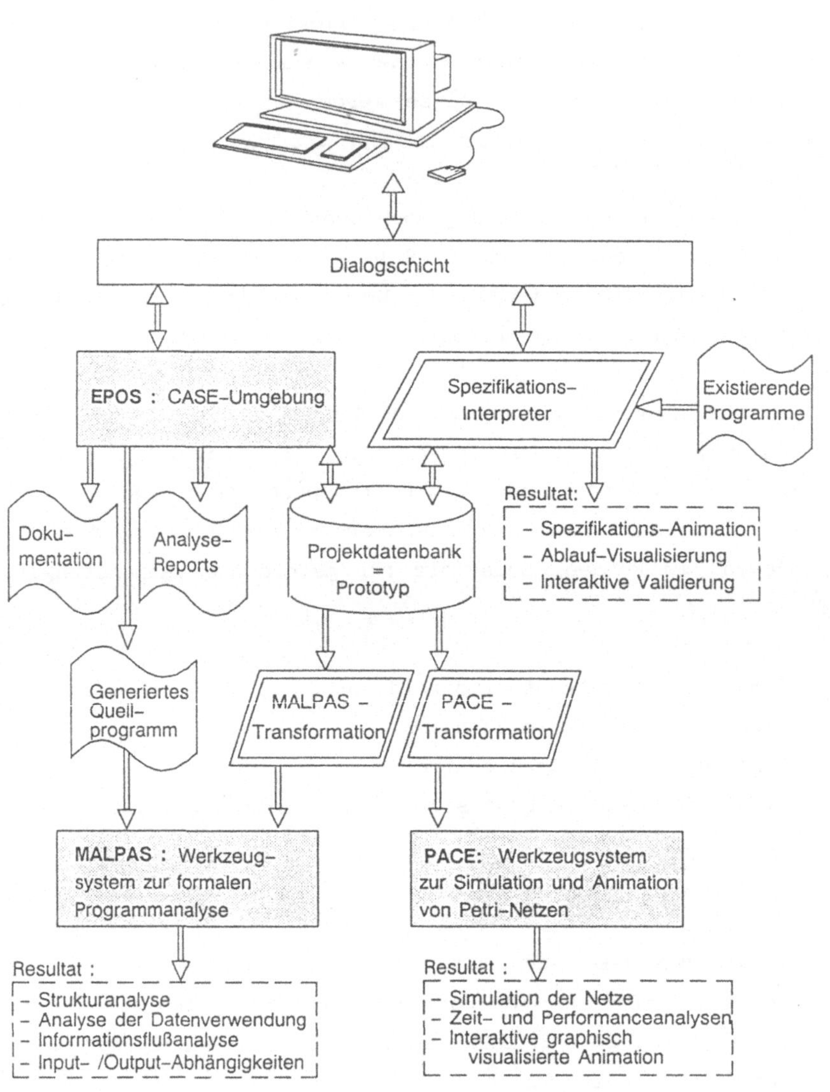

Bild 3:

5.1 Integration externer Werkzeuge

Petri-Netz-Simulation mit MALPAS

Ein wesentlicher Aspekt bei der Systementwicklung ist die Klärung, ob dynamische bzw. zeitliche Anforderungen erfüllt werden. Ein Modell, das solche Betrachtungen erlaubt, sind erweiterte Petri-Netze. Sie eignen sich speziell für zeitbehaftete, dynamische Untersuchungen, da den Transitionen Eigenschaften, wie Zeitverhalten und Schaltwahrscheinlichkeit zugeordnet werden könnnen /10/. Deshalb wurde ein Transformator realisiert, der die dynamischen Ablaufeigenschaften einer EPOS-Spezifikation in die "höheren" Petri-Netze des Werkzeuges PACE umsetzt /11/. PACE ermöglicht dann eine Simulation und interaktive Animation dieser Netze, die eine Verfolgung des Zeitablaufs gewährleistet, Kapazitäts-Mengenbetrachtungen gestattet und die auf Grund der Netzhierarchie die Betrachtung von Teilnetzen, d.h. verschiedener Entwurfsebenen erlaubt.

PACE steht in einer Smalltalk80-Umgebung zur Verfügung. Der Anwender hat damit zusätzlich die Möglichkeit, bestimmte Simulationsaspekte anwendungsspezifisch zu adaptieren. Bild 4 zeigt einen Bildschirmausschnitt während einer PACE-Simulation.

Formale Analyse mit MALPAS

MALPAS ist ein Werkzeug zur formalen Analyse von Spezifikationen oder Programmen und arbeitet auf einer universellen Zwischensprache /12/. Dazu existieren verschiedene Übersetzer, die Quellprogramme (FORTRAN, Pascal, Ada, C) automatisch übersetzen. Diese Quellprogramme können prinzipiell mit den entsprechenden EPOS-Codegeneratoren erzeugt werden. Dies ist jedoch nur ab einem gewissen Fertigstellungsstand einer EPOS-Spezifikation sinnvoll und für eine frühe prototypische Betrachtung in der Regel ungeeignet. Deshalb wurde ein spezieller Übersetzer für EPOS-Spezifikationen realisiert. Damit können dann diese Spezifikationen direkt mit Hilfe von MALPAS analysiert werden:

- Detektion unstrukturierter Abläufe
- Analyse der Datenverwendung und Datenreferenzen
- Abhängigkeiten der Programmpfade von Daten und Bedingungen
- Aufzeigen der exakten Abhängigkeiten zwischen Eingangs- und Ausgangsgrößen für jeden semantisch möglichen Programmpfad.

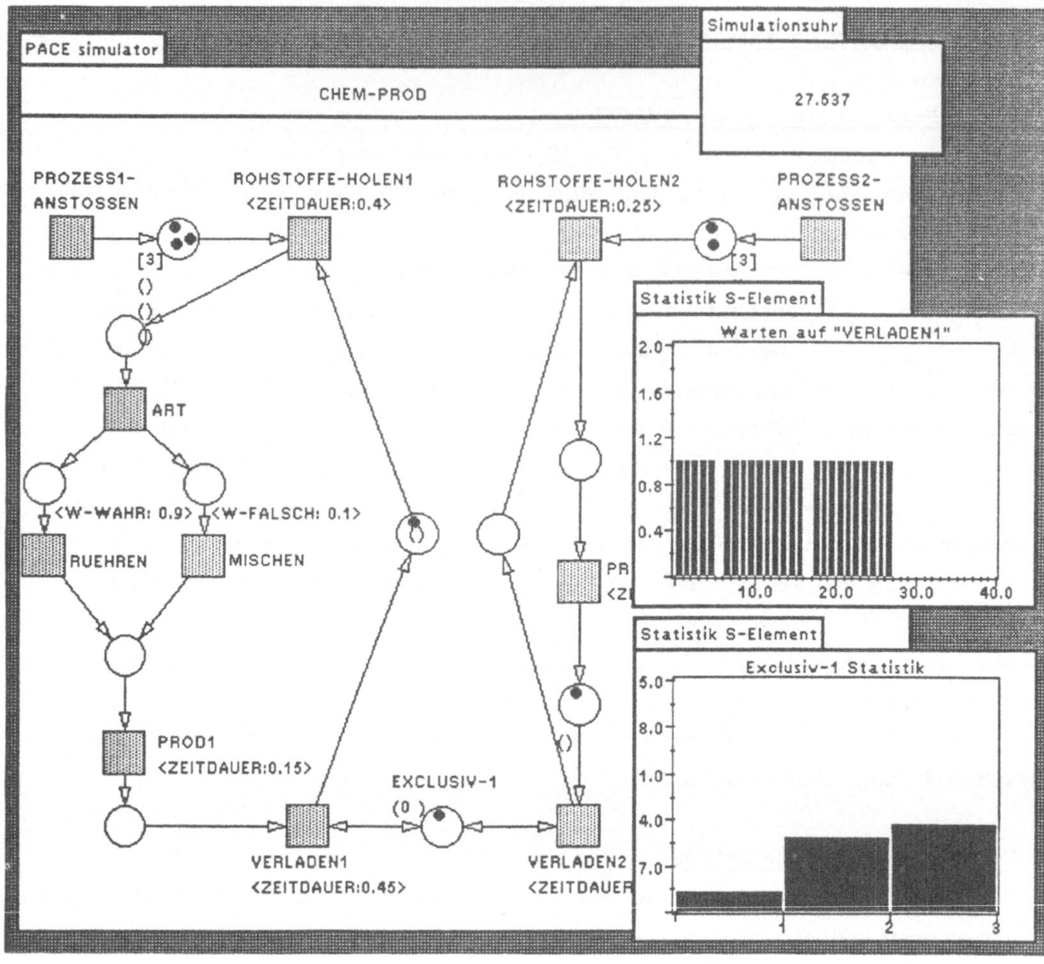

Bild 4: PACE-Simulation

5.2   EPOS-Interpreter

Der EPOS-Interpreter bietet die Möglichkeit, eine EPOS-Systemspezifikation, unabhängig von ihrem Fertigstellungsgrad, jederzeit "ablaufen" zu lassen. Dem Anwender steht damit eine Art "Spezifikationsdebugger" zur Verfügung, der auf (nahezu beliebig) unvollständigen Spezifikationen arbeiten kann.

Wesentliche Eigenschaften sind:
- Wahl des zu interpretierenden Spezifikationsausschnittes
- Wahl verschiedenster TRACE- und WATCH-Funktionen
- Graphische Ablaufdokumentation
- BREAKPOINT- und SHOW-Funktionen
- Reale Ausführung von in der Spezifikation referenzierten Bibliotheksfunktionen.

Besonders erwähnenswert ist die Möglichkeit, beliebige Programmteile aus vom Anwender bereitgestellten Objekt-Bibliotheken in die Spezifikation einzubinden, die dann real ausgeführt werden. Damit können bereits existierende Programme oder z.B. Programme zur Simulation der Systemumgebung (z.B. technische Anlagen) einfach in die Betrachtung mit einbezogen werden.

6.   Literatur

/1/   Hansen, H.-L.: Software-Produktionsumgebungen - Bedeutung - Auswahl - Einsatz, Symposium "Software-Entwicklungssysteme - und -Werkzeuge", TA Esslingen, September 1985, S. 16.3-1 - 16.3-10

/2/   Floyd, C.: A Systematic Look at Prototyping. Approaches to Prototyping. Springer Verlag 1984.

/3/   Spitta, T.: Software-Engineering und Prototyping. Springer-Verlag 1989

/4/   Pomberger, G., Bischofsberger, W., Keller, R., Schmidt, D.:
        Prototypingorientierte Softwareentwicklung. Institut für Infor-
        matik, Universität Zürich 1987.

/5/   GPP mbH: EPOS-Kurzbeschreibung. GPP mbH, Kolpingring 18a, 8024
        Oberhaching, 1990

/6/   Lauber, R. und Lempp, P.: Integrierte Rechnerunterstützung für
        Entwicklung, Projektmanagement und Produktverwaltung mit EPOS

/7/   Göhner, P.: Methoden zur Entwicklung von Realzeitsystemen und ihre
        praktische Anwendung in EPOS, Informatik-Fachberichte Bd. 86
        S. 325-335, Springer-Verlag 1984

/8/   Baur, P.: Fault-Avoidance And Fault-Removal Features of the
        Computer-Aided Development Support System EPOS
        Proceedings 3rd IFAC/IFIP workshop Safety of Computer Control
        Systems SAFECOMP '83, Cambridge U.K., September 1983 (Ed.: J.A.
        Baylis). Pergamon Press, Oxford, New York 1983, S. 103 - 110.

/9/   Baur, P.: Methoden und Werkzeuge zur entwicklungsbegleitenden bzw.
        entwicklungsnachgeschalteten Verifikation/Validation von Software;
        Sichere Software, Heinrich Kersten (Hrsg.)
        Hüthig Buch Verlag, Heidelberg 1990.

/10/   Bruno, G., Marchetto, G.: Rapid Prototyping of control systems
        using high level Petri-Nets. In Proc. 8th Int. Conf. Software
        Engineering, London, August 1985.

/11/   GPP mbH: PACE-Kurzdarstellung
        GPP mbH, Kolpingring 18a, 8024 Oberhaching, 1989.

/12/   RTP Ltd.: MALPAS-Management Guide
        RTP Software Limited 1989, erhältlich bei GPP mbH, 8024 Oberhaching.

Ressourcen-getriebene Konfigurierung
modularer Technischer Systeme

M. Heinrich, E.W. Jüngst
Daimler-Benz AG
Bereich Forschung und Technik
Forschungsinstitut Berlin

Alt-Moabit 91b
D-1000 Berlin 21

Zusammenfassung

Es wird ein Konfigurationsverfahren für modular aufgebaute technische Systeme vorgestellt, das die typischen Nachteile bisheriger Ansätze vermeidet.

Grundlage ist die Modellierung modularer Technischer Systeme auf der Basis von Relationen zwischen Komponenten und den als "Ressourcen" repräsentierten Systemschnittstellen. Eingeführt werden dafür die Standard-Relationen "stellt bereit", "verbraucht" bzw. "nutzt". Diese Modellierung erlaubt ein einfaches iteratives Verfahren zur Auswahl und Positionierung von Komponenten.

Modell und Verfahren bilden die Basis der Expertensystemschale COSMOS, die für Konfigurationsaufgaben bei modularen Technischen Systemen recht universell einsetzbar ist.

Durch das Modell und die Wissenserwerbskomponente von COSMOS wird die Pflege des Wissens entscheidend erleichtert. Der Vorteil unserer Modellierung liegt in der Separation des umfangreichen Katalogwissens vom eigentlichen Expertenwissen, in der Entkopplung des Wissens über einzelne Komponenten voneinander und in der systematischen Strukturierung des Wissens in Taxonomien.

Erste Anwendung von COSMOS ist ein Prototyp eines Expertensystems zur Konfiguration Speicherprogrammierter Steuerungen. Weitere Einsatzfelder werden aufgezeigt.

1. Einleitung

Konfigurieren ist das Entwerfen eines Technischen Systems entsprechend den Anforderungen einer Spezifikation durch Auswählen, Parametrieren und Zusammenfügen von Exemplaren geeigneter Komponententypen eines vorgegebenen Komponentenkatalogs.

Bei der Projektierung von Automatisierungseinrichtungen geht es darum, für die jeweilige Automatisierungsaufgabe unter Beachtung der Kundenforderungen eine individuelle Lösung zu erarbeiten. Betriebswirtschaftlich optimal gelöst wird diese Aufgabe durch Einsatz von Komponenten aus modularen Komponentensystemen. Deren Entwicklungsziel ist, die Komponenten so geschickt zu entwerfen und aufeinander abzustimmen, daß sie im Zusammenwirken ein möglichst weites Anwendungsgebiet abdecken. Solche modularen Komponentensysteme werden aber nicht nur für die informationsverarbeitenden Einrichtungen entworfen, wie Speicherprogrammierte Steuerungen und Prozeßrechner, sondern auch für die Meßwerterfassung und die Stellglieder, wie z.B. Schaltanlagen, Antriebe, Förderzeuge. Die Projektierung besteht dann größtenteils in der Auswahl und dem Zusammenfügen geeigneter Komponenten, also dem Lösen von Konfigurierungsaufgaben.

Für den kostengünstigen Einsatz von modularen Komponentensystemen wird erhebliche Expertise benötigt. Selten gibt es jedoch genügend Experten. Auch birgt die manuelle Lösung komplexer Aufgaben das Risiko, daß Irrtümer, z.B. Inkonsistenzen und vergessene Teile, sich oft erst bei der Inbetriebnahme zeigen. Wie sehr der Einsatz von Konfigurationsexpertensystemen hier eine Kostenreduktion und Qualitätssteigerung bewirken kann, zeigt der wirtschaftliche Erfolg von XCON, dem Expertensystem zur Konfiguration von VAX-Rechnern. Bei XCON stellte sich jedoch auch heraus, daß ein beträchlicher Anteil der erzielten Einsparungen zur ständigen Pflege des Expertensystems aufgewendet werden muß.

Der Aufwand für Erwerb und Pflege des Wissens ist der wesentliche Hemmschuh für den Einsatz von Expertensystemen. Ursache des hohen Aufwands ist die konzeptionelle Schwäche klassischer Expertensysteme, die im allgemeinen nur nachahmen, was ein Experte tut, und kein Wissen darüber besitzen, warum er etwas tut.

Diese Aussage trifft in besonderem Maße auf rein regelbasierte Systeme zu [1, 2]. Frameorientierte Ansätze streben an, das Faktenwissen über einzelne Komponenten zumindest an einer Stelle der Wissensbasis organisatorisch zusammenzufassen [3 - 8]. Andere Ansätze sehen die Architektur des modularen Systems als Variantenraum, wobei die Spezifikation der gewünschten Konfiguration als Constraints formuliert wird. Deren Propagierung führt zur Identifikation einer Lösung im Variantenraum [9 - 12]. Schnittstellen werden dabei als besondere Constraints aufgefaßt [13 - 16]. Andererseits werden Bilanzierungen von Schnittstellen zur Überprüfung von Konfigurationen herangezogen [17].

Der Aufwand für die Erstellung eines Expertensystems inklusive Wissenserwerb und -pflege hängt wesentlich von der gewählten Wissensrepräsentationsform ab. In der Praxis ist ein großer Teil des Wissens in den vorhandenen Komponenten-Katalogen niedergelegt, die damit eine der wesentlichen Quellen der Wissensakquisition sind. Im Katalog werden einerseits die Eigenschaften jeder einzelnen Komponente beschrieben, andererseits Beziehungen zwischen den Komponenten aufgeführt. Diese explizite Nennung der Beziehungen verursacht jedoch große Probleme bei der Wissenspflege, denn beim Hinzufügen oder Entfernen einer Komponente muß stets das Wissen über alle Komponenten durchmustert und, falls zu ihnen Beziehungen bestehen, überarbeitet werden. Dadurch steigt der Pflegeaufwand mit der Zahl der Komponenten schnell an.

Bei der Festlegung der Wissensrepräsentationsform sollte man das in den Katalogen bewährte komponentenorientierte Organisationschema übernehmen. Die Vorteile dieses Ansatzes liegen in der sinnfälligen Zusammenfassung intuitiv zusammengehörender Informationen, ganz abgesehen von der Vereinfachung der Übertragung des Wissens aus dem vorhandenen Katalog in die Wissensbasis. Dabei sollten die Beschreibungen der Komponenten jedoch durch Abstraktion ihrer wechselseitigen Beziehungen voneinander entkoppelt werden. Der Wissenseingabeaufwand kann durch Redundanzvermeidung weiter reduziert und die Pflege wesentlich vereinfacht werden, wenn man berücksichtigt, daß Technische Systeme häufig über Komponenten-Warenkörbe verfügen, in denen viele ähnliche, auf unterschiedliche Anwendungen spezialisierte Komponenten vorkommen. Das Ausnutzen dieser Ähnlichkeiten geschieht zweckmäßigerweise durch Organisation der Komponententypen in einer Taxonomie (Verwandtschaftshierarchie).

Unser Ansatz [18] führt eine neuartige Modellierung ein, die es erlaubt, allein aus den Eigenschaften von Systemschnittstellen ("Ressourcen") und den Beziehungen von Komponenten zu diesen Ressourcen zulässige Konfigurationen durch ein selbstorganisierendes Verfahren herzuleiten.

2. Grundprinzip

Die Abstraktion der Beziehungen zwischen Komponenten als Ressourcenaustausch bildet die Basis eines allgemeinen Modells zur Beschreibung von Konfigurationen Technischer Systeme. Der Prozeß des Konfigurierens erfolgt durch Bilanzieren und Ausgleichen von Ressourcenforderungen.

2.1 Das Ressourcen-Modell

Das Modell orientiert sich an den Bauprinzipien modularer Systeme. In den Design-Richtlinien eines Komponentensystems werden nach der Definition der System-Grundstruktur zuerst die Systemschnittstellen festgelegt, nach denen sich die Konstruktionsentwürfe der einzelnen Komponenten richten müssen. Diese Schnittstellen werden einheitlich als "Ressourcen" beschrieben, wobei es sich um physikalische Ressourcen (z.B. elektrische Energie, Kühlleistung), standardisierte technische Ressourcen (z.B. V24-Schnittstellen, EPROM-Sockel), in-house-Standards (z.B. Bus-Anschlüsse, Software-Dienste) und betriebswirtschaftliche Ressourcen (z.B. Kapital, Zeitvorgabe für die Montage) handeln kann. Diese Systemschnittstellen und ihre Festlegungen überdauern in der Regel mehrere Komponenten-Generationen.

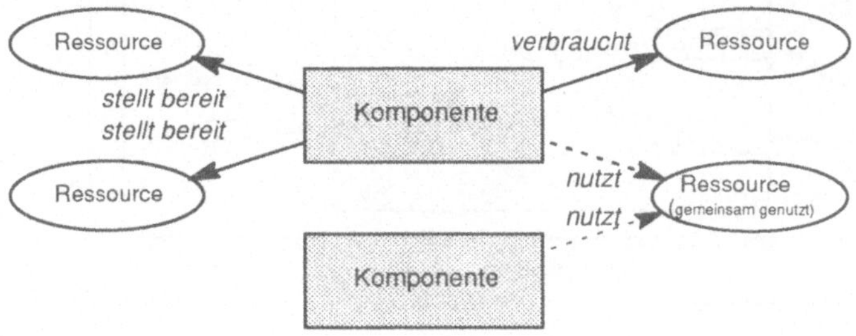

Bild 1: Basis-Beziehungen zwischen
Komponenten und Ressourcen

Als prinzipielle Beziehungen zwischen Komponenten und Ressourcen werden unterschieden (Bild 1):

o Eine Komponente **stellt Ressourcen bereit**
   (z.B. RAM-Speicher bei einer RAM-Speicherkarte)

o Eine Komponente fordert Ressourcen an.
   Dabei werden einige **Ressourcen** von der Komponente **verbraucht** (z.B. Bus-Anschluß)
   Andere **Ressourcen** werden von der Komponente **genutzt**, können aber mit anderen
   Komponenten geteilt werden (z.B. System-Takt, Software–Dienste).

In dieser völlig anwendungsunabhängigen Form lassen sich alle wesentlichen Eigenschaften von Komponen'en als Beziehungen zu Ressourcen ausdrücken. Das gleiche gilt für die Eigenschaften der

Umgebung und für das Technische System als Ganzes (s. Bild 2). Direkte Beziehungen zwischen Komponenten treten in diesem Modell nicht mehr auf, so daß das Hinzufügen und Entfernen von Komponenten durch rein lokale Änderungen der Wissensbasis vorgenommen werden kann.

2.2 Ressourcen-getriebenes Konfigurieren

Das ressourcen–getriebene Konfigurieren sei anhand von Bild 2 erläutert. Es besteht eine Wechselwirkung zwischen Umgebung und Technischem System, wobei die Umgebung einerseits Ressourcen anfordert, die vom Technischen System bereitgestellt werden müssen, ihm andererseits Ressourcen bereitstellt. Um die Anforderungen der Umgebung zu erfüllen, enthält das Technische System geeignete Komponenten, die diese geforderten Ressourcen bereitstellen. Diese Komponenten nutzen oder verbrauchen zu ihrem Betrieb ihrerseits Ressourcen, die wiederum von anderen Komponenten des Technischen Systems oder letzlich von der Umgebung bereitgestellt werden müssen. Für eine korrekte Konfiguration ist es notwendig, daß alle insgesamt geforderten Ressourcen bereitgestellt werden, d.h. die Ressourcenbilanz zumindest ausgeglichen ist.

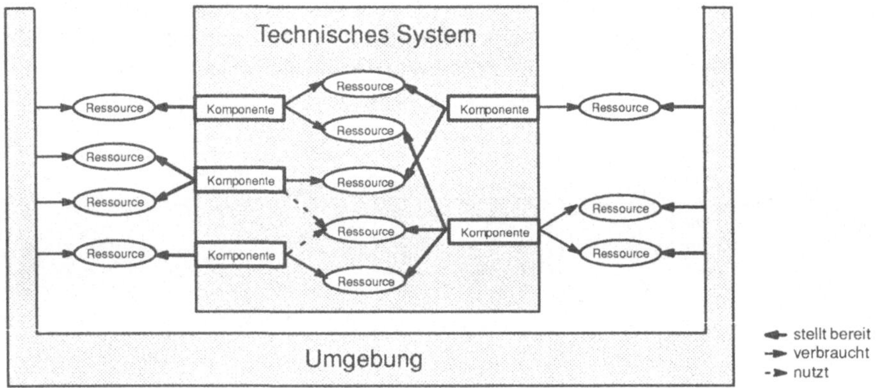

Bild 2: Ressourcen–getriebenes Konfigurieren

3. Die Konfigurationsschale COSMOS

An dieser Modellierung orientieren sich – völlig unabhängig von der jeweiligen Anwendung – Schlußfolgerungsverfahren, Debug-Werkzeuge, Wissensrepräsentation und Wissenserwerbskomponente von COSMOS.

3.1 Schlußfolgerungskomponente

Das iterative Schlußfolgerungsverfahren (Bild 3) beruht auf dem Grundprinzip der ressourcen-getriebenen Konfigurierung. Kernstück ist die sukzessive wissensbasierte Abarbeitung der Ressourcen-Bilanz, in der die Anforderungen und Bereitstellungen aller Ressourcentypen durch die bereits gewählten Komponententypen notiert werden. Aus der Menge der verbleibenden Anforderungen wird die nächste abzudeckende Ressource bestimmt, zu deren Bereitstellung geeignete Komponenten aus dem Katalog ausgesucht werden. Unter ihnen wird die jeweils beste Komponente gewählt und in der Entscheidungs-Liste notiert. Ihre Ressourcen-Bereitstellungen und -Anforderungen werden in die Bilanz eingetragen. Das Verfahren beginnt damit, daß als Systemspezifikation die Umgebung wie eine gewählte Komponente in die Entscheidungsliste eingetragen wird, und endet, wenn alle Anforderungen abgedeckt sind.

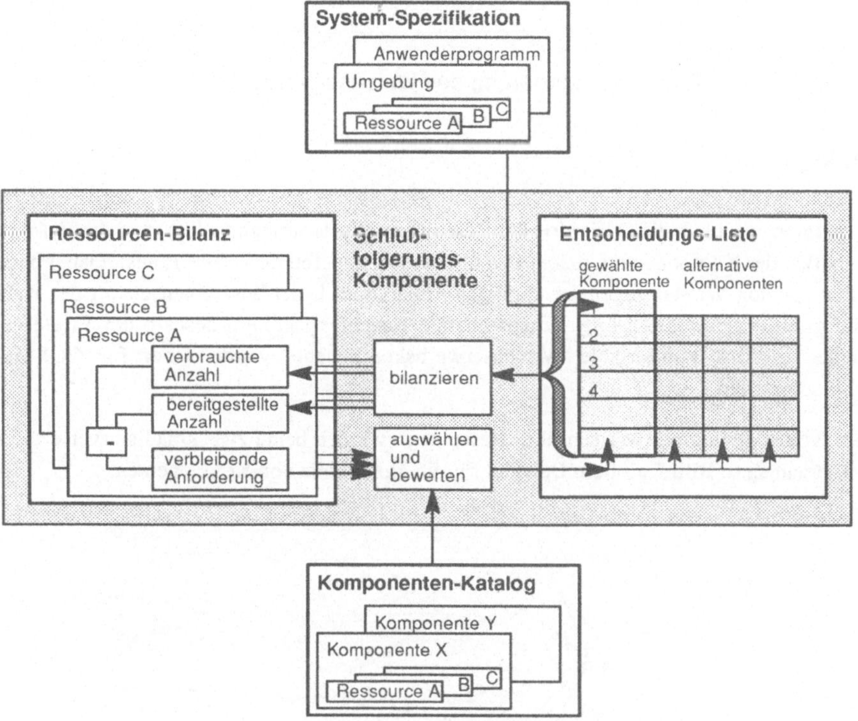

Bild 3: Prinzipieller Aufbau der COSMOS-Schlußfolgerungskomponente

3.2 Strukturierung des Wissens

Das Schlußfolgerungsverfahren wird durch Wissen gesteuert. Unser Modell legt eine Strukturierung in fünf Wissensebenen nahe (Bild 4). Die beiden unteren Ebenen beschreiben Katalogwissen, das in technischen Katalogen und Datenblättern dokumentiert ist. Mit diesem Wissen sollten im Prinzip bereits brauchbare Konfigurationen erstellt werden können. Der Experte verfügt darüberhinaus - dies macht ihn erst zum Experten – über weiteres Wissen, das durch die oberen Ebenen repräsentiert wird.

Beim Katalogwissen wird **Systemwissen** (Wissen über die Systemschnittstellen) und **Komponenten-Katalog** (Wissen über die Komponententypen) unterschieden. Das Expertenwissen kann in **Wissen um Ausnahmen, Bewertungswissen** und **Reihenfolge–Wissen** unterteilt werden. Im folgenden werden diese verschiedenen Wissensbereiche diskutiert.

Bild 4: Strukturierung des Wissens in Ebenen

3.2.1 Katalogwissen

Das Katalogwissen, das heißt die Definition der Ressourcen (Systemschnittstellen) und die Beschreibung der Eigenschaften der Komponenten, macht den überwiegenden Teil des Wissens aus. Dabei bedarf das Komponenten-Katalog-Wissen wegen der ständigen Änderungen in der Zusammensetzung des Warenkorbes am häufigsten der Aktualisierung. Entsprechende Bedeutung kommt deshalb der Effektivität der Wissenspflege zu. Eine komfortable Wissenserwerbskomponente wurde daher für COSMOS als unverzichtbar angesehen.

Wissen über Komponenten und Wissen über Ressourcen werden beide zweckmäßigerweise in eigenen Taxonomien organisiert. Bild 5 gibt ein Beispiel für die Taxonomie von Komponenten.

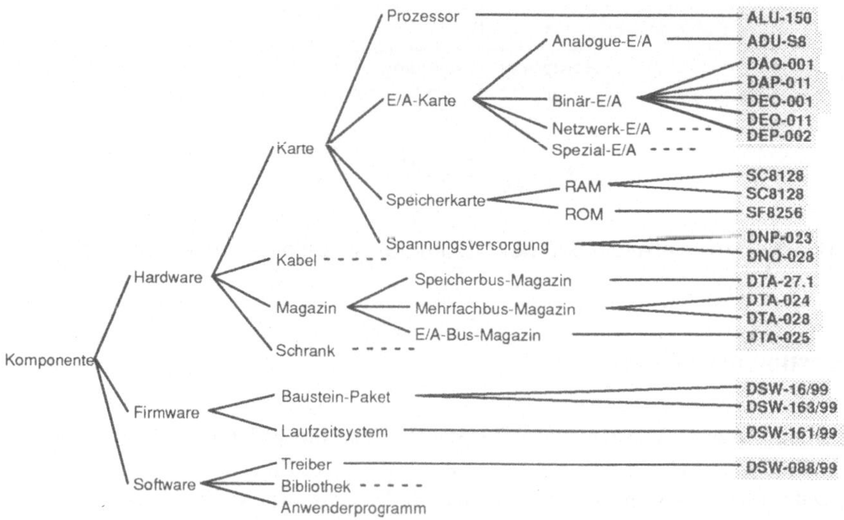

Bild 5: Komponenten-Taxonomie am Beispiel einer SPS-Systemfamilie (Ausschnitt)

In solchen Taxonomien treten abstrakte Komponenten im Sinne von Oberbegriffen zu Komponententypen auf. Die Wissenseingabe wird dadurch rationalisiert, daß Eigenschaften, die mehreren Komponenten gemeinsam sind, nur einmal, und zwar an der zugehörigen abstrakten Komponente, notiert werden. Eine neue

Komponente kann dann mit minimalem Aufwand integriert werden als Spezialfall derjenigen abstrakten Komponente, unter der sie entsprechend der Systematik einzuordnen ist, wobei nur noch die individuellen Eigenschaften der neuen Komponente angegeben werden müssen. Natürlich kann nicht nur Hardware, sondern auch Software als Komponenten in die Taxonomie aufgenommen werden. Analoges gilt für die Ressourcen-Taxonomie (Bild 6).

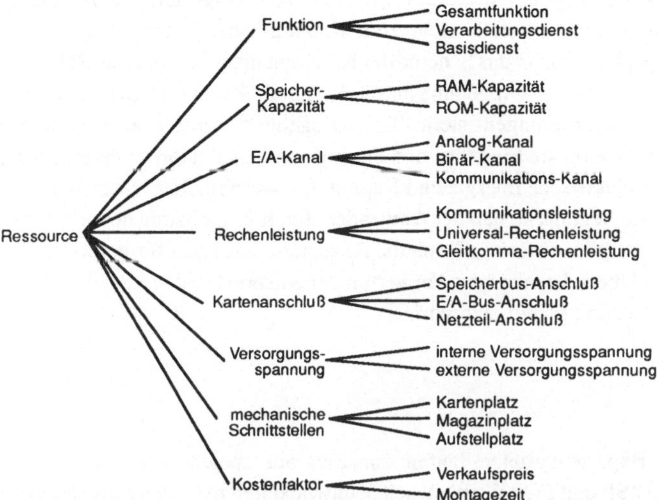

Bild 6: Ressourcen-Taxonomie am Beispiel einer SPS-Systemfamilie (Ausschnitt)

Beide Taxonomien können in der Wissenserwerbskomponente grafisch dargestellt und für die Mensch-Maschine-Schnittstelle genutzt werden. Nach der Festlegung der Wissens–Grundstruktur sind auch Personen, die bisher mit der Redaktion von gedruckten Katalogen befaßt waren, befähigt, die routinemäßige Pflege des Komponenten–Katalogs durchzuführen. Nur solche neuen Komponenten, die prinzipiell andersartige Eigenschaften aufweisen oder deren Repräsentation strukturelle Änderungen erfordern, müssen von einem Experten modelliert und eingegeben werden. Keine dieser Tätigkeiten setzt Kenntnisse der Expertensystemtechnik voraus.

3.2.2. Expertenwissen

Das Expertenwissen ist neben dem **Wissen um Ausnahmen,** d.h. hauptsächlich Unverträglichkeiten zwischen einzelnen Komponenten, vor allem das Wissen über den vorteilhaften Gebrauch des Katalogs.

Hierzu gehört das **Bewertungswissen.** Wenn zur Abdeckung einer Ressourcenanforderung mehrere Komponenten geeignet und mit den bisher ausgewählten Komponenten verträglich sind, soll die günstigste gewählt werden. Die Bewertung des Experten orientiert sich am Nutzen-Kosten-Verhältnis der Komponenten, wobei die jeweiligen Folgekosten überschlägig aus den von der Komponente angeforderten Ressourcen ermittelt werden.

Wesentlich für die Geschwindigkeit der Lösungsfindung und die Güte der ersten gefundenen Lösung ist die Reihenfolge, in der die angeforderten Ressourcen behandelt und durch geeignete Komponenten bereitgestellt werden. Dieses **Reihenfolge-Wissen** steuert den Inferenzverlauf.

Das Expertenwissen kann sehr sinnfällig den Komponenten bzw. den Ressourcen in ihren Taxonomien zugeordnet und mit der COSMOS-Wissenserwerbskomponente effizient gepflegt werden.

3.3. Spezifikation der Konfiguration

Die Spezifikation, d.h. die Aufstellung der Anforderungen an das zu konfigurierende System, gehört zum Wissen über den jeweiligen Anwendungsfall, nicht zum Katalog- oder Expertenwissen. Wie bereits erwähnt, läßt sich auch die Spezifikation in das Schema der Komponenten-Ressourcen-Relationen elegant und vorteilhaft einbringen. Dazu wird die Umgebung als eine spezielle Komponente eingeführt, die an das zu konfigurierende System Anforderungen stellt. Die Umgebung fordert also vom System die Bereitstellung von Ressourcen wie binäre Eingänge und Funktionalität, stellt ihm andererseits Ressourcen zur Verfügung, beispielsweise elektrische Energie und Kapital. Ein wesentlicher Vorteil der Interpretation der Umgebung als Komponente liegt darin, daß der Anwender durch Begrenzung der von der Umgebung bereitgestellten Ressourcen, beispielsweise des Kapitals, Akzeptanz-Kriterien für gefundene Konfigurationen angeben kann, deren Überschreiten zum Verwerfen der entsprechenden Lösung führt, so daß die Suche nach einer zulässigen Lösung fortgesetzt wird.

3.4 Implementation

Auf COSMOS basierende Expertensysteme laufen zur Zeit auf speichererweiterten Rechnern der PC/AT-Klasse mit Common LISP und FOLK, einem selbst entwickelten hybriden Expertensystem-Werkzeugkasten. Da es bereits geeignete Laptops gibt, ist der Einsatz solcher Expertensysteme nicht auf die Projektierung im Büro beschränkt. So können bereits während des Kundengesprächs mehrere Alternativen untersucht, technisch abgeklärte Angebote erstellt und schnellere Abschlüsse getätigt werden.

4. Anwendungsbeispiele

Auf einigen Anwendungsfeldern der Automatisierungstechnik wurde die Eignung der Ressourcengetriebenen Konfigurierung bereits in Applikationsstudien, auch durch Anwender, getestet. Dazu gehören Konfigurationssyteme für Briefverteilanlagen, Zugausrüstungen, elektrische Antriebe, Niederspannungs-Schaltanlagen.

Am weitesten fortgeschritten ist ein Expertensystem-Prototyp zur **Konfigurierung modularer Speicherprogrammierter Steuerungen** der Systemfamilie AEG MODICON A500. Dieser Prototyp beschränkt sich auf eine der beiden Bauformen der Familie (Heckanschlußtechnik) und berücksichtigt über 100 Komponenten bzw. -varianten. Dies sind ca. 50% der Komponenten dieser Bauform. Für Akquisition, Strukturierung und Eingabe dieses Wissens benötigte der Bearbeiter dank der komfortablen Wissenserwerbs- und Testkomponente nur 3 Wochen.

Die Anwenderschnittstelle ist speziell auf die Erfordernisse Speicherprogrammierte Steuerungen zugeschnitten. Dabei wird die Umgebung in drei separaten Spezifikationskomponenten beschrieben. Im Zentrum der Anwenderschnittstelle steht ein grafischer Editor (Browser), der zur Eingabe der hierarchischen Anlagensstruktur aus einzelnen Stationen und ihren Kommunikationsverbindungen dient. Diese Strukturbeschreibung legt dabei nach Spezifikation der Art der Verbindungen die daraus resultierenden Anforderungen an Hardware und Software für die einzelnen Stationen fest, und zwar automatisch durch Parametrierung der ersten der drei Spezifikationskomponenten. In der zweiten Spezifikationskomponente werden alle Anforderungen notiert, die sich aus dem Anwenderprogramm ergeben, wie z.B. RAM-Speicherbedarf. Diese Festlegungen werden sich auch maschinell aus dem Programm gewinnen lassen. In der dritten Spezifikationskomponente werden die übrigen Anforderungen für jede Station explizit durch den Anwender durch Ausfüllen von einfachen Formularen spezifiziert, beispielsweise die Ausführungsformen der Prozeß-Eingabe/Ausgabe.

Nach der Spezifizierung ist der Prototyp in der Lage, innerhalb von ca. 5 min auch umfangreiche Konfigurationen zu erstellen, wofür ein Experte Stunden benötigt. Als Ergebnis wird neben der Stückliste auch die Ressourcen-Bilanz ausgegeben, d.h. eine Gegenüberstellung von spezifizierten und erreichten Eigenschaften wie Preis, Vorgabezeit für die Montage, Gewicht, elektrische Verlustleistung, Auslastung der Netzgeräte. Genauere Informationen über die Herleitung einer gefundenen Lösung bietet die Erklärungskomponente, die auch die Gründe liefert, die zum Verwerfen von Komponenten führten. Manuelle Eingriffe in den Konfigurationsverlauf werden unterstützt.

Ein weiteres Anwendungsfeld von COSMOS liegt in der Komponentenplanung. Hierfür wird vom Marketing bereits vor der Erteilung eines Entwicklungsauftrags die Beschreibung für eine neue oder verbesserte Komponente in die Wissensbasis eingegeben und eine größere Zahl von repräsentativen Altaufträgen testweise noch einmal konfiguriert. Damit kann das Marktpotential der neuen Komponente geprüft und Kostenvorteile verifiziert werden.

Darüberhinaus kann COSMOS bereits in der Phase des Systementwurfs eingesetzt werden, indem die geplanten Schnittstellen und Komponenten in die COSMOS-Wissensbasis eingegeben und entsprechende Ressourcen- bzw. Komponenten-Taxonomien aufgebaut werden. Durch simulierendes Konfigurieren kann die Systemfamilie auf interne Konsistenz und befriedigende Abdeckung des Anwendungsgebiets geprüft werden. Die Beschreibungen der Komponenten dienen später als Spezifikationen für die Entwicklungsabteilung.

5. Fazit

Der COSMOS-Ansatz der Ressourcen-getriebenen Konfigurierung modelliert das statische Zusammenwirken von Komponenten Technischer Systeme auf einer abstrakten Ebene, die genau der Problemstellung der Konfigurierung angemessen ist. Das darauf basierende Schlußfolgerungsverfahren geht von elementaren Prinzipien aus, wie es für Expertensysteme mit "tiefem Wissen" kennzeichnend ist.

Durch diese Modellierung mit der Separation des umfangreichen Katalogwissens vom eigentlichen Expertenwissen, mit der Entkopplung des Wissens über einzelne Komponenten voneinander und mit der systematischen Strukturierung des Wissen in Taxonomien ist ein entscheidender Fortschritt erzielt, der eine rationelle Wissenspflege auch für umfangreiche, komplexe Komponentensysteme verspricht.

Die Allgemeingültigkeit der Modellierung bietet mehrfachen Vorteil. Erstens können mit Hilfe der Expertensystemschale COSMOS Konfigurationssysteme kostengünstig erstellt werden. Zweitens können unterschiedliche Ausrüstungsteile von Anlagen wie Steuerung, Meßwerterfassung und Stellglieder einheitlich modelliert und daher durch eine Anzahl gleichartig zu bedienender und zu pflegender Expertensysteme konfiguriert werden. Drittens können diese Expertensysteme Informationen über die aus dem Zusammenwirken resultierenden gegenseitigen Ressourcenanforderungen austauschen. Dies eröffnet die Aussicht, die Konfigurierung kompletter Automatisierungsanlagen durchgehend zu unterstützen.

Literatur

[1] McDermott, J.   R1: A Rule-Based Configurer of Computer Systems.
                    Artificial Intelligence 19 (1982) 39-88.

[2] Soloway, E.   Assessing the Maintainability of XCON-in-RIME: Coping with the Problems
     et al.   of a VERY Large Rule-Base. Proc. AAAI '87, 824-829.

[3] Haugeneder, H.   Knowledge-Based Configuration of Operating Systems -
     et al.   Problems in Modeling the Domain Knowledge. Proc.
                    GI-Kongreß "Wissensbasierte Systeme" 1985, 121-134.

[4] Borgida, A.   KNOWLEDGE REPRESENTATION AS THE BASIS FOR
     et al.   REQUIREMENTS SPECIFICATION. Computer 18(1985)4, 82-91.

[5] Lan, M.S.   A KNOWLEDGE-BASED APPROACH TO PRINTING PRESS
     et al.   CONFIGURATION. Proc. CAIA '87, 38-43.

[6] Pierick, J.   A KNOWLEDGE REPRESENTATION TECHNIQUE FOR SYSTEMS DEALING WITH
                    HARDWARE CONFIGURATION. Proc. AAAI '86, 991-995.

[7] Lehmann, E.   SICONFEX ein Expertensystem für die Konfiguration eines
     et al.   Betriebssystems. 15. GI-Jahrestagung, 1985, 792-805.

[8] Dias, P.d.C.,   Der Einsatz der frame-basierten Expertensystem-Shell KEN für die Konfiguration
     Bärtschli, M.   nachrichtentechnischer Produkte. 2. Anwenderforum Expertensysteme 1989, Duisburg.

[9] Frayman, F.,   COSSACK: A Constraint-Based Expert System for Configuration Tasks. In:
     Mittal, S.   Sriram, D. and R.A. Adey (eds.), Knowledge-Based Expert
                    Systems in Engineering: Planning and Design, 1987.

[10] Cunis, R.   PLAKON, EIN ÜBERGREIFENDES KONZEPT ZUR WISSENS-REPRÄSENTATION
     et al.   UND PROBLEMLÖSUNG BEI PLANUNGS- UND KONFIGURIERUNGSAUFGABEN.
                    Proc. Expertensysteme '87: Konzepte und Werkzeuge, Nürnberg, 1987, 406-419.

[11] Strecker, H.   Configuration Using PLAKON - An Applications Perspective
                    Proc. 3. Int. GI-Kongreß "Wissensbasierte Systeme", 1989, 352-362.

[13] Mittal, S.,   Towards a generic model of configuration tasks.
     Frayman, F.   Proc. IJCAI '89, 1395-1401.

[12] Baginsky, W.   BASIC ARCHITECTURAL FEATURES OF CONFIGURATION EXPERT SYSTEMS
     et al.   FOR AUTOMATION ENGINEERING. Proc. IEEE/SICE AI'88, 603-607.

[14] Hakkarainen, K.   A KNOWLEDGE-BASED CONFIGURER FOR EMBEDDED COMPUTER
     et al.   SYSTEMS SOFTWARE: REAL-LIFE EXPERIENCE.
                    Proc. IEEE/SICE AI'88, 608-613.

[15] Escherle, A.,   Ein interaktives Werkzeug zur Konfiguration
     Sachs,S.   modularer Rechnersysteme. WIMPEL '88, 322-328.

[16] Eich, E.   CONSTRUCT - ein wissensbasierter Ansatz zur Konfigurierung.
     et al.   2. Anwenderforum Expertensysteme 1988, Duisburg.

[17] Neumann, B.,   Anlagenkonzept und Bilanzverarbeitung. 3. Workshop Planen und
     Weiner, J.   Konfigurieren (1989), Arbeitspapiere der GMD 388(1989), 209-224.

[18] Heinrich, M.   Ein generisches Modell zur Konfiguration Technischer Systeme aus modularen
                    Komponenten, 3. Workshop Planen und Konfigurieren (1989),
                    Arbeitspapiere der GMD 388(1989), 49-57.

ERGONOMISCHE GESTALTUNG VON PROZESSLEIT-
STÄNDEN IM MASCHINENBAU

W. Risch   H. Steinbach
Technische Universität Chemnitz
PSF 964, Chemnitz, 9010/Deutschland

1. Einleitung

Die schrittweise Entwicklung zu rechnerintegrierten Betrieben (CIM-Betriebsstrukturen) im Maschinenbau erfordert die Verwirklichung sozialer und ökonomischer Ziele. Die Realisierung der ökonomischen und sozialen Zielstellungen setzt voraus, daß technisch-technologische und arbeitskraftbezogene Sachverhalte innerhalb der Phase des Systementwurfs /1/ neuer rechnerunterstützter Arbeitsbereiche (flexible Fertigungssysteme, gemischtautomatisierte Produktionsbereiche) komplex zu betrachten und zu gestalten sind. In relativ frühen Phasen der Entwicklung dieser neuen Arbeitsprozesse ist zielgerichtet Einfluß auf die Gestaltung der Arbeitsaufgaben und der Ausführungsbedingungen der Arbeit zu nehmen.

Für die Betriebe ist die Flexibilität der Fertigung, die Qualifikation des Personals und die Gestaltung der Arbeitsbedingungen ein entscheidendes Kriterium für ihre innovative Entwicklung. In den letzten Jahren wurden zunehmend technische Optionen zur Unterstützung menschlicher Arbeit genutzt.
Neben personellen, organisatorischen, qualifikatorischen Aspekten sowie Arbeits- und Umweltschutzproblemen stehen insbesondere Aspekte der menschengerechten Technikgestaltung und -anwendung im Vordergrund.
Das bezieht sich u. a. auf folgende Themen (/2/, /3/):

- Entwicklung und Erprobung neuer Technikkonzepte, z. B. die Gestaltung menschengerechter Software und werkstattorientierter Steuerungskonzepte, die die Beschäftigten aufgabenangemessen in ihrer Arbeit unterstützen und die Gewinnung von Erfahrungen im Arbeitsprozeß fördern.

- Nutzung und Anpassung relevanter neuer informations- und fertigungstechnischer Entwicklungen zur Verbesserung der Arbeitsbedingungen und der Umwelt.

- Entwicklung und Erprobung von CIM-Konzepten und -bausteinen, die menschliche Fähigkeiten und Kompetenzen zum Ausgangspunkt machen und auch dem Bedarf kleiner und mittlerer Betriebe entsprechen.

2. Methodisches Konzept zur Gestaltung der Ausführungsbedingungen

Der Prozeßleitstand als Schnittstelle zwischen Bearbeitungssystem und Produktionsprozeßsteuerung sollte sich künftig ausgehend von der Struktur der Arbeitsaufgaben und den erforderlichen Ausrüstungen weiterentwickeln.

Flexible Fertigungssysteme unterschiedlicher Systemgröße/ -konfiguration und gemischtautomatisierte Fertigungsbereiche fordern arbeitsaufgabenbezogene Gestaltungslösungen.

Erfahrungen beim Aufbau der ersten flexiblen Fertigungssysteme in der DDR zeigten,
daß keine komplexen Gestaltungslösungen erarbeitet wurden. Eine arbeitsfunktions-
bezogene Gestaltung von Arbeitsplätzen, Arbeitsabläufen sowie der Arbeitsumweltfak-
toren erfolgte insbesondere in Prozeßleitständen /4/ nicht.

Es gilt, ausgehend von der Forderung zur Zentralisation von Arbeitsfunktionen /1/
die Methodik zur Bildung von Komplexarbeitsplätzen /4/ für den Entwurf von Prozeßleitständen einzusetzen, die eine Gesamtgestaltung und -bewertung (Persönlichkeitsförderlichkeit, Beeinträchtigungsfreiheit, Ausführbarkeit, Schädigungslosigkeit)
ermöglicht.

Die Methodik zur Bildung von Komplexarbeitsplätzen ist im Rahmen der Entwicklung
neuer flexibler Fertigungssysteme projektbegleitend und schrittweise in Form inte-
grativer Arbeitsgestaltung anzuwenden.

Ausgangspunkt der Gestaltung sind die Arbeitsfunktionen, die aus der Funktionsteilung Mensch-Maschine/Mensch-Rechner resultieren (siehe Bild 1).

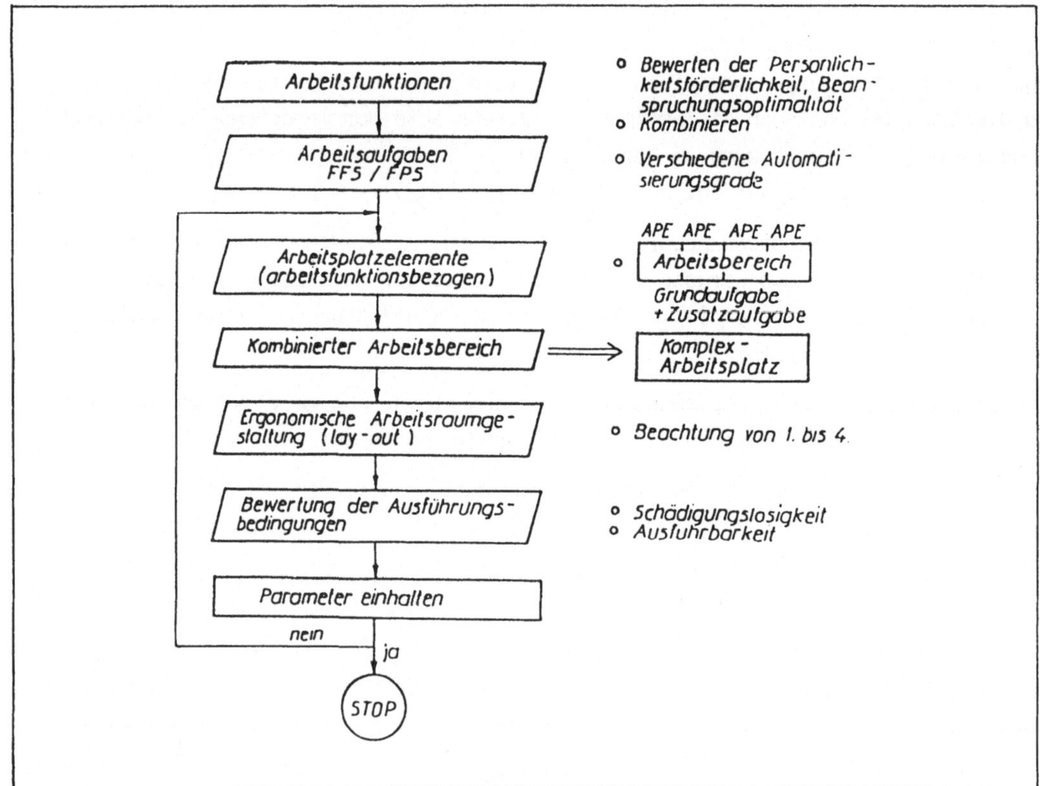

Bild 1: Methode zur Gestaltung von Komplexarbeitsplätzen

In einem ersten Schritt sind die Arbeitsaufgaben zu gestalten, danach können die neuen Orte lebendiger Arbeit entwickelt werden.

Als Ort der Zentralisation ist u. a. neben der Spannarbeitszone der Prozeßleitstand einschließlich benachbarter Räume geeignet.

Für den Entwurf von Komplexarbeitsplätzen ist das Bausteinprinzip, d. h. die Addierbarkeit, Austauschbarkeit und Variabilität der Arbeitsplatzelemente in Form von Grundelementen anzuwenden.

Bestandteil der Methodik zur Bildung von Komplexarbeitsplätzen ist eine Bewertungsmethode für die Ausführungsbedingungen der Fertigungsprozeßsteuerung. Der Einsatz dieser Methode ist besonders für die Neuentwicklung von Bedeutung, um bereits in frühen Phasen des Entwurfs von flexiblen Fertigungssystemen Schädigungslosigkeit und Ausführbarkeit durch Abprüfen entsprechender Parameter zu sichern.

Für die Weiterentwicklung von FMS ist es notwendig, eine Theorie zur Gestaltung von Bedienstellen in Zusammenhang mit dem Entwurf von Komplexarbeitsplätzen zu entwickeln.

Der Gedanke zur Flexibilität von Arbeitsstellen sollte hier Eingang finden.

Pult- und Auftischvarianten kommen diesem Grundsatz bei komfortablen Systembedienstellen mehr entgegen als Schrankvarianten. Flexible Bedienstellen (Monitor, Tastatur, Rechnergrundgerät) setzen sich selbst bei CNC-Bedienterminals durch /5/.

1988/89 vollzog sich in der DDR die Entwicklung von FMS zur Bearbeitung von rotationssymmetrischen bzw. prismatischen Teilen, die mit FMC-Rechnern ausgestattet werden. International sind eine Reihe kleiner Systeme mit dieser Steuerungstechnik bereits schon einige Jahre im Einsatz /7/.

Sie stellen erste CIM-Bausteine für den Anschluß an lokale Netze dar.

Von STEINBACH /4/ wurden unterschiedliche Konzepte für die Gestaltung von Bedien- und Steuerungsbereichen entwickelt.

Fertigungsleitstände/Prozeßleitstände werden in CIM-Betriebsstrukturen zwar ihre volle Berechtigung haben, aber unterschiedliche Formen werden zu gestalten sein:

1. Prozeßleitstand für den Gesamtbetrieb
   (Arbeitsbereich Hauptdispatcher)
2. Fertigungsleitstand für Fertigungssysteme höherer Ordnung
   (Bereichsdispatcher/Meister, Systemtechnologe)
   Erste Raumstrukturen liegen dazu in /4/ sowie in Bild 2 und 3 vor.
3. Fertigungsleitstand bzw. räumlich abgegrenzte Komplexarbeitsplätze für Fertigungssysteme niederer Ordnung
   (komfortable Systembedienstelle für "Freien Bediener")
4. Raumzelle für gemischtautomatisierte Fertigungsbereiche

Die Gestaltung dieser Arbeitsbereiche kann rechnerunterstützt erfolgen /2/. Für den Entwurf am CAD-System müssen die folgenden Voraussetzungen erfüllt werden:
- Die einzusetzende Hardware und übrigen Ausrüstungen sind in Haupt- und Nebenar-

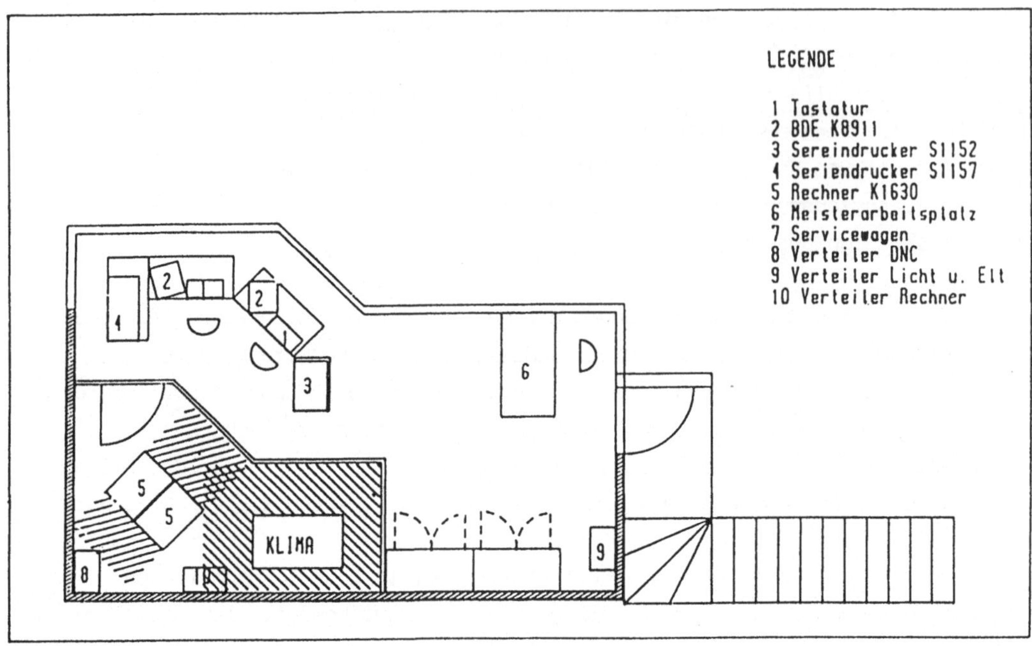

Bild 2: Leitstand für ein kleines FFS oder einen Fertigungsabschnitt

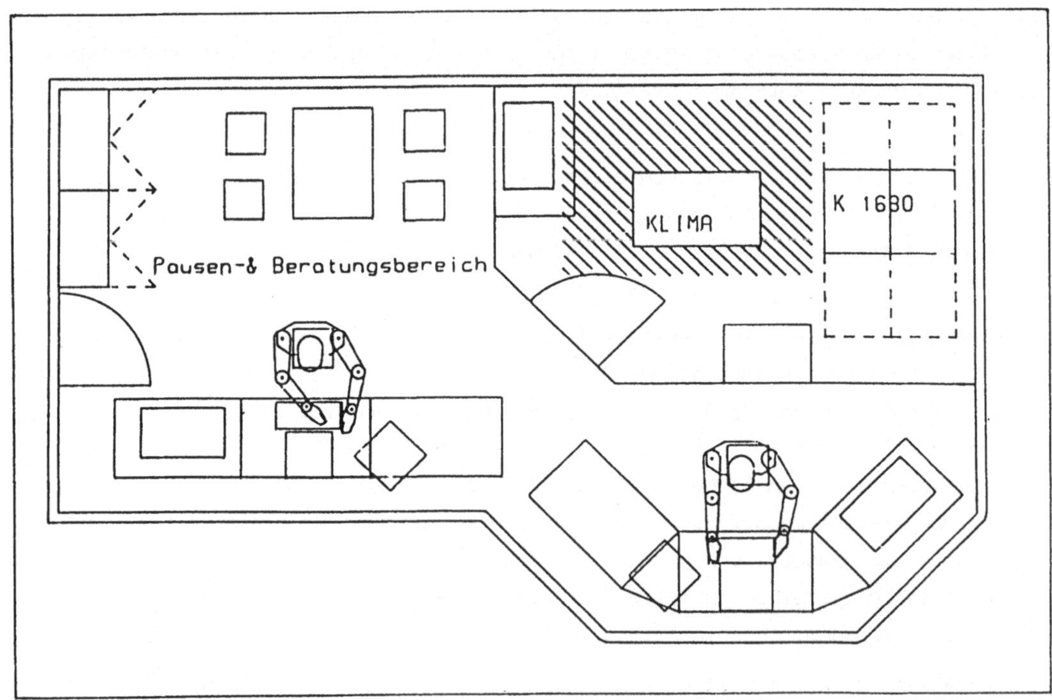

Bild 3: Rechnergestützter Entwurf komplexer Arbeitsbereiche

beitsstellen zu gruppieren.
- Arbeitsplatzelemente (Tische, Segmente) sind arbeitsaufgabenbezogen zu Komplex-
  arbeitsplätzen zusammenzufassen.
- Komplexarbeitsplätze setzen addierbare Ausstattungssysteme voraus.
- Die Ausführung des Prozeßleitstandes in Monolith-, Stahlskelettbauweise oder als
  Container ist in der Arbeitsraumgestaltung zu beachten.
- Die Berechnung der erforderlichen Fläche hat vor der Arbeit am CAD-System zu er-
  folgen bzw. ist in einem gesonderten Programmteil abzuarbeiten.

Die rechnergestützte Layoutgestaltung ist gleichzeitig die Basis für die horizonta-
le Gliederung des Komplexarbeitsplatzes (siehe Bild 3).
Eine grafikorientierte ergonomische Gestaltung und Bewertung von Typenprozeßleit-
ständen ordnet sich in den Gesamtprozeß der rechnergestützten Projektierung von
Fertigungsstätten ein. Die Einbindung arbeitswissenschaftlicher Methoden in diesen
Prozeß ist ein Schritt zur Sicherung eines menschzentrierten Gestaltungskonzeptes
(siehe Bild 4, 5).

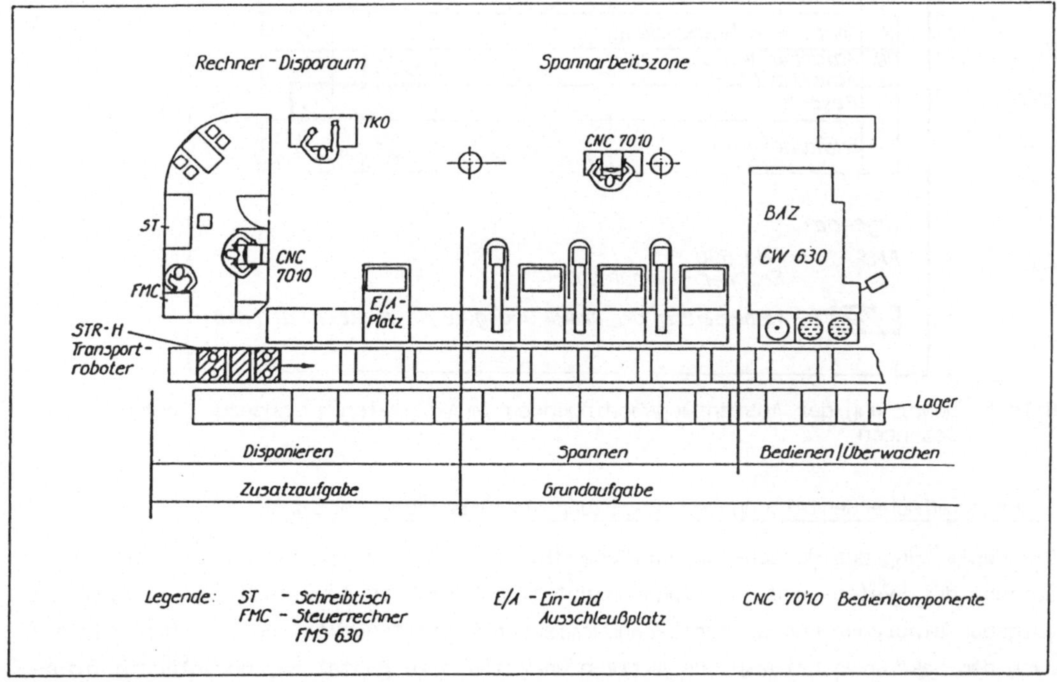

Bild 4: Beispiel einer Gestaltungslösung (FMS 630)

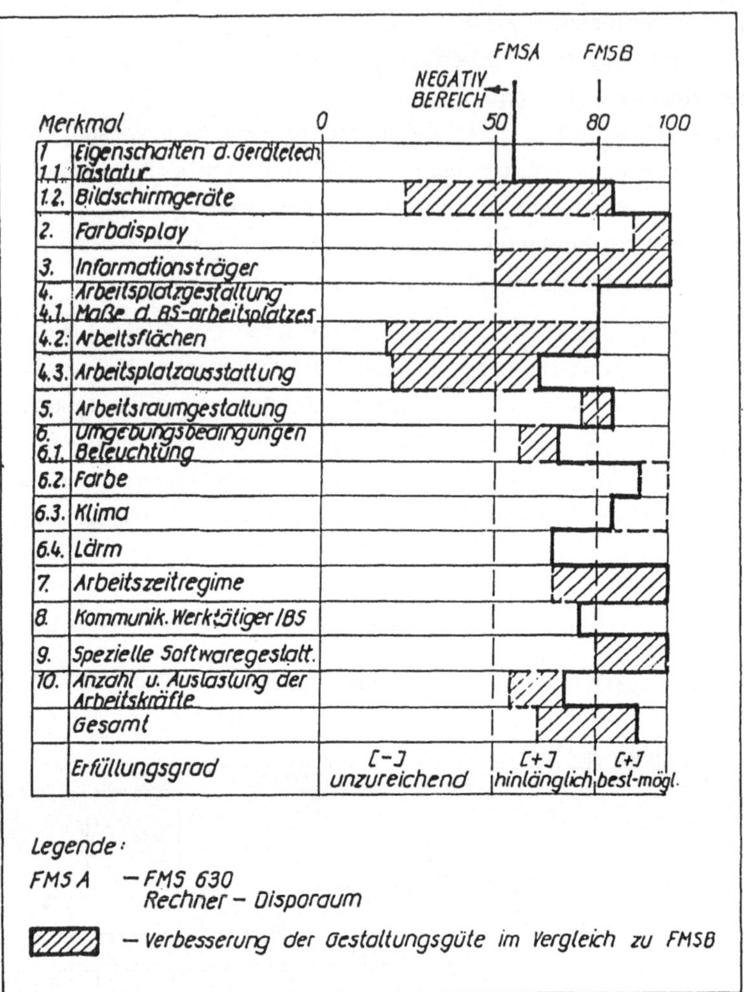

Bild 5: Bewertung der Ausführungsbedingungen im Vergleich zu anderen Gestaltungslösungen

3. Ergonomische Gestaltung und Bewertung von Steuerungssoftware

Die Gestaltung der Mensch-Rechner-Kommunikation im Prozeßleitstand dient der Umsetzung des System- und Funktionskonzeptes, muß sich in Arbeitsabläufe und Arbeitsaufgaben einordnen und software-ergonomischen Gesichtspunkten entsprechen. Sie ist nach dem bekannten Prinzip des aktiven Operateurs zu gestalten, der ständig Informationen über den Fertigungs- und Systemzustand erhält und ihm durch eine entsprechend gestaltete Software Eingriffsmöglichkeiten geschaffen werden. Der Mensch ist zu entlasten von Gedächtnisanforderungen und Routinetätigkeiten. Während komplexe geistige Verarbeitungsoperationen, die seine intellektuellen Fähigkeiten ausschöpfen (infolge ihres aktivierenden und einstellungsfördernden Effekts), beim Menschen in von ihm wählbaren Umfange verbleiben sollten.

Eine für den Menschen ungünstig gestaltete Funktionsteilung Mensch - Rechner - Fertigungssystem beeinflußt den Gestaltungsspielraum zur Arbeitsteilung Mensch - Mensch und schränkt die Erreichung arbeitswissenschaftlich determinierter Zielgrößen spürbar ein. Deshalb sind Fragen der Funktionsteilung beim Systemprojektanten und Softwarehersteller rechtzeitig zu entscheiden, bevor der Systemanwender mit dem Problem der Arbeitsteilung bei der praktischen Überführung konfrontiert wird.

Es sind demzufolge Produktionssteuerungssysteme zu entwickeln, die dem Menschen als Hilfsmittel zur Steuerung und Beherrschung der Produktion dienen.
Dem Menschen (das kann ein Dispatcher sein) dürfen nicht nur Überwachungsfunktionen und die Störungsbeseitigung obliegen, sondern er ist aktiv in die Planung, Leitung und Steuerung der Fertigung einzubeziehen. Hauptagenzien des Menschen sind damit die Entscheidungsfindung und das Verfügen über Kontrollmöglichkeiten. Das verlangt, daß beide Seiten - Mensch und Rechner - über entsprechende Informationen und Informationsverarbeitungsprozeduren verfügen, für die gestalterische Grundlagen erforderlich sind /2/.

Als Mittel dazu werden für erforderlich gehalten:
a) Gestaltungsrichtlinien /-methoden, die dem Softwareentwickler konkrete, möglichst objektbezogene Hinweise geben.

b) Bewertungsmethoden zur Überprüfung der erreichten Qualität innerhalb des Softwareentwicklungsprozesses

c) Bewertungsmethoden zur Bewährungskontrolle bei der Nutzung der Software durch Überprüfung von Belastungswirkungen und Persönlichkeitsförderlichkeit beim Anwender.

Hierzu liegen unterschiedlich differenzierende Lösungen und Teillösungen vor. Hauptgegenstand der Arbeit bildet gegenwärtig die Entwicklung einer kombinierten Gestaltungs- und Bewertungsmethode (GBM) und die Erarbeitung und Zusammenstellung von Instrumentarien zur Bewährungskontrolle/Belastungsanalyse.

Der Lösungsstrategie zur Erarbeitung der GBM liegen folgende Schritte zugrunde:

1. Festlegung einer solchen Softwaretechnologie, die eine ergonomische Beeinflussung des Softwareproduktes ermöglicht.

2. Struktureller Entwurf der GBM in Anpassung an die festgelegte Softwaretechnologie und das notwendige Qualitätssicherungssystem für Software.

3. Aufstellung eines Aufgabenmodells für die Produktionssteuerung in einem flexiblen Fertigungssystem in Form eines idealtypischen Referenzmodells, das alle technologischen Steuerfunktionen und die dazu notwendigen Informationen enthält, unabhängig von der zu realisierenden Funktionsteilung Mensch - Rechner.

4. Ableitung einer arbeitspsychologisch begründeten Funktionsteilung aus dem Aufgabenmodell und Aufbereitung in Form von Gestaltungsforderungen in der GBM.

5. Aufstellung der Gestaltungsforderungen für die Mensch-Rechner-Kommunikation im
   engeren Sinne (Bedienoberfläche, Bediensystem) und Aufbereitung für die GBM:

   . Dialogtechnik
   . Menügestaltung
   . Maskengestaltung
   . Codierung von Informationen
   . Hilfe

6. Entwicklung des Bewertungsteiles zur Überprüfung der erreichten Gestaltungsgüte
   bezüglich

   . Handhabbarkeit
   . Beeinträchtigungsfreiheit
   . Persönlichkeitsförderlichkeit
   . Erlernbarkeit

7. Analyse bestehender Steuerungssysteme mittels Bewertungsmethoden der Bewährungskontrolle zur Ermittlung kritischer Gestaltungsdefizite und deren Umsetzung in
   der GBM parallel zu obigen Bearbeitungsschritten.

4. Zusammenfassung

Für die arbeitswissenschaftliche Gestaltung von Prozeßleitständen liegt ein umfangreiches methodisches Instrumentarium /1/ /2/ /4/ /6/ vor. Es gilt immer stärker eine
arbeitsaufgabenbezogene Gestaltung von Prozeßleitständen mit Hilfe der Methodik zur
Bildung von Komplexarbeitsplätzen zu realisieren, um tayloristische Formen der starken Arbeitsteilung in flexiblen Fertigungssystemen sowie in Fertigungsinseln zu vermeiden. Die entwickelten Instrumentarien sind im Objektbereich des Maschinenbaues
im Rahmen einer Pilotlösung Arbeitsumweltgestaltung in der DDR erprobt. Die Methoden erwiesen sich als unkompliziert in der Handhabung und für interdisziplinäre Ent-
wicklerkollektive anwendbar.
Die Anwendbarkeit der Entwurfsmethoden für Prozeßleitstände in anderen Objektberei-
chen außerhalb des Maschinenbaues gilt es nachzuweisen.

5. Literatur

/1/ Risch, W.; Steinbach, H.: Produktionssteuerung - Systementwurf und Systemgestaltung aus arbeitswissenschaftlicher Sicht.
Vortrag "AUPRO 88", TU Karl-Marx-Stadt, 1988

/2/ Risch, W.: Arbeitswissenschaftliche Gestaltung von Produktionssteuerungssystemen. AWZ-Inf. Nr. 26, Karl-Marx-Stadt, 1989, 97 S.

/3/ Forschungs- und Entwicklungsprogramm Arbeit und Technik. Ma. zum Workshop Fabrikinnovation: BMFT Bonn, 1990

/4/ Steinbach, H.: Ergonomische Gestaltung von Arbeitsbereichen der Fertigungsprozeßsteuerung flexibler Fertigungssysteme.
TU Karl-Marx-Stadt, Fakultät für Maschineningenieurwesen, Sektion Technologie der metallverarbeitenden Industrie, Diss. B, 1988

/5/ Hasecke; Ergonomie - Trends. Industrie - elektrik + elektronik.
Maschinenausrüstung + Automatisierung 33 (1988) 8, Heidelberg, S. 45 - 49

/6/ Kernler, H.: Der elektronisch-grafische Leitstand und die Anbindung relevanter Subsysteme, Kongreß "PPS 88". - Tagungsband, Böblingen 1988

Softwaretechnologien

Ein Verfahren zur Garantie des zeitlichen Verhaltens
in verteilten Echtzeitsystemen

Ralf Agne
Universität Kaiserslautern, Postfach 3049, 6750 Kaiserslautern
e-mail: agne@informatik.uni-kl.de

Kurzfassung

In diesem Aufsatz wird ein Verfahren beschrieben, das durch die Verwendung der globalen zyklischen Schedulingstrategie in verteilten Echtzeitsystemen die Überprüfung von Zeitschranken während der Entwicklungsphase ermöglicht und deren Einhaltung zur Laufzeit garantiert. Softwarewerkzeuge unterstützen die Berechnung von Schedulingplänen. Sowohl die Schedulingstrategie als auch Eigenschaften der unterstützenden Werkzeuge werden diskutiert.

1. Einleitung

Verteilte Systeme werden in zunehmendem Maße im Echtzeitbereich an Stelle von zentralen Prozeßrechnern eingesetzt. Die *korrekte Funktionsweise* eines Echtzeitsystems erfordert nicht nur die Abgabe von *korrekten Daten* an das zu kontrollierende technische System, sondern auch die Einhaltung der von der Umgebung dem Echtzeitsystem vorgegebenen *Zeitschranken*. Zur Berechnung korrekter Ausgabedaten verwalten Echtzeitsysteme *zeitkritische Zustandsdaten* (*Echtzeitdaten*), die den jeweils aktuellen Zustand der Umgebung approximieren und entsprechend der Dynamik des zu kontrollierenden technischen Systems aktualisiert werden müssen.

Wir bezeichnen eine Menge von kommunizierenden Softwareeinheiten, die aufgrund des Auftretens eines bestimmten Ereignisses auszuführen sind, als **Berechnungsvorgang**. Eine Softwareeinheit (SE) muß nicht exklusiv einem Berechnungsvorgang zugeordnet sein, sondern kann in mehreren Berechnungsvorgängen enthalten sein. Jedem Berechnungsvorgang sind *Ausführungszeitschranken* zugeordnet, die bei der Ausführung der SE des Berechnungsvorganges zu beachten sind. Während bei Berechnungsvorgängen mit **harten Echtzeitbedingungen (BhE)** alle zugeordneten Ausführungszeitschranken während der Laufzeit des Echtzeitsystems eingehalten werden müssen, ist bei Berechnungsvorgängen mit **schwachen Echtzeitbedingungen (BsE)** eine Überschreitung zugeordneter Zeitschranken möglich. Es gibt grundsätzlich zwei Möglichkeiten, wodurch die Ausführung von Berechnungsvorgängen ausgelöst werden kann: Beim Erreichen bestimmter Zeitpunkte zur Laufzeit des Echtzeitsystems oder aufgrund asynchron auftretender Ereignisse (z.B. externer Interrupts). Berechnungsvorgänge, die aufgrund bestimmter Zeitpunkte aktiviert werden, werden als **zeitgesteuerte (periodische)** Berechnungsvorgänge bezeichnet, während die sonstigen Berechnungsvorgänge als **ereignisgesteuerte (asynchrone)** Berechnungsvorgänge bezeichnet werden. Asynchrone Berechnungsvorgänge, bei denen die minimale Zeitspanne zwischen dem aufeinanderfolgenden Auftreten des Ereignisses zur Aktivierung des Berechnungsvorganges bekannt ist, werden als **sporadische** Berechnungsvorgänge bezeichnet.

Die Forschungsarbeiten wurden durch ein *Ernst-von-Siemens Stipendium* der Firma Siemens AG gefördert

Neben den BhE und den BsE gibt es in Echtzeitsystemen i.a. noch eine weitere Klasse von Berechnungs-vorgängen (**BiA**), die lediglich in *Ausnahmesituationen* (z.B. beim Drücken eines Notaus-Schalters) aus-geführt werden, um das Gesamtsystem möglichst rasch in einen sicheren und stabilen Zustand zu bringen.

Die korrekte Funktionsweise eines Echtzeitsystems erfordert, daß alle den BhE zugeordneten Zeitschranken während der Laufzeit des Systems eingehalten werden. Deshalb sind Softwarewerkzeuge, die schon während des Entwicklungsprozesses eines Echtzeitsystems Zeitschrankenüberprüfungen bei BhE durchführen von großer Bedeutung.

In diesem Artikel stellen wir ein Softwaresystem zur Unterstützung des Entwicklungsprozesses verteil-ter Echtzeitsysteme vor. Das Softwaresystem besteht aus einer Spezifikationsmethode und einem Verfah-ren zur Berechnung einer globalen zyklischen Schedule. Dieses Softwaresystem kann sowohl im stand-alone Betrieb verwendet als auch in komplexe Softwareentwurfsumgebungen integriert werden. Das Spe-zifikationswerkzeug stellt dem Entwickler eine Schnittstelle zur Festlegung der BhE mit ihren Zeitschran-ken zur Verfügung. Die angegebenen Daten werden aufbereitet und dienen dem Verfahren zur Berech-nung einer globalen Schedule als Eingabe.

Kapitel 2 befaßt sich mit der für verteilte Echtzeitsysteme vorgeschlagenen globalen zyklischen Sche-dulingstrategie. Kapitel 3 beschreibt in groben Zügen unsere Spezifikationsmethode, während Kapitel 4 auf eines unserer Verfahren zur Berechnung einer globalen zyklischen Schedule eingeht. Kapitel 5 schließt diesen Artikel mit einigen Anmerkungen.

2. Die globale zyklische Schedulingstrategie für BhE

Ähnlich den Überlegungen in [1] gehen wir davon aus, daß in einem Echtzeitsystem für die einzelnen Klassen von Berechnungsvorgängen (BhE, BsE, BiA) entsprechend den zu erfüllenden zeitlichen Anfor-derungen unterschiedliche Schedulingstrategien verwendet werden. Jede Schedulingstrategie berücksich-tigt bei ihrer Entscheidung nur die Berechnungsvorgänge ihrer Klasse. Zur Lösung von Schedulingkon-flikten zwischen den einzelnen Klassen dient folgende Prioritätenvergabe: BiA haben Priorität über BhE, die wiederum Priorität über BsE haben.

Wir diskutieren in diesem Kapitel die Verwendung der **globalen zyklischen Schedulingstrategie** für BhE. In [1,2] finden sich Schedulingstrategien, die für BiA und BsE verwendet werden können.

Bei der globalen zyklischen Schedulingdisziplin wird eine Reservierungsstrategie angewendet. Für die SE der BhE werden auf den Rechnern, auf denen sie allokiert sind, Zeitbereiche für ihre Ausführung reser-viert. Kommunikationsvorgänge bedürfen einer besonderen Betrachtung. Zur Durchführung eines Kom-munikationsvorganges wird nicht nur auf das Übertragungsmedium (ÜM) zugegriffen, sondern es müssen auch in Abhängigkeit von der Leistungsfähigkeit der Kommunikationshardware Sende- und Empfangs-protokolle auf den Arbeitsprozessoren der jeweiligen Rechner ausgeführt werden. Deshalb werden sowohl auf dem ÜM Zeitschlitze für die Nachrichtenübertragung reserviert als auch auf den einzelnen Rechnern Zeitbereiche zur Durchführung dieser Protokolle eingeplant.

Für jeden Rechner und für das ÜM existiert je ein Schedulingplan. Diese Schedulingpläne haben alle die gleiche Länge. Ein Schedulingplan für einen Rechner legt die Zeitbereiche fest, in denen die SE bzw. Sende- und Empfangsprotokolle der BhE auszuführen sind. In den nicht reservierten Zeiträumen kann der Rechner für BsE genutzt werden. Alle Rechner des verteilten Echtzeitsystems beginnen gleichzeitig mit der Abarbeitung ihres Schedulingplanes. Wird das Ende des Schedulingplanes erreicht, so wird dieser wie-der von vorne abgearbeitet, wobei auch dieser Wechsel in allen Rechnern gleichzeitig durchgeführt wird.

Die Schedulingpläne der Rechner und des ÜM zusammen bilden einen **globalen zyklischen Scheduling-plan** für das verteilte Echtzeitsystem.

Die Verwendung der globalen zyklischen Schedulingstrategie erfordert eine globale Zeit, die entweder hardwaremäßig oder durch die Realisierung von Synchronisationsalgorithmen als BhE bereitzustellen ist. Weiterhin müssen Angaben über die Ausführungszeiten der Software und die Länge der Nachrichtenzeit-schlitze zur Verfügung stehen. Die Abschätzung solcher Zeiten kann mit geeigneten Werkzeugen automa-tisch durchgeführt werden [3]. Diese Zeiten dürfen zur Laufzeit nicht durch nichtdeterministische Ein-flüsse vergrößert werden. Deshalb gehen wir davon aus, daß im Bereich der BhE solche Einflüsse elimi-niert werden (speicherresidente Software von BhE, etc.). Während der Ausführung der SE von BhE dürfen lediglich Uhrinterrupts und Interrupts zur Aktivierung von BiA Unterbrechungen bewirken. Asynchrone Berechnungsvorgänge, die keine sporadischen Berechnungsvorgänge sind, sind als BhE nicht zulässig. Sporadische Berechnungsvorgänge können entsprechend den Ausführungen in [4] in periodische Berech-nungsvorgänge transformiert werden, so daß letztlich zur Berechnung einer globalen zyklischen Schedule nur periodische BhE vorliegen.

Die Verwendung der globalen zyklischen Schedulingstrategie für BhE bietet eine Reihe von Vorteilen:
- Schon während der Entwicklungsphase kann die Einhaltung von Zeitschranken überprüft werden, so daß zur Laufzeit Zeitschrankenverletzungen ausgeschlossen sind.
- Durch die Reservierung von Zeitschlitzen auf dem ÜM und die Integration von Sende- und Empfangsprotokollen in die Schedulingpläne der Rechner sind zur Laufzeit keine komplexen und zeit-aufwendigen Zugriffsverfahren, Kollisionserkennungs- und Kollisionsbehandlungsverfahren notwen-dig.
- Der synchronisierte Zugriff auf die Echtzeitdaten und die Schnittstellen zum technischen System kann bei der Erstellung der Schedulingpläne berücksichtigt werden, so daß zur Laufzeit keine Synchronisati-onsmaßnahmen wie Semaphore, Monitore oder Sperren notwendig sind.

Der eigentlichen Echtzeitanwendung steht somit ein Maximum an Systemleistung zur Verfügung.

Mit der Verwendung der globalen zyklischen Schedulingstrategie ist das Verhalten des Echtzeitsystems für BhE deterministisch, wie es in [5] gefordert wird. In [11,12] finden sich Überlegungen zur Berechnung zyklischer Schedules für Monoprozessor- und Mehrprozessorsysteme. [6] charakterisiert die Kombination *time-rigid Scheduling* mit *TDMA*-Zugriffsverfahren auf das ÜM als eine Möglichkeit, Determinismus zu erzielen. Diese Kombination, die eine Variante der globalen zyklischen Schedulingstrategie ist, wird im MARS-System [7] eingesetzt. Beim TDMA-Zugriffsverfahren werden auf dem ÜM einheitlich lange Zeitschlitze für die Nachrichtenübertragung reserviert, die in einem round-robin-Verfahren den einzelnen Rechnern exklusiv zugeordnet werden. Für jeden Rechner gibt es einen Schedulingplan, wobei jede SE gemäß ihrer Periode so im Schedulingplan eingeplant wird, daß zwischen zwei aufeinanderfolgenden Ein-planungen einer SE exakt die spezifizierte Periodenzeit der SE liegt. Das Softwaremodell in MARS besteht aus miteinander mittels Datagram-Nachrichten kommunizierenden periodischen SE.

In unserem Softwaremodell unterscheiden wir verschiedene Arten von SE, wobei die Zeitabstände zwi-schen aufeinanderfolgenden Startzeitpunkten einer SE im Schedulingplan i.a. unterschiedlich sind. Die spezifizierte Echtzeitanwendung bestimmt in unserem Modell die Länge und die Lage der reservierten Zeitschlitze auf dem ÜM.

3. Eine Spezifikationsmethode für BhE in verteilten Echtzeitsystemen

Mit Hilfe unserer Spezifikationsmethode können die BhE des Echtzeitsystems durch ihre SE, ihre Kom-munikationsstruktur und ihre Zeitschranken beschrieben werden. Diese Beschreibung wird in eine Dar-stellung transformiert, auf die zur Berechnung einer globalen zyklischen Schedule zugegriffen wird. Das zu entwickelnde Echtzeitsystem besteht aus autonomen Rechnern, die über ein broadcastfähiges ÜM miteinander verbunden sind. Gewisse Rechner haben über Sensor- und Aktorsysteme Zugriff auf das technische System.

Die Gesamtsoftware der BhE wird als Graph dargestellt. Die Knoten sind die SE, die Kanten stellen Kommunikationsvorgänge dar, wobei jede SE dem Rechner zuzuordnen ist, auf der sie ausgeführt wird.

Wir unterscheiden in unserem Softwaremodell verschiedene Arten von SE: SE ohne Kommunikationseingangskanten (periodische SE) und SE mit Kommunikationseingangskanten (UND-SE und ODER-SE). Periodische SE haben keine Vorgänger-SE. SE mit Kommunikationseingangskanten haben Vorgänger-SE und können erst ausgeführt werden, wenn die jeweiligen Nachrichten ihrer Vorgänger-SE zur Verfügung stehen.

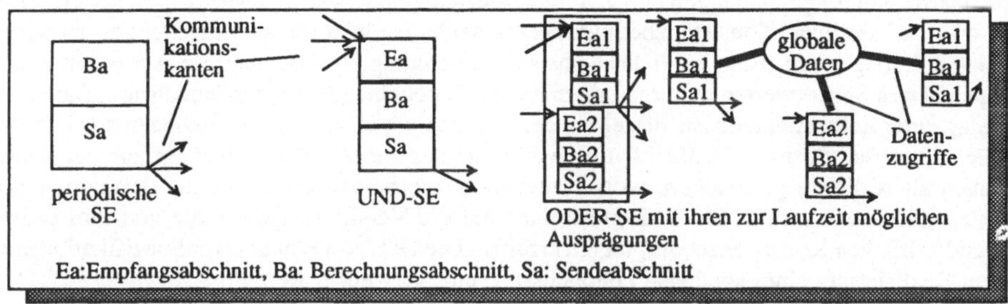

Bild 1: SE-Arten

Periodische SE (Bild 1) bestehen aus zwei Teilen: Einem reinen Berechnungsteil (Berechnungsabschnitt), in dem keine Nachrichten verschickt werden, und einem optionalen Sendeabschnitt. Die Ausführung einer periodischen SE besteht jeweils aus der Abarbeitung des Berechnungsabschnittes und der Abarbeitung des nachfolgenden Sendeabschnittes zur Übergabe der an nachfolgende SE zu verschickenden Nachrichten an die Kommunikationsschicht des Betriebssystems. Jede periodische SE wird durch die Spezifikation einer Aktivierungsperiode P und eines erstmaligen Aktivierungszeitpunktes A relativ zu t_0 charakterisiert; t_0 ist der virtuelle Startzeitpunkt des Echtzeitsystems. Durch P und A werden Anfangszeitpunkte ($t_0+A+k*P$; k=0,1, ...) von Zeitbereichen festgelegt, innerhalb derer die periodische SE jeweils einmal vollständig auszuführen ist. Die jeweiligen Endzeitpunkte dieser Zeitbereiche ergeben sich aus den für die BhE spezifizierten Zeitschranken.

UND-SE (Bild 1) bestehen aus drei Teilen: Einem Empfangsabschnitt, einem Berechnungsabschnitt und einem optionalen Sendeabschnitt. UND-SE haben eine oder mehrere Vorgänger-SE, wobei zur Ausführung einer UND-SE die Nachrichten von allen Vorgänger-SE vorliegen müssen. Der Berechnungsabschnitt und der Sendeabschnitt haben die gleiche Bedeutung wie bei periodischen SE. Der Empfangsabschnitt dient zur Übernahme der benötigten Eingangsnachrichten von der Kommunikationsschicht des Betriebssystems.

Zur Verwaltung der Echtzeitdaten und zum konsistenten Zugriff auf die Schnittstellen zum technischen System enthält unsere Spezifikationsmethode **ODER**-SE (Bild 1). Eine ODER-SE verwaltet globale Daten (Echtzeitdaten, Schnittstellen zum technischen System) und stellt Routinen für den konsistenten Zugriff auf diese Daten zur Verfügung. Jede dieser Zugriffsroutinen besteht aus einem Empfangsabschnitt, einem Berechnungsabschnitt und einem optionalen Sendeabschnitt. Nachrichten an eine ODER-SE werden an die spezifischen Entry-Stellen der gewünschten Zugriffsroutinen geschickt. Eine ODER-SE hat i.a. mehrere Kommunikationseingangskanten, wobei jede ankommende Eingangsnachricht zu einer eigenständigen Aktivierung der ODER-SE zur Abarbeitung der adressierten Zugriffsfunktion führt. Wir bezeichnen diese eigenständigen Aktivierungen einer ODER-SE als **Ausprägungen** dieser SE. Für jede Zugriffsroutine ist ihre Ausführungsart (shared oder exklusiv) anzugeben. Wurde für eine Zugriffsroutine shared angegeben, dann darf mit ihrer Abarbeitung nur dann begonnen werden, wenn sich keine exklusiv-Zugriffsroutine dieser ODER-SE in Berechnung befindet. Mit der Abarbeitung einer exklusiv-Zugriffsroutine darf nur dann begonnen werden, wenn sich keine andere Zugriffsroutine der ODER-SE in Berechnung befindet. Eine Zugriffsroutine befindet sich in Berechnung, wenn mit ihrer Bearbeitung begonnen wurde, diese jedoch noch nicht beendet ist.

Auf den Echtzeitdaten können verschiedene Konsistenzbedingungen definiert sein. *Externe* Konsistenzbedingungen legen fest, inwieweit gespeicherte Daten mit dem Zustand der Umgebung übereinstimmen, definieren also die "Aktualität" der gespeicherten Echtzeitdaten [6,8]. Für die korrekte Funktionsweise eines Echtzeitsystemes ist es wesentlich, daß veraltete Zustandsdaten erkannt werden. Dies ist z.B. dadurch möglich, daß den gespeicherten Echtzeitdaten zeitliche Gültigkeitsbereiche zugeordnet werden. *Interne* Konsistenzbedingungen legen einzuhaltende Beziehungen zwischen den gespeicherten Datenwerten fest, während *zeitliche* Konsistenzbedingungen einzuhaltende zeitliche Beziehungen zwischen den Echtzeitdaten festlegen [8]. Eine zeitliche Konsistenzbedingung ist z.B., daß die Abtastzeitpunkte von zwei gespeicherten Sensorwerten maximal um eine Zeit Δt voneinander abweichen dürfen. Eine mögliche Realisierung einer Echtzeitdatenbasis besteht in der Anwendung eines Versionenkonzepts bei den gespeicherten Echtzeitdaten; für jedes Echtzeitdatum wird nicht nur der aktuellste Wert gespeichert, sondern es werden auch ältere Werte gespeichert, wobei jeder dieser gespeicherten Werte die definierten externen Konsistenzbedingungen erfüllt. Ein BhE greift dann auf die Versionen zu, die die von ihm geforderten internen und zeitlichen Konsistenzbedingungen erfüllen. Die ODER-SE in unserer Spezifikationsmethode können zur Realisierung eines solchen Versionenkonzeptes verwendet werden.

Für jede SE bzw. Zugriffsroutine bei ODER-SE ist ihre Ausführungszeit anzugeben. Weiterhin ist festzulegen, ob die Abarbeitung zur Laufzeit unterbrochen werden kann oder nicht. Hierbei können diese Angaben sowohl für ganze SE als Einheit als auch für die einzelnen Abschnitte gemacht werden.

Neben den SE müssen auch Kommunikationsvorgänge charakterisiert werden. Für alle verwendeten Kommunikationsprotokolle im Bereich der BhE ist ein deterministisches zeitliches Verhalten bei der Nachrichtenübertragung notwendig. Wir unterstützen in unserem Modell nur asynchrone Kommunikationsformen. Sobald die sendende SE eine Nachricht an die Kommunikationsschicht abgegeben hat, kann ihre Abarbeitung fortgesetzt werden. Wir gehen davon aus, daß flexible Kommunikationssysteme verwendet werden, die neben einem reinen Datagram-Service auch Nachrichtenübertragungen mit einer festen, für jeden Kommunikationsauftrag spezifischen Anzahl von Nachrichtenwiederholungen im Fehlerfall ermöglichen. Die Angabe der Wiederholungszahl kann z.B. zur Laufzeit über einen bestimmten Auftragsparameter erfolgen.

Zur Spezifikation von rechnerübergreifenden Kommunikationsvorgängen sind folgende Zeiten anzugeben (Bild 2): Ausführungszeit des Sendeprotokolls (SP), Länge des zu reservierenden Zeitschlitzes auf dem ÜM (ZS), Ausführungszeit des Empfangsprotokolls (EP), zeitlicher Abstand zwischen dem Sendeprotokoll und dem zu reservierenden Zeitschlitz (ASZ) und zeitlicher Abstand zwischen dem Zeitschlitz und dem Empfangsprotokoll (AZE). Sende- und Empfangsprotokoll überlappen sich bei Protokollen mit Nachrichtenwiederholungen im Gegensatz zum reinen Datagram-Service. Deshalb sind die angegebenen ASZ- und AZE-Zeiten bei Protokollen mit Wiederholungen als Fixzeiten zu behandeln, d.h. zwischen dem im globalen Schedulingplan eingeplanten Sendeprotokoll und dem reservierten Zeitschlitz liegt genau die Zeit ASZ, zwischen dem reservierten Zeitschlitz und dem Empfangsprotokoll liegt genau die Zeit AZE. Bei einem Kommunikationsvorgang, bei dem ein Datagram-Service benötigt wird und bei einer entsprechenden Leistungsfähigkeit der Kommunikationshardware können diese Zeiten als Mindestzeiten behandelt werden.

Die Spezifikation von Berechnungsvorgängen erfolgt durch die Angabe von Teilgraphen, wobei jeder Teilgraph nur periodische SE als SE ohne Eingangskanten enthält. Durch die Ausführung von periodischen SE wird die Ausführung von Berechnungsvorgängen aktiviert. Aber nicht jede Ausführung einer periodischen SE bewirkt die Aktivierung aller Berechnungvorgänge, die diese SE enthalten. Für jeden Berechnungsvorgang werden durch die Angabe einer Aktivierungsperiode und eines erstmaligen Aktivierungszeitpunktes relativ zu t_0 Anfangszeitpunkte von Zeitbereichen festgelegt. Nur die Ausführung der periodischen SE des Berechnungsvorganges in diesen Zeitbereichen hat eine Ausführung des gesamten Berechnungsvorganges zur Folge. Weiterhin werden jedem Berechnungsvorgang Zeitschranken relativ zu seinen Aktivierungszeitpunkten zugeordnet, die bei der Ausführung der SE des Berechnungsvorganges einzuhalten sind.

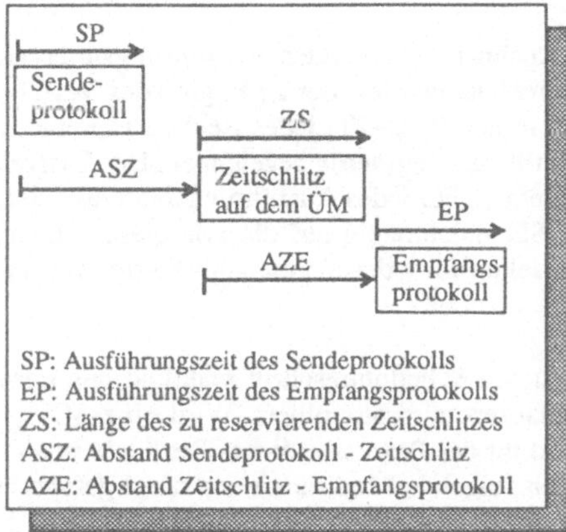

Bild 2: Spezifikation rechnerübergreifender Kommunikationsvorgänge

Bild 3: Initialisierung eines Lastvektors

Ziel unserer Softwareentwicklungsumgebung ist die Berechnung einer globalen zyklischen Schedule der Länge T (T = kgV der spezifizierten Perioden) für die BhE. Wir bezeichnen weiterhin die im Rahmen von Kommunikationsvorgängen einzuplanenden Sende- und Empfangsprotokolle als **Sende-** bzw. **Empfangs-SE**. Das ÜM wird im weiteren auch als "Rechner" aufgefaßt, auf dem Nachrichtenzeitschlitze als **ÜM-SE** "auszuführen" sind, falls nicht explizit auf besondere Maßnahmen beim ÜM eingegangen wird. Nachdem die BhE mit ihren Zeitschranken spezifiziert sind, können für jede SE **Ausführungsintervalle** berechnet werden. Jede SE ist zur Laufzeit jeweils einmal vollständig innerhalb jedes ihrer Ausführungsintervalle auszuführen. Die für die SE berechneten Ausführungsintervalle werden in den Zeitbereich $[t_0, t_0+2T]$ transformiert. Aufgrund dieser Transformation kann es Ausführungsintervalle geben, mit: $LI < t_0 < RI$ bzw. $LI < t_0+2T < RI$ (LI: linke Ausführungsintervallgrenze, RI: rechte Ausführungsintervallgrenze). Deshalb berechnet unser Verfahren eine globale Schedule für den Zeitraum $[t_0, t_0+2T]$, wobei durch spezielle Maßnahmen dafür gesorgt wird, daß es sich bei dem Teil $[t_0+T, t_0+2T]$ um eine zyklische Schedule der Länge T handelt.

4. Ein Verfahren zur Berechnung einer globalen zyklischen Schedule für BhE

Zur Scheduleberechnung wird ein **Schedulingbaum** aufgebaut. Jeder Baumknoten ist mit einem Zeitpunkt markiert, wobei der Wurzel der Zeitpunkt t_0 zugeordnet ist. Ein Pfad von der Wurzel bis zu einem mit einem Zeitpunkt t_i markierten Knoten stellt eine partielle globale Schedule bis zum Zeitpunkt t_i dar. Jeder Baumknoten des Pfades beschreibt einen globalen Schedulingschritt: Durch den Knoten wird für jeden Rechner festgelegt, welche SE im Zeitraum $[t_{i-1}, t_i]$ auszuführen ist (t_{i-1} ist der Zeitpunkt mit dem der

Vaterknoten dieses Knotens markiert ist). Grundsätzlich besteht in jedem solchen Schedulingschritt auch die Möglichkeit, einem Rechner keine SE eines BhE zur Ausführung zuzuordnen. Ziel unseres Berechnungsverfahrens ist das Finden eines Pfades von der Wurzel bis zu einem mit einem mit t_0+2T markierten Knoten.

Bevor wir unser heuristisch gesteuertes Berechnungsverfahren skizzieren, wollen wir zunächst eine Bewertungsfunktion vorstellen, die zur Bewertung eines Schedulingschrittes bzw. eines Schedulingpfades dient. Ein vorliegender, partieller Schedulingpfad soll um einen Schedulingschritt erweitert werden, indem von den Sohnknoten des letzten Pfadknotens ein Knoten zur Pfaderweiterung ausgewählt wird.

Ausgehend von den im Zeitraum $[t_0,t_0+2T]$ für jeden Rechner vorliegenden Ausführungsintervallen, kann für jeden Rechner jeweils ein **Lastvektor** (Bild 3) berechnet werden, der zu Beginn des Verfahrens mit einer "Grundlast" initialisiert wird. Hierzu wird der Zeitraum $[t_0,t_0+2T]$ jedes Rechners in eine feste Anzahl gleichgroßer Zeitabschnitte unterteilt, die wir weiterhin als **Segmente** bezeichnen. Der Lastvektor eines jeden Rechners setzt sich aus diesen Segmenten zusammen. Für jedes Ausführungsintervall wird die Ausführungszeit der in diesem Intervall einzuplanenden SE gleichmäßig auf die von diesem Intervall überdeckten Segmente verteilt. Anschließend wird im Lastvektor für jedes Segment als **Segmentwert** die Summe der zugeordneten Ausführungszeiten gespeichert.

Die Erweiterung eines Schedulingpfades um einen weiteren Schedulingschritt während des Berechnungsverfahrens wird durch Modifikationen aller Lastvektoren mitprotokolliert. Wird eine SE S_k auf einem Rechner für den Zeitraum $[t_{i-1},t_i]$ eingeplant, so wird ihr der Prozessor dieses Rechners in diesem Zeitraum exklusiv zugeordnet. In den Lastvektoren werden folgende Modifikationen durchgeführt: Die noch einzuplanende Ausführungszeit von S_k hat sich verringert. Deshalb wird der Segmentwert der Segmente reduziert, die von diesem Ausführungsintervall der SE S_k ab dem Zeitpunkt t_i noch überdeckt werden. Alle Nachfolger-SE von S_k werden zumindest um die Zeit t_i-t_{i-1} nach hinten verschoben, wodurch die Ausführungszeit dieser Nachfolger-SE in den betroffenen Ausführungsintervallen auf weniger Segmente als bisher zu verteilen ist. Deshalb werden die Segmentwerte der davon betroffenen Segmenten erhöht. Weiterhin werden alle SE, die auf diesem Rechner auch im Zeitraum $[t_{i-1},t_i]$ zu einem Zeitpunkt t (t_{i-1} <= t < t_i) eingeplant werden können und deren Nachfolger zumindest um die Zeit t_i-t verschoben. Dies wird ebenfalls durch eine entsprechende Erhöhung der Segmentwerte der davon betroffenen Segmente berücksichtigt.

Die beschriebenen Maßnahmen werden für jede in einem globalen Schedulingschritt eingeplante SE durchgeführt, so daß die Lastvektoren letztlich die "Auswirkungen" des gesamten globalen Schedulingschrittes beinhalten.

Von den Sohnknoten des letzten Schedulingpfadknotens soll ein Sohn zur Erweiterung der Schedule ausgewählt werden. Zur Bewertung jedes Sohnknotens wird folgendes durchgeführt: Von den Lastvektoren des letzten Pfadknotens werden Kopien angelegt. Für jede SE des durch den Sohnknoten dargestellten Schedulingschrittes wird angenommen, daß sie vollständig und ohne Unterbrechung eingeplant wird. Entsprechend obigen Ausführungen zur Modifikation der Lastvektoren werden die Kopien verändert. Danach wird für jeden Rechner R_j der Differenzvektor zwischen der modifizierten Kopie und dem ursprünglichen Lastvektor gebildet und der Mittelwert der Differenzwerte berechnet, die ungleich null sind. Dieser Mittelwert wird als **mittlere Segmentdifferenz** von R_j bezeichnet. Bei der Mittelwertbildung könnte auch eine zusätzliche Gewichtung eingehen, die den jeweiligen Segmentwert des ursprünglichen Lastvektors berücksichtigt. Ist für jeden Rechner seine mittlere Segmentdifferenz berechnet, so erhält der Schedulingschritt die Bewertungszahl $b=(m_0^2+m_1^2+...+m_n^2)^{1/2}$ (m_0 = mittlere Segmentdifferenz beim ÜM, m_i = mittlere Segmentdifferenz bei Rechner R_i). Einem Pfad im Schedulingbaum wird der Mittelwert über den Bewertungszahlen b der einzelnen Pfadknoten als Bewertungszahl B zugeordnet. Hier kann durch eine gewichtete Mittelwertbildung die Schedulingpfadlänge berücksichtigt werden.

Bild 4 skizziert in groben Zügen unser Berechnungsverfahren. Die Variable "Sched_Pfad" zeigt jeweils auf den letzten Knoten im Schedulingpfad. In jedem Schedulingschritt wird der Baumknoten expandiert, auf den Sched_pfad zeigt. Dieser Baumknoten ist mit t_{i-1} markiert. Seine Sohnmenge stellt alle zur Erweiterung des Schedulingpfades möglichen Schedulingschritte dar. Jeder der Söhne erhält eine Bewertungszahl b. Zur Erweiterung des Schedulingpfades wird der Sohn mit der kleinsten Bewertungszahl b gewählt. Der gewählte Sohn ist mit einem Zeitpunkt t_i zu markieren. Zur Bestimmung des jeweiligen Zeitpunktes wird zunächst für jeden Rechner R_j ein Zeitpunkt $t_{i,j}$ bestimmt: Handelt es sich bei der auf Rechner R_j gemäß des globalen Schedulingschrittes eingeplanten SE S_k um eine ununterbrechbare SE, dann wird $t_{i,j}=t_{i-1}+A$ gesetzt, ansonsten wird $t_{i,j}=min(t_{i-1}+A,t)$ gesetzt. A ist die noch einzuplanende Ausführungszeit von S_k; t bezeichnet den frühesten Zeitpunkt im Zeitraum $[t_{i-1},t_{i-1}+A]$ zu dem eine sonstige SE von R_j erstmalig in ihrem Ausführungsintervall, das diesen Zeitpunkt enthält, eingeplant werden kann. Es handelt sich dabei um eine SE, die keine Vorgänger-SE auf diesem Rechner hat, also um eine periodische SE oder eine SE in einem Kommunikationsvorgang (Empfangs-SE bzw. ÜM-SE). Ist für jeden Rechner R_j der Zeitpunkt $t_{i,j}$ bestimmt, so ist $t_i=min(t_{i,j})$ (j=0,1,..,n) zu setzen. Für den Fall, daß dieser Schedulingschritt zur Laufzeit zu Zeitschrankenverletzungen führen würde, wird dieser Sohnknoten als Möglichkeit zur Erweiterung des Schedulingpfades ausgeschlossen, indem er mit einer "nicht-möglich"-Marke versehen wird. Diese Marke wird entlang des Schedulingpfades "vererbt": Haben alle Söhne eines Knotens eine solche Marke, so wird auch dieser Knoten markiert und der durch die Sohnknoten belegte Speicherplatz kann freigegeben werden. Ist der Wurzel des Schedulingbaumes eine solche "nicht-möglich"-Marke zuzuordnen, so existiert keine globale Schedule. Wurde der gewählte Sohn mit einer "nicht-möglich"-Marke versehen, so wird die Variable Sched_Pfad auf den Knoten mit der kleinsten Bewertungszahl B (s.o.) gesetzt, der nicht im Schedulingpfad liegt und Sohn eines Pfadknotens ist.

```
begin    erzeuge die Wurzel des Schedulingbaumes;
         Sched_Pfad := Wurzel;
         while es wurde noch keine globale Schedule gefunden und es wurde noch nicht festgestellt,
               daß keine globale Schedule möglich ist do
               begin expandiere den Baumknoten, auf den Sched_pfad zeigt;
                     Sched_Pfad := bestimme den Sohnknoten des expandierten Baumknotens,
                                   durch den die partielle Schedule zu erweitern ist
                                   u. markiere ihn mit einem Zeitpunkt;
                  if es wird für die aktuelle Schedule eine Zeitschrankenverletzung
                     festgestellt then
                     begin ordne dem Knoten, auf den Sched_Pfad zeigt eine
                           "nicht-möglich"-Marke zu und "vererbe" diese;
                           Sched_Pfad := bestimme einen Baumknoten, durch den die partielle
                                         Schedule zur Fortsetzung der Berechnung festgelegt wird;
                     end;
               end;
end.
```

Bild 4: Berechnungsverfahren für eine globale zyklische Schedule

Grundsätzlich besteht unter Verwendung der Bewertungszahl B auch die Möglichkeit, einen allgemeinen A^*-Algorithmus [9] zu integrieren. Allerdings gibt es im Berechnungsverfahren einige, den aktuellen Zustand der Scheduleberechnung repräsentierende Mengen. Diese Mengen müßten beim Einsatz eines A^*-Algorithmus entweder jeweils mit abgespeichert oder neu initialisiert werden, was aus Speicherplatzgründen bzw. Laufzeitgründen nicht interessant erscheint. Eine Variante des Algorithmus, bei dem in jedem Schedulingschritt alle Sohnknoten entlang des momentanen Schedulingpfades berücksichtigt werden, ist dagegen vertretbar.

Um das in der Praxis auftretende Problem des beschränkten Speicherplatzes zu lösen, kann z.B. die Kombination des beschriebenen Berechnungsverfahrens mit einem reinen Backtracking oder das in [10] skizzierte, als Iterative-deepening-A* bezeichnete Verfahren eingesetzt werden.

Das bisher skizzierte Verfahren beschreibt die grundsätzliche Idee. Es wird dabei implizit davon ausgegangen, daß die zur Charakterisierung von Kommunikationsvorgängen spezifizierten Versatzzeiten nicht als Fixzeiten, sondern als Mindestzeiten zu behandeln sind. Die Behandlung der Versatzzeiten als Fixzeiten, das Vorhandensein von ODER-SE bzw. ODER-Ausprägungen und das Vorkommen von ununterbrechbaren SE erfordern zusätzliche Maßnahmen, auf die hier nicht näher eingegangen werden kann.

Unser in der Programmiersprache C++ unter dem Betriebssystem UNIX implementierter Prototyp enthält diese zusätzlichen Maßnahmen. Bei Kommunikationsvorgängen wurde der schwierigste Fall angenommen, nämlich die Behandlung aller Versatzzeiten als Fixzeiten. Das Berechnungsverfahren läuft zur Zeit als reines Backtrackingverfahren. Steuerungsheuristiken im Sinne der beschriebenen Bewertungsfunktion werden gerade integriert. Wir implementieren nur Approximationen der Bewertungsfunktion, da die beschriebene Bewertungsfunktion zu laufzeitaufwendig ist. Gleichzeitig werden Maßnahmen integriert, um Speicherplatzengpässe zu behandeln.

5. Anmerkungen

Neben dem skizzierten zentralen Berechnungsverfahren zur Berechnung einer globalen zyklischen Schedule haben wir auch ein verteiltes Verfahren entwickelt und als Prototyp implementiert. Dieses verteilte Verfahren besteht aus einer Menge kommunizierender Prozesse zur Berechnung einer globalen Schedule. Für jeden Rechner und für das ÜM gibt es einen Prozeß, der für diesen Rechner bzw. für das ÜM eine lokale Schedule berechnet. Synchronisationsmaßnahmen durch Nachrichtenkommunikation gewährleisten, daß die berechneten lokalen Schedulingpläne zusammen eine globale zyklische Schedule bilden. Dieses verteilte Berechnungsverfahren wurde in der Programmiersprache C unter dem Betriebssystem UNIX als miteinander über Socket-Verbindungen kommunizierende Prozesse implementiert. Auch hier werden zur Zeit Steuerungsheuristiken integriert. Desweiteren planen wir eine Portierung des verteilten Verfahrens auf Transputer.

In Echtzeitsystemen mit unterschiedlichen Betriebsphasen steht i.a. zur Laufzeit nicht nur eine einzige zyklische Schedule zur Verfügung, sondern es kann grundsätzlich für jede Betriebsphase eine zyklische Schedule bereitgestellt werden.

Literatur

1. A. Damm, *"Kernel Aspects of the Distributed Real-Time Operating System of MARS"*, Research Report Nr. 6/87, Institut für Technische Informatik, TU Wien

2. E.D. Jensen, C.D. Locke, H. Tokuda, *"A Time-Driven Scheduling Model for Real-Time Operating Systems"* IEEE 1987 Real-Time Systems Symposium, Proceedings

3. P. Puschner, Ch. Koza, *"Calculating the Maximum Execution Time of Real-Time Programs"*, The Journal of Real-Time Systems, Vol.1, No.2, 1989

4. A. Mok, *"Fundamental Design Problems of Distributed Systems for the Hard-Real-Time Environment"*, Ph.D. Thesis, Massachsetts Institute of Technology, 1983

5. R.G. Herrtwich, *"Echtzeit"*, Informatik-Spektrum (1989) 12

6. H. Kopetz, *" Scheduling in Distributed Real-Time Systems"*, INRIA Advanced Seminar on Real-Time Local Area Networks, France, 1986, Proceedings

7. H. Kopetz u.a., *"Distributed Fault-Tolerant Real-Time Systems: The Mars Approach"*, IEEE Micro, February 1989

8. K.-J. Lin, M.-J. Lin, *"Enhancing Availability in Distributed Real-Time Databases"*, SIGMOD RECORD, Vol.17, No.1, 1988

9. A. Barr, E.A. Feigenbaum, *"The Handbook of Artifical Intelligence"*, Volume I, William Kaufmann Inc., Los Altos, California, ISBN 0-86576-005-5

10. R.E. Korf, *"Depth-First Iterative-Deepening: An Optimal Admissible Tree Search"*, Artifical Intelligence 27 (1985)

11. J.Y.-T. Leung, M.L. Merrill, *"A Note on Preemptive Scheduling of Periodic, Real-Time Tasks"*, Information Processing Letters, Vol.11, No.3, 1980

12. E.L. Lawler, C.U. Martel, *"Scheduling Periodically Occuring Tasks on Multiple Processors"*, Information Processing Letters, Vol.12, No.1, 1981

Analysing PEARL Programs for Timely Schedulability and Executability

(Analyse zeitgerechter Zuteilbarkeit und Ausführbarkeit von PEARL-Programmen)

Alexander D. Stoyenko
Department of Computer and Information Science
New Jersey Institute of Technology
University Heights
Newark, New Jersey, 07102 USA
alex@vienna.njit.edu

Wolfgang A. Halang
Department of Mathematics and Computing Science
University of Groningen
P.O. Box 800
9700 AV Groningen, The Netherlands
halang@cs.rug.nl

Abstract

PEARL is a widely-used, standardised, industrial process-control programming language. The programs, that are written in PEARL, are typically expected to operate in a predictable, timely, reliable manner. Developing such programs is no easy task. Yet, with the invent of the experimental programming language Real-Time Euclid and its *schedulability analyser,* a tool that is used to predict whether real-time software will adhere to its critical timing constraints, it has been demonstrated that this task may be made considerably easier.

In this paper, we describe how PEARL can be modified to be better suitable for real-time application writing, and to accommodate schedulability analysis. We propose a design for a schedulability analyser for PEARL. We feel confident that the analyser can be applied successfully to PEARL programs, thus aiding greatly in the process of developing predictable, timely and reliable software.

I. Introduction

In [1, No. 9.2.11], real-time operation of a computer system is defined as: "The operating mode of a computer system in which the programs for the processing of data arriving from the outside are permanently ready, so that their results will be available within predetermined periods of time; the arrival times of the data can be randomly distributed or be already a priori determined depending upon the different applications." Among the examples of real-time environments or applications are windtunnel control [2], astronomical data acquisition [3], plane control [4], space shuttle control [5], overcurrent and high impedance fault relaying [6], radar applications [7], and amusement park ride control [8].

The emphasis of the above definition is on the system performing all of the above functions *sufficiently quickly*. Moreover, it is clear that only when the system performs its timely activities reliably and in a predictable, ascertainable manner, may the users of the system be satisfied.

Owing to the growing number and sophistication of real-time applications, there is also a growing need for real-time software to control them. Real-time software must be verified to adhere to its critical timing constraints prior to its use. We refer to this verification process as *schedulability analysis* [9][1]. Schedulability analysis is done by applying an automatic tool, called the *schedulability analyser*, to real-time software. The tool reports to the programmer to what extent the software satisfies its critical timing constraints, and what corrections may be made to guarantee that all constraints are met. The schedulability analyser aids greatly in the development of reliable and predictable real-time software. We hope that one day the benefits of schedulability analysis will be recognised to the extent that no real-time software will be developed without the use of the schedulability analyser.

Our experience with schedulability analysis and schedulability analysable languages is based on the development of the experimental language Real-Time Euclid [9, 10] and its schedulability analyser [10, 11]. A descendant of the Pascal [12], Euclid [13] and Concurrent Euclid [14] line of languages, Real-Time Euclid is the first language specifically designed with provisions for schedulability analysis built-in. Thus, all Real-Time Euclid programs are schedulability analysable.

The same ideas that went into the design of Real-Time Euclid and its schedulability analyser can be applied to PEARL [15], a widely-used, standardised, industrial process-control language. While PEARL is not, in its current form, entirely schedulability analysable, it actually comes the closest to that among all industrial languages. Thus, only moderate modifications are needed to make PEARL schedulability analysable.

The body of this paper is organised as follows. In Section II, we suggest modifications to PEARL, to ensure that PEARL programs can be analysed for guaranteed schedulability. In Section III we present a set of schedulability analysis techniques, that are applied to PEARL. In Section IV we report on the state of our research and its evaluation. Still in Section IV we summarise the main points brought up in this paper and conclude with directions for future research.

II. PEARL and Suggested Modifications to PEARL

So far, the scope of this work has included Basic PEARL only. PEARL is a modular, procedural and block-oriented language, widely-used in industrical process-control applications. Like in Real-Time Euclid, tasks (processes) are declared statically. Unlike in Real-Time Euclid, however, task activation information is not associated with each task statically, but is specified by schedules at runtime.

Task control statements are used to activate, suspend, resume or terminate tasks. Task control operation may specify a schedule, detailing the conditions of the operation. The conditions may be temporal - or event-driven in nature, like in Real-Time Euclid. Tasks are synchronised and shared data are protected via generalised semaphores. There is an unstructured exception handling facility based on software signals and **ON** conditions (as in PL/I). PEARL defines a generalised data-flow model of I/O through **DATION**s or **DAT**a sta**TION**s. A dation defines the topology of a device, and the data and control specifications of device access.

---

[1]The English term "schedulability" can be translated into German as "zeitgerechte Zuteilbarkeit und Ausführbarkeit", or, literally, "timely schedulability and executability". Since this paper is in English, we shall always refer to the term simply as "schedulability".

A. *Adherence to Critical Timing Constraints*

As this paper is not intended to be either a language survey or a language requirements wish-list, we shall not turn it into such. For a fuller treatment of the subject, the reader is referred to [10, 16, 17]. However, we should like to motivate our proposed modifications of PEARL, before we introduce them.

A real-time application imposes critical timing constraints on the software that controls it. The time-critical nature of a real-time application makes it necessary that to be considered reliable and predictable, the controlling software must adhere predictably to these constraints. The timing constraints are typically described in terms of timing schedules that govern task activation, execution and termination. Each task activation is expected to perform its computations successfully before a deadline, explicitly specified in its schedule, expires.

Since deadlines and other such timing constraints directly correspond to explicit application requirements, the computation model of a task activation has no need to include any abstractions that may prevent the execution from adhering to the schedule. Thus, any abstraction that takes arbitrarily long to execute is unnecessary. From the theory of decidability of computation models, it follows that for any analysis, that attempts to ascertain a priori the adherence of a set of real-time task executions to their schedules, to succeed, any computation abstraction that takes arbitrarily long to execute must, in fact, be forbidden.

If we apply this reasoning to a high-level real-time language, then it follows that the language should allow the expression of timing schedules and real-time tasks whose activation, execution and termination are controlled by schedules. On the other hand, the language should not allow any construct that may take arbitrarily long to execute to be used in task construction.

Thus, a number of the familiar conventional constructs, such as, for instance, a general while-loop, should be banned from real-time languages. Indeed, an iterative computation that has a deadline must be expressable in terms of a constant-bounded loop. Otherwise, the computation may miss its deadline and thus fail. If, on the other hand, unbounded iteration is called for in the real-time application, then it must be the case that the iterative computation has no inherent deadline associated with it. Then, a proper way to implement the computation is to make every single iteration (or a bounded number of iterations) a task activation.

Once a language is modified to describe each program as a set of real-time tasks whose activation, execution and termination is governed by explicit schedules, and to forbid any constructs that may take arbitrarily long to execute, expressing real-time applications in the language becomes quite natural. Moreover, every program in the language becomes schedulability analysable.

Despite what we consider an apparent need for ensuring that real-time languages are schedulability analysable, a survey of the current languages actually used in real-time programming [18] indicates that none make sufficient provisions for schedulability analysis. The state of affairs is well summarised by Berry: "... paradoxically, one can verify that the current so-called "real-time programming languages" do not provide any explicit means of correctly expressing time constraints. A fortiori, they provide no insurance that the constraints would be respected when executing the program." [19]

B. *Making PEARL Schedulability Analysable*

All temporal conditions used in a schedule must be in terms of compile-time constants. Every task is associated with a frame size, expressed as a compile-time constant. Once a task is activated, it must complete by the time computed as the sum of the current time and the frame size. Usually, the scheduler will detect well in advance whether the task will be able to complete by that time, that is, its deadline. Special signals are raised if this is impossible or the deadline is actually missed. These restrictions ensure that the schedules are only used to express explicit timing constraints, as imposed by the applications.

In [20-22] and in the sequel PEARL is slightly modified and restricted in order to enable schedulability analysis.

Semaphores are removed from the language and replaced by the structured **LOCK** mechanisms for the formulation of critical regions. Endowed with a timeout constant to bound the queue waiting time, an alternative action to be taken in the case of unsuccessful waiting, and an execution time bound for the body of the **LOCK**, it is possible to determine, already at compile-time, a global **NOLONGERTHAN** parameter for each **LOCK** statement. The I/O operations, viz. **PUT, GET, READ, WRITE**, SEND and **TAKE**, are realised as runtime system procedures with known upper execution time bounds. If accessing devices are shared between several tasks, they must be used within the framework of **LOCK** statements. The above provisions ensure that no synchronisation or I/O operation takes arbitrarily long to execute.

Every signal event declaration is associated with a compile-time limit on the number of times the event may be raised in a single activation of a task causing and handling the exception. The same limit applies, naturally, to all specifications (imports) of the same event throughout the program. The minimum duration between two subsequent occurrences of the same interrupt, which is equivalent to its maximum arrival frequency, is specified in the interrupt declaration. These restrictions ensure that there can be no arbitrarily long chains of raised signals or interrupts.

To prevent unpredictable iteration, **WHILE** clauses are no longer permitted in **REPEAT** statements. Moreover, every **REPEAT** must specify a maximum compile-time number of iterations. The **GOTO** statement is removed from the language. To avoid arbitrarily long chains of invocations, (mutually) recursive calls of PEARL procedures are no longer allowed.

In the remainder of this paper, when we say "PEARL", we mean PEARL with the modifications, as described in this Section, incorporated.

III. A Schedulability Analyser for PEARL

The schedulability analyser consists of a front end and a back end.

A. Front End of the Schedulability Analyser

The front end of the schedulability analyser is incorporated into the coder, and its task is to extract timing information and calling information from each compilation unit.[2]. Clearly, the front end of the analyser is compiler- and machine architecture-dependent. The output of the front end, however, is free of such dependencies.

---

[2]By "compilation unit" we mean anything that is compilable, that is, a module, a task, or a procedure

A.1 Front End Segment Trees

As statements are translated into assembly instructions in the coder, the front end of the analyser builds a tree of segments for each procedure or task. The various types of segments are as follows.

(1)   A simple segment contains the amount of time it takes to execute an explicit section of code.

(2)   An internal-call-segment contains the name of a called internal routine, and will eventually be resolved by substituting the amount of time it takes for that routine to execute. External- and kernel-call segments are similar to an internal-call-segment.

(3)   An if-segment splits the tree into two subtrees where one subtree corresponds to the then-clause, and the other to the else-clause of the correspondent if-statement. An if-segment will eventually be resolved and only the subtree which takes longer to execute will stay. Case-statements result in (a sequence of) if-segments, similar to the way if-statements do.

A loop does not by itself cause a tree sub-branching. If the body of the loop is all pure explicit code, then the loop just results in a simple segment incorporating the amount of time it takes to execute a single iteration times the number of iterations. If the body of the loop contains a call- or an if-segment, then the loop is unwound into a segment chain, that is, its body is repeated the number of iterations times.

A.2 Instruction-Level Timing Information Extraction

As an assembly instruction is generated, the front end of the analyser records its execution time. The time is computed as a function of the opcode, the operands and addressing modes. A procedure explaining how to compute such timing information for a particular machine architecture is typically given in a machine instructions manual such as [23]. The time is added to the current simple segment being built.

A.3 Critical Section, Routine and Task Records

Apart from building segment trees, the front end of the analyser also records information on the organisation of critical sections as well as calling information. A critical section is defined as the body of a **LOCK** statement. We are contemplating adding structured time-bounded monitors, as in Real-Time Euclid, to PEARL, but are only addressing critical sections at the moment.

For every critical section its start and end are recorded, as well as which routines and tasks use it. For every routine it is recorded who calls it, whom it calls, what critical sections it uses, and what I/O operations it executes. Finally, for each task its frame and activation information are recorded as well as whom the task calls, what critical sections it uses, and what I/O operations it executes.

B. Back End of the Schedulability Analyser

The back end of the schedulability analyser is a separate program, and its task is to resolve all information gathered by the front end, and predict guaranteed response times and other schedulability information and diagnostics for the entire real-time application. The back end constructs a map from a PEARL program, or more precisely, from segment trees built by the front end, to an instance of a schedulability analysable model. The instance is then solved (analysed) for guaranteed schedulability. The back end of the analyser is independent of the compiler, the machine architecture and so on.

B.1 Resolving Segment Trees

The back end of the analyser starts off by resolving segment trees built by the front end. All individual segment trees and critical section, routine and task records are concatenated into one file, including the segment trees and records pertinent to predefined kernel and I/O routines. All non-kernel calls are recursively resolved by substituting the corresponding routine segment tree in place of each call-segment. As many if-segments as possible are resolved.

B.2 Converting Task Trees

After all non-kernel calls are resolved, only task segment trees are left. The trees are converted to different segment trees, where each possible segment is one of the following.

(1)   An interruptible segment is one containing the amount of time it takes to execute a section of code in the absence of contention. The execution can be interrupted, though.

(2)   A non-interruptible segment is one containing the amount of time it takes to execute a section of code in the absence of contention. The execution cannot be interrupted.

(3)   A specified-delay segment corresponds to a delay specified in a **NOLONGERTHAN-**clause or in a schedule in the program.

(4)   A queue-delay segment corresponds to a delay while waiting to enter a critical section.

(5)   A non-interruptible-code delay segment corresponds to a delay while waiting for another ask to complete executing a non- interruptible section of code.

Once segment tree conversion is complete, the tasks are ready to be mapped to an instance of a schedulability analysable model.

B.3 A Schedulability Analysable Model

The exact form of the model depends on the topology of the system on which the real-time application is intended to be run. For the purpose of this discussion, it suffices to consider a model with a single CPU, local memory and devices. All inter-task communication takes place via critical sections (that is, LOCK synchronisation). Each resource is protected by either a DATION (that is treated like a critical section), or a critical section. Processor allocation as well as critical section queueing are carried out using the earliest-deadline- first task scheduling discipline. The following system of equations describes the high level of the model.

$$GT_i = I_i + N_i + ID_i + CD_i + ND_i + SD_i \leq F_i \qquad (1)$$

for $i \in \{1 ,..., \#T \}$, where
$\#T$ is the number of tasks in the application,
I_i is the total CPU requirement for executing interruptible parts of task T_i, in the absence of contention,

N_i is the total CPU requirement for executing non-interruptible parts of T_i, in the absence of contention,

ID_i is the worst case amount of time T_i spends waiting due to other tasks executing their interruptible parts,

CD_i is the worst case amount of time T_i spends in critical section queues,

ND_i is the worst case amount of time T_i spends waiting due to other tasks executing their non-interruptible parts,

SD_i is the worst case amount of time T_i spends in specified delays,

GT_i is the guaranteed response time of T_i, and

F_i is the specified frame of T_i .

I_i, N_i and SD_i are easily derived by adding up interruptible, non-interruptible and specified-delay segments, respectively. To derive CD_i, ID_i and ND_i, more work than simply adding up the segments is necessary.

B.3.1 Critical Section Queue and Non-interruptible Delays

A critical section is interruptible, but is necessarily in between a queue-delay segment (waiting to enter the critical section) and a non-interruptible segment (updating critical section queues upon leaving the section).

In non-real-time systems, queueing for resources[3] has traditionally been estimated via statistical modeling, commonly referred to as *queueing network modeling*. In real-time systems, this statistical, averaged approach is insufficient, because it is not enough to know that most deadlines are met most of the time. We must establish whether all deadlines are met all the time. Thus, real-time systems must be analysed via *worst-case* schedulability analysis.

Many schedulability analysis techniques reported in the literature [24-35] have dealt with resource queueing by means of solutions that do not take into account segments' relative positions and inter-segment distances on the time line [25] when estimating delays. These solutions are typically closed-form or polynomial, and thus work very quickly. However, they result in overly pessimistic worst-case queue delays, and ultimately in overly pessimistic guaranteed response times [18]. For example, a well-known solution by Leinbaugh [24] computes CD_i as:

$$CD_i = \sum_{j=1}^{\#C} CD_{ij}, \quad CD_{ij} = \sum_{\substack{t=1 \\ t \neq i}}^{\#T} I_{tj} \times \left(\left\lceil \frac{F_i}{F_j} \right\rceil + 1 \right)$$

where there are #C critical sections, labeled 1,2 ,..., #C, CD_{ij} is the jth critical section's contribution to CD_i, and I_{tj} is the jth critical section's contribution to I_t. This solution assumes that every critical section segment blocks every other critical section segment every time: an overly pessimistic assumption.

We use *frame superimposition* [10] to derive CD_i's and ND_i's. In fact, these delays are computed at once and *the sum* of CD_i and ND_i is maximised. We maximise the sum because the sum is a part of GT_i, and GT_i is the entity whose upper bound we are ultimately interested in. Frame superimposition simply means fixing a single T_i's frame at time t_0 and positioning frames of other tasks along the time line in such a way as to maximise the amount of critical section contention T_i incurs. Frame superimposition uses relative positions of

---

[3]Traditionally, hardware resource queueing has been measured, but with the increased complexity of system software, software resource queueing has become important. In PEARL we confine all software and hardware resource access to critical sections.

segments and distances between segments on the time line. The algorithm shifts frames exhaustively, for every time unit, for every task, for every combination of frames possible.

The frame superimposition algorithm is clearly exponential. In fact, finding the optimal worst-case bound for resource contention in the presence of deadlines is NP-complete, given that even the most basic deadline scheduling problems are [36, 37]. However, so is compilation in general. Moreover, this part of the analysis operates on segment trees, there are many instructions, even statements reduced to an averaged segment, and relatively few task combinations are legal. Finally, better delay bounds are derived by our algorithm, than by closed-form or polynomial algorithms. Thus we feel justified in using frame superimposition.

B.3.2 Interruptible Delays

ID_i is the delay caused by other tasks executing their interruptible segments while T_i is ready. If T_i is in a critical section, then, it cannot be pre-empted by another task trying to get into the same critical section. On the other hand, T_i can be pre-empted by tasks not trying to use the same critical section. We thus compute ID_i as

$$ID_i = \sum_{k=0}^{\#C} ID_{ik}, \quad ID_{ik} = s_{ik}WID_{ik} \tag{2,3}$$

where ID_{ik} is the kth critical section's contribution to ID_i. Critical sections $1,2,...,\#C$ are real critical sections, while the critical section 0 corresponds to the non-critical section interruptible segments.[4]

WID_{ik} is the worst-case interruptible interference for the kth critical section, computed as

$$WID_{ik} = \begin{cases} k \neq 0 & \displaystyle\sum_{\substack{j=1 \\ j\neq i}}^{\#T}\sum_{\substack{t=0 \\ t\neq k}}^{\#C} I_{jt} \\[3ex] k = 0 & \displaystyle\sum_{\substack{j=1 \\ j\neq i}}^{\#T}\sum_{t=0}^{\#C} I_{jt} \end{cases} \tag{4}$$

s_{ik} is the slow-down rate for the kth critical section, which is further constrained by equations (5) and (6).

$$0 \leq s_{ik} \leq 1, \quad \sum_{i=1}^{\#T}(1-s_{ik}) \leq 1 \tag{5,6}$$

B.3.3 Overall Solution Algorithm

If we are only interested in verifying a given set of frames, then we only have to solve the set of systems of equations (1) through (6) once. However, if we want to find out how short response times we can guarantee, then we instead fixed-point iterate over the following system of constrained equations.

$$GT_i^{<x+1>} \leq F_i^{<x>}$$

where x is the iteration count,
$F_i^{<x>}$ is given or guessed,

---

[4]We refer to these segments as "critical section segments" to keep our notation unified.

$GT_i^{<x>}$ is obtained by solving (1) through (6),

and $F_i^{<x+1>}=GT_i^{<x>}$

Our set of systems of equations is underdetermined. To make it determined, we can make a number of constructive assumptions.

In some systems, tasks naturally group into families. Each family is responsible for performing synchronised, related tasks. Often, system engineers define frame ratios for each task family. As a result, we would then add the following system of equations to our system.

$$\frac{F_i^{<x>}}{F_j^{<x>}} = \frac{F_i^{<0>}}{F_j^{<0>}} \tag{8}$$

for $i,j \in \{1 ,..., \#T \}$, for all x.

Unless there is a specific reason to do otherwise, we can assume that for each task, slow-down rates are the same for all critical sections, that is:

$$s_{i0} = s_{i0} = , ... , = s_{i\#C}$$

for $i \in \{1 ,..., \#T \}$.

Once we determine (1) through (9), we iterate until we get no more reduction in GT_i's, or alternatively until we get no more reduction in the sum of GT_i's. We then report the GT_i and s_i for each task, as well as timing information for every routine, call, critical section access, even every statement. The programmer may then study the results, and decide whether the program should be run as it exists, or whether it should be modified.

If the purpose of the analysis was to verify a given set of frames, then in the case of a positive solution of our set of systems of equations the programmer need not do any changes to the program to achieve guaranteed schedulability. Similarly, if the purpose was to guarantee as tight a set of response times as possible, and the resulting set of times is satisfactory, no changes to the program are needed.

If, however, the guaranteed response times obtained are too large, the supplementary timing information, that is, specific routine, call, statement and other such information, is of help to determine where the tasks can be streamlined.

IV. A Last Word

A detailed description of the evaluation of this work is outside the scope of this paper, but can be found in [10]. For completeness, the reader is also referred to [38-41]. In this section we describe briefly the evaluation of the work so far, state the current status and give indications for future activities.

A. *Implementation*

No implementation or evaluation of modified PEARL or its schedulability analyser has taken place yet. However, the ideas used to modify PEARL and design its schedulability analyser are very similar to those used in Real-Time Euclid.

At the University of Toronto, we have designed and built a Real- Time Euclid compiler. The assembly code generated is for the NS16000/32000 microprocessor family. To support Real-Time Euclid programs at run-time, we have designed and built a two-layer Real-Time Euclid kernel. The upper layer, referred to as the Time Manager, keeps track of time in real-time units. The Time Manager maintains the status of each task, next activation times, frames and other timing information. The Time Manager also handles timer interrupts. The lower kernel layer, referred to as the Basic Kernel, is responsible for setting up the hardware, transferring timer interrupts to the Time Manager, handling device interrupts, maintaining various task queues, and basic earliest-deadline-first scheduling.

A prototype schedulability analyser for Real-Time Euclid has been implemented. The front end of the analyser is embedded in the coder pass of the compiler. While assembly code is generated, this part of the schedulability analyser extracts timing information, builds list, and segment trees. The back end of the analyser acts similarly to the way described of Section III. Real-Time Euclid programs are mapped to a more complicated model, though. The model reflects the topology of the hardware (see below), thus assuming distributed nodes, with a CPU and resources at each node. Inter-node communication is done via a centralised mailbox.

The hardware on which Real-Time Euclid has been evaluated is configured as a distributed system, consisting of three nodes and an independent megabyte memory board, all connected by a Multibus. Each node has a NS32016 CPU, a megabyte of local memory, and a terminal. The CPUs can communicate via the independent board.

B. Evaluation

Teams of senior undergraduate students have written medium size (under 5K lines of code) Real-Time Euclid programs. Each program does data acquisition via simulated sensors and keyboards, processes, analyses and reduces the data, and outputs commands via simulated output 13 devices and terminal screens.

We have used the schedulability analyser to predict response times and compute other schedulability information for these programs. We have run these programs, recorded the observed actual response times and other statistics, and compared the actual statistics to those guaranteed by the analyser.

We have found that in the absence of overload (that is, missed deadlines) the worst-case guaranteed statistics differ very little (usually, under 10%, and never over 26%) from those observed. In the presence of missed deadlines, we have recorded a 25% to 44% discrepancy. In contrast, when the forementioned Leinbaugh method had been used to predict performace of the same programs, that method has predicted bounds that differed by 40% to 200% from the observed statistics.

C. Summary and Future Directions

By schedulability analysis [9, 10], a real-time application can be, to a significant extent, guaranteed to perform in a predictable, timely manner. Schedulability analysis is thus an important part of the real-time application development process.

To be used for writing real-time software, a programming language should be schedulability analysable. In this paper, we have suggested modifications to PEARL, a real-time language widely used in industrial process control. The modifications ensure that every PEARL program can be analysed for schedulability.

Based on our experience with Real-Time Euclid, we have proposed a schedulability analyser for PEARL. The analyser uses frame superimposition to derive guaranteed, accurate bounds of runtime statistics. While we have not yet evaluated PEARL's schedulability analyser expermentally, our successful experience with Real-Time Euclid and its analyser leads us to believe that the PEARL evaluation too is going to be a success. We intend to undertake an experimental evaluation of our proposal. In the course of the evaluation, we may extend our ideas also to Distributed PEARL [42]. Owing to the positive experience with the schedulability analyser for Real-Time Euclid applications, which were of distributed nature, we do not anticipate major problems in migrating to Distributed PEARL.

Another direction for future research we are interested in pursuing has to do with the design of the schedulability analysis itself. Frame superimposition makes extensive use of timing information to predict guaranteed performance statistics. However, no semantic (that is, variable state) information has been used so far. We intend to investigate to what extent semantic information may be used in frame superimposition, from both the effectiveness and efficiency points of view.

References

[1] DIN 44 300 A2, Informationsverarbeitung. Berlin: Beuth-Verlag 1985.

[2] C. D. Williams, "The Data Acquisition, Data Reduction and Control System (DARCS) for the NRCC 2x3m Windtunnel," Proceeding of the IEEE 1984 Real-Time Systems Symposium, December 1984, pp. 89-94.

[3] P. W. Kelton, "Distributed Computing for Astronomical Data Acquisition at McDonald Observatory," Proceeding of the IEEE 1984 Real-Time Systems Symposium, December 1984, pp. 83-88.

[4] G. Kaplan, "The X-29: Is it coming or going?," IEEE Spectrum, June 1985, pp. 54-60.

[5] G. D. Carlow, "Architecture of the Space Shuttle Primary Avionics Software System," Communications of the ACM, Vol. 27, No. 9, September 1984, pp. 926-936.

[6] B. M. Aucoin, R. P. Heller, "Overcurrent and High Impedance Fault Relaying using a Microcomputer," Proceedings of the 7th Texas Conference on Computing Systems, November 1978, pp. 2.5-2.9.

[7] E. T. Fathi, N. R. Fines, "Real-Time Data Acquisition, Processing and Distribution for Radar Applications," Proceeding of the IEEE 1984 Real-Time Systems Symposium, December 1984, pp. 95-101.

[8] V. P. Nelson, H. L. Fellows, Jr., "A Microcomputer-Based Controller for an Amusement Park Ride," IEEE Micro, August 1981, pp. 13-22.

[9] A. D. Stoyenko, E. Kligerman, "Real-Time Euclid: A Language for Reliable Real-Time Systems," IEEE Transactions on Software Engineering, Vol. SE-12, No. 9, September 1986, pp. 940-949.

[10] A. D. Stoyenko, A Real-Time Language with A Schedulability Analyzer, Ph.D. Thesis, Department of Computer Science, University of Toronto, 1987, available as Computer Systems Research Institute Technical Report CSRI-206.

[11] A. D. Stoyenko, "A Schedulability Analyzer for Real-Time Euclid," Proceedings of the IEEE 1987 Real-Time Systems Symposium, December 1987, pp. 218-225.

[12] N. Wirth, "The Programming Language Pascal," Acta Informatica 1, 1971, pp. 35-63. 15 [13] R. C. Holt et al., "Euclid: A language for producing quality software," Proceedings of the National Computer Conference, Chicago, May 1981.

[14] J. R. Cordy, R. C. Holt, Specification of Concurrent Euclid, Technical Report CSRG-133, Computer Systems Research Group, University of Toronto, August 1981.

[15] DIN 66 253, Programming Language PEARL (Programmiersprache PEARL), Teil 1 Basic PEARL, 1981; Teil 2 Full PEARL, 1982. Berlin: BeuthVerlag.

[16] W. A. Halang, "On Real-Time Features Available in High-Level Languages and Yet to be Implemented," Microprocessing and Microprogramming, Vol. 12, 1983, pp. 79-87.

[17] W. A. Halang, A. D. Stoyenko, "Comparitative Evaluation of High Level Real Time Programming Languages," Real-Time Systems, Vol. 2, No. 4, 1990.

[18] A. D. Stoyenko, The State of the Art in Hard-Real-Time Modeling and Languages, Ph.D. Qualifying Paper, Department of Computer Science, University of Toronto, May 1986.

[19] G. Berry, S. Moisan, J.-P. Rigault, "Esterel: Towards a Synchronous and Semantically Sound High Level Language for Real-Time Applications," Proceedings of the IEEE 1983 Real- Time Systems Symposium, December 1983, pp. 30-37.

[20] W. A. Halang, "A Proposal for Extensions of PEARL to Facilitate the Formulation of Hard Real-Time Applications," Informatik- Fachberichte 86, pp. 573-582, Berlin-Heidelberg-New York-Tokyo: Springer-Verlag 1984.

[21] W. A. Halang, R. Henn, "Additional PEARL Language Structures for the Implementation of Reliable and Inherently Safe Real-Time Systems," Proceedings of the 15th IFAC/IFIP Workshop on Real Time Programming, Oxford: Pergamon Press 1988, pp. 35-42.

[22] W. Ehrenberger, "Softwarezuverlassigkeit und Programmiersprache," PEARL-Rundschau, Vol. 3, No. 2, 1982, pp. 49-55.

[23] Series 3200 Instruction Set Reference Manual, National Semiconductor Corporation, Santa Clara, June 1984.

[24] D. W. Leinbaugh, "Guaranteed Response Times in a Hard-Real-Time Environment," IEEE Transactions on Software Engineering, Vol. SE- 6, No. 1, January 1980, pp. 85-91.

[25] D. W. Leinbaugh, M.-R. Yamini, "Guaranteed Response Times in a Distributed Hard-Real-Time Environment", Proceedings of the IEEE 16 1982 Real-Time Systems Symposium, December 1982, pp. 157-169.

[26] C. L. Liu, J. W. Layland, "Scheduling Algorithms for Multiprogramming in a Hard-Real-Time Environment", JACM, Vol. 20, No. 1, January 1973, pp. 46-61.

[27] A. K. Mok, "The Design of Real-Time Programming Systems Based on Process Models", Proceedings of the IEEE 1984 Real-Time Systems Symposium, December 1984, pp. 5-17.

[28] A. K. Mok, M. L. Dertouzos, "Multiprocessor Scheduling in a Hard Real-Time Environment", Proceedings of the 7th Texas Conference on Computing Systems, November 1978, pp. 5.1-5.12.

[29] K. Ramamritham, J. A. Stankovic, "Dynamic Task Scheduling in Distributed Hard Real-Time Systems", Proceedings of the IEEE 4th International Conference on Distributed Computing Systems, May 1984, pp. 96-107.

[30] K. Ramamritham, J. A. Stankovic, S. Cheng, "Evaluation of a Flexible Task Scheduling Algorithm for Distributed Hard Real-Time Systems", IEEE Transactions on Computers, Vol. C-34, No. 12, December 1985, pp. 1130-1143.

[31] P. G. Sorenson, A Methodology for Real-Time System Development, Ph.D. Thesis, Department of Computer Science, University of Toronto, 1974.

[32] P. G. Sorenson, V. C. Hamacher, "A Real-Time System Design Methodology", INFOR, Vol. 13, No. 1, February 1975, pp. 1-18.

[33] A. D. Stoyenko, Real-Time Systems: Scheduling and Structure, M.Sc. Thesis, Department of Computer Science, University of Toronto, 1984.

[34] T. J. Teixeira, "Static Priority Interrupt Scheduling", Proceedings of the 7th Texas Conference on Computing Systems, November 1978, pp. 5.13-5.18.

[35] W. Zhao, K. Ramamritham, "Distributed Scheduling Using Bidding and Focused Addressing", Proceeding of the IEEE 1985 Real-Time Systems Symposium, December 1985, pp. 103-111.

[36] M. R. Garey, D. S. Johnson, "Complexity Results for Multiprocessor Scheduling under Resource Constraints", SIAM Journal on Computing, Vol. 4, No. 4, December 1975, pp. 397-411.

[37] J. D. Ullman, "Polynomial complete scheduling problems", Proceedings of the 4th Symposium on OS Principles, 1973, pp. 96-101.

[38] E. Kligerman, A Programming Environment for Real-Time Systems, M.Sc. Thesis, Department of Computer Science, University of Toronto, 1987.

[39] C. Ngan, Implementing the Real-Time Euclid Compiler, Student Project Report, Department of Computer Science, University of Toronto, August 1986.

[40] S. A. Thurlow, Simulation of a Real Time Control System Using the Real-Time Euclid Programming Language, Student Project Report, Department of Computer Science, University of Toronto, April 1987.

[41] G. Parnis, Simulation of Packet Level Handshaking in X.25 Using the Real-Time Euclid Programming Language, Student Project Report, Department of Computer Science, University of Toronto, April 1987.

[42] DIN 66 253, Programmiersprache PEARL - Mehrrechner-PEARL, Teil 3, Berlin: Beuth-Verlag 1989.

<u>Einsatz von zeitbewerteten Fuzzy-Petri-Netzen in Expertensyste-</u>
<u>men zur operativen Führung komplexer Produktionssysteme</u>

H.-P. Lipp
TU Chemnitz
Sektion Informatik

1. Problemstellung

Moderne just-in-time Produktionskonzepte erfordern leistungsfähige
Entscheidungssysteme, die in Störungssituationen durch die Entwick-
lung gültiger Steuerhandlungen maßgeblich zur Stabilisierung des
Produktionsablaufes beitragen. Solche, bisher von einem Dispatcher
oder Schichtleiter im instationären Betriebsgeschehen ermittelten
Entscheidungen, beinhalten alle operativen Steuerhandlungen, welche
das System in einen aus übergeordneten Planungsebenen festgelegten
Zielzustand transformieren. Anders als in off-line geführten CAD-
Prozessen ist jedoch dieser Zustandswechsel nur erreichbar, wenn die
Steuereingriffe in allen Teilsystemen unter Beachtung ihrer strukturellen Abhängigkeiten gleichzeitig, gut aufeinander abgestimmt
ausgeführt werden. In einem mehrstufigen Entscheidungsprozeß muß
deshalb zu jedem Taktzeitpunkt aus der verfügbaren Handlungsvielfalt
ein Vektor von Steuerregeln gebildet und zu der gesuchten Steuerstrategie zusammengestellt werden. Diese Steuereingriffe können als
Regelwerk eines Problemlösungsprozesses betrachtet werden, dessen
Ergebnis eine Steuerstrategie aus mehrdimensionalen Komplexregeln
liefert, die zuvor aus den momentan verfügbaren Einzelregeln gebildet werden müssen.
Durch den Einfluß von Prozeßstörungen ist die elementare Regelmenge
unterschiedlich verfügbar. Das System ist damit in seiner Steuerbar-
keit mehr oder weniger stark eingeschränkt. Die Lösungen haben
deshalb eine begrenzte Gültigkeit, so daß sie wiederholt ermittelt
werden müssen, wobei möglicherweise unterschiedliche Problemlöser-
strategien eingesetzt werden.
In diesem Entscheidungsprozeß müssen die Lösungen in einer von der
Prozeßdynamik abhängigen Zeit ermittelt werden. Unterstützt durch
Erfahrungen von Experten werden deshalb kompakte Modelle benötigt,
so daß adäquate Lösungen im Echtzeitbetrieb entwickelt werden
können. Der vom Prozeß bedingte Handlungszwang gestattet dabei kein
Zurückstellen von Entscheidungen. Oft müssen sogar vorläufige Lösun-

gen an vereinfachten Modellen ermittelt werden. Die Regeln können dabei wegen ihrer fehlenden direkten Umkehrbarkeit nicht durch einfaches Probieren am Prozeß bestimmt werden, so daß der Einsatz von Simulationssystemen notwendig wird, welche zur Unterstützung des operativen Entscheidungsprozesses die zu erwartenden Auswirkungen prozeßnah dargestellen.

Um den Dispatcher wirkungsvoll in seinem operativen Entscheidungsprozeß zu entlasten, werden Expertensysteme benötigt, die auf speziellen für diese Aufgabenklasse geeigneten Konzepten aufbauen. In den nachfolgenden Ausführungen wird ein Konzept eines Expertensystems beschrieben, das den Forderungen des Echtzeitprozesses entsprechen soll. Seine Wissensbasis hat wegen der Parallelität und Asynchronität der zu aktivierenden Steuerregeln die Struktur eines Fuzzy-Petri-Netzes/1/. Das Fuzzy-Petri-Netz wird durch entsprechende Echtzeitkopplungen mit Prozeßwerten versorgt, so daß ein Problemlöser eine Lösungsstrategie aus parallel einzusetzenden Steueroperationen entwickeln kann.

2. Eine Modellstruktur für den operativen Entscheidungsprozeß in komplexen Systemen

2.1. Das Fuzzy-Petri-Netz-Konzept/1/

Eine Beschreibung von Handlungsfolgen zur Steuerung komplexer Systeme ist durch Petri-Netze anschaulich immer dann möglich, wenn der Arbeitsfortschritt durch das asynchrone Zusammenwirken mehrerer paralleler Teilprozesse bestimmt wird. Dabei ist vorteilhaft, daß bei der Darstellung unterschiedlicher Prozeßzustände die Systemstruktur im Modell erhalten bleibt. Nachteilig ist jedoch das zweiwertige Verhalten der Plätze und Transitionen, wodurch Prozeßbedingungen und Steueraktionen nur in den Zuständen "zulässig/unzulässig" oder "Ein/Aus" abgebildet werden können. Ein für reale Systeme aber durchaus übliches mehr oder weniger gutes Schalten bei nicht ganz exakt erfüllten Bedingungen, läßt sich in diesem "scharfen" Petri-Netz nur mit einem sehr hohen Modellaufwand ausdrücken. Um solche durch unvollständige Systeminformationen bedingten oder auf Erfahrungswissen begründete Handlungsfolgen dennoch in einem operativen Entscheidungsprozeß berücksichtigen zu können, wurde das Fuzzy-Petri-Netz-Konzept entwickelt.

Das Fuzzy-Petri-Netz wird durch Verunschärfung der Plätze und Transitionen aus "scharfen" Petri-Netzen abgeleitet. Die Aufweichung von Bedingungskomplexen und Aktionen wird dabei mit dem Ziel vorgenommen, um unvollständige Informationen und subjektives Expertenwissen bei der Modellierung von Entscheidungssituationen in komplexen Systemen zu berücksichtigen. Die scharfen Netzelemente werden unscharfen Mengen zugeordnet, wobei, da viele für bestimmte Situationen gültige Petri-Netze zu einem Fuzzy-Petri-Netz zusammengefaßt werden, eine Modellvereinfachung erreicht wird. Die scharfen Netzkomponenten entsprechen dann mit unterschiedlicher Zugehörigkeit den Elementen im unscharfen Netz. In den Plätzen werden dabei bewußt viele Marken zugelassen, um so den Erfüllungsgrad der Prozeßbedingungen möglichst fein unterscheiden zu können. In Abhängigkeit von der Güte dieser Bedingungen wird ein mehr oder weniger starkes Reagieren der unscharfen Transitionen ausgelöst. Für das Fuzzy-Petri-Netz gilt:

$$FPN = (P, T, F, S, G, M, m_0)$$

$P = \{\tilde{p}\}$ — Menge der unscharfen Plätze
$$\tilde{p} = \{ (m(\tilde{p}); \mu_{\tilde{p}}(m(\tilde{p}))) \}$$

$T = \{\tilde{t}\}$ — Menge der unscharfen Transitionen
$$\tilde{t} = \{ (t; \mu_{\tilde{t}}(t)) \}$$

$F : (PxT) \cup (TxP) \longrightarrow [0,1]$ — Überführungsfunktion

$S : (PxT) \longrightarrow [0,1]$ — Startbewertung

$G : (TxP) \longrightarrow [0,1]$ — Zielbewertung

$M : (P) \longrightarrow N$ — zulässige Markierungen

$m_0 \in M$ — Anfangsmarkierung

Unscharfe Plätze stellen fuzzy Mengen über Platzmarkierungen dar. Ihre unterschiedlich bewerteten Markenbelegungen dokumentieren dabei den mehr oder weniger guten Erfüllungsgrad von Prozeßbedingungen. Die unscharfen Transitionen werden aus abgeschlossenen Teilsystemen oder Regeln abgeleitet, die Markenströme von Eingangsplätzen in einem bestimmten Verhältnis auf Ausgangsplätze übertragen. Sie repräsentieren die Menge von Einstellungen, Prozeßrealisierungen oder Regelausführungen, die unterschiedlich gut die Steuerabsicht eines Anlagenexperten umsetzen. Diese Realisierungen werden von einer bestimmten Variabilität ihrer Markenströme begleitet, so daß eine flußabhängige Anpassung der Platzmarkierungen möglich ist.

Der Wechsel von einer Transitionseinstellung in eine andere wird als unscharfes Schalten betrachtet. Gegenüber einer scharfen Transition, die nur die Zustände "kein Markenstrom" und "Schalten mit einer bestimmten Markenanzahl" beinhaltet, kann bei ihr ein mit unterschiedlicher Markenanzahl n-wertiges Schalten realisiert werden. Es liegt eine sichere Schaltregel vor, wenn die unscharfe Transition in eine Einstellung wechselt, die eine maximale Verbesserung der Platz- und Transitionsbewertungen bewirkt.

In welcher Reihenfolge Transitionen an der Verbesserung angrenzender Platzmarkierungen beteiligt werden, wird aus lokalen und globalen Strukturkenntnissen abgeleitet und in der Startbewertung $S(p \times t)$ der Kanten berücksichtigt.

Die Zielbewertung $G(t \times p)$ des unscharfen Netzes dient zur Auswahl von unscharfen Schaltfolgen an den Transitionen unter Beachtung des augenblicklichen Netzzustandes. Abhängig vom Anwendungsfall soll sie Lösungen ermöglichen, die neben einem minimalen Zielabstand auch ein gutes dynamisches Lösungsverhalten garantieren.

In einem Fuzzy-Petri-Netz stellt eine Schaltfolge das wiederholte unscharfe Schalten der Transitionen dar, die das Petri-Netz von seinem Ist- in einen Zielzustand überführt. Die Schaltfolge der Transitionen wird von einem Problemlöser bestimmt, der den Schaltzeitpunkt und die Größe der Transitionsänderungen situationsbedingt so festlegt, daß bei möglichst guten Platzbedingungen der Zielmarkenstrom erreicht wird. Das Ergebnis dieses mehrstufigen Suchprozesses ist eine Lösungstrajektorie, die zu jedem Taktzeitpunkt die notwendigen Transitionsrealisierungen zur Verbesserung der Zielbewertung des Netzes beinhaltet. Der Problemlöser ist dabei durch die unscharfen Netzbewertungen in seinem Lösungsverhalten einstellbar.

2.2. Zeitbewertete Fuzzy-Petri-Netze

Schaltzeiten, Totzeiten, Verzögerungszeiten u.a. müssen in vielen Entscheidungen unbedingt berücksichtigt werden. Mit der Entwicklung von zeitbewerteten Petri-Netzen wurde deshalb eine Modellstruktur geschaffen, in der das bedingungs- und zeitabhängige Verhalten eines Systems nebeneinander in einem Modell beschrieben werden kann. In zeitbehafteten Petri-Netzen setzt sich der Schaltvorgang einer Transition aus der Initialisierungs-, Ausführungs- und Beendigungsphase zusammen. In "scharfen" Petri-Netzen erfolgt das Schalten der Transition nach festen Zeitmaßen, wobei nach Beendigung des Vorganges ideale Schaltfolgebedingungen gesetzt werden. Es gibt aber durch Bearbeitungs- oder Wartezeiten gekennzeichnete Prozeßabläufe, deren Dauer zusätzlich durch situationsbezogene Bedingungen zeitvariabel

begrenzt werden, und bei denen in Abhängigkeit von der tatsächlich
benötigten Prozeßzeit Folgebedingungen so gesetzt werden, daß mit
unterschiedlichen Reaktionen der Vorgang fortgesetzt werden muß. Mit
einer unscharfen zeitbewerteten Transition ist eine Netzstruktur
entwickelt worden, die solche variablel begrenzten Aktionen in einem
Entscheidungsprozeß berücksichtigen kann.

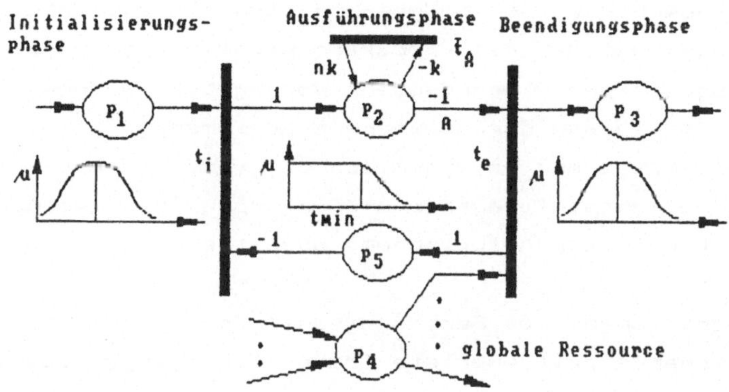

Bild 1: Unscharf zeitbehaftete Petri-Netz-Struktur

In Bild 1 ist die Netzstruktur einer zeitbewerteten Transition mit
unscharfen Prozeßeigenschaften angegeben. Bei vorhandenen Ausgangs-
bedingungen startet ihre Initialisierungstransition τ_i einen unscharf bewerteten Zeitzähler (τ_A, $\tilde{p}_2$), dessen einseitig abfallende
Zugehörigkeitsfunktion bei Überschreitung einer minimalen Zeitgrenze
t_{min} durch eine schlechter werdende Bewertung die Abschlußtransition
τ_e immer mehr zum Schalten anregt. Wenn die Voraussetzungen ($\tilde{p}_4$) für
das Schalten der Abschlußtransition erfüllt sind, wird der Zeitzähler gestoppt und die Ausführungsphase τ_A wird beendet. Bei zunehmender zeitlichen Verzögerung der Ausführungsphase wirkt sich die
unscharfe Zeitbewertung des Zählers mit einer immer größer werdenden
Triebkraft auf die Arbeitsfähigkeit der Abschlußtransition aus, so
daß diese gegenüber anderen vorrangig schalten kann. Mit dieser
unscharfen Eigenschaft verfügt die zeitbewertete Transition über
eine variable Ausführungszeit, wobei mit zunehmender Verzögerung
unterschiedliche Reaktionen auf globale Schaltbedingungen ausgeführt
werden können.

2.3. Unscharfe Petri-Netz-Modelle für die dynamische Produktionsablaufplanung in flexiblen Fertigungssystemen

In diesem Abschnitt wird die Anwendung des Fuzzy-Petri-Netz-Konzeptes in einem Expertensystem beschrieben, das zur Entwicklung von dynamischen Produktionsablaufplänen in der Leitebene flexibler Fertigungsstrukturen eingesetzt wurde. In seiner Wissensbasis enthält es in Form von unscharfen Petri-Netz-Modellen die abzuarbeitenden Fertigungsaufträge, welche als Produktionsziel von der administrativen- und dispositiven Planungsebene vorgegeben wurden. Bei Störungen oder bei Einflüssen durch extern existierende konventionelle Fertigungsabschnitte muß es dynamisch die verfügbaren Fertigungsmittel den Bearbeitungsaufträgen zuordnen, wobei der Zeitpunkt, der Ort und die Art der technologischen Operationen festzulegen sind.

Die Bearbeitungsoperationen der Fertigungsaufträge werden in dem Petri-Netz durch unscharfe zeitbewertete Transitionen abgebildet. In unscharfer Weise kann damit ihre situations- und zeitbezogene Einflußnahme auf die Systemressourcen im Produktionsplan berücksichtigt werden. Die Initialisierungs-, Ausführungs- und Abschlußphase der unscharf zeitbewerteten Transitionen entsprechen dabei Abschnitten der Fertigungsoperation, die durch ihren technologischen Inhalt und die Belegung bestimmter Fertigungshilfsmittel gekennzeichnet sind. Jede Fertigungsoperation wird deshalb unterteilt in Bereitstellungs-, Bearbeitungs- und Abschlußoperationen dargestellt.

Die Bereitstellungsoperation schafft dabei alle Voraussetzungen für die Ausführung der nachfolgend ablaufenden Bearbeitungsphase. Sie schließt damit neben dem Transport des Werkstücks, der Werkzeuge und anderer Fertigungshilfsmittel auch den Rüstvorgang in der Maschine ein.

Die Bearbeitungsoperation schließt sich an die Bereitstellungsphase an. Sie beinhaltet alle technologischen Prozesse am Werkstück, wobei das Transportsystem in dieser Zeit für andere Aufträge wieder frei zur Verfügung steht.

Die Abschlußoperation beendet den Aufenthalt des Werkstückes in der Station. Sie kann frühestens aufgerufen werden, wenn die Bearbeitung abgeschlossen wurde und das Transportsystem das Werkstück von der Maschine befördern kann. Mit dem Abtransport des Werkstückes aus der Fertigungszelle werden die benutzten Systemressourcen für den weiteren Einsatz bereitgestellt.

Um neben der Maschine die zeitlich begrenzte Verfügbarkeit des

Transportsystems und anderer Fertigungsressourcen im Produktionsab-
laufplan berücksichtigen zu können, wird in Bild 2 ein erweitertes
Netzmodell für die Fertigungsoperation angegeben. Ihre Bereitstel-
lungs- und Abschlußoperationen sind ebenfalls wie der Bearbeitungs-
abschnitt mit einer endlichen Zeit versehen und durch unscharfe
zeitbewertete Transitionen abgebildet. Die Fertigungsoperation
stellt hier eine Folge von mehreren Zeitabschnitten dar, die unmittelbar hintereinander durchlaufen werden. In Abhängigkeit davon, ob
das Werkstück nach seiner Bearbeitung in ein Zwischenlager transportiert wird, ob es direkt in die freie Folgestation gebracht wird
oder ob es nach einem Umrüstvorgang in der gleichen Maschine weiter
bearbeitet wird, für all diese Fälle ist in dieser Darstellung eine
spezielle Abschlußoperation angegeben. Jede dieser Abschlußphasen
wird durch die jeweils benötigte Zeit, einem Unschärfegrad und durch
die Belegung bzw. Freigabe ihrer global genutzten Hilfseinrichtungen
charakterisiert.

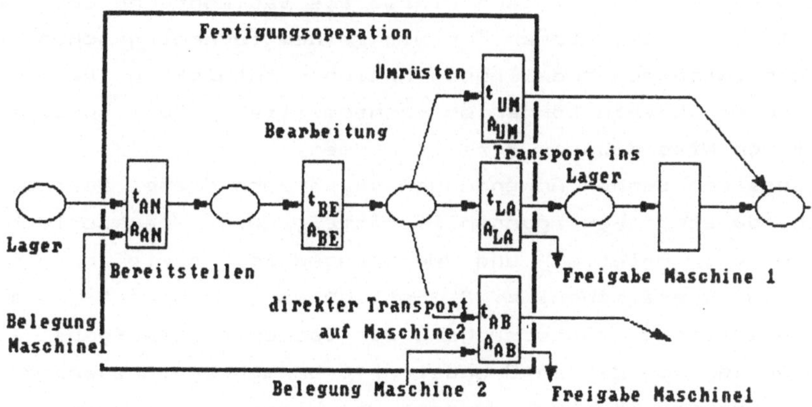

Bild 2: Modell einer Fertigungsoperation mit Berücksichtigung
ihrer Bearbeitungs- und Transportzeiten

Neben den bisher beschriebenen lokalen Zeitbewertungen der Fertigungsoperationen können auch globale, auf den gesamten Fertigungsauftrag bezogene Zeitmaße, in das unscharfe Petri-Netz aufgenommen
werden. Mit diesen Netzstrukturen können dann in unscharfer Weise
Eilaufträge vorrangig im Produktionsablauf eingesetzt werden. Im
Zusammenspiel mit den lokalen Operationsbewertungen wirken diese Bewertungen erst im fortgeschrittenen Produktionsstadium, während die
lokalen kapazitätsabhängigen schon zu Beginn der Auftragsabarbeitung
auf die Verteilung der Betriebsmittel Einfluß nehmen.

3. Entwicklung von Petri-Netz-Modellen

Die Wissensbasis eines Expertensystems unterliegt, wenn es zur operativen Führung flexibler Fertigungssysteme eingesetzt wird, wegen der hohen Produktflexibilität und der sich ständig ändernden Auftragslage starken strukturellen Änderungen. Anders als in herkömmlichen Expertensystemen muß deshalb gleichzeitig mit der sich verändernden Auftragslage seine Wissensbasis neu erstellt werden. Da vom Nutzer keine tiefen Kenntnisse über ihren Aufbau vorausgesetzt werden können, muß eine Wissenserwerbskomponente bereitgestellt werden, welche die Entwicklung der Wissensbasis alleine auf der Grundlage technologischer Prozeßkenntnisse ermöglicht.

Abgeleitet aus den fuzzy Netzstrukturen wurde deshalb eine blockorientierte Sprache entwickelt, die Teilnetze zu prozeßabhängigen Objekten zusammenfaßt und somit als Modellbausteine für die Konstruktion der Wissensbasis zur Verfügung stellt. Diese Objekte werden bezogen auf eine Aufgabenklasse als Basissprache dem Nutzer bereitgestellt, so daß dieser nur mit seinem technologischen Fachwissen die notwendigen Modellkonstruktionen entwickeln kann. Beim Austausch dieser Objekte können so fachspezifische Basissprachen für den Entwurf von Wissensbasen erzeugt werden.

In fertigungstechnischen Anwendungen verkörpern diese Basisobjekte die verschiedensten Fertigungshilfsmittel, wie Transportroboter, Fertigungszellen, Paletten und Werkzeugsätze, sowie die unterschiedlichsten Operationen, z. B. Fertigungs-, Transport-, Umrüstoder Spannoperationen. Entsprechend der Bearbeitungsaufgabe können durch den Nutzer von diesen Grundstrukturen mehrere Instanzen bzw. komplexe Strukturen angelegt werden. Die Wissensbasis ergibt sich dann aus der technologisch sinnvollen Verkopplung dieser Komponenten zu einem das Produktionsregime beschreibenden Gesamtmodell des Systems.

Jedes Objekt der Basissprache verfügt über einen sichtbaren- und nicht sichtbaren Bereich. Der unsichtbare Objektteil enthält alle unabhängig vom Einsatzfall konstant bleibenden Netzbestandteile. Der sichtbare Objektteil beinhaltet dagegen alle notwendigen Informationen zur Spezifizierung der Objektinstanzen. Dazu gehören Parametereinstellungen, der Anzeigeteil und das Interface für den Anschluß an andere Blöcke. Mit dem Anzeigeteil können Netzelemente angegeben werden, die im Dialog mit dem Nutzer angezeigt werden sollen oder als Meß- bzw. Stellwerte mit dem Prozeß in Verbindung stehen. Jeder Block verfügt weiterhin über Vererbungseigenschaften, die bei der Bildung abgeleiteter Objekte interne Strukturen, Parameter bzw. das Interface in die Konstruktion einbringen.

In Bild 3 ist ein Ausschnitt eines Netzmodells für zwei Fertigungsaufträge abgebildet, die jeweils aus den zwei Fertigungsoperationen Bohren und Schlichten bestehen und gemeinsam die zwei Bearbeitungsstationen Maschine 1 und 2 benutzen. Als Basisblöcke werden hier die Objekte Fertigungsoperation, Lager und Bearbeitungsmaschine verwendet. Von diesen Objekten wurden mehrere Instanzen angelegt, die hier mit ihren Namen Bohren, Schlichten, Maschine1, Maschine2 usw. gekennzeichnet wurden. Als einzigstes Objekt besitzt die Fertigungsoperation einen nicht sichtbaren Bereich, der hier schraffiert dargestellt wurde. Das zur Verkopplung dienende Interface ist als Transition bei den Operationen oder als Platz bei den Lagern oder Maschinen in entsprechenden Feldern sichtbar angegeben. Ebenfalls sichtbar sind für jedes Objekt seine Parameter (Fertigungszeit(TF), Werkstückanzahl(WST) und Anzahl der Bearbeitungsstationen(ANZ)) in entsprechenden Fenstern angezeigt.

Als abgeleitete Blockstruktur sind in diesem Beispiel die zwei Fertigungsaufträge dargestellt. Von ihren Basisobjekten erben sie das Interface zu den Maschinen und die jeweiligen Parameter. Die Teilelager werden zu internen Elementen und deshalb nach außen hin nicht sichtbar.

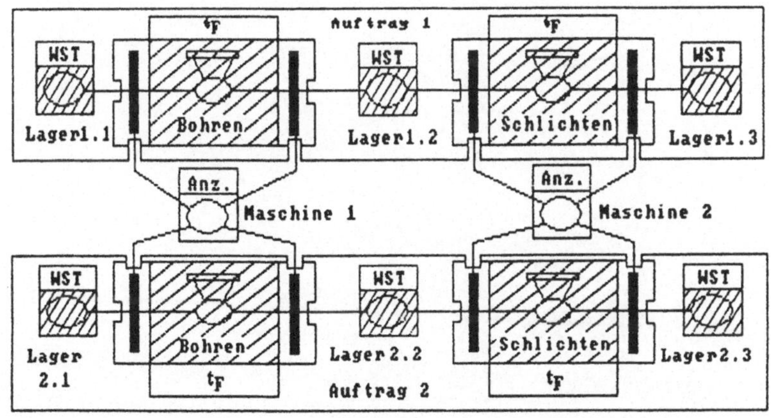

Bild 4: Blockstruktur eines Fuzzy-Petri-Netzes

4. Literatur

/1/   Lipp, H.-P.; Günther, R.; Sonntag, P.
        Unscharfe Petri-Netze - Ein Basiskonzept für
        computerunterstützte Entscheidungssysteme in
        komplexen Systemen
        Wiss. Schriftenreihe der TU Karl-Marx-Stadt 7/1989

/2/   Peterson, J.L.
        Petri-Net-Theory and modelling of systems
        Prenice-Hall Inc. 1981

/3/   Starke, P. H.
        Petri-Netze
        Deutscher Verlag der Wissenschaften Berlin 1980

/4/   Zadeh, L. A.
        Fuzzy Sets
        Inf. control 1965, 8, S. 338-353

Objektorientierter Entwurf und Programmierung von
Echtzeitsystemen

W. Mala, C. Grein
Elektronik System GmbH, München

1.   Überblick

Dieser Vortrag gibt einen Überblick über die Erfahrungen mit der
objekt-orientierten SW-Entwurfsmethode HOOD in Echtzeitsystemen.
Es werden Problemstellungen und Lösungen beim Übergang von
informellen SW-Anforderungen bis zum ablauffähigen ADA-Kode
aufgezeigt. Die vorliegenden Erfahrungen mit der Methode HOOD und
den verfügbaren Entwicklungs-Werkzeugen werden bewertet.

2.   HOOD Methode

HOOD (Hierarchical Object Oriented Design) ist eine Software
Entwicklungsmethode, die von einem französischen
Firmenkonsortium für die Entwicklung von komplexen ADA
Programmen entwickelt wurde.

HOOD zerlegt ein Vorhaben in Objekte und unterstützt die
schrittweise Verfeinerung eines Entwurfes durch hierarchische
Aufgliederung in kleinere Einheiten auf zwei Arten:

-       Senior-Junior Relation
        Senioren geben Aufträge an Junioren weiter, die von diesen
        bearbeitet werden.
-       Parent-Child Relation:
        Aufteilung eines Objektes in Unterobjekte.

Die Vorgehensweise des SW-Entwurfes mit HOOD ist streng
formalisiert und basiert auf folgenden Grundlagen:

-       einer vollständigen Top down System-(bzw. Problem-)
        Beschreibung
-       der Identifizierung der Objekte mit Hilfe des OOD-
        Verfahrens [1]
-       den in der HOOD Designmethode festgelegten
        Arbeitsschritten:

        *   Objekt Definition in zwei Schritten und auf zwei
                Ebenen
        *   Entwurf der Prozess-Hierarchie (Design Process Tree)

Durch sein stufenweises und rekursives Vorgehen ist HOOD gut
geeignet, komplexe Software-Systeme in kleine übersichtliche
Einheiten zu zerlegen.

Bild 1 gibt einen Überblick über die HOOD-Entwurfsschritte.

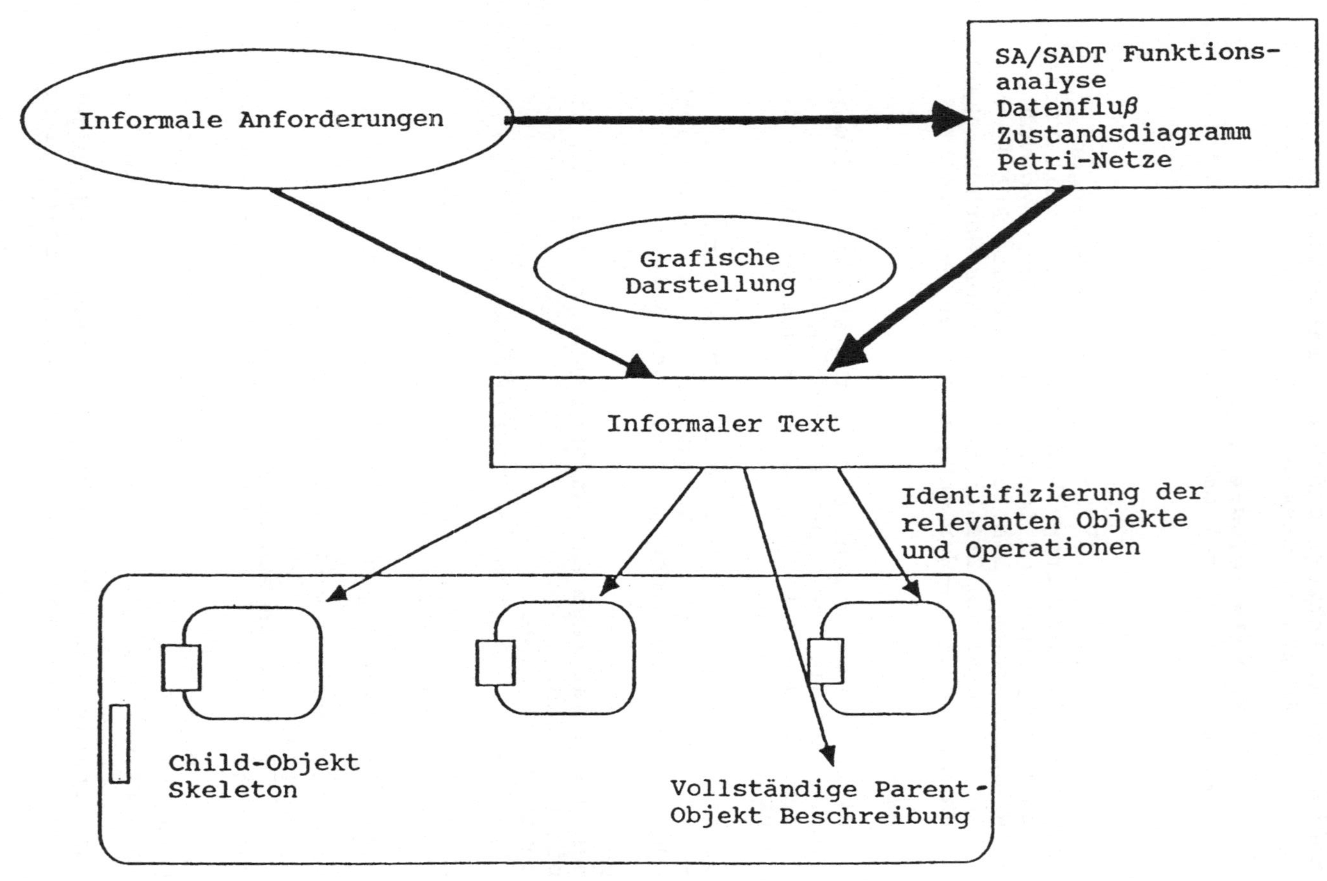

Bild 1: Ablauf der Entwurfsschritte

Im folgenden werden die Erfahrungen mit HOOD bei der Entwicklung
eines Realzeit-Programms dargestellt. Auf die Methode selbst wird
nur insoweit eingegangen, wie es zum Verständnis notwendig ist.
Für Details wird auf das HOOD Manual der European Space Agency
[2] verwiesen.

2.1 Objekte

Der Unterschied zwischen HOOD und den traditionellen Software-
Entwicklungsmethoden besteht in der Objektorientiertheit. Objekte
sind nur scheinbar ein neuer Begriff. Im Bereich der künstlichen
Intelligenz wurden objektorientierte Konzepte schon längere Zeit
verwendet, bevor der Begriff Objekt geprägt wurde. Schon SIMULA
67 führte den Klassenbegriff ein, der zur Zusammenfassung von
Zustandsinformation und Prozessen diente.

Im folgenden werden kurz die von BOOCH [1] definierten
wesentlichen Eigenschaften eines Objektes zusammengefaßt.

Ein Objekt ist einem Ding der realen Welt zugeordnet, das heißt,
es ist eine Abstraktion eines Teils der Problemebene. Ein Objekt
der Problemebene besteht natürlicherweise aus Daten und
Operationen.

Ein Objekt hat einen Zustand, der zeitlichen Änderungen
unterworfen ist.

Ein Objekt wird einerseits charakterisiert durch Operationen, die
es durch andere Objekte erleidet, andererseits macht es selbst
Annahmen darüber, wie es andere Objekte handhaben kann. Die
ersten Operationen sind die vom gegebenen Objekt zur Verfügung
gestellten, die letzteren die von ihm von anderen Objekten
verlangten.

Ein Objekt kann aktiv oder passiv sein. In den tradionellen
Sprachen gibt es nur den sequentiellen Kontrollfluß: Ein Objekt
ruft den Dienst eines anderen auf, der direkt ausgeführt wird;
nach Beendigung wird die Kontrolle an das aufrufende Objekt
zurückgegeben. Diese Art der Programmausführung definiert passive
Objekte. Aktive Objekte agieren nebenläufig.

Ein Objekt gehört einer (möglicherweise anonymen) Klasse an. Das
Wort Klasse wird hier verwendet, um die Einheit von Daten und
Operationen zu bezeichnen.

Ein Objekt hat einen Namen, der nur beschränkte Gültigkeit und
Sichtbarkeit besitzt.

Ein Objekt kann von innen oder von außen betrachtet werden. Die
Außenansicht eines Objektes ist abstrakt und verbirgt die zur
Implementation nötigen Details, die nur von innen sichtbar sind.

2.2 HOOD Chapter Skeleton

HOOD schreibt für jedes zu entwickelnde Objekt die Ausführung von
vier Schritten vor, die von einer allgemeinen Darstellung des zu
lösenden Problems bis zum detaillierten Entwurf mit ADA als
Programm-Entwurfssprache (PDL) führen.

Zielsetzung ist es, daß die formalen Spezifikationen der Objekte
zwar schon im Hinblick auf den späteren ADA-Kode erstellt werden,
die Festlegung auf die ADA-Syntax jedoch erst in einem
fortgeschrittenen Stadium erfolgt. Daher schreibt die HOOD-
Methode eine eigene Syntax vor (siehe Object Definition Skeleton
- ODS).

Die vier Schritte werden in einem HOOD Chapter Skeleton (HCS)
zusammengefaßt:

H1   Definition des Problems.
H2   Erarbeitung einer informellen, textuellen Lösung.
H3   Formalisierung der Strategie.
      1.   Wahl der Objekte durch Schlüsselhauptwörter in der
           textuellen Lösung.
      2.   Wahl der Operationen durch Schlüsselverben in der
           textuellen Lösung.
      3.   Gruppierung der Objekte und Operationen.
      4.   Graphische Darstellung.
H4   Formalisierung der Lösung.
      1.   Formale Definition der Objekte durch Object Definition
           Skeletons.
      2.   Übersetzung in ADA.

Diese Vorgehensweise ist sehr natürlich und übersichtlich. Daher
nur einige Anmerkungen zur Strategie bei der Auswahl der Objekte
und Operationen, zur graphischen Darstellung und zu den Objekt
Definition Skeletons.

2.3 Auswahl der Objekte und Operationen

Im HOOD-Manual wird der Vorschlag gemacht, die informelle,
textuelle Lösung des Problems so zu formulieren, daß hieraus
durch Unterstreichen der Hauptwörter und Verben die Objekte und
Operationen gewonnen werden können.

Dieser Vorschlag kann leicht zu Problemen führen, wenn er
wörtlich genommen wird. Es scheint eher umgekehrt zu sein: Sind
die richtigen Objekte und Operationen erst einmal gefunden, ist
es relativ einfach, einen Text so abzufassen, daß sie als
Hauptwörter und Verben vorkommen.

Irvin et al. stellen in [3] fest, daß die Wahl der richtigen
Objekte und Operationen eine äußerst schwierige, iterative
Aufgabe ist, die möglichst mit der gesamten Arbeitsgruppe
gemeinsam durchgeführt werden soll.

In diesem Schritt kann also die Methode keine besondere
Hilfestellung bieten, hier hilft nur ein sehr tiefes Verständnis
der Anforderungen an das zu entwickelnde System und Erfahrung im
objektorientierten Programmieren.

2.4 Objekt-Typen

Die wesentlichen Objekt-Typen in HOOD sind: aktive und passive
Objekte.

Ein aktives Objekt hat einen inneren Zustand und ändert diesen
Zustand von sich aus ohne Anstoß durch andere Objekte.

Ein passives Objekt dagegen bietet Dienstleistungen an, es wird
nur auf Anstoß von anderen Objekten hin tätig und beendet seine
Tätigkeit mit Abschluß des Auftrages.

Ein aktives Objekt wird durch ein ADA-Paket mit einem
eingeschlossenen Prozeß dargestellt. Die OBJEKT CONTROL
STRUCTURE im ODS dient zur Darstellung dieses Prozesses.

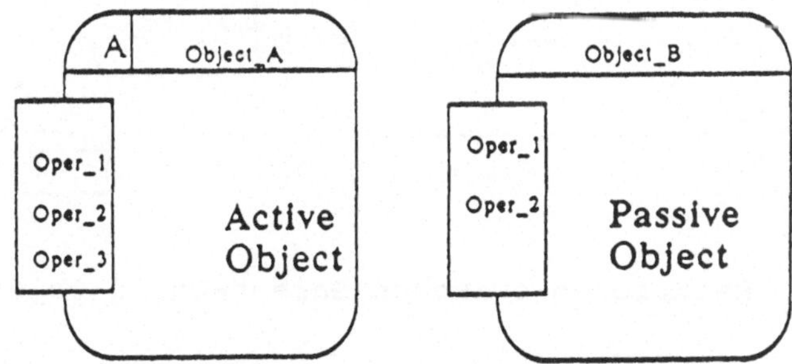

Bild 2:   Graphische Darstellung für Objekte

2.5 Graphik

Von großem Nutzen ist der graphische Teil der HOOD-Methode, der
es gestattet, die wechselseitigen Abhängigkeiten der
unterschiedlichen Objekte des Software-Entwurfes darzustellen:

-       Die Senior-Junior-Relation wird durch sogenannte 'Use'-
        Pfeile dargestellt. Datenflüsse und Ausnahmen entlang
        dieser Pfeile können ebenfalls dargestellt werden.

-       Die Parent-Child-Relation wird durch Schachtelung der
        Graphiken dargestellt.

-       Operationen und ihre Implementierung in Child-Objekten
        werden durch sogenannte 'Implemented by'-Pfeile
        dargestellt.

-       Anforderungen an die Ausführungsart (parallel, sequentiell,
        timed) werden durch sogenannte 'Trigger'-Pfeile dargestellt.

Bild 3 und 4 zeigen typische HOOD Diagramme.

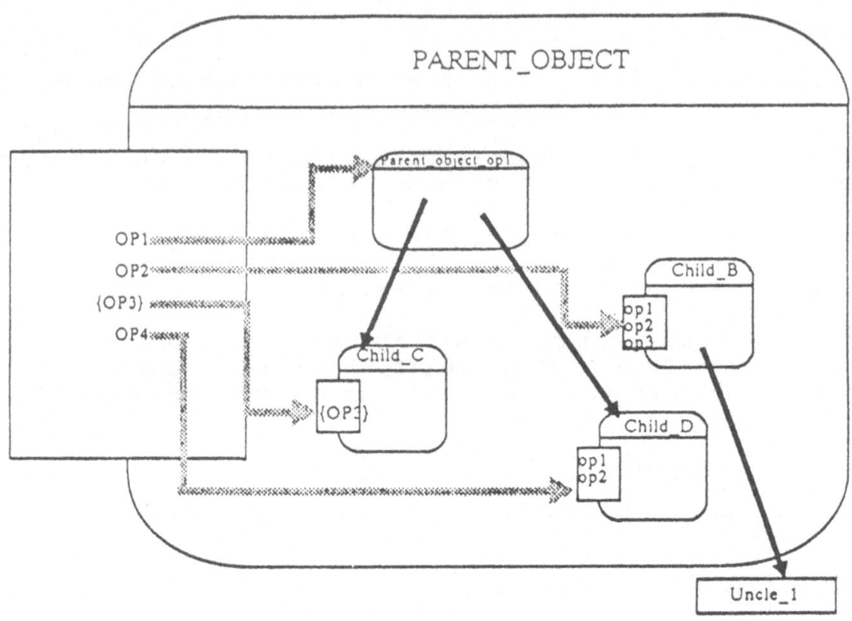

Bild 3:   Relationen zwischen Parent-Child Objekten

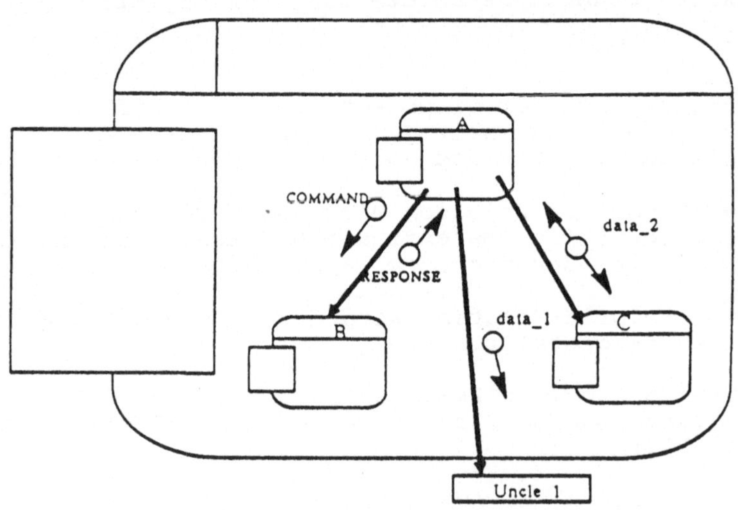

Bild 4: Darstellung des Datenflusses zwischen Objekten

HOOD-Diagramme sind allerdings nicht geeignet, Zustandsänderungen und Kontrollflüsse darzustellen. Hier besteht die
Gefahr, daß die Graphiken mit diesen Informationen überladen
werden und sie ihre Übersichtlichkeit und damit ihre Aussagekraft
verlieren.

2.6 Object Definition Skeleton (HOOD Version 2.2)

Jedes Objekt wird im vierten (und letzten) HOOD-Entwurfsschritt
mittels einer Schablone, dem sogenannten Object Definition
Skeleton (ODS) formal spezifiziert:

```
OBJECT Name [:CLASS [(Parameter_List)] IS
            DESCRIPTION
            IMPLEMENTATION OR SYNCHRONIZATION CONSTRAINTS
            PROVIDED INTERFACE
                TYPES
                DATA
                OPERATIONS
                EXCEPTIONS
            REQUIRED INTERFACE
                USED OBJECTS
                TYPES
                DATA
                OPERATIONS
            INTERNALS
                INCLUDED OBJECTS
                TYPES
                DATA
                OPERATIONS
                OBJECT CONTROL STRUCTURE
                OPERATION CONTROL STRUCTURES
END Name;

Object_Name IS NEW Class_Name [(Parameter_List)];
```

Die einzelnen Einträge in diese Schablone sind eine Mischung aus
ADA-Deklarationen und freiem Text. Unter der Überschrift
OPERATION CONTROL STRUCTURES wird ADA-Kode für die Operationen
aus den Teilen PROVIDED INTERFACE und INTERNALS eingegeben, unter
OBJECT CONTROL STRUCTURE (nur für aktive Objekte) der Kode für
die Steuerung innerhalb des Objektes.

Dieser formale Teil ist etwas gewöhnungsbedürftig, weil über der
ADA-Syntax die HOOD-Syntax liegt und es am Anfang nicht immer
offensichtlich ist, wo und in welcher Form die gewünschte
Information (wenn überhaupt) untergebracht werden kann. Hinzu
kommen Unterschiede in den Ausprägungen dieser Schablone bei den
verschiedenen HOOD-Werkzeugen.

Der Übergang von den ODS zum fertigen ADA-Kode ist relativ
einfach.

3.   Bewertung

3.1 Bewertung der HOOD-Methode

Vom Ansatz ist HOOD sehr gut für den Software-Entwurf von
komplexen Echtzeit-Softwaresystemen geeignet. Die HOOD-Methode,
Version 3.0, verfügt über ausreichende Entwurfsmittel für den SW-
Entwurf typischer Echtzeitsysteme (nebenläufige und verteilte
Prozesse).

Alle angebotenen HOOD-Werkzeuge enthalten Prüfmechanismen, die
die Konsistenz und Vollständigkeit des Entwurfes sichern sollen.

Die HOOD-Methode befindet sich, wie alle anderen OOD-Methoden
auch, noch in der Entwicklung. Die Methode wird von der 'HOOD
Working Group' kontinuierlich weiter entwickelt.

Die HOOD-Methode kann schon jetzt für ADA-Software-
Entwicklungen eingesetzt werden. Sie formalisiert ein Vorgehen,
das natürlich ist. Man sollte sich jedoch nicht zu sehr an die
Reihenfolge der Schritte im HCS klammern. Ihre Abfolge ist eine
Idealisierung; bei realen Entwicklungen läßt es sich kaum
vermeiden, zu früheren Entwurfsebenen zurückzukehren und sie zu
ändern.

Ein weiterer wesentlicher Aspekt ist die gute Qualität der SW-
Entwurfsdokumentation. Sie ist übersichtlich und durch die
strenge Formalisierung der Methode auch vollständig.

3.2 HOOD-Werkzeuge

Die Unterstützung der HOOD-Methode durch Werkzeuge ist derzeit
sehr unterschiedlich, da alle untersuchten Werkzeuge noch nicht
vollständig den letzten Entwicklungsstand der Methode (Version
3.0, September 1989) unterstützen.

Die folgenden HOOD-Werkzeuge wurden untersucht:

-    ADA-Nice von Intecs Sistemi [4]
-    HOOD-SF von Systematica Limited [5]
-    HOOD-Toolset von Software Sciences Limited [6]

Vom grundsätzlichen Aufbau her gesehen, sind sie alle
gleichwertig: Alle arbeiten mit Maus und bieten graphische und
textuelle Editoren, wobei graphische und formale textuelle
Eingaben auf wechselseitige Konsistenz gemäß den HOOD-Regeln
überprüft werden. Kode-Generatoren selektieren aus den formalen
Spezifikationen der ODS ADA-Kode, der natürlich noch verfeinert
werden muß.

Die Unterschiede liegen in der Bedienungsfreundlichkeit für den
Benutzer und in der Ausführlichkeit der Dokumentation.

Literatur

---

[1]   Software Engineering with ADA
       Grady Booch
       1986

[2]   European Space Agency
       HOOD Reference Manual Issue 3.0
       September 1989

[3]   Malcom Irving, John Lee, David Upcott
       Getting Started With Object Oriented Design
       ADA UK 8th International Conference September 1989
       ADA User Vol 10 Supplement, 1989

[4]   Intecs Sistemi s.p.a
       ADA-Nice, The Integrated HOOD Toolset
       User Manual Version 1.2 Field Test, September 1989

[5]   Systematica Limited
       HOOD-SF (Version 0)
       User Manual, 1989

[6]   Software Sciences
       An Overview of the HOOD Toolset, 1989

Eine kritische Bewertung der Software-Entwurfsmethode HOOD

T. Tempelmeier, München

Zusammenfassung: Nach einer kurzen Übersicht über die Software-Entwurfsmethode HOOD, Version 3.0, wird diese Methode hinsichtlich ihrer Einsatzfähigkeit in der Praxis bewertet. Hierbei ergeben sich positive Gesichtspunkte, z.B. die Anleitung zu einem planmäßigen Vorgehen, aber auch negative Aspekte, z.B. das Code-Erzeugungsschema oder die Definition der aktiven Objekte. Insgesamt wird HOOD als brauchbar eingestuft. Der Einsatz von HOOD muß jedoch mit sorgfältiger Schulung einhergehen, nicht zuletzt auch, um die Schwächen und Problemkreise in HOOD durch anwendungsspezifische Richtlinien umgehen zu können.

Stichwörter: HOOD, OOD, Ada, Hierarchical Object Oriented Design, Software-Entwurf, Software-Engineering, Luft- und Raumfahrt.

1. Einleitung

Die Software-Entwurfsmethode HOOD (Hierarchical Object Oriented Design) wurde seit 1986 im Auftrag der European Space Agency (ESA) entwickelt. Ihr Einsatz wird für die geplanten Raumfahrtprojekte der ESA und für weitere Großprojekte im Luftfahrtbereich verbindlich vorgeschrieben. HOOD wird über den Bereich der Großunternehmen hinaus auch alle Zulieferbetriebe mit Software-Anteilen betreffen.

HOOD lehnt sich stark an die Programmiersprache Ada an. Einige Konzepte von Ada wurden jedoch in HOOD durch tatsächlich oder vermeintlich bessere ersetzt. Anders als bei der Definition von Ada erfolgte aber die Definition von HOOD weitgehend innerhalb der ESA und der betroffenen Großunternehmen, ohne Beteiligung der internationalen wissenschaftlichen Gemeinde. Durch diesen Beitrag soll unter anderem versucht werden, eine offene wissenschaftliche Diskussion anzustoßen, die der zu erwartenden Verbreitung und Bedeutung der Methode HOOD gerecht wird.

In diesem Beitrag wird ein allgemein anerkannter und paxisrelevanter Stand der Technik vorausgesetzt [1], der in etwa durch die folgenden Schlagwörter charakterisiert ist:
• Tasks und Module (als Bausteine eines Echtzeit-Softwaresystems),
• Modultechnik, insbesondere abstrakte Datentypmodule,

• Objektorientierung beim Entwurf.

Die im folgenden enthaltene Kritik richtet sich demnach in keiner Weise gegen objektorientierte Ansätze und die Modultechnik im allgemeinen, sondern nur speziell gegen bestimmte Ausprägungen der Methode HOOD.

Im folgenden wird zunächst die Methode HOOD kurz erläutert und ein Bezug zu ähnlichen Methoden hergestellt. Anschließend werden die Stärken und die möglichen Problemkreise von HOOD dargestellt. Zum Abschluß wird eine Gesamtbeurteilung der Methode HOOD aus der Sicht des Autors abgegeben.

2. Die Software-Entwurfsmethode HOOD

HOOD wird hier nur in der aktuellen Version 3.0, die teilweise deutlich von der weithin bekannten Version 2.2 aus dem Jahr 1987 abweicht, betrachtet. Für eine ausführlichere Beschreibung wird auf [2-5] verwiesen.

2.1 Konzepte und grafische / textuelle Repräsentationsformen

Als Grundbausteine für den Software-Entwurf stellt HOOD *aktive und passive Objekte* zur Verfügung. Diese entsprechen in erster Näherung den Tasks und Paketen in Ada. Die grafische Repräsentation von Objekten ist aus Bild 1 ersichtlich. Objekte werden durch Rechtecke mit abgerundeten Ecken dargestellt, wobei aktive Objekte zusätzlich den Buchstaben A in der linken oberen Ecke tragen. Die Schnittstelle der

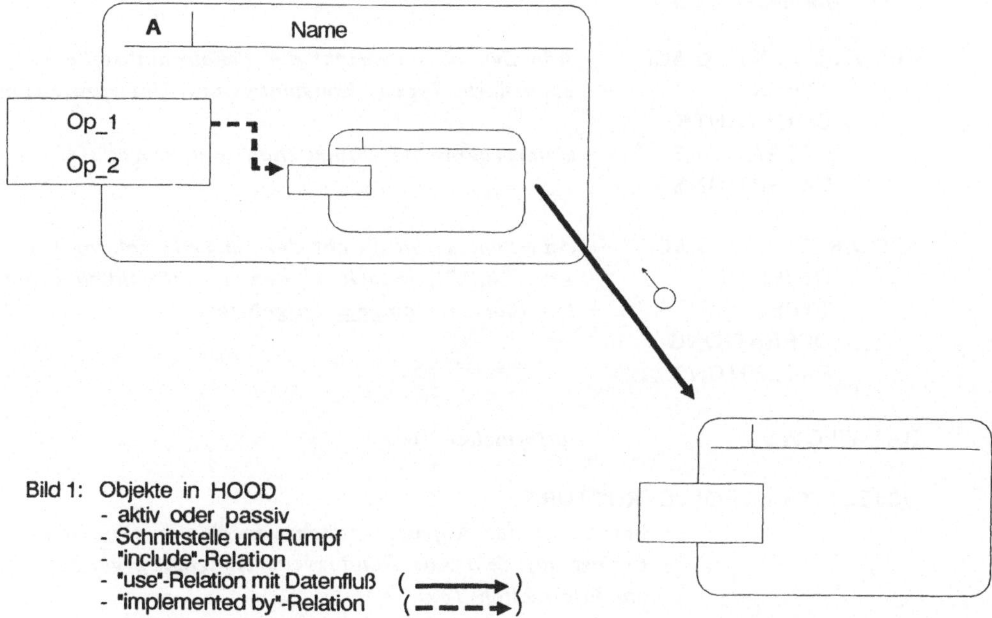

Bild 1: Objekte in HOOD
- aktiv oder passiv
- Schnittstelle und Rumpf
- "include"-Relation
- "use"-Relation mit Datenfluß (——→)
- "implemented by"-Relation (— — —→)

Objekte nach außen, in HOOD *Provided Interface* genannt, wird durch ein vorstehendes Rechteck symbolisiert, in welches die exportierten Operationen des Objekts eingetragen werden.

Objekte können andere Objekte enthalten; HOOD spricht diesbezüglich von einer *Parent-Child-* oder *Include-Relation*. Diese Beziehung entspricht der von Ada und anderen Programmiersprachen bekannten Verschachtelung. Falls Operationen der Schnittstelle direkt durch ein enthaltenes Objekt realisiert werden, so definiert HOOD dies als *Implemented_by-Relation*.

Werden durch ein Objekt die Ressourcen benützt, die ein anderes, nicht im ersten Objekt enthaltenes Objekt zur Verfügung stellt, so spricht HOOD von der *Use-Relation*. Diese Beziehung entspricht der üblichen Import-Beziehung (in Ada: "with").

Zu jedem Objekt gibt es eine textuelle Beschreibung, das sogenannte *Object Description Skeleton (ODS)*, welches alle Informationen zum Objekt in einer Ada-ähnlichen Syntax noch einmal detailliert enthält. Durch dieses ODS kann eine Überfrachtung der grafischen Darstellungen vermieden werden. Das ODS ist wie folgt aufgebaut:

```
OBJECT  name  IS   active_or_passive

    DESCRIPTION
      -- informeller Text

    IMPLEMENTATION_OR_SYNCHRONISATION_CONSTRAINTS
      -- informeller Text

    PROVIDED_INTERFACE      -- Ada-Syntax; entspricht der Paketschnittstelle;
        TYPES               -- exportierte Typen, Konstanten und Unterprogramme.
        CONSTANTS
        OPERATIONS          -- Unterprogramme (Funktionen und Prozeduren).
        EXCEPTIONS

    REQUIRED_INTERFACE      -- Ada-Syntax; entspricht der Importbeziehung (in
        OBJECTS             -- Ada: "with"), jedoch werden die tatsächlich benötig-
        TYPES               -- ten Elemente einzeln aufgelistet.
        OPERATIONS
        EXCEPTIONS

    DATAFLOWS                   -- informeller Text

    OBJECT_CONTROL_STRUCTURE
                        -- Beschreibt das Aufrufverhalten von aktiven Objekten.
                        -- Basiert auf dem Ada-Rendezvous-Konzept, ergänzt
                        -- um informellen Text.
```

```
INTERNALS                -- Ada-Syntax; entspricht dem Paketrumpf.
     OBJECTS
     DECLARATIONS
     OPERATIONS          -- Unterprogramme (Funktionen und Prozeduren).

     OPERATION_CONTROL_STRUCTURE
                         -- Ada-Pseudocode aller Funktionen und Operationen.
                         -- Entspricht den Ada-Unterprogrammrümpfen.


END_OBJECT  name ;
```

Mit Hilfe des ODS besteht auch in eingeschränktem Maße die Möglichkeit, generische Pakete (HOOD-Bezeichnung: *Class*) zu definieren. Auf die Beschreibung einiger weiterer Objekttypen, die mit der Version 3.0 in HOOD eingeführt wurden, kann hier verzichtet werden, da sie keine wesentlichen Neuerungen beinhalten.

2.2 Anleitung für ein planmäßiges Vorgehen und Beurteilung der Entwurfsgüte

Der Begriff "Methode" beinhaltet unauflöslich auch Regeln für ein planmäßiges Vorgehen oder eine Verfahrensweise. In HOOD wird in diesem Zusammenhang eine hierarchische Top-Down-Zerlegung der Objekte vorgeschlagen (vgl. Bild 2). Dies führt entsprechend der *Include-Relation* zum sogenannten *Design Process Tree*.

HOOD beinhaltet ferner für die einzelnen Zerlegungsschritte des Entwurfsprozesses (*Basic Design Steps*) eine sehr ausführliche Anleitung [5]. Es handelt sich dabei um eine Weiterentwicklung der Ideen von BOOCH [6,7].

Bezüglich einer Güte-Beurteilung der erstellten Entwürfe stützt sich HOOD auf allgemein anerkannte Kriterien wie hohe Bindung, geringe Kopplung, großes Fan-in, kleines Fan-out, usw. ab.

Insgesamt wird HOOD seinem Anspruch, eine <u>Methode</u> zu definieren, voll gerecht. HOOD geht über eine bloße Repräsentationsform für Software-Entwürfe hinaus und bietet auch eine systematische Verfahrensweise sowie Gütekriterien an.

2.3 Tools

Gegenwärtig sind dem Autor drei brauchbare Toolsysteme, die die Konzepte, Repräsentationsformen und die Vorgehensweise von HOOD unterstützen, bekannt. Die Toolsysteme sind auf VMS- und Unix-Workstations ablauffähig. Eines dieser Toolsysteme wird direkt durch die Firma Digital Equipment Corporation vermarktet.

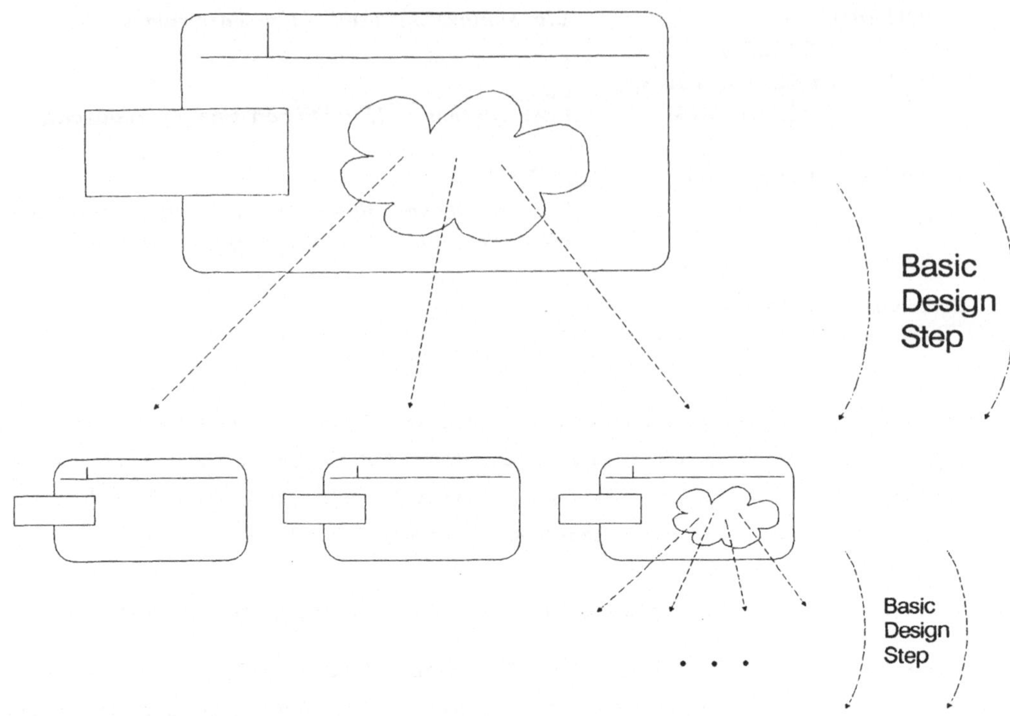

Bild 2: Der "Basic Design Step" in HOOD
und der resultierende "Design Process Tree"

3. Bezug zu ähnlichen Methoden

Es gibt eine Reihe von Methoden, die ebenfalls für die Programmiersprache Ada geeignet sind, insbesondere:

- OOD oder Object-Oriented Development [6,7],
- die Methode von Buhr [8],
- OOSD oder Object-Oriented Structured Design [9] sowie
- die Methode nach Nielsen / Shumate [10].

Ein Vergleich dieser Methoden mit HOOD zeigt zunächst, daß hinsichtlich der zugrunde gelegten Konzepte und der angebotenen Repräsentationsformen nur geringe Unterschiede bestehen. Die aufgeführten Methoden enthalten aber im Gegensatz zu HOOD nur Konzepte, die praktisch identisch zu den Konzepten von Ada sind, und vermeiden somit teilweise die in Kapitel 5 für HOOD aufgezeigten Probleme. Die augenfälligen Ähnlichkeiten der grafischen Darstellungsformen dieser Methoden untereinander und mit HOOD sind in [1, 2] dargestellt. Die feststellbaren Unterschiede sind praktisch vernachlässigbar. (Kurzformel: "Grafikelemente mit abgerundeten Ecken versus Grafikelemente mit eckigen Ecken").

Einige neuere Arbeiten [11, 12, 13] bestätigen durch ihre ähnlichen bzw. übereinstimmenden Konzepte die durch HOOD vorgegebene Linie.

Ein Methodenvergleich sollte statt der Repräsentationsformen aber eher die angebotenen planmäßigen Vorgehensweisen vergleichen. Hier ergeben sich stärkere Unterschiede: OOSD erhebt schon vom Titel der Erstveröffentlichung her gar nicht den Anspruch, eine bestimmte Vorgehensweise vorzugeben. OOD muß als weniger detaillierter Vorgänger von HOOD angesehen werden. Bei Buhr überwiegt ebenfalls die Betonung der Konzepte und Repräsentationsformen.
Nielsen / Shumate konzentrieren sich fast ausschließlich auf eine systematische, praxisorientierte Verfahrensweise; das Fehlen eines werbewirksamen Methodennamens und ansprechender Grafikelemente dürften die einzigen "Nachteile" dieser sehr fundierten Methode sein.

Insgesamt ist nach Meinung des Autors eine Trennung in Repräsentationsformen (z.B. OOSD oder Buhr) und in umfassende Anleitungen für ein planmäßiges Vorgehen (z.B. Nielsen / Shumate) der richtige Ansatz.

4. Stärken von HOOD

Die Stärke von HOOD ist die ausführliche Festlegung eines planmäßigen Vorgehens für den Entwurfsvorgang ([5] und Bild 2):

- Der gesamte Entwurfsvorgang wird in eine Anzahl von einzelnen Entwurfsschritten (*Basic Design Steps*) aufgeliedert, woraus sich der *Design Process Tree* und auch bereits ein Gliederungsschema für das Entwurfsdokument ergeben.

- Die einzelnen Entwurfsschritte (*Basic Design Steps*) werden noch einmal in 9 Einzelaktivitäten aufgegliedert (Appendix D in [4]).

- Prinzipien für die Objektbildung sind klar (wenn auch zu knapp) als Bestandteil der Methode ausgewiesen.

Gerade für weniger erfahrene Entwerfer ist die systematische Führung durch den Entwurfsprozeß eine wichtige Hilfe, auch wenn sie kein Ersatz für eine fundierte Software-Engineering-Ausbildung sein kann. HOOD ist damit den meisten anderen Entwurfsmethoden weit überlegen.

Andererseits ist HOOD nach gegenwärtiger Definition noch weit vom Niveau und der Praxisrelevanz der Methode nach Nielsen / Shumate entfernt. HOOD ist aber offen für methodische Ergänzungen, so daß z. B. die Leitlinien von Nielsen / Shumate auch in firmen- oder anwendungsspezifischer Form eingebracht werden können.

5. Problembereiche in HOOD

5.1 Übersicht

Nach Meinung des Autors enthält HOOD folgende Schwächen und Problembereiche, geordnet nach ihrer Bedeutung:
* Definition der aktiven Objekte
* Schema für die Erzeugung von Ada-Code
* Fehlen von methodischen Leitlinien zum Auffinden von aktiven Objekten
* Datenflüsse (*Dataflows*) in unüblicher / unnützer Definition
* Fragwürdigkeit der streng hierarchischen Gesamtstruktur der mit HOOD entworfenen Systeme
* Unterschiede zum Scoping in Ada
* Zusätzliche Regeln zu Ada
  - weitere, spezielle Objekttypen (*Environment, Virtual Node, OP_Control*)
  - zusätzliche Pragmas zur Steuerung der Ada-Codeerzeugung (*Server_Task, FIFO*)
  - Spezialformen für die Task-Kommunikation (*ASER, LSER, HSER, TOERR*).

Der zusätzliche Mindest-Lernaufwand für die formalen Anteile von HOOD ist aus folgenden Angaben ersichtlich:
* 7 Seiten BNF-Syntax, 63 Produktionsregeln
* 37 reservierte Worte (zusätzlich zu Ada!)
* 52 verbal gefaßte Regeln.

Zur Vermeidung von Mißverständnissen sei betont, daß ein zusätzlicher Lernaufwand für eine Methode voll zu akzeptieren wäre, wenn dadurch bekannte Schwachstellen in Ada umgangen werden könnten oder wenn dadurch leistungsfähigere Konzepte eingeführt würden.

Im Rahmen dieses Beitrags wird die weitere Behandlung auf die beiden ersten Punkte beschränkt.

5.2 Definition der aktiven Objekte

Die Definition des Begriffs *Active Object* in HOOD ist unklar und schwer verständlich. Einige Zitate aus dem Glossar zu [4] zeigen dies sehr deutlich :

"Active Object: An Object in which the →OBCS controls the flow of control between all →contrained operations of the object"

→ "Object Control Structure - OBCS: Part of the ODS that defines the control flow between constrained operations"

→ "Constrained Operation: An operation of an active object that is constrained by the status of the object as defined in the OBCS"

Insgesamt dürften mit aktiven Objekten diejenigen gemeint sein, die ähnlich einer Ada-Task einen eigenen internen Ablauf haben. Die Abgrenzung gegenüber Ada-Tasks liegt darin, daß aktive Objekte auch mehrere, parallele interne Abläufe haben können. Verwunderlich bleibt dabei, daß sowohl in aktiven als auch in passiven (!) Objekten aktive (Kind-)Objekte enthalten sein können.

Die überspitzten, in der Praxis nutzlosen Unterschiede zwischen aktiven und passiven Objekten in HOOD seien durch ein kleines Beispiel beleuchtet. Zwei Objekte, **a_stack** und **p_stack**, die sich nur in ganz geringem Maße unterscheiden, müssen nach HOOD einmal als aktives Objekt und einmal als passives Objekt betrachtet werden.

- **OBJECT a_stack IS ACTIVE**

```
. . .
provided_operation POP (elem : out elem_type) is
begin
    stack_task.pop (elem);
      -- let stack_task be included in object a_stack
      -- caller may be suspended, if stack is empty
end pop;
```

- **OBJECT p_stack IS PASSIVE**

```
. . .
provided_operation POP (elem : out elem_type; error : out error_type) is
begin
  select
    stack_task.pop (elem);
      -- let stack_task be included in object p_stack
    error := none;
  else
    error := stack_empty_or_busy;
  end select;
end pop;
```

Für den Anwender in der Praxis ist eine Umdeutung des Konzepts der aktiven Objekte hilfreich, derart, daß er es (Mengen von) Ada-Tasks gleichsetzt. Das bewährte Konzept "Task" wäre insgesamt dem unklar definierten und nicht hilfreichen Konzept der aktiven Objekte vorzuziehen.

5.3 Schema für die Erzeugung von Ada-Code

Im Standardfall werden aus einem (!) HOOD-Objekt drei Ada-Pakete erzeugt. Dies führt bereits ohne Berücksichtigung von *use*-Relationen zu anderen Objekten auf eine Vielzahl von "with"-Klauseln vor den Paketteilen.

Das Codeerzeugungs-Schema für aktive Objekte ist noch komplizierter. Der erzeugte Ada-Code ist in diesem Fall kaum mehr handhabbar.

In der Feinentwurfs- und in der Kodierphase muß mit Ada-Entwurfseinheiten und in der Ada-Denkwelt gearbeitet werden. Das wird auch in HOOD so gesehen [4,5]. Durch die standardmäßig erzeugte Flut von Paketen und die generierten, unübersichtlichen Rendezvousmechanismen wird diese Arbeit aber unnötig und nahezu unverantwortlich erschwert.

6. Fazit

Negative Aspekte:

- HOOD enthält problematische Konzepte, wie etwa das der aktiven Objekte. Eine anwendungsspezifische Umdeutung dieser Konzepte, z.B. eine Gleichsetzung der aktiven Objekte mit (Mengen von) Ada-Tasks, kann die praktische Arbeit erleichtern.

- Der zusätzliche Lernaufwand für HOOD ist größer als von der Sache her nötig und gerechtfertigt wäre.

- Das Ada-Codeerzeugungsschema in der gegenwärtigen Form erzeugt eine Flut von Paketen, durch welche die Feinentwurfs- und Kodierungsphase erheblich – nahezu unverantwortlich – erschwert wird.

- Die Anwendung von HOOD bedeutet nicht, daß für die Entwerfer damit eine fundierte Software-Engineering-Ausbildung überflüssig geworden wäre.

Positive Aspekte:

- HOOD beinhaltet ein planmäßiges und methodisches Vorgehen für den Entwurfsvorgang.

- Weitergehende und überlegene Methoden (z.B. Nielsen / Shumate) können in HOOD über die Software-Engineering-Ausbildung der HOOD-Anwender oder über firmen- oder projektspezifische Anwendungsrichtlinien eingebracht werden.

- Für HOOD ist eine brauchbare Toolunterstützung verfügbar.

- HOOD ist trotz aller Problematiken eine akzeptable, vielleicht sogar eine der besseren Software-Entwurfsmethoden. Es ist bedauerlich, daß der gute Grundansatz unter einigen verschrobenen Konzepten leidet.

Handlungsbedarf:

• Der Übergang von HOOD Version 2.2 nach Version 3.0 ist eher als Irrweg denn als Fortschritt anzusehen. Alle von HOOD Betroffenen sollten entsprechend ihrer Kompetenz Einfluß auf die zukünftige Entwicklung von HOOD nehmen.

Abschließend sei nochmals betont, daß die hier vorgebrachten Kritikpunkte nur gewisse Ausprägungen der Methode HOOD, insbesondere in der aktuellen Version 3.0, betreffen. Die grundsätzliche Überlegenheit des objektorientierten Ansatzes im Sinne der Bildung von Datenstruktur- und abstrakten Datentypmodulen wird vom Autor voll bejaht.

Literaturangaben

[1] Tempelmeier, T.: *Praxis des Entwurfs von Echtzeit-Softwaresystemen.*
atp Automatierungstechnische Praxis, 31 (1989) 11, 532-538.

[2] Tempelmeier, T.: *Eine Übersicht über die Software-Entwurfsmethode HOOD.*
In: W. Gerth, P. Baacke (Hrsg.): PEARL 90 - Workshop über Realzeitsysteme. 11. Fachtagung des PEARL-Vereins e.V. Boppard, November 1990. Erscheint im Springer-Verlag, Berlin, 1990, in der Reihe der Informatik-Fachberichte.

[3] Mala, W., Grein, C.: *Objektorientierter Entwurf und Programmierung von Echtzeitsystemen.* Im vorliegenden Tagungsband.

[4] HOOD Working Group: *HOOD Reference Manual. Issue 3.0.*
European Space Agency, Noordwijk, The Netherlands, Sept. 1989.

[5] HOOD Working Group: *HOOD User Manual. Issue 3.0.*
European Space Agency, Noordwijk, The Netherlands, Dec. 1989.

[6] Booch, G.: *Software Engineering with Ada.*
Benjamin/Cummings, Menlo Park, Ca., 1986.

[7] Booch, G.: *Object-Oriented Development.*
IEEE Transactions on Software Engineering, SE-12 (1986), 211-221.

[8] Buhr, R.J.A.: *System Design with Ada.* Prentice-Hall, Englewood Cliffs, N.J., 1984.

[9] Wasserman, A.I., Pircher, P.A., Muller, R.J.: *The Object-Oriented Structured Design Notation for Software Design Representation.* Computer, 23 (1990) 3, 50-63.

[10] Nielsen, K., Shumate, K.: *Designing Large Real-Time Systems with Ada.*
Intertext Publications & McGraw-Hill, New York, N.Y., 1988.

[11] Jalote, P.: *Functional Refinement and Nested Objects for Object-Oriented Design.* IEEE Transactions on Software Engineering, SE-15 (1989), 264-270.

[12] Seidewitz, E.: *General Object-Oriented Software Development: Background and Experience.* Journal of Systems and Software 9 (1989), 95-108.

[13] Wolf, L.W., Clarke, L.A., Wileden, J.C.: *The AdaPIC Tool Set: Supporting Interface Control and Analysis Throughout the Software Development Process.* IEEE Transactions on Software Engineering, SE-15 (1989), 250-263.

Anschrift des Autors:

Dr. Theodor Tempelmeier, Elfenstraße 39, D - 8000 München 83, Telefon 089 / 602 622, Email: tempelmeier @ lrz-extern.uni-muenchen.dbp.de. Der Autor ist Professor für Informatik an der Fachhochschule Rosenheim und befaßte sich früher bei der Firma MBB und seither im Rahmen verschiedener Beratungstätigkeiten mit HOOD und ähnlichen Vorgehensweisen.

Neuartige Entwicklung wissensbasierter Systeme mit DIWA

Gabriele Schmiedel
Siemens AG, Otto-Hahn-Ring 6, 8000 München 83

1 Einleitung / Definitionen

Wissensbasierte Systeme (oder Expertensysteme) sind ein Teil der sogenannten *Künstlichen Intelligenz*, der bisher am weitesten den Einzug in die praktischen Anwendungen und damit in die Industrie geschafft hat. Hierzu gehören insbesondere die wissensbasierten Diagnosesysteme im Bereich der heuristischen Diagnose.

Wodurch unterscheidet sich ein wissensbasiertes System von einem konventionellen Software-System? Keineswegs durch die Wissensbasiertheit in dem Sinne, daß diese Systeme intelligenter reagieren, sondern durch ihre Architektur. Diese Architektur ermöglicht es, daß das *Anwendungswissen* explizit und vom *Wissen über die Abarbeitung* des Anwendungswissens getrennt repräsentiert werden kann. Gegenüber dieser Wissensrepräsentation liegen die beiden Arten von Wissen in den konventionellen Systemen vermischt vor. In Abbildung 1 ist die Verteilung des Wissens stark schematisiert dargestellt. Die schraffierten Flächen stellen das Anwendungswissen dar.

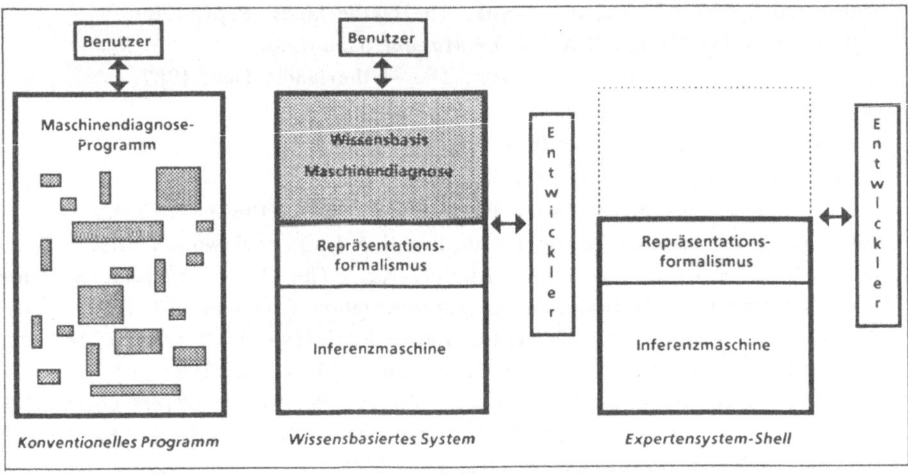

Abb. 1: Konventionelles Programm, wissensbasiertes System und Expertensystemshell

Neben der expliziten Darstellung des enthaltenen Wissens hat diese Architektur der wissensbasierten Systeme den Vorteil, daß für eine weitere, ähnliche Anwendung, z.B. die

Diagnose einer anderen Maschine, nur das Anwendungswissen neu implementiert werden muß.

Ein wissensbasiertes System besteht somit aus mindestens 3 Komponenten:

- Einem Repräsentationsformalismus für das Wissen *(Regeln und Fakten, logische Formeln, Objekte, ...)*,
- einem Interpretationsalgorithmus, der zum Repräsentationsformalismus gehört *(Regelinterpreter, Theorembeweiser, Methoden, ...)* und
- einer Wissensbasis, die Wissen in Form des Repräsentationsformalismus enthält.

Eine Expertensystemshell ist in dem obigen Sinne ein wissensbasiertes System ohne Wissensbasis. In der Praxis wird noch zwischen Shells unterschieden, die nur einen Repräsentationsformalismus anbieten und solchen, die mehrere, miteinander verknüpfbare Formalismen anbieten. Die letzteren nennt man hybride Shells.

Um ein wissensbasiertes Diagnosesystem für die Diagnose einer komplexen Maschine zu entwickeln, gibt es verschiedene Vorgehensweisen.

2 Traditioneller Weg

Bei der traditionellen Entwicklung eines wissensbasierten Systems wird der Experte, dessen Wissen und Erfahrung über die zu diagnostizierende Maschine und über das Vorgehen bei der Diagnose Gegenstand der Wissensbasis sein soll, von einem sogenannten *Wissensingenieur* befragt. Dieser soll das Wissen, über welches der Experte berichtet, in eine strukturierte Form bringen. Wesentliche Aufgaben hierbei sind:

- Die intuitiven Handlungen und das hierin implizit enthaltene Wissen explizit zu machen und
- das beim Experten vorhandene Wissen, das in irgendeiner, nicht näher beschriebenen Form vorliegt, in eine strukturierte Form zu bringen.

Dieser traditionelle Weg I ist in Abbildung 2 (mit einer Linie aus Punkten und Strichen) dargestellt.

Beispiel: Der Experte kommt zu einer defekten Maschine, schaut sich wenige Meßwerte an und kann sofort die möglichen Ursachen auf ein bis zwei Möglichkeiten einschränken. Befragt man ihn, wie er zu diesen Schlüssen kommt, muß er sich überlegen, welche Zwischenschritte für diese Schlußfolgerung nötig waren. Je mehr Erfahrung ein Experte hat, umso eher führt er diese Zwischenschritte implizit aus.

Der Wissensingenieur entwickelt aus dem strukturierten Wissen dann ein wissensbasiertes System, indem er das strukturierte Wissen in einen geeigneten Repräsentationsformalismus, z.B. in Regeln, übersetzt.

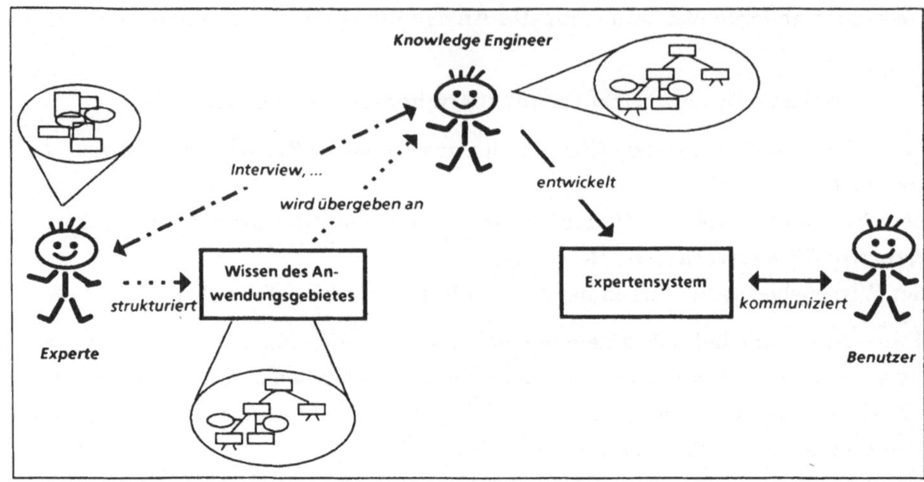

Abb. 2: Traditioneller Weg I und II

Prototyp

Auf diesem traditionellen Weg haben wir einen Prototypen eines wissensbasierten Systems für die Wartung und Instandhaltung in der Chipfertigung realisiert.

Maschinenausfälle in der Chipfertigung sind äußerst kostenintensiv, vor allem wenn sie zu längeren Ausfallzeiten führen. Sie sind besonders in der Spät- und Nachtschicht problematisch, wenn wenige Techniker eine ganze Produktionshalle zu überwachen haben. Hier treten verstärkt Probleme auf, die nicht ursächlich technologisch bedingt sind.

Während der Schicht am Tage sind Experten für die einzelnen Fertigungsmaschinen anwesend, die bei einem Maschinenstillstand zu Rate gezogen werden. Im allgemeinen ist jedoch in der Spät- und Nachtschicht *der Experte* für die gerade ausgefallene Maschine nicht anwesend. Obwohl die auftretenden Fehler in den meisten Fällen relativ einfach zu beheben wären, dauert die Fehlerlokalisierung länger und verursacht teure Maschinenausfallzeiten und damit Produktionsausfall.

Eine Analyse der typischen Fehler zeigt, daß viele davon mit nur etwas 'Erfahrungswissen' über die jeweilige Maschine auch von weniger erfahrenen Technikern behoben werden könnten. Dieses Erfahrungswissen sollte durch ein wissensbasiertes System den Technikern zur Verfügung gestellt werden. Dieses enthält einen Teil des Expertenwissens über den Plasmaätzer, eine der Chipfertigungsmaschinen.

Der Prototyp wurde unter realen Bedingungen mit potentiellen Benutzern getestet. Die verschiedenen vom Experten eingebauten Fehler konnten von den Testpersonen alle mithilfe des wissensbasierten Systems beseitigt werden (siehe SCHMIEDEL / WINKELMANN 89).

Bei der Wissensstrukturierung kamen wir nach einer gewissen Zeit zu der Feststellung, daß die Struktur, die dem Expertenwissen zugrunde liegt, sich nicht mehr änderte, auch wenn noch weiteres Wissen des Experten strukturiert wurde. Daraufhin haben wir ausgehend vom Basismodell der Problemzerlegung das *Diagnosemodell der defekten Maschine* entwickelt, mit dem das Expertenwissen strukturiert werden kann. Dieses Modell und seine graphische Repräsentation sind weiter unten beschrieben.

Nach einer Einarbeitungszeit, in der der Wissensingenieur das Wissen mit dem Experten zusammen strukturierte, konnte der Experte sein Wissen selbständig mithilfe dieses Diagnosemodell strukturieren und beschreiben. Die so entstandenen Wissensstrukturen wurden nochmals gemeinsam durchgesprochen. Der Wissensingenieur erstellte daraus ein ablauffähiges wissensbasiertes System. Dieser traditionelle Weg II ist in Abbildung 2 gepunktet skizziert.

Diagnosemodell der defekten Maschine

Das Diagnosemodell der 'defekten Maschine' beschreibt die Zerlegung des globalen Problems *defekte Maschine* in Teilprobleme. Diese Zerlegung ist hierarchisch und endet bei den zur Fehlerursache korrespondierenden Reparaturanleitungen oder den Wartungsausgängen. Mit Wartungsausgang wird eine Situation bezeichnet, in der die Rufbereitschaft oder der Experte zu Rate gezogen wird, weil das Expertensystemwissen nicht mehr ausreicht.

Graphisch läßt sich dieses Modell als Baum darstellen (siehe auch Abbildung 3). Das globale Problem *defekte Maschine* ist die Wurzel, die Teilprobleme sind die Knoten und die Reparaturanleitungen bzw. die Wartungsausgänge die Blätter des Baumes.

Die Übergänge von einem Teilproblem zum anderen werden als Kanten bezeichnet. Diese sind beschriftet, um die Fehlerlokalisierung durch möglichst frühzeitige Einschränkung der Suche im Baum auf einen Teilbaum zu beschleunigen. Es gibt zwei Typen von Beschriftungen, sogenannte **Indizien**, wie z.B. Meßwerte, bestimmte Schalterstellungen oder deren Verknüpfungen *(in Ovalen dargestellt)* und ganze Zahlen, die eine **Abarbeitungsreihenfolge** für die möglichen Ursachen eines Problems definieren *(als Zahlen an den Kanten dargestellt)*. Diese Abarbeitungsreihenfolge wird vom Experten zur Entwicklungszeit aufgrund seiner Erfahrung (z.B. Reparaturkosten, Auftrittshäufigkeit, Überprüfungsaufwand) festgelegt, wenn es kein geeignetes Indiz gibt.

Die zu einem Problem gehörenden Erläuterungen, die u.a. vor Messungen ausgegeben werden, enthalten z.B. Anleitungen zum Ausbau bestimmter Maschinenteile, so daß die Messung überhaupt durchgeführt werden kann *(mit doppeltem Rand dargestellt)*.

Desweiteren gibt es einen besonderen Problemknoten, den sogenannten **Rücksprungknoten**. Dieser ist dadurch gekennzeichnet, daß er an der nachfolgenden Kante eine Indizbeschriftung hat und es eine Indizbedingung zu diesem Indiz gibt, die das Nichtvorhandensein des zum Rücksprungknoten gehörenden Problems charakterisiert *(durchgestrichene Verbindung vom Oval zum Knoten)*.

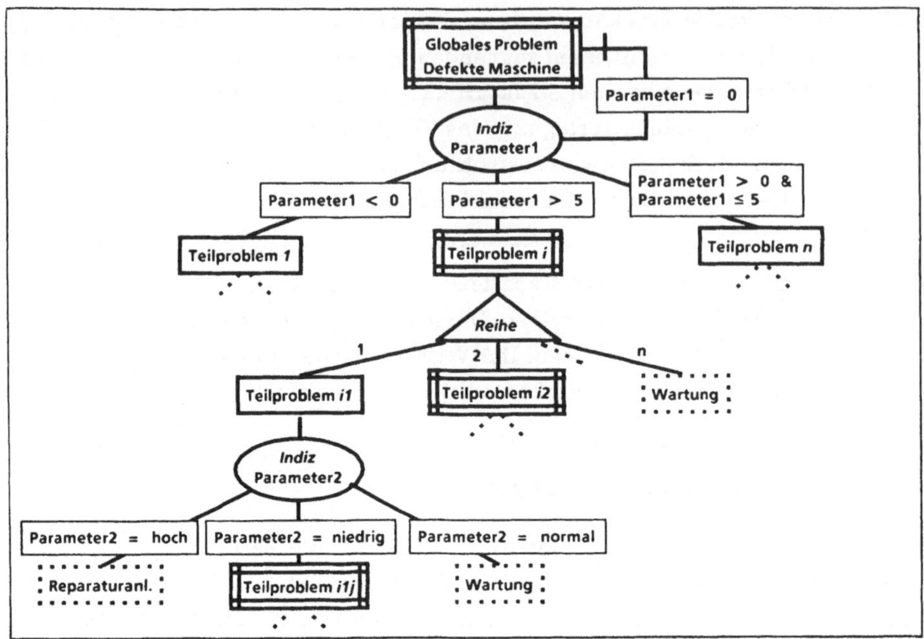

Abb. 3: Diagnosemodell der defekten Maschine

3 Neuartiger Weg

Beim oben beschriebenen zweiten Weg zeichnet der Experte für eine reale Anwendung recht große graphische Strukturen, die über mehrere Papierseiten verteilt sind und die bei Änderungen, insbesondere Einfügungen mitten in der Struktur, schnell unübersichtlich werden.

Der Wissensingenieur auf der anderen Seite hat nach der Einarbeitung des Experten in das Diagnosemodell nur noch eine reine Routineaufgabe. Er erstellt aus den vom Experten erstellten Wissensstrukturen ein ablauffähiges wissensbasiertes System.

Wir entwickeln das Werkzeug **DIWA** (*DI*agnosewerkzeug mit graphischer *W*issens*A*kquisitionskomponente), welches eine neuartige Entwicklung eines wissensbasierten Systems ermöglicht und den traditionellen Entwicklungsweg drastisch verkürzt (siehe Abbildung 4).

DIWA stellt dem Experten eine graphische Wissensakquisitionskomponente für das Diagnosemodell der 'defekten Maschine' zur Verfügung. Hiermit kann der Experte sein mit dem Diagnosemodell strukturiertes Wissen interaktiv in graphischer Form eingeben (siehe Abbildung 5).

Für das DIWA-System wurde das Diagnosemodell der defekten Maschine erweitert. Es wurden folgende Konzepte neu hinzugenommen oder überarbeitet:

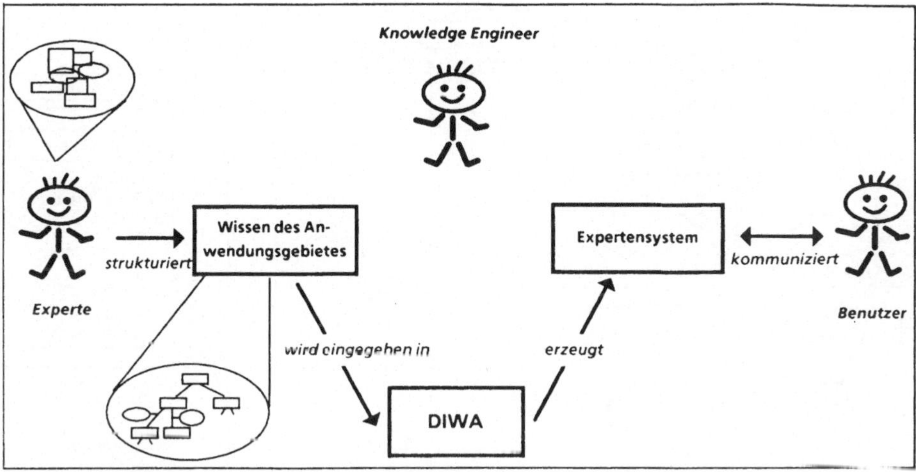

Abb. 4: Neuartiger Entwicklungsweg mit DIWA

- Wissensbasen
- Wissensmodule
- M-Wissensmodule
- Sprache für die Indizbedingungen

Eine Wissensbasis ist eine abgeschlossene Wissensstruktur, die z.B. für die Diagnose einer Maschine notwendig ist. Wissensmodule und M-Wissensmodule dienen der hierarchischen Modularisierung von großen Wissensstrukturen. Sie können selbst wiederum Wissensmodule und M-Wissensmodule enthalten. Die M-Wissensmodule sind im Gegensatz zu den Wissensmodulen mehrfach einzusetzen, d.h. es kann von verschiedenen Stellen in der Wissensstruktur oder auch aus verschiedenen Wissensbasen auf ein und denselben M-Wissensmodul verwiesen werden. Die Sprache für die Indizbedingungen der Indizien beschreibt die Verknüpfungsmöglichkeiten von Parametern und Konstanten zu Indizbedingungen und wurde durch eine kontextfreie Grammatik festgelegt.

DIWA übernimmt die oben beschriebene Routinearbeit des Knowledge Engineers und bietet dem Experten eine komfortable Oberfläche für die Eingabe der Wissensstrukturen an.

Nach der Wissenseditierung erzeugt DIWA aus den eingegebenen Wissensstrukturen automatisch eine Wissensbasis, die sofort mit dem zugehörigen DIWA-Ablaufsystem eingesetzt werden kann.

Das DIWA-System, bestehend aus Entwicklungs- und Ablaufsystem, entspricht im oben definierten Sinne, einer auf die Anwendungsklasse spezialisierten Shell. Das DIWA-Ablaufsystem enthält den zum Diagnosemodell der defekten Maschine gehörigen Interpretationsalgorithmus und eine entsprechende Bedienoberfläche.

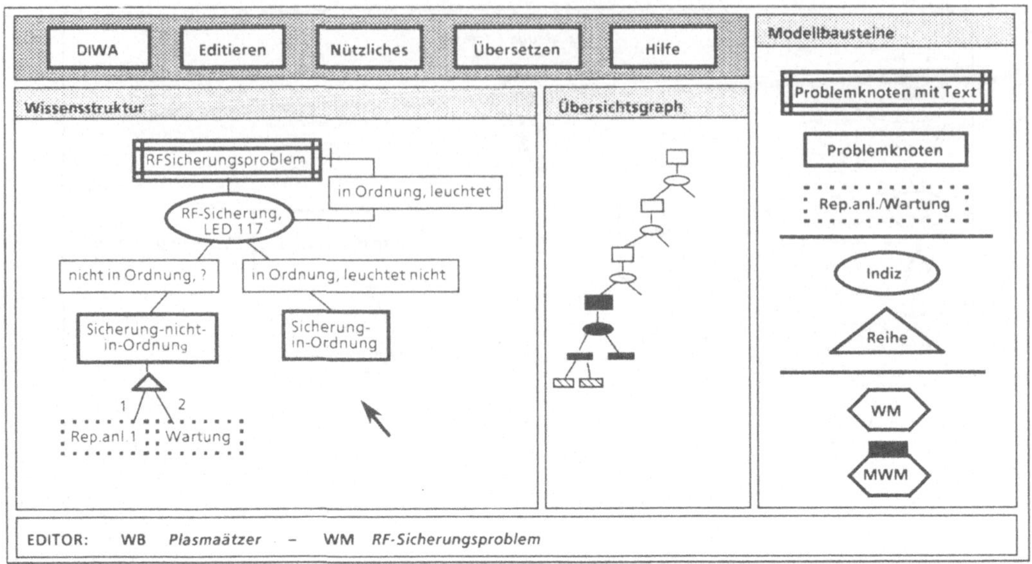

Abb. 5: Skizze der graphischen Wissensakquisitionskomponente von DIWA

Mit DIWA wird

- der traditionelle Weg der Entwicklung eines wissensbasierten Systems drastisch verkürzt.
- der Experte die Erstellung, Pflege und Weiterentwicklung der Wissensbasis selbständig vornehmen können, ohne die Hilfe des Knowledge Engineers.
- der Einsatz des mit DIWA entwickelten wissensbasierten Systems weitaus früher möglich sein als bei der traditionellen Entwicklung.
- der Experte aufgrund der graphischen Wissenstrukturen das eingegebene Wissen leichter validieren können.

DIWA wird derzeit auf einer WS 30 Workstation (Apollo) unter UNIX mit Common Lisp und dem zugehörigen Objektsystem CLOS entwickelt.

4 Ausblick

DIWA ist in der ersten Version als ein OFF-LINE-System konzipiert. Eine ON-LINE-Kopplung zu Datenbanken, Automatisierungsystemen oder auch zu Berechnungsroutinen (z.B. in C) ist für eine weitere Version vorgesehen. Diese Erweiterung betrifft insbesondere die Indize des Diagnosemodells mit ihren Indizbedingungen und Parametern. So kann der Wert für einen Parameter dann nicht mehr nur vom Benutzer erfragt, sondern auch von einer Datenbank, einem Automatisierungssystem kommen oder durch eine Berechnung ermittelt werden.

Desweiteren könnte man sich eine Rückkopplung des Diagnoseergebnisses an ein Automatisierungssystem vorstellen.

5 Zusammenfassung

Um ein wissensbasiertes Diagnosesystem für die Diagnose einer komplexen Maschine zu entwickeln, gibt es verschiedene Vorgehensweisen.

Bei der traditionellen Entwicklung eines wissensbasierten Systems wird der Experte von einem Knowledge Engineer befragt. Dieser bringt das Wissen, über welches der Experte berichtet, in eine strukturierte Form und entwickelt daraus dann ein wissensbasiertes System.

Aufgrund der Erfahrungen bei der Wissensstrukturierung während der beschriebenen Prototyp-Entwicklung war es möglich, das Diagnosemodell der 'defekten Maschine' zu entwickeln, mit dem der Experte nach einer Einarbeitungszeit sein Wissen selbständig strukturieren konnte.

Eine neuartige Entwicklung eines wissensbasierten Systems ist mit dem DIWA-System möglich. Dieses stellt dem Experten eine graphische Wissensakquisitionskomponente für das Diagnosemodell der defekten Maschine zur Verfügung. Hiermit kann der Experte sein mit dem Diagnosemodell strukturiertes Wissen interaktiv in graphischer Form eingeben.

Nach der Wissenseditierung erzeugt DIWA aus den eingegebenen Wissensstrukturen automatisch eine Wissensbasis, die sofort mit dem zugehörigen DIWA-Ablaufsystem eingesetzt werden kann.

In einer weiteren Version von DIWA ist eine ON-LINE-Kopplung zu Datenbanken, Automatisierungssystemen oder Berechnungsroutinen vorgesehen.

6 Literatur

SCHMIEDEL / WINKELMANN 89:
   Ein wissensbasiertes Diagnosesystem in der Halbleiterfertigung. Erfahrungsbericht. KI 3/89 (GI), Oldenbourg 1989.

Ein Agentensystem für eine flexible Fertigungssteuerung

Klaus Fischer*
Institut für Informatik der Technischen Universität München
Postfach 202420
D-8000 München 2
E-mail: fischer@lan.informatik.tu-muenchen.dbp.de

Zusammenfassung

In diesem Artikel wird ein Konzept zur Steuerung einer flexiblen Fertigung vorgestellt. Das Konzept stützt sich auf ein verteiltes Agentensystem, das über einer objektorientierten Wissensbasis arbeitet. Die Agenten können dynamisch entstehen und sich über Kommunikationsverbindungen Nachrichten zusenden; dabei werden gegenseitig Dienstleistungen erbracht und in Anspruch genommen. Das globale Faktenwissen, auf das alle Agenten Zugriff haben, ist in der objektorientierten Wissensbasis in einer Klassenhierarchie strukturiert. Zusätzlich ist dort auch prozedurales Systemwissen in Form von Regelmengen abgelegt. Die Agenten können anhand dieser Wissensbasis Fakten- und Regelwissen erlernen. Die Konzepte und ihre Umsetzung in ein ablauffähiges System werden anhand eines Modells für eine Fertigungsanlage erläutert.

1 Einleitung

CIM (Computer-Integrated Manufacturing) ist heute ein Schlagwort, dem durch die aktuelle Forschung echte Inhalte verliehen werden. In Laborumgebungen ist es bereits möglich, einen vollautomatischen Produktionsprozeß für einen Fertigungsauftrag durchzuführen, der vom Produktlayout bis zur Fertigung des Endprodukts reicht. Anfänglich waren in der Forschung vor allem Erfolge auf Einzelgebieten, z.B. im CAD-Bereich erreicht worden. Heute ist es der Gesichtspunkt der Integration, der im Mittelpunkt der aktuellen CIM-Forschung steht [9]. Dabei stand am Anfang das Problem der Kommunikation zwischen Einzelkomponenten im Vordergrund. Die immer weitergehende Vernetzung der Einzelsysteme und die Definition von Kommunikationsprotokollen haben große Fortschritte gebracht. Man kann sich heute auf die eigentlichen Probleme der Fertigungssteuerung konzentrieren. Der vorliegende Artikel stellt ein Konzept für den Aufbau eines Fertigungssteuerungssystems vor. Die Probleme die dabei betrachtet werden werden in der Literatur meist unter dem Kürzel CAM (Computer-Aided Manufacturing) zusammengefaßt. Gegenstand der Untersuchung ist hier vor allem die Steuerung des Material- und Informationsflusses in einer flexiblen Fertigung. Bei der Steuerung des Materialflusses wird festgelegt, wann und von wem ein Werkstück bzw. ein Werkzeug von einem Ort zu einem anderen transportiert werden soll. Die Steuerung des Informationsflusses dagegen muß sicherstellen, daß die Produktionsdaten, z.B. Art der Verarbeitung, rechtzeitig mit dem physikalischen Transport verfügbar sind. Das im weiteren vorgestellte Konzept ist eine Verallgemeinerung von Ergebnissen, die im Rahmen des SFB 331 *Informationsverarbeitung in autonomen mobilen Handhabungsystemen* im Teilprojekt A2 *Aufgabenorientierte Programmierung* entwickelt wurden.

---

*Der vorliegende Artikel wurde von der Deutschen Forschungsgemeinschaft im Rahmen des Sonderforschungsbereichs (SFB) 331 *Informationsverarbeitung in autonomen mobilen Handhabungssystemen,* Teilprojekt A2, gefördert.

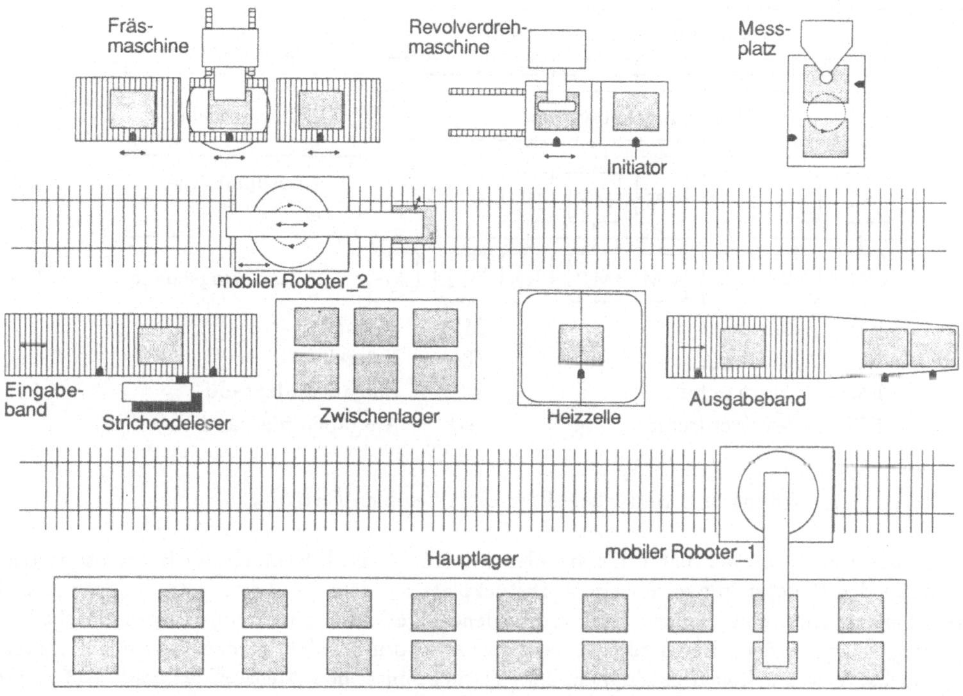

Abbildung 1: Funktionsmodell der Fertigungsanlage.

2 Beschreibung des Modells der Fertigungsanlage

Als Anwendungsbeispiel für die im weiteren vorgestellten Konzepte dient ein 140x120 cm^2 großes Modell einer Fertigungsanlage, das aus Fischer-Technik-Baugruppen aufgebaut ist (Abb. 1). Zwei mobile Roboter, die jeweils 4 Achsen besitzen, führen die Handhabungs- und Transportaufgaben aus. Das Hauptlager kann nur von Roboter_1, die drei Maschinen auf der anderen Seite können nur von Roboter_2 bedient werden. Im Bereich des Eingabeförderbandes, des Zwischenlagers, der Heizzelle und des Ausgabeförderbandes überlappt der Arbeitsraum der beiden Roboter. Deshalb ist es notwendig, den Robotern Strategien für eine Kollisionsvermeidung vorzugeben. Bei der Beschickung der Maschinen und beim Beladen der Heizzelle müssen die Bewegungen der Roboter mit den Aktionen der Maschinen koordiniert werden.

Die Werkstücke gelangen über das Eingabeförderband, auf dem ein Strichcodeleser zur Werkstückidentifizierung angebracht ist, in den Bearbeitungsprozeß. Anhand von induktiven Sensoren (Initiatoren), wie sie auch auf den Werkstückplätzen der Maschinen und beim Ausgabeförderband angebracht sind, kann die Anwesenheit eines Werkstücks festgestellt werden. Das Hauptlager dient als Aufnahmeplatz für Teile, die gerade nicht bearbeitet werden können oder auf eine Auslieferung warten. Das Zwischenlager wird als Übergabebereich zwischen den beiden Robotern verwendet.

Die Heizzelle in der Mitte der Fabrik kann von beiden Robotern beschickt werden. Ihr Innenraum ist durch ein Dach abgeschlossen, das geöffnet sein muß, wenn ein Werkstück eingelegt oder entnommen werden soll. Bei der Bearbeitung wird durch einen Regelungsprozeß mit Hilfe von Heizdrähten und Ventilatoren ein vorgegebenes Temperaturprofil eingehalten. Die Fräsmaschine kann von zwei Seiten her beladen werden. Die Werkstücke werden mit Hilfe von Förderbändern von links und rechts auf den Bearbeitungsplatz geschoben und von dort wieder auf den Eingabeplatz entladen. Bei der Bearbeitung folgt der Fräskopf einer durch ein Programm vorgegebenen Bahn. Auch die Revolver-

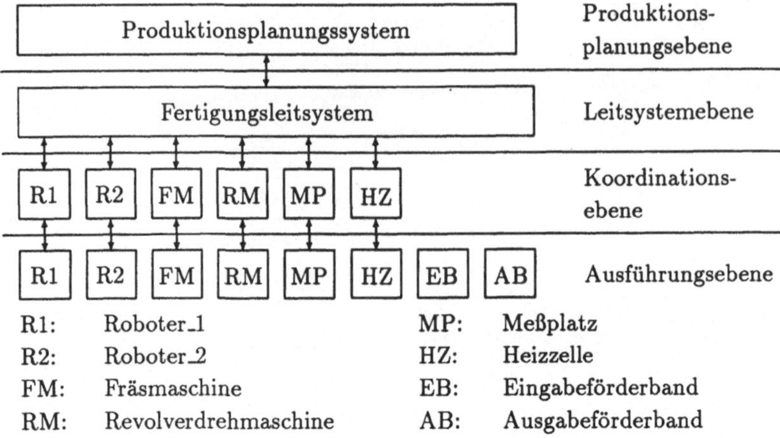

Abbildung 2: Strukturdefinition einer flexiblen Fertigungsteuerung.

drehmaschine kann von zwei Seiten her beladen werden. Je nach Stellung des Bearbeitungstisches ist dabei für den Roboter immer nur ein Werkstückplatz auf dem Bearbeitungstisch zugänglich. Bei der Bearbeitung kann eine Sequenz aus verschiedenen Werkzeugen und zugehörigen Bearbeitungszeiten vorgegeben werden. Der Meßplatz kann nur von vorn beladen werden. Für den Roboter ist immer nur der Werkstückplatz zugänglich, der sich nicht unter dem Meßkopf befindet. Auf zu große Abweichungen vom vorgegebenen Sollwert muß die Fertigungssteuerung entsprechend reagieren, z.B. durch die Wiederholung eines Bearbeitungsvorgangs.

3 Eine hierarchische Ebenenstruktur für eine flexible Fertigungssteuerung

Die komplexe Aufgabe der Planung und Steuerung des Produktionsablaufs läßt sich auf mehrere Ebenen so verteilen (Abb. 2), daß jeder einzelnen Ebene ein klar definierter Kompetenzbereich zukommt. Bisher herrscht in der Literatur [4],[7],[8] nur bei der Funktionalität der oberen beiden Ebenen weitgehende Einigkeit. Die tieferen Ebenen werden sehr unterschiedlich und auf die speziellen Hardwaregegebenheiten zugeschnitten definiert. Bei dem im weiteren vorgestellten Ansatz wird die Funktionalität der einzelnen Ebenen am Beispiel des in Abschnitt 2 beschriebenen Modells erläutert. Bei der Definition der tieferen Ebenen steht der Gesichtspunkt im Vordergrund, daß die Roboter und Maschinen ihre Aufgaben autonom bearbeiten können sollen.

Eine Aufteilung des Planungs- und Steuerungsvorgangs auf mehrere Ebenen bringt außerdem den Vorteil mit sich, daß unter Umgehung der übergeordneten Ebenen direkt auf tiefere Ebenen zugegriffen werden kann. Damit kann auf Ausführungsebene ein Einrichtbetrieb realisiert werden, auf Koordinationsebene kann ein menschlicher Operateur eine Maschine mit einem Werkstück beschicken und es können Aufträge unter Umgehung der Produktionsplanung direkt an das Fertigungsleitsystem eingelastet werden, was z.B. für die Behandlung von Eilauftägen interessant sein kann. Wichtig ist, daß bei Aktionen, bei denen eine der tieferen Ebenen direkt angesprochen wird, der korrekte Ablauf in den oberen Ebenen nicht beeinträchtigt wird.

Das Produktionsplanungssystem ist die oberste planende Instanz der flexiblen Fertigungssteuerung. Hier können Produktionsaufträge zur Herstellung von Endprodukten aus einer vorgegebenen Produktpalette erfaßt werden. Die Aufträge werden unter Einbeziehung der aktuellen Auslastung terminiert und an das Fertigungsleitsystem weitergegeben. Es darf dabei nur eine bestimmte Anzahl von Fertigungsaufträgen gleichzeitig in das Fertigungsleitsystem eingelastet werden. Für die Fertigung eines Endprodukts ist ein Rohteile aus einer bestimmten Rohteilgruppe notwendig. Bei der Einlieferung der Rohteile auf dem Eingabeförderband werden die Rohteile identifiziert und damit automatisch einer bestimmten Rohteilgruppe zugeordnet. Falls die für einen Auftrag notwendigen Rohteile im Lager nicht verfügbar sind, wird eine Bestellung für diese Rohteile auf einem Ausgabegerät ausgegeben.

Das Fertigungsleitsystem überwacht die Ausführung der von der Produktionsplanung terminierten Fertigungsaufträge. Hier erfolgt die Zuordnung eines Werkstücks zu einem Fertigunsauftrag. Die Stationen, die zur Fertigung eines bestimmten Endprodukts von einem Werkstück durchlaufen werden, sind als Teil des Fertigungsauftrags in einer Liste von Bearbeitungsschritten festgelegt; diese Liste wird auch Arbeitsplan genannt [7]. Im Arbeitsplan speziell markierte Schritte sind permutierbar, wenn dadurch der nächste Arbeitsschritt eines Werkstücks früher ausgeführt werden kann. Wenn das Werkstück, dessen nächster Bearbeitungsschritt als nächstes ausgeführt werden soll, geschickt ausgewählt wird, wird ein optimaler Materialfluß der Werkstücke durch die Fertigungsanlage erzielt. Optimalitätskriterien sind eine maximale Maschinenauslastung oder minimale Durchlaufzeiten der Werkstücke durch die Fertigungsanlage.

Auf der Koordinationsebene werden alle Komponenten der realen Fertigungsumgebung als autonome Einheiten betrachtet, die eine Aufgabe selbständig bearbeiten und eigenständig entscheiden, welche Einzelaktion als nächste auszuführen ist. Im Beispiel des Modells der Fertigungsanlage sind folgende Komponenten autonome Einheiten: die beiden Roboter, die Fräsmaschine, die Revolverdrehmaschine, der Meßplatz und die Heizzelle. Für jede autonome Einheit existiert auf der Koordinationsebene ein Planer, der die korrekte Ausführung einer Einzelaufgabe für diese autonome Einheit plant und überwacht [3]. In vielen Fällen müssen bei solchen Einzelaufgabe autonome Einheiten kooperieren. Als Beispiel soll hier das Einlegen eines Werkstücks in die Heizzelle dienen. Das Dach der Heizzelle muß geöffnet werden, bevor einer der Roboter das Werkstück hineinlegen kann. Erst wenn der Roboter mit seinem Arm den Arbeitsraum der Heizzelle verlassen hat, darf das Dach der Heizzelle geschlossen werden und der Bearbeitungsvorgang kann beginnen.

In der Ausführungsebene werden unmittelbar die Motoren der Komponenten der Fertigungsanlage angesprochen. In dieser Schicht findet ein wesentlicher Teil der Sensordatenverarbeitung statt. Hierarchische Sensor-Aktor-Netze ermöglichen eine sensorgestützte Ausführung der Einzelaktionen [6]. In der Ausführungsebene gewonnene Informationen werden, soweit dies sinnvoll ist, in einer global zugänglichen Wissensbasis abgelegt, so daß diese Informationen den höheren Planungsschichten zugänglich sind, z.B. der Code der vom Strichcodeleser erkannt wurde.

4 Definition des Agentensystems

Dieser Abschnitt stellt ein allgemeines Schema zur einheitlichen Beschreibung der logischen Teilsysteme einer flexiblen Fertigungssteuerung vor. Dieses Schema läßt sich für die einzelnen Ebenen um zusätzliche Konzepte erweitern, die der Aufgabenstellung auf den einzelnen Ebenen angepaßt ist. Es liegt nahe, die zum Aufbau der Struktur der Fertigungssteuerung beschriebenen logischen Teilsysteme, auf einem abstrakten Niveau als eigenständige Objekte zu betrachten; dazu wird der Begriff des Agenten eingeführt.

Ein Agent A ist ein Tripel $A := (\mathcal{F}, \mathcal{R}, \mathcal{D})$, mit

$\mathcal{F}$ ist eine Menge von Fakten, die das lokale Faktenwissen des Agenten repräsentiert. Ein Faktum ist dabei als Eingenschaftenliste zu verstehen, die bestimmten Bezeichnern symbolische Werte zuweist.

$\mathcal{R}$ ist eine Menge von Regeln, die Strategien für das allgemeine Verhalten des Agenten vorgeben.

$\mathcal{D}$ ist eine Menge von Diensten, die der Agent als Schnittstelle nach außen anbietet.

Für einen Agenten A soll durch $A_{\mathcal{F}}$, $A_{\mathcal{R}}$ und $A_{\mathcal{D}}$ die jeweilige Menge angesprochen werden können. A kann Botschaften empfangen, die entweder die lokale Faktenmenge $A_{\mathcal{F}}$ verändern oder ein Dienst der Menge $A_{\mathcal{D}}$ aktivieren. Jeder Dienst eines Agenten besteht aus einer Menge von Regeln, die genau dann aktiv sind, wenn der zugehörige Dienst durch das Empfangen einer Botschaft aktiviert wurde. Bei Ausführung eines Dienstes des Agenten A, kann sich sein Faktenwissen $A_{\mathcal{F}}$ und sein Regelwissen $A_{\mathcal{R}}$ verändern. Der Agent lernt also, indem er Dienstleistungen erbringt, wobei das Lernen auch darin bestehen kann, hinfällig gewordene Fakten und Regeln zu löschen. Weiter kann durch das Ausführen eines Dienstes des Agenten A eine Menge von neuen Agenten entstehen. Die bei einem Agenten eingehenden Dienstbotschaften werden nach einer bestimmten Strategie abgearbeitet (siehe Abschnitt 6). Wichtig ist zunächst nur, daß in einem Agenten immer nur ein Dienst aktiv sein kann. Die anderen Dienste werden sofern vorhanden in einer Warteschlange verwaltet und erst aktiviert, wenn der aktuelle Dienst beendet ist.

Zur Modellierung des die Problemlösung beschreibenden Fakten- und Regelwissens wird eine objektorientierte Wissensbasis verwendet [1], weil sich hier das Wissen kompakt strukturieren läßt. Objekte sind die Basiseinheit, aus denen die Wissensbasis aufgebaut wird. In ihrer Struktur gleiche Objekte werden zu Klassen zusammengefaßt. Die Beschreibung einer Klasse ist damit ein prototypisches Objekt, das angibt, wie die Elemente (wir nennen diese Elemente Instanzen) dieser Klasse strukturiert sind. Die Eigenschaften der Instanzen werden durch Attribute beschrieben, die ihrerseits durch Facetten beschrieben werden. Die Klassen sind in einer Hierarchie angeordnet, in der Attribute von übergeordneten, allgemeinen Klassen an untergeordnete, spezielle Klassen vererbt werden. Es lassen sich Integritätsbedingungen formulieren, die objektübergreifende Konsistenz zusichern können, weil sie bei Veränderungen automatisch geprüft werden.

Zur Modellierung der Problemlösung für eine bestimmte Aufgabenstellung wird eine Menge von kooperierenden Agenten spezifiziert. Die statische Beschreibung der Problemlösung ist dabei durch das Prototypensystem $\mathcal{S}_{prot} := (\mathcal{A}_{prot}, W_{init})$ gegeben, mit

$\mathcal{A}_{prot}$ ist die Menge der prototypischen Agenten; das sind die Agenten, die dynamisch im Agentensystem entstehen können, d.h. die als Problemlöser dem System potentiell zur Verfügung stehen. Es können dabei mehrere Inkarnationen von einem bestimmten prototypischen Agenten in das System eingebracht werden.

W_{init} beschreibt den Initialzustand der Wissensbasis.

Die Aufgabenbearbeitung beginnt mit dem Initialsystem $\mathcal{S}_0 := (\{A_1, ..., A_n\}, W_{init})$, mit $A_1, ..., A_n \in \mathcal{A}_{prot}$. Bei der Problemlösung erbringen die Agenten Dienste und nehmen Dienste anderer Agenten in Anspruch. Dabei kooperieren die Agenten, indem sie sich gegenseitig Botschaften zusenden. Neue Agenten können dynamisch in das System eingebracht und wieder aus dem System entfernt werden. Die Identität von Agenten kann Inhalt von Botschaften sein, so daß sich beliebige Kommunikationsstrukturen aufbauen lassen. Das Wissen der Wissensbasis ist allen Agenten zugänglich. Damit kann ein Agent seine Identität allgemein bekannt machen, indem er sie in die Wissensbasis einträgt.

5 Abbildung der hierarchischen Ebenenstruktur auf ein Agentensystem

Nun soll gezeigt werden, wie die in Abschnitt 3 eingeführte hierarchische Ebenenstruktur für die Fertigungssteuerung auf ein Agentensystem abgebildet werden kann. Aus Platzgründen kann dabei nicht die vollständige Funktionalität aller Agenten angegeben werden; es wird vielmehr jeweils der wesentliche Kern der Dienstleistungen beschrieben, die ein bestimmter Agent verfügbar macht.

Die Modellierung der Benutzerschnittstelle

Eigentlich ist die Benutzerschnittstelle die oberste Schicht in der Planungshierarchie; sie wurde in Abschnitt 3 nicht explizit aufgeführt, weil sie in der Fertigungssteuerung keine Planungsinstanz ist. Es wurde schon erwähnt, daß der Benutzer nicht nur mit dem Produktionsplanungssystem kommunizieren können soll, sondern, daß es auch sinnvoll ist, wenn der Benutzer Zugriff auf die tieferen Planungsebenen hat. Die Benutzerschnittstellen werden jeweils als eigene Agenten repräsentiert, deren Aufgabe es ist, die gewünschte Funktionalität auf einem Ein-/Ausgabegerät verfügbar zu machen und die Ergebnisse der Dienste übersichtlich darzustellen. Auf diese Weise lassen sich Benutzerschnittstellen problemlos vervielfältigen, wenn dort, wo ein exklusiver Zugriff notwendig ist, durch die Verwendung der Wissensbasis, die notwendigen Reservierungsmechanismen bereitgestellt werden.

Die Modellierung der Produktionsplanungsebene

Für die Produktionsplanungsebene wird ein Produktionsplanungsagent eingeführt. Neben den Diensten zur Erfassung und Stornierung von Produktionsaufträgen, werden Dienste zur Abfrage von Statusinformationen angeboten. Dabei kann abgefragt werden, wieviele Aufträge aktuell vorliegen, wie weit die Bearbeitung fortgeschritten ist und inwieweit die Terminvorgaben eingehalten werden können. Während der gesamten Laufzeit werden statistische Daten gesammelt, z.B. durchschnittliche Durchlaufzeiten von Werkstücken, durchschnittliche Abweichungen von den Terminvorgaben etc. Als allgemeine Verhaltensregeln sind Strategien zur Bestimmung der voraussichtlichen Endtermine vorgegeben.

ERFASSE AUFTRAG <Endprodukt> <Anzahl> [<Anfangstermin>]
> Durch diesen Dienst wird ein Auftrag in das Produktionsplanungssystem eingelastet. Der Parameter <Endprodukt> gibt an, von welcher Art das Endprodukt sein soll. Die <Anzahl> legt fest, wieviele Endprodukte durch diesen Auftrag gefertigt werden sollen. Optional kann noch der früheste <Anfangstermin> des Fertigungsauftrags angegeben werden. Wird ein Auftrag im Produktionsplanungssystem erfaßt, so erhält er automatisch eine Auftragsnummer zugeteilt.

STORNIERE AUFTRAG <Nr>
> Der Auftrag mit der Auftragsnummer <Nr> wird abgebrochen, d.h. er wird aus dem Planungssystem gestrichen und die bereits an das Fertigungsleitsystem eingelasteten Fertigungsaufträge werden storniert.

Die Modellierung der Leitsystemebene

Auch für das Fertigungsleitsystem wird ein Agent eingeführt, der Fertigungsleitagent. Als Dienste werden das Einlasten und Löschen von Fertigungsaufträgen, sowie Abfragedienste für die komplexen internen Zustände des Fertigungsleitagenten bereitgestellt. Abgefragte Informationen werden in einem Fenster eines Graphiksichtgeräts ausgegeben; solange die Informationen angezeigt werden, wird automatisch mit den internen Zustandsänderungen auch die graphische Darstellung verändert. Die Informationsdienste liefern die Belegung der Lager und der Maschinen, die Liste der Fertigungsaufträge und den Bearbeitungsstand einzelner Fertigungsaufträge. Die allgemeinen Verhaltensregeln legen die Strategien zur Steuerung des Materialflusses fest; im wesentlichen geht es dabei darum, Fahraufträge für die Roboter in einer optimalen Reihenfolge zu erzeugen.

FERTIGE <Id> <Endprodukt> <Endtermin> <Priorität>
> Mit diesem Dienst wird ein Fertigungsauftrag in das Fertigungsleitsystem eingelastet; dabei wird ein für die Herstellung von <Endprodukt> geeignetes Rohteil ausgewählt und für diesen Fertigungsauftrag reserviert. <Id> wird zur Identifizierung des Fertigungsauftrags verwendet und muß ein eindeutiger Bezeichner sein. Durch <Endprodukt> kann anhand der Wissensbasis die Reihenfolge der Bearbeitungsschritte bestimmt werden. Über die Parameter <Endtermin>

und <Priorität> kann Einfluß darauf genommen werden, wie schnell ein Werkstück bei der Bearbeitung zur nächsten Bearbeitungsstelle befördert wird.

STORNIERE FERTIGUNGSAUFTRAG <Id>

Der Fertigungsauftrag <Id> wird storniert. Wenn die Bearbeitung bereits begonnen wurde, wird sie bis zum Endprodukt vollständig ausgeführt. Das Werkstück wird aber am Ende nicht ausgeliefert, sondern im Hauptlager abgelegt; es kann für neu eingehende Fertigungsaufträge als Endprodukt verwendet werden. Wurde die Bearbeitung des Werkstücks noch nicht begonnen, wird der Fertigungsauftrag gelöscht und das Werkstück freigegeben, das für den Fertigungsauftrag reserviert war.

Die Modellierung der Koordinationsebene

Jede autonome Einheit wird auf der Koordinationsebene als eigenständiger Agent modelliert. Alle diese Agenten bieten die gleichen Dienste an und haben bei ihrer Entstehung die gleichen Mengen von Fakten und allgemeinen Verhaltensregeln. BEARBEITE ist der wichtigste Dienst, den diese Agenten anbieten; er erhält als Parameter eine Aufgabenspezifikation. Anhand dieser Aufgabenspezifikation sucht der Agent in der Wissensbasis nach einem Verhaltensmuster für die ihm gestellte Aufgabe. Verhaltensmuster bestehen aus einer Menge von Regeln, die vorgeben, wie eine bestimmte Aufgabe gelöst werden kann. Das Konzept der Verhaltensmuster wurde für die Planung von Handhabungsaufgaben eines autonomen Roboters in einer Fertigungsumgebung entwickelt; nähere Informationen dazu sind in [2] und [3] zu finden. Wichtig ist nur, zu bemerken, daß die Ausführung von Regeln in einem Koordinationsagenten eine minimale Zeitspanne benötigen sollten, damit der Agent immer auf externe Ereignisse reagieren kann. Dies wird dadurch erreicht, daß die effektive Ausführung von Aktionen der autonomen Einheiten von den Ausführungsagenten überwacht wird; vom Koordinationsagent erfolgt also nur der Anstoß zur Ausführung der Aktion. Müssen von einem Koordinationsagenten langwierige Berechnungen ausgeführt werden, z.B. Wegfindungsalgorithmen, so müssen diese Berechnungen in einen eigenen Agenten ausgelagert und von diesem als Dienstleistung angefordert werden. Durch weitere Dienste kann die Aufgabenausführung unter- bzw. abgebrochen und der interne Zustand des Koordinationsagenten abgefragt werden.

BEARBEITE <Aufgabenspezifikation>

Erhält ein Koordinationsagent diesen Dienst, so prüft er, ob in seinen allgemeinen Verhaltensregeln bereits ein Verhaltensmuster enthalten ist, das die gestellte Aufgabe lösen kann. Ist dies nicht der Fall, so wird in der Wissensbasis anhand der Aufgabenspezifikation ein Verhaltensmuster gesucht, das die gestellte Aufgabe löst. Ist dort kein solches Verhaltensmuster vorhanden, erfolgt eine Fehlermeldung; anderenfalls wird es in die Menge der allgemeinen Verhaltensregeln eingelesen und automatisch ausgeführt, d.h. die Regeln werden aktiv. Dabei können weitere Verhaltensmuster aktiviert werden, die falls sie noch nicht lokal verfügbar sind, aus der Wissensbasis nachgeladen werden.

Die Modellierung der Ausführungsebene

Für jede Komponente der Fertigungsanlage wird ein Ausführungsagent eingeführt. Eine Komponente ist dabei ein Teilsystem der Fertigungsanlage, das rechnergestützt angesteuert werden kann und eine sinnvolle Funktionseinheit bildet. In der Regel wird man für jede auf der Koordinationsebene modellierte autonome Einheit einen Ausführungsagenten spezifizieren, der die Einzelaktionen dieser autonomen Einheit ausführen und überwachen kann. Es gibt aber auch Komponenten, die auf Koordinationsebene nicht als autonome Einheiten modelliert sind, für die aber trotzdem ein Ausführungsagent angegeben werden muß; z.B. das Ein- und das Ausgabeförderband. Die Funktionalität hängt stark vom Aufbau der einzelnen Komponenten ab. Exemplarisch soll nur ein Dienst des Ausführungsagenten für die Fräsmaschine angegeben werden.

BELADE FRAESMASCHINE VON <Ort>

Für <Ort> kann LINKS oder RECHTS angegeben werden, wobei die Ortsangabe aus der Sicht des Roboters zu interpretieren ist, der die Handhabungsaufgabe ausführt. Der Dienst wird nur ausgeführt, wenn der Platz, von dem das Werkstück weggenommen werden soll, besetzt ist und der Bearbeitungsplatz frei ist. Die Zustandsänderungen werden in der Wissensbasis bekannt gemacht, so daß sie auf der Koordinationsebene berücksichtigt werden können.

6 Ein verteiltes System zur parallelen regelorientierten Programmierung

Das bisher beschriebene Agentensystem ist mit Hilfe von OPS5 implementiert worden. Dies gelang, weil OPS5 so erweitert werden konnte, daß es auf komfortable Weise möglich ist, asynchrone Ereignisse in effizienter Weise zu verarbeiten [2],[3]. Die Fähigkeit auf asynchrone Ereignisse in effizienter Weise zu reagieren, ist vor allem auf der Leitsystem- und Koordinationsebene von grundlegender Bedeutung. Durch die Regelprogrammierung ist es nur notwendig, die Situation zu beschreiben, in der bestimmte Aktionen ausgelöst werden sollen. Zeitliche Abhängigkeiten müssen nur dort berücksichtigt werden, d.h. in den Regeln repräsentiert werden, wo sie durch die Aufgabenstellung vorgegeben sind. Die Implementierung eines ersten Prototypen des vorgestellten Agentensystems wurde in [10] beschrieben. Durch [5] wurde die Funktionalität dieses Prototypen so erweitert, daß die Agenten auf beliebig viele Knoten in einem VAX/VMS-LAN-Netz verteilt werden können.

Die objektorientierte Wissensbasis ist im Teilprojekt A4 des SFB 331 entwickelt worden [1], im folgenden wird sie kurz mit WB bezeichnet; sie existiert im ganzen System genau einmal. Auf jedem Knoten des LAN-Netzes gibt es einen speziellen Agenten (GFA), dessen Identität alle anderen Agenten auf diesem Knoten kennen. Die GFAs aller Knoten haben eine Verbindung zur WB und sind untereinander vollständig durch Kommunikationsverbindungen vernetzt. Für die Agenten eines Knotens fungiert der GFA als Schnittstelle zur WB. Der GFA versucht dabei, alle Anfragen zunächst aus seiner lokalen Faktenmenge zu beantworten, nur wenn hier keine Information vorliegt, greift er auf die WB zu. WB-Objekte werden dabei auf Fakten der Form

(LITERALIZE *Globales-Faktum Klasse Attribut Instanz Wert*)

abgebildet. Hat ein GFA einmal ein solches Faktum aus der WB gelesen, so erhält er jedesmal, wenn der Wert in der WB neu geschrieben wird, diesen Wert automatisch zugesandt; der GFA verteilt dann diesen Wert an die Agenten auf seinem Knoten, von denen er weiß, daß sie die alten Versionen dieser Fakten bereits erworben hatten. Abbildung 3 zeigt eine mögliche Konfiguration für die Implementierung des in Abschnitt 3 vorgestellten Schichtenmodells; dabei wurden die Ausführungsagenten aus Gründen der Übersichtlichkeit fortgelassen. Bei der Implementierung der Agenten als OPS5-Prozesse, entspricht die Faktenmenge eines Agenten dem Inhalt des OPS5-Faktenspeichers. Die allgemeinen Verhaltensregeln entsprechen einer Menge von Regeln, die nur für eine bestimmte Aufgabenstellung speziell strukturiert sind. Jeder Dienst wird durch eine Menge von Regeln realisiert, die alle in ihrer ersten Bedingung die existenz eines Faktums folgender Faktenklasse abprüfen:

(LITERALIZE *Auftrags-Kontext Dienst-Name Auftrags-Id Auftraggeber-Id Status*)

Diese Fakten entstehen asynchron im OPS5-Faktenspeicher eines Agenten, wenn ein Dienst mit Hilfe von SENDE-AUFTRAG (siehe unten) aufgerufen wird. Ein Dienst ist aktiv, wenn ein entsprechendes *Auftrags-Kontext*-Faktum mit ˆ*Status*=AKTIV existiert. Es ist immer nur ein Dienst aktiv. Den einzelnen Diensten sind dabei Prioritäten zugeordnet, so daß immer der Dienst mit höchster Priorität aktiv ist, wobei das älteste der entsprechenden *Auftrags-Kontext*-Fakten verwendet wird, falls es mehrere geben sollte. Dadurch wird für die Dienste ein prioritätengesteuerte FIFO[1]-Bearbeitungsstrategie erreicht. Der bisher beschriebene Teil der Auftragsverwaltung ist durch eine Menge von Regeln in OPS5 implementiert worden. Zur Prozeßkommunikation stehen eine Reihe

---

[1] first-in-first-out

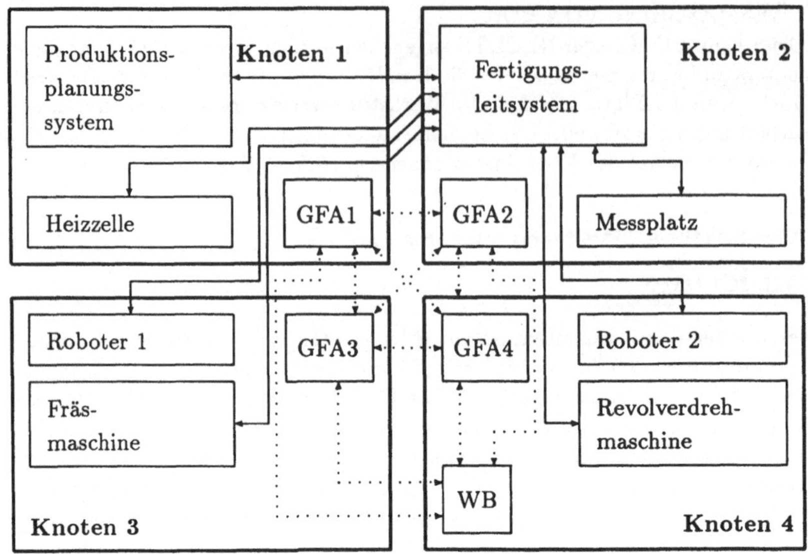

Abbildung 3: Konfiguration eines Agentensystems für die Steuerung der Fertigungsanlage

von Externprozeduren zur Verfügung, die mit Hilfe der OPS5-Anweisung CALL im Aktionsteil einer Regel aufgerufen werden können. Die wichtigste Prozeduren sollen nun kurz erklärt werden.

ERZEUGE-AGENT <Name> <Knoten>

> Mit Hilfe dieser Prozedur kann ein neuer Agent erzeugt werden. Der Erzeuger erhält ein Faktum in seinen lokalen Faktenspeicher eingetragen, das die Identität des neu entstandenen Agenten enthält. Damit kann der Erzeuger immer sofort mit dem neu entstandenen Agenten kommunizieren. <Name> gibt den Namen der Datei an, die das ausführbare Programm für den neu entstandenen Agenten enthält. Durch <Knoten> kann angegeben werden, auf welchem Knoten im Netz der Agent erzeugt werden soll. Es darf auch LOKAL und NICHT-LOKAL angegeben werden; der Agent entsteht im ersten Fall auf dem gleichen Knoten wie der Aufrufer bzw. im zweiten Fall nach einer bestimmten Strategie auf einem anderen Knoten im Netz.

SENDE-AUFTRAG <Agent-Id> <Dienst-Name> <Modus> <Faktenspez.>

> Mit Hilfe dieser Prozedur kann ein Dienst des Agenten <Agent-Id> aufgerufen werden. Falls <Modus>=WARTEN, so wird in der Prozedur gewartet, bis die erste Antwort von Agent <Agent-Id> erhalten wird; anderenfalls kann der Aufrufer weiter arbeiten ohne auf eine Rückmeldung von <Agent-Id> zu warten. Als Ergebnis erhält der Aufrufer ein Faktum in seinen Faktenspeicher eingetragen, das die Auftragsidentifikation AUFTRAGS-ID enthält, die automatisch erzeugt wird. Beim Agenten <Agent-Id> entsteht daraufhin asynchron ein *Auftrags-Kontext*-Faktum mit ^*Dienst-Name*=<Dienst-Name>, ^*Auftrags-Id*=AUFTRAGS-ID und ^*Auftraggeber-Id*=Identität des Aufrufers[2]. Außerdem wird das durch die Symbolfolge <Faktenspez.> spezifizierte Faktum, bei der Dienstaktivierung, in den Faktenspeicher von <Agent-Id> eingetragen.

SENDE-FAKTUM <Agent-Id> <Faktenspez.>

> Durch den Aufruf dieser Prozedur ist es einem Agenten möglich, dem Agenten mit der Identität <Agent-Id> asynchron ein Faktum in den Faktenspeicher eintragen zu lassen. <Faktenspez.>

---

[2]In der PASCAL-Schnittstelle wird bei der Initialisierung die Identität des Aufrufers festgehalten.

gibt als Symbolfolge an, welche Gestalt das zu verschickende Faktum haben soll.

SENDE-ANTWORT <Auftraggeber-Id> <Auftrags-Id> <Faktenspez.>
Durch den Aufruf dieser Prozedur kann ein Agent, der gerade einen Dienst ausführt, ein Antwortfaktum an den Auftraggeber zurücksenden. <Auftraggeber-Id> und <Auftrags-Id> werden dabei aus dem *Auftrags-Kontext*-Faktum übernommen. <Faktenspez.> gibt als Symbolfolge an, welche Gestalt das Antwortfaktum haben soll.

7 Zusammenfassung und Ausblick

In dem vorliegenden Artikel wurde ein Konzept für ein Agentensystem zur Steuerung eines flexiblen Fertigungssystems vorgestellt. Der Implementierung des Agentensystems liegt eine Erweiterung von OPS5 zu einem verteilten System zur parallelen regelbasierten Programmierung zu Grunde. Die Erfahrungen mit bisherigen Implementierungen von Agentensystemen für Planungsaufgaben haben gezeigt, daß sich die Planung und Überwachung von Abläufen mit asynchron auftretenden Ereignissen auf symbolischer Ebene sehr gut mit den vorgestellten Konzepten beschreiben läßt. Die vorwärtsgerichtete Schlußfolgerungssemantik von OPS5 kann hier optimal eingesetzt werden. Aus den bei der Implementierung des vorgestellten Agentensystems für die Fertigungsanlage gewonnenen Erfahrungen sollen weitere Konzepte für die einzelnen Planungsebenen abgeleitet werden. Daß die Erweiterung von OPS5 zu einem verteilten System nur auf das OPS5-System aufgesetzt und nicht wirklich in das OPS5-System integriert werden konnten, machte sich vor allem beim Testen und Überwachen der Abläufe in den Agenten negativ bemerkbar. Die Testhilfen, die das OPS5-System anbietet, konnten aus diesem Grund nicht optimal verfügbar gemacht werden und reichen damit für ein komfortables Arbeiten nicht aus. Daher wird versucht einen Regelinterpreter zu implementieren, der die OPS5-Schlußfolgerungssemantik beibehält, in den aber die in diesem Artikel vorgestellten Konzepte voll integriert sind.

Literatur

[1] Bocionek S.: Dynamic Flavors. TU München, Institut für Informatik, Report Nr. TUM-I8708, Juni 1987.

[2] Fischer K.: Regelbasierte Synchronisation zwischen Roboter und Maschinen. TU München, Institut für Informatik, Report Nr. TUM-I8816, Dezember 1988.

[3] Fischer K.: Knowledge-Based Task Planning for Autonomous Mobile Robot Systems. Proc. of the 2nd Inter. Conf. on Intelligent Autonomous Systems, Amsterdam, Dezember 1989.

[4] Groha A.: Universelles Zellenrechnerkonzept für flexible Fertigungssysteme. Springer-Verlag, Berlin 1988.

[5] Knittl R., Ruckdäschel R., Sing C.: Erweiterung der Funktionalität von POPE zu einem Netzwerkfähigen System. TU München, Institut für Informatik, Fortgeschrittenenpraktikum, erscheint im Dezember 1990.

[6] Hagg E.: Logische Sensoren und Aktoren, ein Ansatz zur Entwicklung von Multisensoranwendungen für Fertigungsumgebungen. 12. DAGM Symposium Mustererkennung, September 1990.

[7] Hirschberg G.: Produktionsautomatisierung und Flexibilität aus der Sicht des Informatikers. VDI Bericht NR. 723, 1989.

[8] Kupec T.: Integration of Autonomous, Mobile Robots in Flexible Manufacturing Systems. Proc. of the 2nd Inter. Conf. on Intelligent Autonomous Systems, Amsterdam, Dezember 1989.

[9] Scheer, A.-W.: CIM - Der computergesteuerte Industriebetrieb, 4. neu bearbeitete und erweiterte Auflage, Springer Verlag 1990.

[10] Weikert P.: Parallelisierung von OPS5 durch Nachrichtenübermittlung. TU München, Institut für Informatik, Diplomarbeit, Februar 1989.

<u>PROZESSDIAGNOSE AUF DER GRUNDLAGE EINER</u>

<u>PRÄDIKATENLOGISCHEN PROZESSBESCHREIBUNG</u>

Jan Lunze

Zentralinstitut für Kernforschung Rossendorf
Postfach 19, D-08051 Dresden
Tel. (051) 591 2461

Zusammenfassung. Es wird eine prädikatenlogische Darstellungsform für kausale dynamische Systeme definiert und ein Algorithmus beschrieben, mit dem Diagnoseaufgaben gelöst werden können. Aufgrund der Struktur des Algorithmus ist dieser für Echtzeitanwendungen geeignet.

Schlagwörter. Prozeßdiagnose, wissensbasierte Systeme, modellbasierte Diagnose, Echtzeitexpertensysteme.

Einleitung

Wichtige Aufgaben der Prozeßleittechnik sind die Überwachung des Betriebes einer technologischen Anlage, die Anzeige anormaler Betriebszustände und die Ermittlung der Ursachen für fehlerhaftes Verhalten. Während die Anzeige untypischer bzw. gefährlicher Betriebszustände in Form von Alarmen zum klassischen Funktionsumfang von Leitsystemen gehört, ist der Umfang der von der Leittechnik selbsttätig gelösten Diagnoseaufgaben noch gering[9]. Ein wichtiger Grund dafür liegt in der Komplexität der für Diagnoseaufgaben im allg. verwendeten Ereignis- bzw. Diagnosegraphen, die aufgrund der Ursache-Wirkungs-Beziehungen der Anlage aufgestellt werden und als Grundlage für die Lösung der Diagnoseaufgaben dienen.

Die Verwendung logischer Darstellungsformen für das Prozeßverhalten kann dazu beitragen, den Funktionsumfang moderner Leitsysteme wesentlich zu erweitern. Dafür sprechen zwei Gründe: Erstens bietet die prädikatenlogische Prozeßbeschreibung gegenüber dem Ereignisgraphen wesentliche Vorteile bezüglich der Zahl von Ereignissen und Randbedingungen, die in die Diagnose einbezogen werden können. Es muß nicht der Ereignisgraph bzw. eine damit vergleichbare Darstellung des Anlagenverhaltens aufgestellt werden, in der die einzelnen Ursache-Wirkungs-Beziehungen ineinander verschachtelt sind, sondern es kann von Aussagen ausgegangen werden, die unmittelbar an die ingenieurtechnische Beschreibung der Anlage anschließen. Im Gegensatz zu den in [8] zusammengefaßten systemtheoretisch begründeten Methoden wird hier kein analytisches

Modell, sondern eine auf Ereignisfolgen beruhende Prozeßbeschreibung verwendet. Zweitens können durch die Behandlung der Diagnoseaufgabe als Wissensverarbeitungsproblem die im Fachgebiet künstliche Intelligenz entwickelten modernen softwaretechnologischen Konzepte für die Leittechnik nutzbar gemacht werden.

Im folgenden wird ein Verfahren für die Prozeßdiagnose vorgestellt, das auf einer prädikatenlogischen Darstellung der Ursache-Wirkungs-Beziehungen innerhalb der Anlage beruht. Es ermittelt aus den in zeitlicher Folge eingehenden Alarmmeldungen den die Betriebsstörung verursachenden Fehler. Unter Ausnutzung spezifischer Eigenschaften des Diagnoseproblems wird ein Wissensverarbeitungsalgorithmus entwickelt, der aufgrund seiner Struktur für Echtzeitanwendungen besonders geeignet erscheint. Im Vergleich zu den in der künstlichen Intelligenz betrachteten allgemeinen Diagnosestrategien, z.B. [2], [10], [11], wird hier von einer für die Prozeßdiagnose zugeschnittenen Wissensrepräsentation ausgegangen und ein Verfahren entwickelt, mit dem die Diagnoseaufgabe vollständig gelöst werden kann (obwohl allgemeine prädikatenlogische Probleme nicht entscheidbar sind). Im Sprachgebrauch der künstlichen Intelligenz handelt es sich um ein modellbasiertes Verfahren. Für die mit der Einbindung der wissensbsierten Prozeßdiagnose in die Leittechnik verbundenen Probleme wird auf [3] verwiesen.

1. Die Diagnoseaufgabe

Die Auswertung von Alarmen ist eine typische Diagnoseaufgabe, bei der das Prozeßverhalten nach außen hin durch das Auftreten diskreter Ereignisse (Alarmmeldungen) erkennbar wird, unabhängig davon, ob die zu überwachende Anlage kontinuierlich oder diskret arbeitet. Deshalb werden Steuerungen und Prozeßzustände durch globale, im Zeitraum der Alarmauswertung feststehende Größen U_i bzw. Z_i qualitativ beschrieben, z.B.

$$U_1 = \text{"Ventil 1 ist geöffnet"}, \quad Z_1 = \text{"Pegel 2 ist niedrig"}.$$

In ähnlicher Weise werden Fehler F_i, Alarmmeldungen M_i und charakteristische Ereignisse K_i definiert. Die Mengen aller dieser Ereignisse heißen $\underline{U}$, $\underline{Z}$, $\underline{F}$, $\underline{M}$ bzw. $\underline{K}$.

Die Alarmmeldungen werden zu bestimmten Zeitpunkten t_i aktiviert, wobei der Zeitpunkt $t = 0$ auf den Moment des Auftretens des ersten Alarms festgelegt ist. Die bis zur Zeit t erhaltenen Meldungen sind in der Menge

$$\underline{M}_t = \{(M_k,0),(M_1,t_1),..,(M_m,t_m)\} \quad \text{für } M_i \in \underline{M}, \; t_i < t \qquad (1)$$

zusammengefaßt, wobei jedes Paar (M_i, t_i) den Namen M_i der Meldung und den Zeitpunkt t_i des Auftretens beschreibt. Weitere Voraussetzungen sind die folgenden:

Voraussetzung 1: Die Alarme sind durch einen einzigen Fehler F_t verursacht, der nach seinem Eintreten bestehen bleibt. (Erweiterung auf Mehrfachfehler sind möglich, können hier aber nicht behandelt werden).

Voraussetzung 2: Es wird der Zeitraum T nach dem Störeintritt betrachtet, in dem sich alle kurzfristigen Störauswirkungen in Alarmmeldungen wiederspiegeln, d.h. es wird angenommen, daß nach der Zeitraum T kein weiterer Alarm aktiviert wird.

Gesucht wird der die Betriebsstörung verursachende primäre Fehler:

Definition 1: Ein Fehler F_i heißt **primäre Alarmursache**, wenn die Ausbreitung der durch ihn erzeugten Störung des Betriebszustandes der Anlage innerhalb der Zeitspanne T auf eine Menge $\underline{M}(F_i)$ von Alarmen M_i führt, die identisch mit der Alarmmeldung $\underline{M}_T$ ist: $\underline{M}(F_i) = \underline{M}_T$.

Diese Definition bezieht sich auf $t \geq T$. Die Diagnose soll jedoch bereits nach der ersten Alarmmeldung einsetzen (vgl. Abschn. 3):

Diagnoseaufgabe: Nach Eintreten von Alarmmeldungen soll die Menge $\underline{F}_t$ aller Fehler F_i bestimmt werden, die aufgrund der Meldung $\underline{M}_t$ nicht als primäre Alarmursache ausgeschlossen werden kann.

Für $t > T$ ist die Lösung dieser Diagnoseaufgabe die Menge $\underline{F}_T$ aller primären Alarmursachen. Diese Menge enthält also alle Fehler, von denen jeder allein die Alarmmeldung $\underline{M}_T$ erzeugen kann.

2. Prädikatenlogische Formulierung der Diagnoseaufgabe

In der prädikatenlogischen Beschreibung der Ursache-Wirkungs-Beziehungen werden für die nicht zeitabhängigen Ereignisse einstellige Prädikate eingeführt:

$$\text{Steuerung}(U), \quad \text{Fehler}(F), \quad \text{Zustand}(Z), \tag{2}$$

wobei U, Z und F Variable sind, die für Steuerereignisse, Fehler bzw. Zustände stehen. Die zeitabhängigen Ereignisse und Alarmmeldungen sind durch dreistellige Prädikaten beschrieben, von den das zweite Argument einen Zeitpunkt t und das dritte Argument eine Unsicherheit d angibt, mit der dieser Zeitpunkt beschrieben ist:

$$\text{Alarm}(M,t,d) \qquad \text{für } M \in \underline{M}$$

$$\text{Ereignis}(K,t,d) \qquad \text{für } K \in \underline{K}. \tag{3}$$

Das heißt, daß der tatsächliche Auftrittzeitpunkt t_1 im Intervall

$$t - d \leq t_1 \leq t + d \tag{4}$$

liegt. So bedeutet

    Ereignis("Behälter 2 ist leer", 3, 1),

daß das Ereignis "Behälter 2 ist leer" im Zeitintervall

    $t \in [2, 4]$

eintritt. Diese Unsicherheit bzgl. des Auftrittszeitpunktes eines Alarms ermöglicht die Lösung der Diagnoseaufgabe auch dann, wenn sich Alarmmeldungen
"überholen". Weitere Prädikate wie z.B. SUM oder kleiner_als werden zur Darstellung von Zeitrelationen gebraucht.

Aufgrund der Kausalität sind für die Darstellung des Prozessverhaltens Formeln
folgender Art ausreichend:

    Ursache => Ereignis(K,t,d)

    Ursache => Alarm(M,t,d)

Dabei steht "Ursache" für die konjunktive Verknüpfung von Literalen mit den
Prädikaten Steuerung, Fehler, Zustand und Ereignis. Die in der Form (5)
notierten Beziehungen sind in vielfältiger Weise untereinander abhängig. Darauf muß – im Gegensatz zur Darstellung in einem Ereignisbaum – bei der Aufstellung der Prozeßbeschreibung keine Rücksicht genommen werden.

Für die aktuelle Alarmmeldung (1) wird das zweistellige Prädikat

    Meldung(M,t)   mit $M \in \underline{M}$, t > 0   (6)

eingeführt. Eine Meldung (6) tritt in den durch Alarm(M,t,d) beschriebenen
Zeitintervall auf, wenn die Beziehung (4) gilt:

    $\text{Alarm}(M,t,d)\ \&\ \text{SUM}(t,d_1,t_1)\ \&\ \text{kleiner\_als}(\text{abs}(d_1),d)$

    $\Rightarrow \text{Meldung}(M,t_1)$ (7)

Die Alarmmeldung $\underline{M}_T$ (Gl.(1) für t = T) kann entsprechend (6) als

    $\text{Meldung}(M_k,0)\ \&\ \text{Meldung}(M_1,t_1)\ \&...\&\ \text{Meldung}(M_m,t_m)$ (8)

geschrieben werden.

Das Diagnoseproblem besteht in dem Beweis, daß aus der Systembeschreibung (5),
den Definitionen der verwendeten Hilfsprädikate sowie einem durch

    $\text{Fehler}(F_i)$ für ein fester $F_i \in \underline{F}$ (9)

beschriebenen Fehler die Alarmmeldung (8) gefolgert werden kann.

Gegeben: 1. Systembeschreibung (5)
        2. Definition der Hilfsprädikate SUM, kleiner_als usw.
        3. Beschreibung von Steuerung und Zustand entsprechend (2)
        4. Fehler(F_i) für ein festes $F_i \in \underline{F}$

Gesucht: Beweis, daß die aktuelle Alarmmeldung (8) aus den gegebenen Formeln erzeugt werden kann. Jeder Fehler F_i, für den das möglich ist, ist eine primäre Alarmursache.

3. Lösungsweg

Um diese Aufgabe algorithmisch lösen zu können, wird im allgemeinen auf die Resolutionsmethode zurückgegriffen. Dabei wird die Behauptung (8) in negierter Form zu den gegebenen Aussagen hinzugefügt und aus der erweiterten Formelmenge ein Widerspruch erzeugt. Unter Verwendung von Resolutionsregel und Faktorisierung als Inferenzregeln sowie einer geeigneten Suchstrategie können korrekte und widerspruchsvollständige Beweisverfahren aufgestellt werden (siehe z.B. [6]). Dieser Lösungsweg ist jedoch nicht unter Echtzeitbedingungen gangbbar, da bei realistischen Anwendungen sehr viele Formeln der Form (5) vorliegen und folglich der Suchraum für das Beweisverfahren sehr groß ist.

Im folgenden wird eine Lösung angegeben, bei der in einem Off-line-Teil die Suche ausgeführt (Algorithmus 1) und damit die On-line-Verarbeitung (Algorithmus 2) auf wenige Schritte reduziert wird.

Algorithmus 1: Umformung der Prozeßbeschreibung (5)

Für gegebene Störungen und Zustände können die durch (5) beschriebenen Wirkungsabläufe in direkte Beziehungen zwischen Fehlern und Alarmen umgeformt werden

Fehler(F_i) =>

$$\text{Alarm}(M_1,t_1,d_1) \ \& \ \text{Alarm}(M_2,t_2,d_2) \ \& \ldots\& \ \text{Alarm}(M_q,t_q,d_q), \qquad (10)$$

wobei aufgrund der Eindeutigkeit des Störablaufes für jeden Fehler $F_i \in \underline{F}$ genau eine Implikation entsteht. Die Überführung von (3), (5) in (10) geschieht folgendermaßen [5]:

1. Aufgrund ihrer Struktur können die Formeln (5) für vorgegebene Steuerungen und Zustände auf die Form

verkuerzte_Ursache => Ereignis(K,t,d)   (11)

verkuerzte_Ursache => Alarm(M,t,d)

gebracht werden, wobei "verkuerzte_Ursache" konjunktive Verknüpfungen von Literalen mit den Prädikaten Fehler und Ereignis bezeichnet.

2. Für einen vorgegebenen Fehler F_i werden die Formeln (11) in

$$\text{Ereignis}(K,t,d)$$

$$\text{Ereignisse} \Rightarrow \text{Ereignis}(K,t,d) \tag{12}$$

$$\text{Ereignisse} \Rightarrow \text{Alarm}(M,t,d)$$

überführt, wobei "Ereignisse" konjunktive Verknüpfungen von Ereignis-Prädikaten darstellen. Tritt der Fehler F_i negiert oder treten andere als der gegebene Fehler F_i in Formeln (11) auf, so werden diese Formeln gestrichen.

3. Durch Vorwärtsverkettung der Formeln (12) kann die zum Fehler F_i gehörende rechte Seite von (10) abgeleitet werden. Dieser Schritt enthält das eigentliche Suchproblem, das aber aufgrund der Kausalität der Prozeßbeschreibung (5) lösbar ist [5]. Das Ergebnis hat die Form (10). Der Algorithmus kann vor dem Eintreten von Alarmmeldungen ausgeführt werden.

Algorithmus 2: Alarmauswertung

Die rechten Seiten der Formeln (10) beschreiben Alarmfolgen, bei denen

$$\text{Alarm}(M_i, t_i, d_i)$$

das Eintreten der Alarmmeldungen M_i im Zeitintervall

$$t \in [t_i - d_i, t_i + d_i]$$

kennzeichnet. Tritt die entsprechende Meldung gar nicht oder außerhalb dieses Intervalls auf oder erfolgt eine in (10) nicht verzeichnete Meldung, so ist der Fehler F_i keine primäre Alarmursache (Def. 1). Das Diagnoseproblem kann also am einfachsten dadurch gelöst werden, daß in allen Formeln (10) der rechte Teil unter Beachtung der Relation (7) zwischen den "Alarm"- und "Meldung"-Prädikaten verglichen wird.

Es soll jedoch bereits vor Ablauf der Zeitspanne T nach dem Auftreten des ersten Alarms die Menge der möglichen Fehler bestimmt werden. Da zu einem Zeitpunkt

$$t_1 : 0 \leq t_1 \leq T$$

nicht bekannt ist, welche zusätzliche Alarme noch eintreffen werden, soll wie in [4] die Menge E_{t_1} aller Alarme bestimmt werden, die aufgrund der bis zur Zeit t_1 gemeldeten Alarme nicht als primäre Alarmursache ausgeschlossen werden können. Diese Erweiterung der Aufgabenstellung bringt für die logische Darstellung und Lösung zwei eng miteinander verknüpfte Probleme:

* Nichtmonotonie des Schließens: Schlußfolgerungen, die die in $\underline{F}_{t1}$ zusammengefaßten Störursachen betreffen, müssen bei Be kanntwerden neuer Alarme überprüft und gegebenenfalls zurückge zogen werden, denn nicht alle Schlußfolgerungen in $\underline{F}_{t1}$ sind auch Bestandteil von $\underline{F}_{t2}$.

* Default-Logik: Bevor nicht die gesamte Alarmmeldung $\underline{M}_T$ vor liegt, kann aus dem Modell (10) mit der klassischen Logik kein Fehler geschlossen werden. Die Schlußweise muß so erweitert werden, daß F_i auch dann eine gültige Folgerung ist, wenn nur einige der in (10) geforderten Alarme eingegangen sind.

Bisher gibt es kein einheitliches theoretisches Fundament, um diese Probleme zufriedenstellend zu lösen [1]. In [5] wurde ein Algorithmus abgeleitet, der die folgenden zusätzlichen Auswerteregeln verwendet:

Regel 1: Zum Zeitpunkt t ist die Aussage

$$\text{Alarm}(M_i, t_i, d_i)$$

wahr, wenn keine Aussage

$$\text{Meldung}(M_i, t_j)$$

existiert, aber $t < t_i + d_i$ gilt.

Regel 2: Die rechte Seite von (10) ist nicht erfüllt, wenn

$$\text{Meldung }(M_i, t_i)$$

mit einer nicht auf der rechten Seite von (10) auftretenden Meldung M_i in (8) enthalten ist.

Der Algorithmus 2 beinhaltet die Auswertung der Formeln (10) entsprechend Gl. (7) und dieser beiden Regeln. Alle Fehler, deren Gültigkeit dabei nachgewiesen werden können, bilden die Menge $\underline{F}_t$.

Eine wesentliche Vereinfachung des Algorithmus 2 ergibt sich aufgrund der in [4] bewiesenen Monotonieeigenschaft der Schlußfolgerung $\underline{F}_t$. Da mit fortschreitender Zeit mehr Alarme zur Verfügung stehen, wird die Menge $\underline{F}_t$ stets nur verkleinert;

$$\underline{F}_{t2} \subseteq \underline{F}_{t1} \quad \text{für } t_2 \leq t_1, \ \underline{M}_{t1} \neq 0 \tag{13}$$

Deshalb kann der Algorithmus auf die noch möglichen Fehler und den entsprechenden Teil der Wissensbasis fokusiert werden [5].

Struktur des Diagnosesystems

Das Diagnosesystem besteht aus zwei Teilen (Bild 1). Im Off-line-Teil erfolgt die Umformung der Wissensbasis nach jeder Änderung der Steuerung bzw. des Prozeßzustandes vor der eigentlichen Diagnose. Dieser Schritt ist deshalb unkritisch bezüglich der notwendigen Rechenzeit. Der On-line-Teil besteht aus dem Algorithmus 2. Die vom Algorithmus 1 vorbereiteten Regeln werden anhand der aktuellen Alarmmeldungen $\underline{M}_t$ abgearbeitet, wobei die Menge der möglichen Fehler $\underline{F}_t$ ermittelt wird. Das erfordert nur wenige Verarbeitungsschritte und ist deshalb unter Echtzeitbedingungen auch für umfangreiche Systeme mit vielen möglichen Alarmen durchführbar.

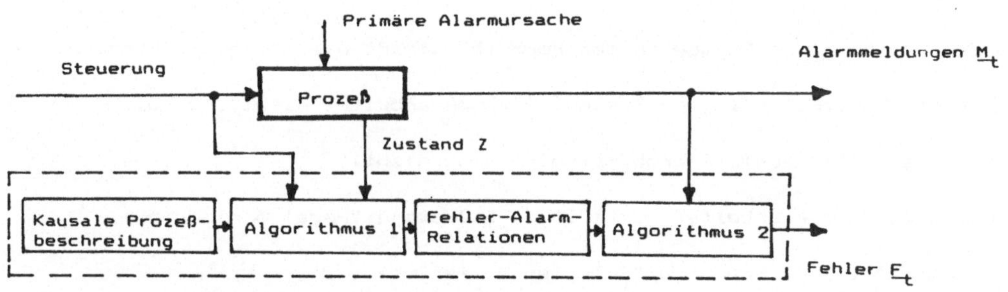

Bild 1: Kopplung des Diagnosealgorithmus mit dem Prozeß

4. Beispiel

Zur Veranschaulichung der betrachteten Vorgehensweise wird die im Bild 2 gezeigte Anlage betrachtet, die aus drei durch Rohrleitungen verbundenen Wasserbehältern besteht. Durch eine Pumpe soll der im Bild angedeutete Höhenunterschied zwischen dem zweiten und dritten Behälter überwunden werden. Regelkreise, die auf ein Ventil bzw. die Pumpe zugreifen, sorgen im Normalbetrieb für Füllstände in den Behältern 2 und 3 , die von der Wasserentnahme unabhängig sind. Zu den möglichen Alarmmeldungen gehören

Alarm 2H – Überlaufen des Behälters 2

Alarm 3L – Leerlaufen des Behälters 3

Alarm 4L – zu niedriger Wasserdruck beim Verbraucher.

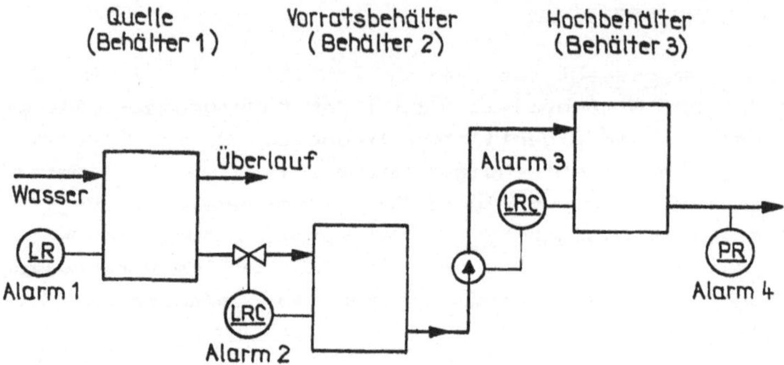

Bild 2: Wasserversorgungssystem

Die Beschreibung der Anlage in der Form (5) umfaßt u.a. folgende Formeln

    Fehler("Stromausfall") => Ereignis("Pumpe arbeitet nicht", 0, 0)

(sofort nach Stromausfall arbeitet die Pumpe nicht)

    Ereignis("Pumpe arbeitet nicht",t,d) & Zustand("Pegel 3 niedrig")

    & SUM($t,4,t_1$) => Ereignis("Behälter 3 leer", t_1, 2)

(bei niedrigem Füllstand im Behälter 3 fällt nach Ausfall der Pumpe z.Z. $t \pm d$ der Wasserstand innerhalb von 4 ± 2 Zeiteinheiten unter die Minimalhöhe)

    Ereignis("Behälter 3 leer", t, d) => Alarm("Alarm 3L", t, d)

(Alarm 3L wird sofort nach Unterschreiten der Minimalhöhe ausgelöst). Durch Algorithmus 1 wird aus der Wissensbasis u.a. die Formel

    Fehler("Stromausfall") => Alarm("Alarm 3L",0,2) & Alarm("Alarm 4L",0,3)

abgeleitet, wobei sich der Zeitpunkt 0 für die Auftrittszeit aus der o.a. Definition von t = 0 ergibt. Algorithmus 2 liefert folgendes Ergebnis:

    Meldung("Alarm 4L", 0)

führt auf zwei mögliche Alarmursachen

    F_0 = {"Stromausfall", "Rohrleitung verstopft"}.

Tritt anschließend

159

    Meldung("Alarm 3L", 1)

auf, bleibt eine einzige Alarmursache

    $\underline{F}_1$ = {"Stromausfall"}

übrig.

Literatur

[1] Brewka, G.: Nichtmonotone Logiken. KI (1989) H. 2, S. 5-12.

[2] de Kleer, J.; Williams, B.C.: Diagnosing multiple faults, AI 32 (1987)
    No. 1.

[3] Kerndlmaier, M. (Hrsg.) *Technische Expertensysteme für Prozeßführung und
    Diagnose.* Oldenbourg-Verlag, München 1989.

[4] Lunze, J.: Ein Verfahren zur Prozessdiagnose auf der Grundlage der Aussa-
    genlogik. msr 33 (1990) H. 12.

[5] Lunze, J.; Serbezow, T.: Logikbasierte Diagnose unter Berücksichtigung der
    Prozeßdynamik msr (erscheint 1991).

[6] Lunze, J.; Schwarz, W.: *Künstliche Intelligenz.* Verlag Technik und Sprin-
    ger-Verlag, Berlin 1990.

[7] Narayanan, N.H.; Viswanadham, N.: A methodology for knowledge acquisition
    and reasoning in failure analysis of systems. IEEE Trans-SMC-17 (1987)
    No. 2, pp. 274-288

[8] Patton, R.; Frank, P.M.; Clark, R.: *Fault Diagnosis of Dynamic Systems.*
    Prentice-Hall, London 1988.

[9] Polke, M.; Will, B.: Neue Qualität der Führung verfahrenstechnischer Pro-
    zesse. ATP 31(1989)H. 9, S.115-120.

[10] Puppe, F.: Diagnostisches Problemlösen mit Expertensystemen. Springer-
    Verlag, Berlin 1986

[11] Reiter, R.: A theory of diagnostics from first principles, AI 32 (1987)
    No. 1, pp. 57-95.

Wissensbasierte Unterstützung der Musterzulassung

Ein Beitrag zur Verifikation und Validation
von Avionik-Software

Klaus-Peter Aupperle
Dr. Reinhard Rößler

ESG Elektronik-System-Gesellschaft mbH
Vogelweideplatz 9

8000 München 80

1. EINFÜHRUNG

Durch die steigenden funktionellen Anforderungen an zukünftige Luftfahr-
zeuge, die in zunehmendem Maße den Einsatz von software-kontrollierten
Systemen erfordern, werden auch wesentliche Avionik-Funktionen, die
Einfluß auf die Flugsicherheit besitzen, durch Software-Komponenten übernommen.

Gerade bei Avionik-Software kommt daher der Verifikation, nämlich der
Prüfung der Korrektheit der Software gegen die Spezifikation, und der Validation, nämlich der Prüfung, ob die Software den Absichten des Nutzers
entspricht, besondere Bedeutung zu: Avionik-Software betrifft in hohem
Maße flugkritische Funktionen, deren Ausfall zu Beschädigungen des Luft-
fahrzeuges und zu Verlust von Menschenleben führen kann.

In diesem Beitrag wird ein Studienergebnis vorgestellt, mit dem ein Teilaspekt
dieses umfassenden Themas unterstützt wird, nämlich der Musterzulassung als
dem formalen und juristisch relevanten Prozeß der Zulassung eines Flugzeug-
musters mit einer speziellen Software-Version.

2. ZIELSETZUNG UND PROBLEMSTELLUNG

Der Prozeß der Musterzulassung ist dadurch gekennzeichnet, daß der Muster-
zulasser eine Systementwicklung begleitet und dabei das erforderliche
Vertrauen gewinnt, um die Zertifizierung zu erteilen.
Dabei müssen Entscheidungen häufig auf der Grundlage unvollständiger und
unsicherer Information getroffen werden, die auf Grund der Erfahrung des
Musterzulassers bewertet werden.

Die Basisinformationen umfassen dabei
- Wissen über die Systemanforderungen
- Wissen über die Vorschriften
- Wissen über den Entwicklungsprozeß von Software.

Den grundsätzlichen Ablauf der Entwicklung von (Avionik-)Software zeigt Bild 2-1.

Hinzu kommen in hohem Maße die Erfahrungen des Musterzulassers.

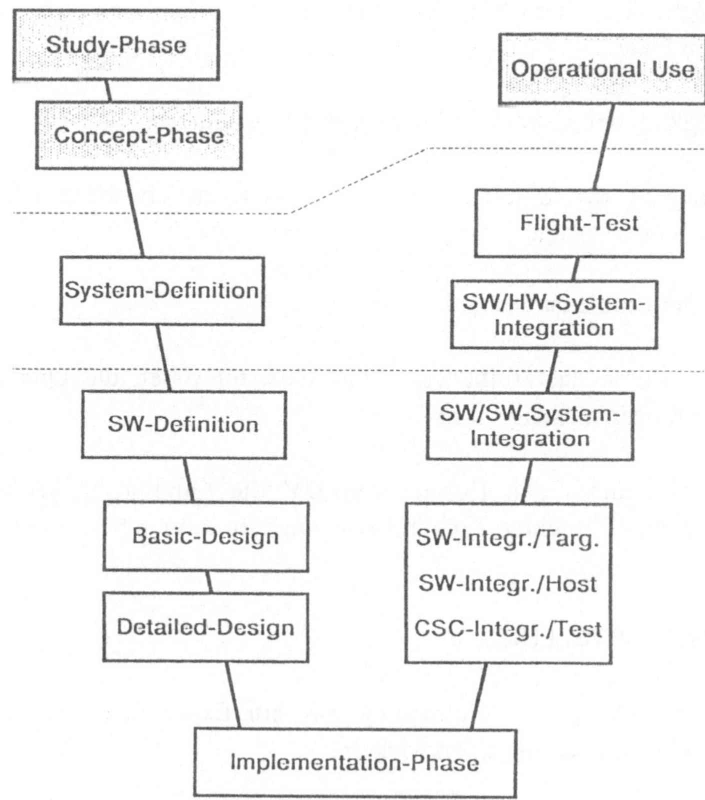

Bild 2-1: Phasenmodell Software-Entwicklung

3.   LÖSUNGSANSATZ

Die Problemstellung zeigt auf, daß das gesetzte Ziel mit Hilfe der Technologie wissensbasierter Systeme erreichbar ist. Allgemein zeichnen sich derartige Systeme aus durch:

- eine Trennung des Wissens über die generellen Lösungsstrategien von dem Faktenwissen des jeweils vorliegenden Problems

- die Möglichkeit der Berücksichtigung von unsicheren bzw. unvollständigen Daten

- die Fähigkeit, den gewählten Lösungsweg zu erläutern

- die Fähigkeit des Systems, weiteres Wissen zu erwerben, ohne neu strukturiert zu werden.

Zielsetzung der Studie ist deshalb,

- einerseits durch das Aufbereiten von Expertenwissen die Qualität von Entscheidungen zu steigern und

- andererseits durch den Einsatz von DV die Qualität in großen und umfangreichen Projekten auch in Belastungssituationen beizubehalten.

4.   AUFGABENSTELLUNG

Für die Untersuchung der Zielsetzung war ein Expertensystem mit einer konkreten Aufgabenstellung zu entwickeln.

Hierzu waren folgende Leistungen gefordert:

- Dokumentation der Systemanforderungen, insbesondere der Anforderungen, die sicherheitskritische Bereiche berühren

- Kenntnis der für Verifikation, Validation und Zulassung relevanten Vorschriften und allgemeinen Zusammenhänge (Erfahrung des Zulassers)

- Dokumentation der Vorgehensweise bei der Software-Erstellung, d.h. Festhalten von Projektzustand, Ergebnisse von Reviews, Tests usw.

- Beratung des Zulassers durch
  * Anfragen des Zulassers (XPS ist passiv; Informationretrieval)
  * Hinweise an den Zulasser (XPS ist aktiv; Prüfungen, Bewertungen)

Das Expertensystem unterstützt im Rahmen der Musterzulassung

- die Vereinnahmung und Prüfung von zertifizierungsrelevanten Informationen aus Dokumenten,

- die Beratung des Musterzulassers hinsichtlich Abweichungen von Vorgehensweise und Richtlinien, fehlender Information und Bewertung der Produktqualität,

- die Verfolgung von Änderungen und

- die Dokumentation der Vorgehensweise bei der Software-Entwicklung.

5. DURCHFÜHRUNG

5.1. Wissenserfassung

Die Wissenserfassung gliederte sich in folgende Teile:

- Erarbeitung des Grundlagenwissens aus den vorhandenen Vorschriften

- Einbeziehung des konkreten Erfahrungswissens von Musterzulassern und Entwicklern von Avionik-Software

Hierzu wurden nach der Erarbeitung des Vorschriftenwissens Mitarbeiter der zuständigen Dienststellen und Mitarbeiter aus Avionik-Software-Entwicklungsteams befragt. Während zwei Validierungstests in der Mitte und zum Ende der Realisierung mit Experten aus dem Behördenbereich wurde das Erfahrungswissen ergänzt, detailliert und strukturiert.

5.2. Spezifikation

Im Rahmen der Spezifikation wurden folgende Teilaufgaben bearbeitet:

- Abgrenzung des Anwendungsbeispiels

- Erarbeitung des Systemkonzepts

  Das System unterstützt den Zulasser bei der Auswertung und Prüfung der im Verlauf des Projektes entstandenen Dokumente (siehe Bild 5-1). Mittels der dabei gewonnenen Informationen berät es ihn dann bei der im Rahmen der Zertifizierung zu treffenden Zulassungsaussage.

- Definition der Nutzerschnittstelle

- Strukturierung der Wissensbank und Erarbeitung erster Wissensinhalte

- Konzipierung der Schlußfolgerungsmechanismen und der generellen Lösungsverfahren

- Festlegung von Anforderungen an die Wissenserwerbskomponente

- Festlegung der Entwicklungsumgebung für die Realisierung

Für die Entwicklung wurde als Ziel- und Entwicklungsumgebung für den Prototypen ein PC AT 386 mit der Expertensystem-Shell ART-IM ausgewählt.

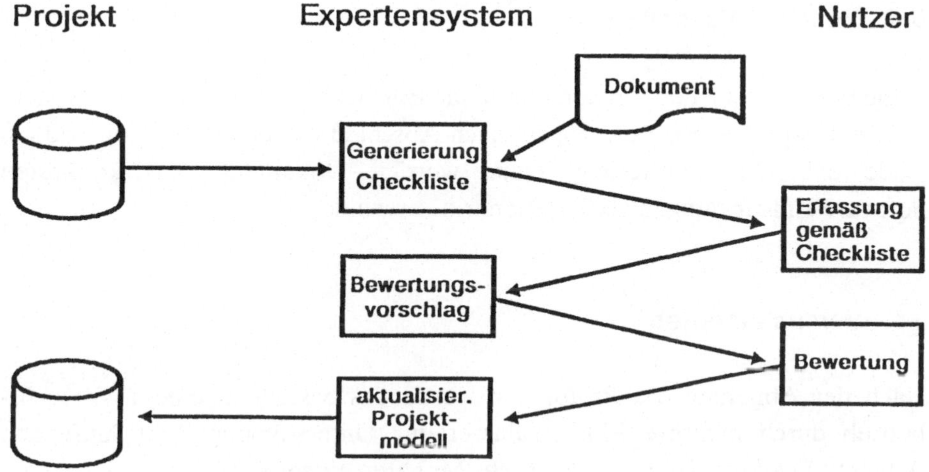

5.3. Realisierung

Die Realisierung erfolgte stufenweise in folgenden Schritten:

- Realisierung eines Systemkerns, der bereits alle vorgesehenen Komponenten als Platzhalter enthält, sodaß die Kommunikation und das Zusammenspiel der Systemkomponenten von Beginn der Entwicklung gesichert war.

- Realisierung von zwei aufeinander aufbauenden Versionen.

Jede Version war in sich abgeschlossen und lauffähig und lieferte jeweils vollständigere Ergebnisse. Mit der Realisierung der Folgeversion wurde erst begonnen, wenn die aktuelle Version stabil lief. Jede der zwei Versionen bot die Möglichkeit von Zwischentests des Gesamtsystems.

5.4. Test und Validation

Neben den Tests jeder Version durch das Entwickler-Team wurde das System zweimal Expertentests unterzogen, nach Abschluß der ersten Version und am Ende der Implementierung. Dazu wurden Dokumente aus konkreten Software-Entwicklungen als Testszenario verwendet.

5.5. Versuchsbetrieb

Nach der Abnahme des Prototypen wurde das System in einem Versuchs-betrieb durch mehrere Musterzulasser zur Unterstützung ihrer laufenden Arbeiten in einem Zeitraum von sechs Monaten eingesetzt.

Damit wurde eine Bewährung des Prototypen im Feldtest erreicht und gleich-zeitig die Akzeptanz bei einer größeren Gruppe von Endnutzern erworben.

6. ERGEBNISSE

Bereits der Prototyp kann die Aufgaben des Musterzulassers im Rahmen der Zertifizierung wirksam unterstützen und damit die Qualität von Entschei-dungen über die Zulassung von Avionik-Software dadurch steigern, daß

- noch ungeübte Nutzer mit Expertenwissen unterstützt werden,

- der Nutzer entlastet wird und dadurch mehr Zeit für wesentliche Entscheidungen hat und

- die Bewertung der Produktqualität auf dem aktuellen Projektstand beruht, der insbesondere bei sehr großem Informationsaufkommen auch die Details und Verknüpfungen enthält, die der Musterzulasser in der Regel derzeit nicht berücksichtigen kann (Mengenproblem).

Gleichzeitig wird durch den Einsatz von DV-Gerät die Qualität in Belastungssituationen erhalten.

Der von Fa. ESG realisierte Prototyp beinhaltet alle Komponenten der geforderten Systemunterstützung und stellt eine Basis für weiteren Ausbau zur Verfügung.

Auszubauen sind im Wesentlichen:

- die Wissensbank, um die Einschränkungen der Aufgabenstellung aufzufüllen.

  Dabei ist die Struktur der Wissensbank mit derselben Vorgehensweise zu erweitern, wie sie zur Erstellung des Prototypen eingeschlagen wurde. Das abgelegte Wissen ist dabei durch den ständigen Kontakt mit dem Nutzer zu ergänzen.

- die Wissenserwerbskomponente, um die Software-Pflege und -Änderung durch den Endnutzer selbst zu ermöglichen und

- die Hardware, um akzeptable Reaktionszeiten des Systems und die Bearbeitbarkeit umfangreicher Projekte zu erhalten.

  Hierbei sind neben systemtechnischen Anforderungen für das Expertensystem die künftige DV-Ausstattung der mit der Musterzulassung betrauten Dienststelle, das damit verbundene Datenbankkonzept sowie die Anbindung an Systeme anderer Institutionen wie beispielsweise der örtlichen Güteprüfstellen oder der Herstellerfirmen zu berücksichtigen.

7. REFERENZDOKUMENTE

/1/ AQAP - 13
NATO SW Quality Control System Requirements

/2/ AFISC SSH 1-1
Software System Safety

/3/ BDLI - Studie
Prüfung und Zulassung von softwareabhängigen Systemen in Luftfährzeugen

/4/   BMVg Org 1 - Az: 61-01
      Richtlinie für SW Dokumentation von DV-Anteilen in Wehrmaterial
      SWDokWM

/5/   BWB - ML: Luftfahrt-Tauglichkeits-Richtlinie-Software
      Entwicklung, Prüfung und Zulassung von SW für Rechner in Luftfahrzeugen
      der Bundeswehr
      LTR 7030 - 001

/6/   BesAnLwUKdo 006/81
      Besondere Anweisung des Luftwaffenunterstützungskommandos, Verfahren
      für die Softwarepflege und -änderung von DV-Anteilen in fliegenden
      Waffensystemen

/7/   DoD-Std 2167A
      Defense System Software Development

/8/   DoD-Std 2168 (Draft)
      Software Quality Evaluation

/9/   BWB FE VI 2
      Software-Entwicklungsstandard der Bundeswehr
      Vorgehensmodell, Version 2.0

/10/ MIL-Std 1521
      Technical Reviews and Audits for Systems and Computer Programs

/11/ MIL-Std 1574A
      System Safety Program for Space and Missile Systems

/12/ RTCA/DO - 178 A
      Software Considerations in Airborne Systems and Equipment Certification

/13/ ZDV 19/1
      Das Prüf- und Zulassungswesen für Luftfahrtgerät der Bundeswehr

/14/ Jude E. Franklin u.a.: Expert System Technology for the Military, Proceedings
      of the IEEE, Oktober 1988

/15/ The Military Software Market in the US, Frost & Sullivan Inc., 1979

/16/ Robert Balzer u.a.: Software Technology in the 1990's, Using a New
      Paradigm, IEEE Computer, November 1983

BETRIEBSSYSTEM-FUNKTIONEN IN EINEM FEHLERTOLERANTEN ECHTZEITSYSTEM

M. Bayer, K. Gresser
Lehrstuhl für Prozeßrechner, Prof. Dr. G. Färber
TU München

Zusammenfassung

Dieser Bericht stellt die Architektur und das Betriebssystem eines fehlertoleranten Echtzeitsystems vor. Neben einem funktionsfähigen System wurde am Lehrstuhl für Prozeßrechner der TU München ein leistungsfähiges Simulationswerkzeug entwickelt, mit dem es möglich ist, das funktionelle und zeitliche Verhalten verschiedener Betriebssystemvarianten in Verbindung mit Applikationslösungen zu untersuchen. Verschiedene Betriebssystem-Funktionen werden gegenübergestellt.

Das Projekt wird von der Deutschen Forschungsgemeinschaft im Rahmen eines Schwerpunkt-programms gefördert.

1. Einleitung

Der zunehmende Einsatz von Mehrrechnersystemen auch auf relativ prozeßnahen Hierarchiestufen bei Prozeßleitsystemen bietet eine gute Möglichkeit, die vorhandene Parallelstruktur nicht nur zur Steigerung der Systemleistung einzusetzen, sondern sie auch zur Implementierung von Fehlertoleranz zu nutzen.

FUTURE ("Fault tolerant and Userfriendly system with Taskspecific Utilisation of REdundancy") ist ein Echtzeitsystem mit software-implementierter Fehlertoleranz, das auf dem Konzept taskspezifischer Zuverlässigkeitsklassen basiert [2]. Die Verwendung mehrerer Standardrechner dient sowohl der Leistungssteigerung als auch der Zuverlässigkeitserhöhung. Dazu werden vom Betriebssystem Tasks mit variablen Zuverlässigkeitsklassen (variable Anzahl von Replikaten) verwaltet, so daß eine Anpassung des Mehrrechnersystems an die unterschiedlichen Zuverlässigkeitsanforderungen einzelner Prozeßpunkte möglich ist. Außer der Angabe der Taskklasse bleibt die Fehlertoleranz für den Anwender unsichtbar.

Ziel dieses am Lehrstuhl für Prozeßrechner durchgeführten Projekts ist zum einen die Untersuchung der Funktionen des fehlertoleranten Betriebssystems, zum anderen die Unterstützung bei der Erstellung von Anwendungen.

2. Systemarchitektur und Untersuchungswerkzeuge

2.1 FUTURE im Überblick

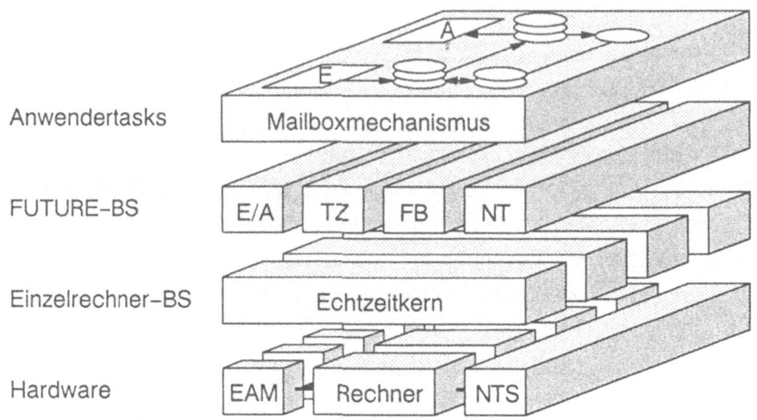

Bild 1 : FUTURE Modell

Die unterste Ebene des in Bild 1 gezeigten Modells zeigt die Hardware. Mehrere Standard-Rechnereinheiten sind über ein Nachrichten-Transport-System miteinander gekoppelt und über Ein/Ausgabe-Modulen an den Prozeß angeschlossen. Auf jedem Rechner läuft ein Echtzeitkern, auf dem das rechnerübergreifende FUTURE-Betriebssystem aufsetzt. Die Funktionen Ein/Ausgabe, Taskzuteilung, Fehlerbehandlung und Nachrichtentransport werden später vorgestellt. Auf oberster Ebene liegen die Anwendertasks, die über einen Mailbox-Mechanismus (SEND, RECEIVE) miteinander kommunizieren. Die taskklassenspezifische Votierung der Nachrichten ist Aufgabe des Betriebssystems. Bei der Codierung von Anwendertasks braucht die Fehlertoleranz nicht berücksichtigt werden.

FUTURE ist am Lehrstuhl für Prozeßrechner als reales System und als Simulator aufgebaut. Der Simulator ermöglicht kurze Entwicklungszeiten für die Betriebssystem-Funktionen, die am realen System verifiziert werden. Die Zeiten werden am realen System gemessen und in den Simulator zurückgeführt.

2.2 realisiertes System

Bild 2 zeigt den Aufbau des realisierten Systems. Vier Rechner sind über ein Kommunikationssystem gekoppelt. Ein fünfter Rechner dient als Entwicklungseinheit; er ermöglicht Download, Modula-Source-Level-Debugging und Terminalemulation für den Mehrrechnerbetrieb.

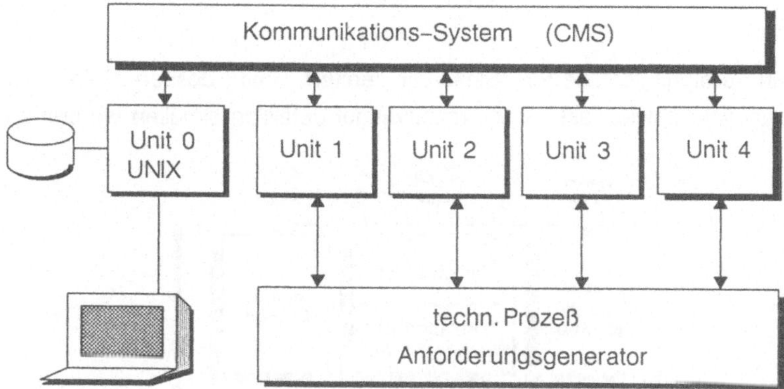

Bild 2 : realisiertes System

Auf dem realen System ist ein vollständiges fehlertolerantes Betriebssystem implementiert, wobei man sich hier auf nur jeweils eine Betriebssystem-Variante beschränkt hat [12,14]. Als Lastgenerator dient ein eigener Rechner, der bis zu 64 Prozeßgrößen simuliert (Bild 3) [6].

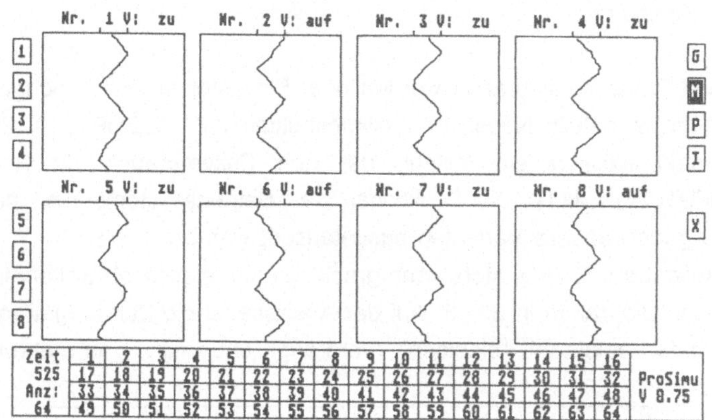

Bild 3 : Lastgenerator

Aus der Sicht des Rechnersystems verhält sich dieser Lastgenerator wie ein Feldbus, der bis zu 64 Ein/Ausgabemodule adressieren kann. Damit ist es möglich, die Ein/Ausgabekonzepte zu implementieren und zu testen.

Zum Test und zur Veranschaulichung der Fehlertoleranzeigenschaften injiziert ein Fehlergenerator mit variabler statistischer Verteilung Fehler durch Verändern von Speicherinhalten. Die Reaktion des Systems auf Fehler kann so beobachtet werden. Der Lastgenerator zeigt dann graphisch, daß nur diejenigen Prozeßpunkte einen Fehler überleben, die von fehlertoleranten Tasks gesteuert werden.

2.3 Simulator

Es wurde ein ereignisgesteuerter Simulator erstellt, mit dessen Hilfe Hardware- und Betriebssystemstruktur sowie Task- und Prozeßverlauf detailliert simuliert werden können (Bild 4).

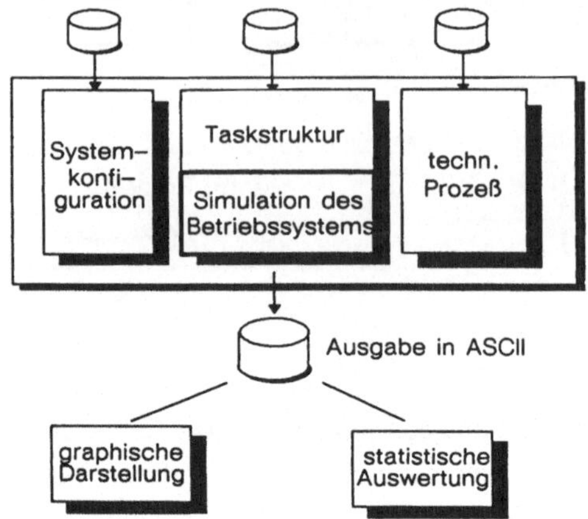

Bild 4 : FUTURE – Simulator

System-, Task- und Prozeßkonfiguration werden dem Simulator in einer hochsprachenähnlichen Beschreibungssprache mitgeteilt. Neben Kontrollkonstrukten ('IF', 'LOOP', ...), Zuweisungen und Betriebssystemaufrufen existiert ein Befehl, der den Zeitverbrauch für Rechenoperationen beschreibt ('COMPUTE( )') [11]. Die Funktionen des Betriebssystems sind programmiert und können als Module gegen andere Varianten ausgetauscht werden.

Die tabellarische Ausgabe des Simulators kann graphisch und statistisch weiterverarbeitet werden. Beispiel hierfür ist das Diagramm in Bild 5. Auf der Abszisse ist die Zeit aufgetragen; die Ordinate zeigt die Taskzustände 'fehlerhaft', 'blockiert', 'verdrängt' und 'aktiv' von mehreren Tasks.

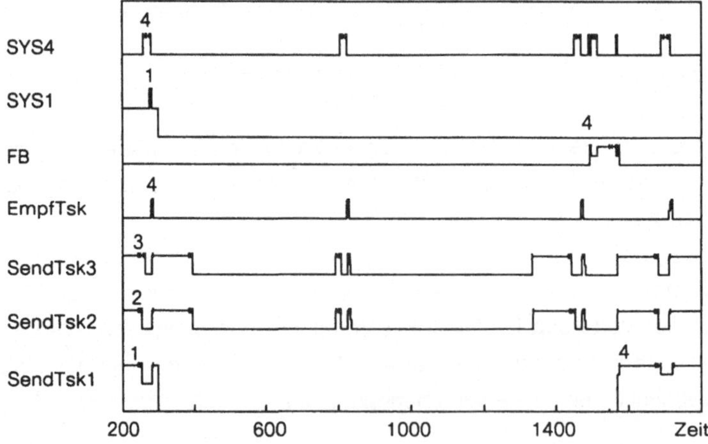

Bild 5 : Beispiel für die Simulations-Aufbereitung

3. Betriebssystem

Das entwickelte Betriebssystem ist in die Blöcke Nachrichtentransport, Taskzuteilung, Fehlerbehandlung und Ein/Ausgabe untergliedert.

3.1 Nachrichtentransport

Der Nachrichtentransport wird von dem SEND-Aufruf initiiert. Die rechnerübergreifende Kommunikation ist Aufgabe eines intelligenten Nachrichtentransport-Systems [7]. Wegen der Forderungen nach Rückwirkungsfreiheit, Redundanz und Broadcast-Unterstützung wurde die im Bild 6 gezeigte Netz-Topologie gewählt [13].

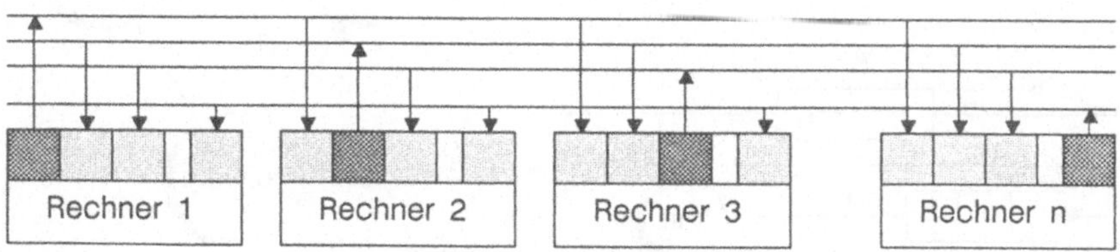

Bild 6 : Topologie des Nachrichtentransports

Jeder Rechner besitzt einen Kommunikationsspeicher, der in Bereiche unterteilt ist. Schreibzugriffe auf den privaten Speicherbereich werden in die entsprechenden Speicherfenster aller anderen Rechner automatisch übertragen. Dazu besitzt jeder Rechner einen Bus mit alleinigem Schreibrecht. Somit sind Systeminformationen global verfügbar und die Gefahr einer Verfälschung durch defekte Einheiten ausgeschlossen [4]. Die Daten werden in den Empfängern votiert, in die Nachrichten-Warteschlange eingetragen und der Hauptprozessor über die eingetroffene Nachricht informiert. Auch das Synchronisieren von Cotasks und Starten der Fehlerbehandlung bei erkannten Fehlern sind Aufgaben des Nachrichten-Transports.

3.2 Taskzuteilung

Es wurden Betriebssystem-Funktionen für asynchrones lokales und globales Scheduling untersucht. Asynchron bedeutet hier, daß die Cotasks nicht gleichzeitig auf den verschiedenen Rechnern zugeteilt werden (Zeitscheiben), sondern daß der Scheduler jedes Rechners unabhängig von den anderen die nächste lauffähige Cotask bestimmt. Untersuchungen von synchronem Scheduling [9] zeigten hohe Verluste durch ungenutzte Zeitscheiben und hohen Kommunikationsaufwand, so daß es hier nicht behandelt wird. Bei den asynchronen Zuteilungsverfahren erfolgt die Synchronisierung zusammengehörender Cotasks an natürlichen Synchronisationspunkten, den SEND-Aufrufen. An diesen Punkten wird über die gesendete Nachricht votiert.

3.2.1 asynchrones lokales Scheduling

Bei lokalem Zuteilungsmechanismus führen die Rechner unabhängig voneinander – jedoch nach gleichen Kriterien – Scheduling durch. Es wird immer die lauffähige Cotask mit der höchsten Priorität zugeteilt. Die Rechnerzuteilung der Tasks ist im fehlerfreien Fall statisch, wobei auf jedem Rechner ein eigener Zuteilungskreislauf stattfindet (Bild 7). Die Taskzustände 'running', 'blocked' und 'ready' entsprechen den Zuständen herkömmlicher Echtzeit-Betriebssysteme.

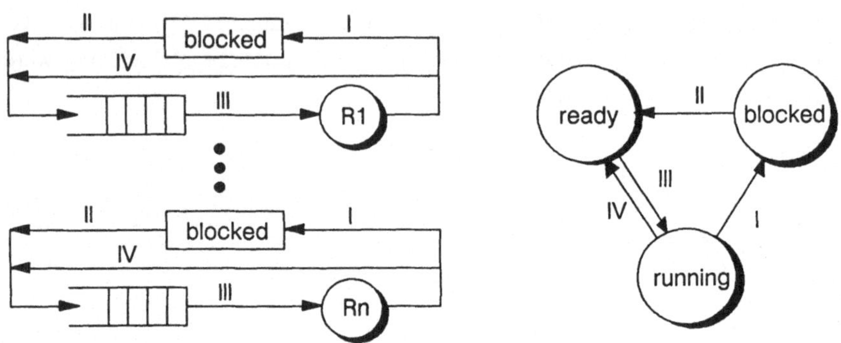

Bild 7: Zustandsübergänge bei lokalem Scheduling

3.2.2 asynchrones globales Scheduling

Im Gegensatz zur Taskzuteilung mit lokalen Warteschlangen wird die Taskzuteilung auf die Rechner nicht festgelegt. Vielmehr werden Cotasks aus einer übergeordneten, globalen (redundant verteilten) Warteschlange den gerade freien Rechnern zugeteilt. Zur Ermöglichung der Fehlerlokalisation und Vermeidung der Fehlerausbreitung müssen jedoch zwei Einschränkungen beachtet werden:

- Eine Cotask darf nur dann den Rechner wechseln, wenn deren Fehlerfreiheit verifiziert wurde, und
- ein Rechner darf jeweils nur eine Cotask einer Taskgruppe bearbeiten

Als Folge dieser Einschränkungen existiert auf jedem Rechner neben der globalen eine lokale Warteschlange, in der die Cotasks solange verweilen, bis sie die Voraussetzungen für den Rechnerwechsel erfüllen (Bild 8).

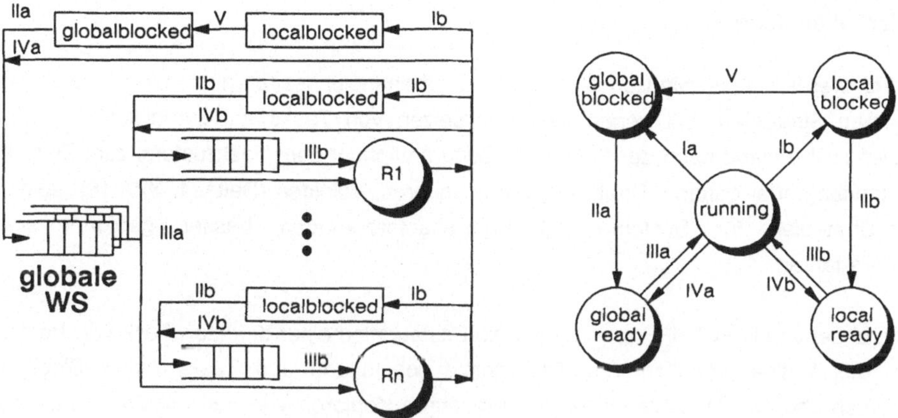

Bild 8 : Zustandsübergänge bei globalem Scheduling

Bei diesem Modell ist die (kommunikationsaufwendige) Verwaltung der globalen Warteschlange nicht immer nötig; vielmehr werden Tasks, die ohnehin den Rechner momentan nicht wechseln dürfen, lokal behandelt.

Ausgewählt wird nun die lauffähige Cotask mit der höchsten Priorität, die nicht gegen o.g. Einschränkungen verstößt. Die bisherigen Taskzustände sind nun nicht mehr ausreichend; es mußten die neuen Zustände 'global blocked' und 'global ready' definiert werden [15].

3.2.3 Bewertung

Das globale Scheduling ist komplizierter als das lokale. Aus diesem Grund ist der Betriebssystem-Overhead beim globalen Scheduling größer.

Der offensichtlichste Vorteil des globalen Schedulings ist die gleichmäßige Rechnerauslastung.

Ein weiterer Vorteil von globalem Scheduling besteht in der besseren Reaktionsfähigkeit des gesamten Systems, da im allgemeinen die Wartezeiten der Tasks durch die gleichmäßige Aufteilung auf die Rechner geringer sind. Auch die Synchronisationszeiten zusammengehörender Cotasks sind kürzer als bei lokalem Scheduling.

Da die Umverteilung von Tasks bei einem Rechnerausfall vom globalen Scheduling automatisch durchgeführt wird, müssen nicht - wie beim lokalem Scheduling - zusätzliche Rekonfigurations-zeiten mitberücksichtigt werden. Nicht zu vergessen ist die größere Anwenderfreundlichkeit der globalen Variante der Taskzuteilung, da die Verteilung der Task auf die Rechner automatisch erfolgt. Das gleiche gilt auch für die Neuverteilung bei einem Rechnerausfall.

3.3 Fehlerbehandlung

In FUTURE werden ausschließlich Vorwärts-Fehlerbehandlungsstrategien angewendet. Darunter versteht man Strategien, bei denen kein Rücksetzen von Tasks auf zurückliegende Zustände erfolgt. Der Taskzustand nach der Fehlerbehebung entspricht dem Taskzustand zum Zeitpunkt der Fehlererkennung oder danach. Erfahrungen aus anderen Projekten (Delta-4, XPA [8]) zeigen, daß Vorwärts-Strategien für Systeme mit Echtzeitanforderungen besser geeignet sind als Rollback-Verfahren.

Die Fehlerbehebung in FUTURE sieht zuerst die Restauration einer defekten Task vor. Restauration bedeutet das Wiederherstellen des Taskkontext (bestehend aus Task-Kontroll-Block, Stack, lokalen Daten und den taskeigenen Mailboxen) durch Kopieren von den intakten Cotasks. Dabei geht man von transienten Fehlern aus. Gelingt die Restauration nicht, oder weist die Art des Fehlers oder die Fehlergeschichte auf einen permanenten Fehler hin, wird die Rekonfiguration eingeleitet; sie besteht in der schrittweisen Migration der betroffenen Cotasks auf andere Rechner.

Von besonderer Bedeutung ist der Zeitpunkt der Fehlerbehebung. Sollen Fehler sofort bei Erkennung auf höchster Priorität behoben werden, müssen die Zeiten dafür bei Berechnung der Echtzeitbedingungen berücksichtigt werden. Das kann zu niedrigen Rechnerauslastungen führen, da für Reaktionen auf mögliche Fehler Reserven vorzusehen sind. Nutzt man natürliche Blockierzeiten oder Idle-Zeiten für die Fehlerbehebung, so verringert sich bzw. verschwindet der negative Einfluß auf das Echtzeitverhalten, jedoch steigt durch das temporäre Zurückstufen der Zuverlässigkeitsklasse die Gefahr eines Doppelfehlers, der zum Systemzusammenbruch führt. Folgende Tabelle stellt 6 Strategien gegenüber, die sich im Zeitpunkt und in der Priorität der Fehlerbehebung unterscheiden [5]:

| Strategie | Zeitpunkt der Fehlerbeheb. | Priorität der Fehlerbeh. | Auswirkungen auf | | | Zeit zwischen Fehlerentdeckung und -Behebung |
			höherere Priorität	fehlerhafte Task	niedere Priorität	
1	sofort	System	ja	ja	ja	minimal
2	sofort	Task	nein	ja	ja	abh. von höher-prioren Tasks
3	nächstes Blockieren	System	ja	ja	ja	abh.von fehlerh. Task
4	nächstes Blockieren	Task	nein	ja	ja	abh.von fehlerh. und höherprioren Tasks
5	ausreichend langer Blockier-zustand	Task	nein	nein	ja	abh.von fehlerh. und höherprioren Tasks
6	ausreichend langer Idlezustand	Idle	nein	nein	nein	abh. von allen Tasks

3.4 Prozeß Ein- und Ausgabe

Zum Anschluß von Prozeßperipheriegeräten erhalten Feldbusse eine große Bedeutung [3]. Die Problematik der Prozeßankopplung an fehlertolerante Rechnersysteme wird in [1] behandelt. Feldbusse für diesen Einsatzzweck müssen echtzeitfähig, multimasterfähig und redundant ausgelegt sein. Sensoren und Aktoren sind dabei über intelligente E/A-Module an den Bus angeschlossen.

Die hier durchgeführten Untersuchungen stützen sich auf das oben genannte Modulkonzept. Die Modulen erhalten die Aufgabe des Votierens, womit sich der Wirkungsbereich der Fehlertoleranz weit bis an den Prozeß erstreckt.

Die Ein/Ausgabe wird als eigene Klasse-3-Task ausgeführt. Ausgabe- und Einleseaufträge werden von beliebigen Tasks über den Aufruf 'SEND' an die E/A-Task geschickt. Eingelesene Werte werden mit 'RECEIVE' empfangen, optional auch Rückmeldungen von Ausgabeaufträgen. Ereignisse, die von den Modulen gemeldet werden, gibt die E/A-Task ebenfalls mit dem Mailbox-Mechanismus an die Anwendertasks weiter [10].

Durch die Verwendung einer E/A-Task ist es möglich, auf die bereits in den Betriebssystemaufrufen 'SEND' und 'RECEIVE' enthaltenen Funktionen der Fehlerbehandlung und des Schedulings aufzusetzen, ohne diese neu entwickeln zu müssen. Ebenso bereitet die Überbrückung von Rechnergrenzen bei der Ein/Ausgabe keinerlei Mehraufwand.

4. Ausblick

Das FUTURE-System mit dem dazugehörigen Simulator ermöglicht es, Aussagen über den durch Fehlertoleranz verursachten Systemoverhead und den daraus resultierenden veränderten Reaktionszeiten in Abhängigkeit verschiedener Betriebssystemvarianten zu ermitteln. Zur Zeit wird an einer weiteren Taskzuteilungsvariante, dem reaktionszeitgesteuerten Scheduling gearbeitet, das wiederum in einer lokalen und globalen Version entwickelt wird.

Literatur

[1]   Endl H., 'Prozeßankopplung in fehlertoleranten Rechnersystemen', Dissertation, Fakultät für Elektrotechnik, TU München, 1988

[2]   Färber G., 'Task-Spezific Implementation of Fault Tolerance in Process Automation Systems', in Dal Cin, Dilger (Eds.), Workshop on Self-Diagosis and Fault-Tolerance, Attempto Verlag Tübingen, 1981

[3]   Färber G., 'Entwicklungstendenzen der Mikroelektronik und der Informationstechnik', atp 31 (1989) 9, S.400 ff, 1989

[4]   Gresser K., 'Broadcast Communication in Fault Tolerant Multicomputer Systems', occam user group, 13-th technical meeting, York, England, 1990

[5]   Hirte J., 'Analyse und Simulation der Fehlerbehandlung im FUTURE-System', Diplomarbeit, Lehrstuhl für Prozeßrechner, TU München, 1989

[6]   Klingler M., 'Simulation und Steuerung eines technischen Prozesses', Studienarbeit, Lehrstuhl für Prozeßrechner, TU München, 1990

[7]   Fischbacher P., Gresser K., Koller G., Stein A., Triller M., 'Basisfunktionen für die UNIX- Implementierung in einem fehlertoleranten Multimikrorechnersystem', Abschlußbericht des DFG-Forschungsvorhabens Fa 109/8, Lehrstuhl für Prozeßrechner, TU München, 1989

[8]   Powell D. (Editor), 'Delta 4, Overall System Specification', Delta-4 Project Consortium, BULL-SA, Echirolles, Frankreich, 1988

[9]   Ries W., 'Prozessorzuteilungsverfahren in fehlertoleranten Mehrrechnersystemen', Dissertation, Fakultät für Elektrotechnik, TU München, 1985

[10]   Stürmer A., 'Fehlertolerante Prozeßsteuerung', Diplomarbeit, Lehrstuhl für Prozeßrechner, TU München, 1990

[11]   Taubert R., 'Simulation des Zeitverhaltens von Anwendertasks im fehlertoleranten Mehrrechnersystem FUTURE', Diplomarbeit, Lehrstuhl für Prozeßrechner, TU München, 1989

[12]   Thielen H., 'Fehlerbehandlung im fehlertoleranten Echtzeitsystem FUTURE – Implementierung im realen System', Diplomarbeit, Lehrstuhl für Prozeßrechner, TU München, 1990

[13]   Wiemann B., 'Kommunikationsfunktionen für fehlertolerante Mehrrechnersysteme mit busartiger Verbindungsstruktur', Dissertation, Fakultät für Elektrotechnik, TU München, 1986

[14]   Wildenburg R., 'Implementierung des FUTURE – Nachrichtenmechanismus auf Basis des CMS', Diplomarbeit, Lehrstuhl für Prozeßrechner, TU München, 1989

[15]   Wittmann P., 'Untersuchung von unterschiedlichen Schedulingstrategien', Diplomarbeit, Lehrstuhl für Prozeßrechner, TU München, 1989

Die Realisierung eines Echtzeit-UNIX für den Einsatz in der Automatisierungstechnik

Roland Weigel
Siemens AG
Bereich Automatisierungstechnik
Geschäftsgebiet Steuerungs- und Prozeßregelsysteme
Microcomputersysteme
Siemens AG AUT V151
Postfach 4848
D-8500 Nürnberg 1

1. Einleitung

In den bisherigen Echtzeitanwendungen dominieren vorwiegend kleine, auf schnelle Reaktionen angepaßte Kernel als sogenannte Echtzeitbetriebssysteme. Die Anwendungen, die auf diesen Systemen ablaufen sind meist sehr effizient programmiert und der jeweiligen Aufgabe recht gut angepaßt. Der Wunsch der Anwender nach Portabilität der Software ist mit diesem Ansatz jedoch nicht zu erfüllen. Diese Anforderung läßt sich am besten mit einem als Standard anerkanntem Betriebssystem erreichen. Als der Standard bei Betriebssystemen hat sich UNIX V durchgesetzt. Aus Effizienzgründen ist jedoch der Einsatz eines normalen UNIX V im Echtzeitbereich nicht möglich. Dies liegt vor allem an der Realisierung des Betriebssystem-Kernels. Mit Sorix wird hier einen Ansatz vorgestellt bei dem UNIX und Echtzeit in einem System vereinigt worden sind.

2. Hardware Komponenten

Das Herzstück eines Microcomputersystems SX ist der Single-Board-Computer (SBC) mit dem 32 bit RISC Prozessor 80960 von Intel mit integrierter Gleitkommaarithmetik. Die Baugruppe enthält den Hauptspeicher mit 4 oder 16 Mbyte RAM-Speicher, eine EPROM-Bank mit 256 kbyte für den Boot-Monitor, zwei serielle Schnittstellen, sowie einen Steckplatz für ein

Kommunikationsmodul. Weitere zwei Kommunikationsmodule können über eine Erweiterungsbaugruppe gesteckt werden. Folgende Kommunikationsmodule sind derzeit verfügbar:

- SCSI-Modul für den Anschluß von Festplatten, Magnetbandlaufwerken, Floppy-Laufwerken und optischen Platten
- LAN-Modul für den Zugriff auf LAN über TCP/IP und ISO-Protokolle
- RS232-TTY-Modul für den Anschluß von bis zu sechs interruptgesteuerten seriellen Schnittstellen
- HDLC-Modul für WAN-Zugang
- Grafik-Modul für den Betrieb von X Windows
- MMC-Modul für den Zugriff auf MMC 216 und SIMATIC S5 Baugruppen über Parallelbuskopplung

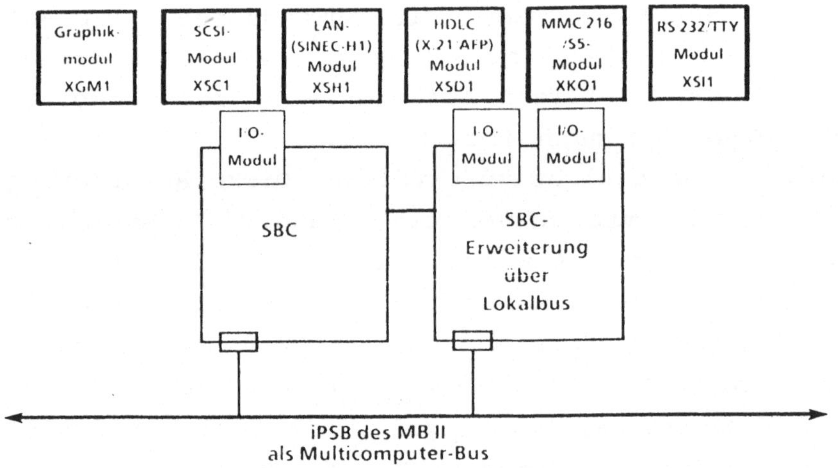

Abbildung 1: I/O-Module für SX-Baugruppen

Als Systembus verwendet das Mikrocomputersystem SX den genormten Multibus II. Über diesen Systembus können sowohl die zur Verfügung stehenden MBII Signalformerbaugruppen angesprochen werden und Multicomputing mit mehreren SBC betrieben werden, aber auch mit anderen MBII-Baugruppen kommuniziert werden.

Im Mikrocomputersystem SX werden ein Baugruppenträger mit 11
MBII Steckplätzen für den industriellen Einsatz und ein
Tischgehäuse mit 5 MBII Steckplätzen angeboten.

3. Das Echtzeit-Unix Betriebssystem SORIX

Wie bereits in der Einleitung erwähnt ist der Einsatz eines
Standard UNIX V Systems in der Echtzeitanwendung nicht
möglich. Das liegt vor allem an der fehlenden Deterministik,
die für normales UNIX typisch ist. Es kann nicht garantiert
werden, daß ein wichtiges Programm in einer vorhersagbaren
Zeit seine Abarbeitung beenden kann. Dies kann in der
Prozeßrechentechnik, wo Reaktionszeiten von wenigen Milli-
oder sogar Mikrosekunden erwartet werden, fatale Folgen
haben.
Um die gegensätzlichen Eigenschaften von UNIX und Echtzeit zu
vereinbaren gibt es verschiedene Lösungsansätze. Ein Weg ist
das Einbringen von Preemption-Points in einen Unix Kernel, um
so die Unterbrechbarkeit gewährleisten zu können. Die andere
Möglichkeit ist das Zusammenarbeiten eines UNIX und eines
Echtzeitkernels. Bei SORIX wurde ein völlig neuer Weg
beschritten.
SORIX wurde vom Design her als Echtzeitbetriebsystem
entwickelt. Bei SORIX ist der Kernel prinzipiell immer
unterbrechbar, mit Ausnahme von einigen nicht unterbrechbaren
Teilen von höchstens 20 Mikrosekunden Dauer Daher können mit
diesem Ansatz wesentlich kürzeren Interruptreaktionszeiten
erreicht werden. Trotzdem gewährleistet SORIX die volle
UNIX V.2 Kompatibilität. Neben der Verbesserung der Reaktion
auf externe Ereignisse wurden noch eine Reihe weiterer neue
Funktionen in SORIX eingebracht, die den Einsatz in der
Automatisierungstechnik erst ermöglichen. Diese SORIX-
Erweiterungen werden in den folgenden Abschnitten
vorgestellt.

3.1 Scheduling

Bei SORIX existieren insgesamt 32 Prioritäten, von denen 16 für Anwenderprozesse und 16 für Interruptprozeduren zur Verfügung stehen. Der Anwender hat die Möglichkeit eine Grenze zwischen Echtzeitprozessen und normalen UNIX Prozessen festzulegen. Alle Prozesse deren Priorität über dieser Schwelle liegt, werden als Echtzeitprozesse mit fester Priorität behandelt. Liegt die Priorität eines Prozesses unterhalb dieser Schwelle ist dieser Prozess ein UNIX-Prozeß und unterliegt den für UNIX typischen Prioritätenveränderungen (Fair Scheduling).

Für das Scheduling von Echtzeitprozesen können die Hardware Eigenschaften des Prozessors ausgenutzt werden ohne daß ein Software-Scheduler zum Einsatz kommen muß. Dies führt zu den sehr schnellen Prozeßwechselzeiten bei SORIX von unter 100 Mikrosekunden.

Weitere Prozeßattribute, die SORIX von UNIX unterscheiden, sind die Eigenschaften hauptspeicherresident, preemptive und ohne Zeitscheibe. Hauptspeicherresidente Programme unterliegen nicht dem SORIX Demand Paging und garantieren somit schnellere Reaktionszeiten, da der möglicherweise notwendige Eintransfer von ausgelagerten Seiten entfällt. Das Attribut Preemptive dient dazu um höherpriore Prozesse in die Lage zu versetzten sofort einen niederprioren Prozess zu verdrängen ohne auf dessen Unterbrechung warten zu müssen.

Die Prioritäten und Prozeßeigenschaften können mit Hilfe eines neuen Systemaufrufs "mkproc" an das Programm übergeben werden. Dieser Aufruf hat die Funktionalität der beiden Aufrufe "fork" und "exec" vereinigt und erreicht neben einer eleganteren Programmierung auch eine deutliche Performance-Steigerung, da das Umkopieren des Prozeßabbildes beim Fork entfällt.

3.2 File-System

Bei der Implementierung des File-Systems sind eine Reihe von
Optimierungen für Echtzeit-Anforderungen in SORIX eingebracht
worden. Eine Möglichkeit Kopf-Positionier-Zeiten zu sparen
sind zusammenhängende Files (contiguous Files). Hier liegen
im Gegensatz zu Standard-UNIX-Filesystemen die Datenblöcke
eines Files physikalisch unmittelbar nebeneinander.
Die Funktionen der Inodes bei UNIX werden bei SORIX durch
einen Index-Record übernommen. Dieser ist größer als ein UNIX
Inode und kann damit mehr Verweise auf Datenblöcke aufnehmen.
Hierdurch wird erreicht, daß das SORIX-Filesystem auf größere
Filelängen optimiert ist, die bei Echtzeit-Anwendungen häufig
vorkommen.
Eine weitere Eigenschaften von SORIX ist der sog. RAW-I/O,
also ein direkter Zugriff auf Platte unter Umgehung der
Hauptspeicher-Pufferung, nicht nur auf Platten-Partition-
Ebene wie bei Standard UNIX-Implementierungen, sondern auch
file-spezifisch möglich. Hierdurch kann der Anwender bei
wichtigen Files sicher sein, daß geschriebene Daten auch
tatsächlich sofort auf der Platte stehen, wenn dies aus
Sicherheitsgründen wichtig ist. Zusätzlich können mit Hilfe
des RAW-I/O größere Puffer als ein Plattenblock mit einem
Treiber-Auftrag transferiert werden. Als zusätzliche Er-
weiterung kann bei SORIX auf Raw-I/O bearbeitete Files
asynchron zugegriffen werden.
Als echtzeitorientiertes Datenhaltungspaket wird zu SORIX
X/ISAM angeboten. X/ISAM bietet aufsetzend auf SORIX Files
eine Library für einen index-sequentiellen, satzorientierten
Zugriff auf Daten. Für X/ISAM Dateien sind variable
Satzlängen möglich.

3.3 Interprozeßkommunikation bei SORIX

Ursprünglich enthielt UNIX als Mittel für die Interprozeßkommunikation nur Pipes und Signals. Da dies selbst für

Standard UNIX Anwendungen nicht ausreichend ist, wurden in der System V Kernel-Extension Messages, Semaphore und Shared Memory neu aufgenommen. Für Echtzeitanwendungen ist dies aber immer noch nicht ausreichend. Deshalb wurde in SORIX die Interprozeßkommunikation um Events und Quicksemaphore (schnelle Semaphore) erweitert.

Events stellen eine erhebliche Erweiterung der bekannten UNIX-Signals dar. Events können aber, im Gegensatz zu diesen,

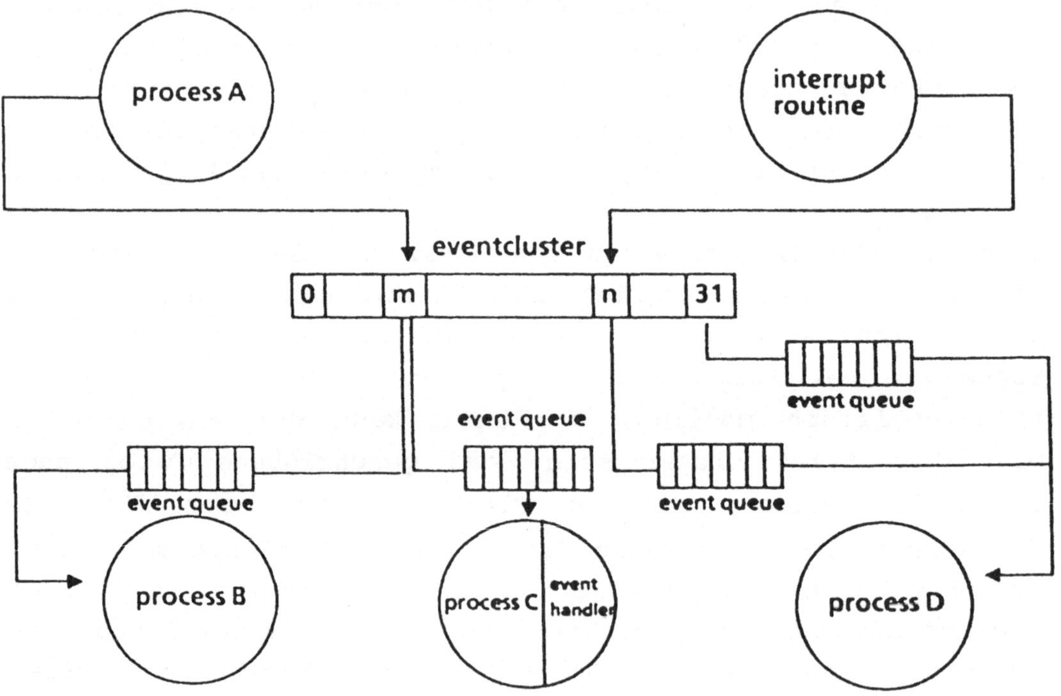

Abbildung 2: SORIX Event-Cluster

zwischen beliebigen Prozessen verschickt werden, ohne daß die Prozeß-ID des Empfängers bekannt ist. Die SORIX-Events können in Warteschlangen eingereiht werden, was sicher stellt, daß keine Events verloren gehen können. Außerdem kann mit einem Event eine 4-Byte Message übermittelt werden, was eine erhebliche Geschwindigkeitssteigerung beim Verschicken sehr kurzer Nachrichten gegenüber der Message-Funktion bringt.

Die zählenden Quicksemaphore sind ein sehr effektives Synchronisationsmittel zwischen Prozessen, mit geringstem

Systemoverhead, da hier direkt die Hardware des Prozessors benutzt werden kann. Die Zeit für eine Signal-Operation an eine Quicksemaphore beträgt 12usec.

3.4 Die SORIX Zeitfunktion

Bei Echtzeitaufgaben werden hohe Anforderungen an die Zeitbearbeitung sowohl was Funktionalität als auch was Genaugigkeit betrifft. Die Zeitgranularität beträgt bei SORIX 10 usec. Die speziellen SORIX-Zeitfunktionen wirken dabei nicht direkt auf Prozesse, sonder auf die Interprozeßkommunikation. Damit hat der Anwender erheblich mehr Freiheitsgrade zur Verfügung. So können zeitabhängig Events verschickt oder Quicksemaphoroperationen durchgeführt werden. Die Zeitangabe kann sowohl in Relativ- als auch in Absolutzeit erfolgen. Die Zeitaufträge können sowohl einzeln oder zyklisch durchgeführt werden.

3.5 Das SORIX Server-Konzept

Besonders im Bereich der Automatisierungstechnik gibt es viele Anwendungsfälle bei denen Treiberleistungen des Betriebssystems modifiziert werden müssen. Einer dieser Anwendungsfälle ist zum Beispiel die Implementierung eines neuen Protokolls an einer seriellen Schnittstelle. Derartige Probleme des Anwenders lassen sich mit Hilfe des Server-Konzepts in Sorix sehr elegant lösen. Der Server ist grundsätzlich ein normaler Anwenderprozeß, der wie jeder andere Pozeß gestartet wird. Erst durch einen speziellen Betriebssystemaufruf wird dieser Prozeß zum Server. Mit diesem Aufruf verbindet sich der Server mit einem als Gerätedatei gekennzeichneten Knoten im Filesystem. Alle Filesystem-Operationen (z.B. open, read, write) anderer Prozesse, die an diesen Knoten gerichtet sind werden vom

Betriebssystem an den Server weitergeleitet. Um diese Operationen für die Anwenderprogramme (Client-Prozesse) bearbeiten zu können ist der Server berechtigt in die Datenbereiche dieser Prozesse zu schreiben bzw. daraus lesen zu können.

Geräte-Zugriff über Server-Prozeß

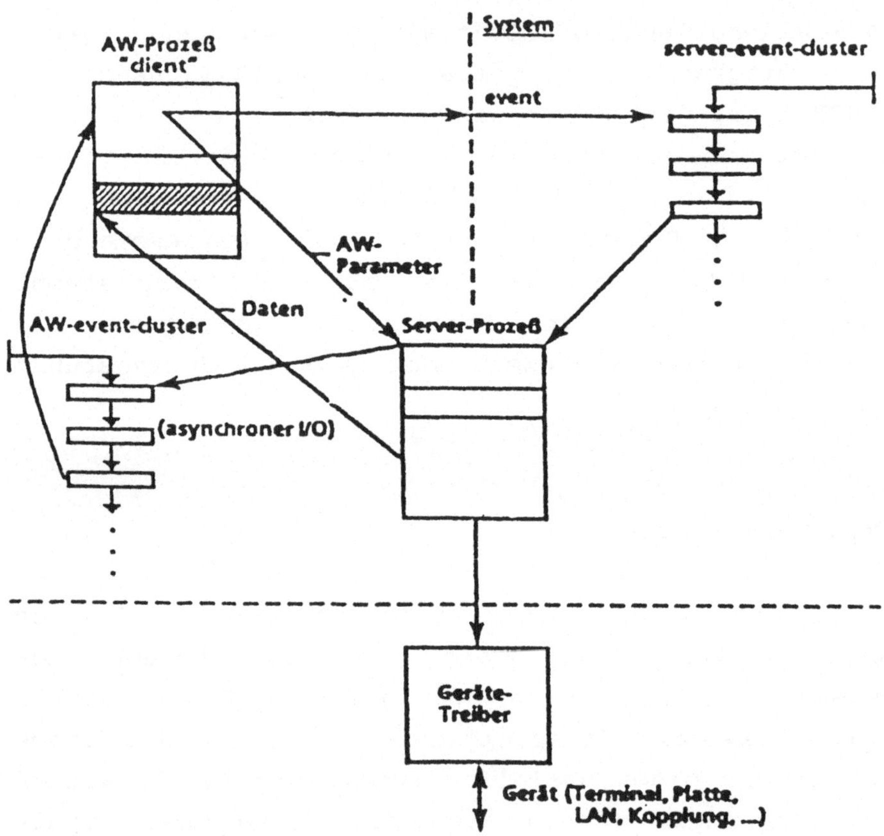

Abbildung 3: Geräte Zugriff über Server Prozeß

Eine weitere vorteilhafte Eigenschaft des Server-Konzepts ist die Realisierung von asynchronen I/O-Operationen. Das heißt der Client-Prozeß kann einen Auftrag an einen Server absetzen und sofort weiterlaufen ohne auf die Beendigung des Auftrags

zu warten. Er wird dann vom Server mittels Event von der Beendigigung des Auftrags informiert.
Wichtige Anwendungen des SORIX-Server-Konzepts gibt es vor allem in der Kommunikation bei der Implementierung der Schichten 3 und 4 der TCP/IP und ISO Protokolle.

3.6 Multicomputing Funktionen

Das Mikrocomputersystem SX verwendet, wie bereits erwähnt, als Rückwandbus den Multibus II von Intel. Damit können bis zu 20 Baugruppen mit Hilfe des Message-Passing-Konzepts miteinander kommunizieren. Auf dieses Konzept setzen die SORIX Multicomputing Funktionen auf. Multicomputing bedeutet in diesem Zusammenhang die Aufteilung einer Anwendung auf mehrere selbsständige Funktionsrechner mit eigenem Betriebssystem und unterschiedlichem Hardware Ausbau. Im Gegensatz zu Multiprozessing legt bei Multicomputing der Anwender selbst fest auf welchem Prozessor ein Programm ablaufen wird. Dadurch ist die bei Echtzeitanwendungen notwendige Deterministik besser gewährleistet.

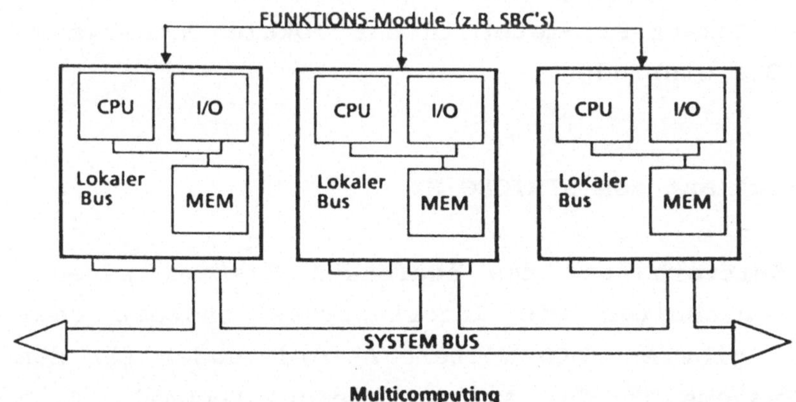

Abbildung 4: Multicomputing

Die Multicomputing Dienste von SORIX zeichnen sich dadurch aus, daß die am MBII verteilten Funktionen transparent von der Anwendung benutzt werden können. Eine Anwendung aus mehreren Prozessen, die auf einem SBC läuft, ist ohne Änderung auf mehrere SBCs verteilbar. Folgende Multicomputing-Funktionen werden von SORIX unterstützt:

- verteilte Events
- verteilte (UNIX-)Messages-Queues
- verteilter Server
- verteiltes Filesystem

Bei verteilten Events und verteilten Message-Queues wird die Transparenz dadurch hergestellt, daß diese grundsätzlich global definiert sind. Durch einen speziellen Parameter können diese lokal definiert werden und so eine höhere Performance erreichen.
Mit Hilfe des verteilten Servers ist es möglich die Dienste eines Servers auf einem anderen SBC zu nutzen.
Das verteilte File-System ist funktionell dem RFS (remote file system) ähnlich. Über einen Advertise Aufruf kann ein SBC Teile seines Dateibaums anderen SBCs zur Benutzung anbieten. Bei diesem Aufruf kann das Benutzungsrecht SBC-spezifisch vergeben werden. Die SBCs, die die angebotenen Directories verwenden wollen, müssen diese über einen Mount-Aufruf mit einem Zusatz-Parameter in das lokale Dateisystem des jeweiligen SBCs einbinden.

4. Booting und Hochlauf des SIMICRO SX

Neben für das Zeitverhalten des laufenden Systems wesent-lichen Systemeigenschaften wie Interruptreaktionszeit oder Prozeßwechselzeit gibt es noch weitere Anforderungen für den Einsatz eines Systems in der Automatisierungstechnik. Eine dieser Anforderungen ist ein, gegenüber einem normalen Unix-System, in wesentlichen Punkten verbessertes Hochlauf-verhalten. Dabei sind vor allem hohe Flexibilität und ein schneller Hochlauf, auch nach einem Spannungsausfall, die wichtigsten Eigenschaften. Die Flexibilität zeigt sich bei

SORIX besonders in der hohen Anzahl von verschiedenen Boot-Geräten. Ein Booten des Systems kann von Festplatte, Floppy-Disk, Streamer, EPROM, LAN oder MBII erfolgen, wobei die benötigten Boot-Parameter aus dem EEPROM ausgelesen werden. Diese verschiedenen Boot-Geräte können besonders flexibel genutzt werden, da eine Ersatzgerätestrategie benutzt werden kann. Die Uhrzeit wird beim Systemhochlauf am MBII verteilt und kann jederzeit zwischen den einzelnen SBC synchronisiert werden.

Während des Hochlaufs werden die Betriebsystemparamter aus dem EEPROM ausgelesen und das System läuft mit der entsprechenden Konfiguration hoch. Dies erleichtert für den Anwender die Änderung der Betriebssystemkonfiguration da ein Editieren der Datei "conf.c" entfällt.

Neben der Flexibilität ist es auch wichtig ein System schnell hochlaufen zu lassen, besonders dann wenn es gilt nach einem Spannungsausfall möglichst schnell wieder hochzufahren ohne Daten zu verlieren. Um diese Anforderungen zu erfüllen gibt es bei SORIX eine Warmstartfunktionalität, die den gepufferten Hauptspeicher des SX ausnützt. Dabei wird bei Spannungswiederkehr zunächst auf den alten Buffer-Cache ein Sync-Aufruf durchgeführt der die Filesysteme auf der Platte wieder konsistent macht. Damit wird die möglicherweise ziemliche lange Zeit für die "Reparatur" des Filesystems mit dem Utility "fsck" abgekürzt. RAM-Disks bleiben auch nach dem Warmstart erhalten. Außerdem besteht die Möglichkeit, Shared Memory Bereiche, in denen häufig für das gesamte System wichtige Anwenderdaten hinterlegt sind, zu retten. Dadurch kann die Anwendung nach einem Spannungsausfall sehr schnell wieder aufsetzen.

Des Herunterfahren des Systems kann zentral von einer Baugruppe aus vorgenommen werden. Dabei ist es auch möglich nur einen einzelnen SBC rückzusetzen. Außerdem gibt es noch einige spezielle Funktionen, wie das Abschalten der Stromversorgung und das zeitgesteuerte Wiedereinschalten.

5. Kommunikation

Um die vielfältigen Anforderungen im Bereich der Kommunikation zu erfüllen gibt es im Mikrocomputersystem SIMICRO SX sowohl die Protokollfamilie TCP/IP als auch SINEC H1 mit ISO-OSI als Transportprotokoll. Beide Transportsysteme sind über die Standardschnittstellen Berkley-Sockets (TCP) bzw. XTI (SH1) ansprechbar. Mit dem LAN-Modul können beide Protokollfamilien auf einer Anschaltung gleichzeitig betrieben werden. Bei der Einbindung der Protokolle in das SORIX-Konzept wurde besonders auf eine hohe Performance geachtet und ein Durchsatz von 400 Paketen pro Sekunde erreicht. Eine große Palette von Diensten, die auf den unteren Protokollschichten aufsitzen erweitern das Angebot an Kommunikationsmöglichkeiten.

Die Terminalemulation TELNET ermöglicht die Fernbedinung eines anderen Rechners über das Netzwerk. Die Datentransferprotokolle FTP (file transfer protocol) und TFTP (trivial file transfer protocol) ermöglichen den netzweiten Austausch von Text und Binärdateien. Das SMTP (simple mail transer protocol) stellt einen Electronic-mail-Dienst zur Verfügung. Das NFS (network file system) wird ein netzweites Filesystem bieten. Die AP-Protokolle (automation protocols) bieten den Zugang zu den höheren Schichten der Siemens Automatisierungsprotokolle. Mit Hilfe des RTDS (remote test and debug system) ist ein Test einer Aplikationen über das Netzwerk möglich. Außerdem ist das Booten des Systems mit dem Protokoll TFTP von einem übergelagerten Rechner möglich.

6. X WINDOW

Um grafische Ausgaben vorzunehmen wird beim Mikrokomputersystem SX das X Window System verwendet. X WINDOW wurde schon beim System-Design als verteiltes System konzpiert. Das X WINDOW-System besteht aus zwei grundsätzlichen Instanzen, die Client-Ebene und die Server-Ebene. Client-Prozesse geben,

eventuell über LAN, Grafik-Aufträge an den sog. X SERVER ab,
der diese am Bildschirm darstellt. Eine X LIBRARY (X LIB)

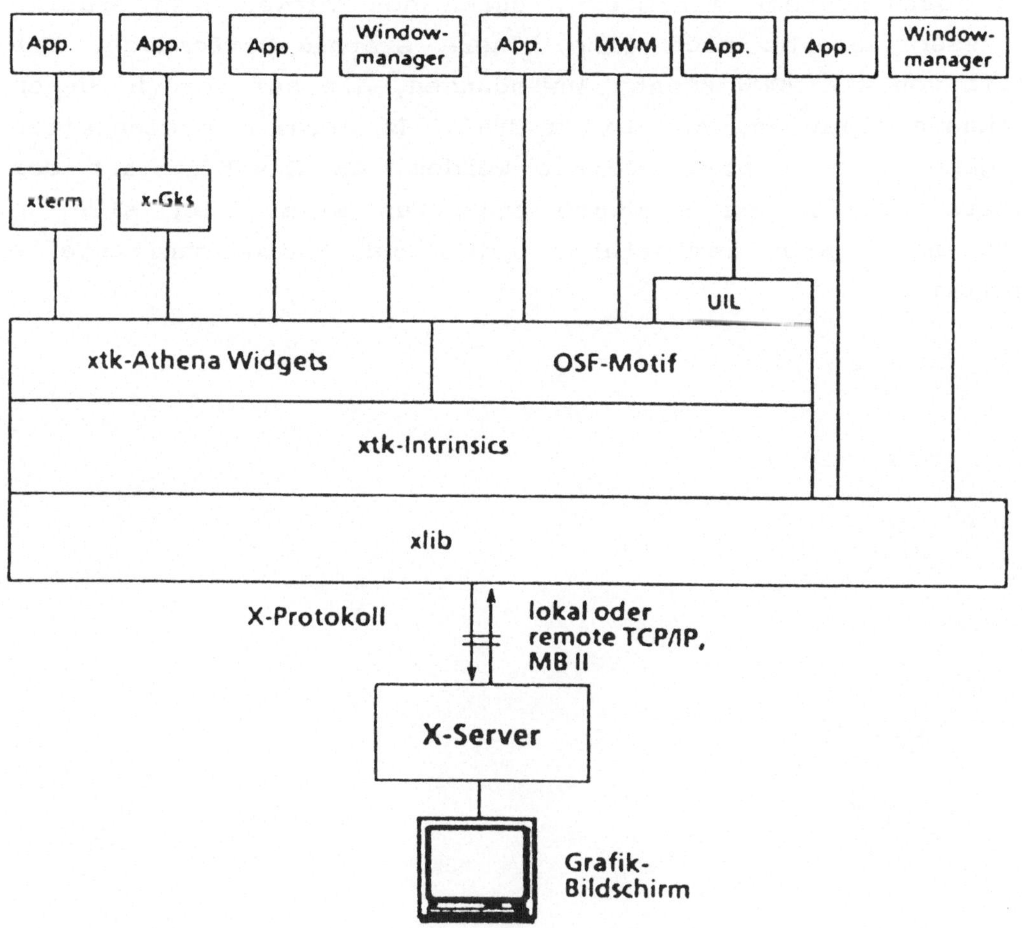

Abbildung 5: X WINDOWS

stellt für die Client-Ebene Graphik- bzw. Basis-Window-
Aufrufe zur Verfügung. Aufsetzend auf der X LIBRARY
existieren eine Reihe von Toolkits, die dem Anwender
komplexere Funktionen bieten wie z.B. MOTIF von OSF und OPEN
LOOK von AT&T.Für das Mikrocomputersystem SX steht die X LIB
als allgemeine Graphik-Schnitstelle und der X SERVER für das
Grafik-Modul zur Verfügung.

Das Graphik-Modul bietet dadurch, daß es lokal angesprochen
werden kann, eine höhere Performance. Da hierbei der X SERVER
auf dem Modul selbst abläuft, wird die Performance-Belastung
des Hosts-Systems minimiert. Durch den verwendeten Grafik-
Prozessor ergibt sich eine höhere Graphik-Perfomance. Mit
Hilfe von X WINDOW können Anwendungen, die auf verschiedenen
Rechnern ablaufen, auf einem Grafik-Bildschirm gleichzeitig
Ausgaben machen bzw. bedient werden. Da X WINDOW auf der
unterschiedlichsten Hardware angeboten wird, läßt sich so
auch eine sehr komfortable grafische Bedienschnittstelle
realisieren.

<u>THE IMPACT OF FORCED DIVERSITY</u>
<u>ON THE FAILURE BEHAVIOUR</u>
<u>OF MULTI-VERSION SOFTWARE</u>

F. Saglietti
Gesellschaft für Reaktorsicherheit (GRS) mbH
Forschungsgelände
D-8046 Garching

<u>Abstract:</u> The intention of this paper is to study the fault-tolerance improvement expected to be achieved by introducing forced diversity during the software development process by means of dissimilar methodologies. In particular, this impact will be analyzed with respect to the majority and the granularity of the system voter.

<u>Keywords:</u> Software fault-tolerance, multi-version software, forced diversity, development methodology, voter majority, voter granularity

1. INTRODUCTION

The principal means to achieve software fault-tolerance is certainly represented by the use of diversity, where the "independent" (in the sense of separate) development of more versions aiming to provide the same service is intended to randomly distribute the unavoidable errors onto the diverse programs, thus permitting them to be detected and tolerated by the output comparison of a voter.

Unfortunately, we know by theoretical and experimental investigations that the randomness of the error occurrence is strongly affected by the specific characteristics of the underlying problem to be solved and of the techniques adopted to solve it, so that in general we have to expect a number of common bugs caused by some intrinsic difficulties in the problem solution, but also generated and enabled to persist by the commonalities in the development processes.

This obviously leads to a dependent failure behaviour of parallel versions with increasing probabilities of simultaneous failures.

On the other hand, fault-tolerance would be best achieved by forcing the errors occurring in each program to affect disjoint input subsets, thus yielding the best possible failure behaviour, even much better than the originally desired independence.

As the problem complexity is essentially determined by the task to be performed by the system, we may try to decrease the common failure probability by extending the original mere product diversity to the more general concept of process diversity, ensuring dissimilarity of the development methodologies, e.g. with respect to aspects as:

- specification language
- implementation language
- algorithms
- data structures
- tools
- personnel
- testing methods.

In this case some particular classes of errors will be preferably produced or remain undetected as a result of a particular process rather than of another one, and even coincident errors will probably have different effects on the final result, allowing their detection.

The intention of this paper is to study the failure behaviour improvement expected by the additional diversity introduced into the fault-tolerant system by dissimilar methodologies, in particular with respect to the majority and the granularity of the voter.

The second section summarizes the already known theoretical results of Littlewood and Miller, which will be successively analyzed and confirmed by use of calculations performed on the basis of published experimental results.

Section 3 proposes an extension of the existing theory taking also into account the granularity of the voter, which will allow the interpretation of further experimental data.

2. <u>COMMON FAILURE BEHAVIOUR OF FORCED AND UNFORCED DIVERSE SYSTEMS WITH RESPECT TO THE VOTER MAJORITY</u>

2.1 Theoretical Results of Littlewood and Miller

In /5/ Littlewood and Miller propose a generalization of the work presented by Eckhardt and Lee in /3/.

The key measure in the original work was represented by the intensity funtion $\theta(x)$ indicating the probability that a program, which is randomly chosen out of a population of versions intended to satisfy the same set of requirements, will fail for a particular input x. Because of the differing difficulty in processing different inputs, Eckhardt and Lee came to the convincing conclusion that the random variable θ will generally take different values for different randomly chosen inputs x, and on this basis they proved the failure dependence of parallel versions.

This concept was extended by Littlewood and Miller, considering for each available development methodology the corresponding set of programs to be produced on the basis of a given specification. Within each particular methodology the situation is exactly the one previously described, so that we may distinguish for the different development techniques A, B, C, ... considered the corresponding random variables θ_A, θ_B, θ_C, ... representing the failure intensity within each method.

The following main result of Littlewood and Miller is essentially based on the indifference assumption between the various methodologies considered, supposing that a multi-version system developed within a methodology A can be expected to be as good as one resulting by means of another technique B.

This assumption is obviously very idealistic and in practical cases not to be realistically verified; in praxis it may only represent a statement about our subjective indifference, which is mainly based on lack of knowledge; anyway, applying the mathematical conclusions we should always be conscious of the difference between the required indifference and the mere ignorance mostly replacing it.

In case of a 1-out-of-2 system, succeeding when at least one of both component versions succeeds, this theory results in the following rule:

If one is indifferent between a randomly chosen AA system and a randomly chosen BB system, it would be recommendable to build instead a randomly chosen AB system.

Littlewood and Miller generalize this result for an arbitrary 1-out-
of-n system failing if no of n versions provides the correct output;
the analogous assumption requires indifference between designs which
involve merely permutations of methodologies. They show then that the
best design is the one which uses all the available methodological diversity and spreads it as evenly as possible among versions.

Of course, these results simply refer to the average system behaviour
both concerning the program space and the input space. Nonetheless,
being as usually ignorant about the detailed conditions of the situation considered, the above average inequalities can usefully assist in
decision making in terms of providing advice and rules for design
choice.

In /6/ the same authors extend these concepts in an even wider sense,
in that they allow more levels of decision. This means that each overall methodology can be interpreted as the final result of more design
decisions related to different development aspects.

Assuming the single design choices to be independent in the sense that
they will not influence each other, it can again be proved that the
more diverse methodologies are preferable, i.e. the ones introducing
dissimilarity at a higher number of levels.

Of course, all the reported results following from the summarized theo-
ries are not particularly surprising, basically confirming our intuitive expectations.

The more astonishing is the fact that the above statements about
1-out-of-2 systems (resp. 1-out-of-n systems) do not analogously hold
for configurations with 2-out-of-3 (resp. (n+1)-out-of-(2n+1)) majority.

In fact, for these architectures Littlewood and Miller show in /5/ that
diverse design may be worse than homogeneous design, thus coming to the
surprising conclusion that forced diversity may not be the most desirable strategy.

2.2 Experimental Results of Kelly and Avizienis

This section is devoted to the data analysis of an experiment on forced
diversity described in /4/ by Kelly and Avizienis. It is our intention
to propose an interpretation of the results in the light of the theoretical considerations previously presented. The experiment considered
was performed at UCLA, where 18 programs were produced on the basis of
a specification written in one of the following 3 different specification languages:

- the formal language OBJ (7 versions)

- the program design language PDL (5 versions)

- the natural language English (6 versions).

A test consisting of 100 input transactions was developed to uncover as
many bugs as possible. The outputs were classified as

- good points including correct outputs and cosmetic errors

- detected points representing wrong results rejected by the respective self-checking mechanism with which the program was instrumented

- undetected points determined by incorrect outputs which were not identified as failures by the corresponding error indicator mentioned above.

This distinction allows 2 possible interpretations of failure occurrence:

a) the system is considered to fail only when it cannot identify an erroneous output

   or

b) the system fails each time it does not correctly provide the service requested.

Thus the first definition concerns primarily the output reliability, whereas the second one requires additionally system availability. With respect to these measures we can now look at the failure behaviour within program classes with the same specification language, obtaining table 1:

| | Stand Alone Test Data | | | | | average failure probability | |
Version	OK Points	Cosmetic Errors	Good (OK+Cos)	Detected Errors	Undetected Errors	a	b
OBJ1	73	0	73	2	25		
OBJ2	71	18	89	8	3		
OBJ3	67	11	78	4	18		
OBJ4	69	3	72	8	20	23.6	27.7
OBJ5	67	12	79	0	21		
OBJ6	46	0	46	0	54		
OBJ7	52	17	69	7	24		
PDL1	59	2	61	1	38		
PDL2	54	2	56	32	12		
PDL3	95	0	95	4	1	15.8	24.2
PDL4	45	28	73	0	27		
PDL5	94	0	94	5	1		
ENG1	74	12	86	0	14		
ENG2	67	27	94	0	6		
ENG3	97	1	98	0	2	22.7	28.3
ENG4	30	5	35	25	40		
ENG5	55	6	61	0	39		
ENG6	53	3	56	9	35		

<u>Table 1:</u> Stand alone test data (mainly from /4/)

In spite of a slight difference in favour of the PDL class, we will regard the deviations on the whole as neglectable and assume in the following methodology indifference, in order to permit a first evaluation about the applicability of the theoretical results in real situations.

On the basis of the data published in /4/ and regarding the behaviour of triplets of homogeneous type OOO, PPP, EEE and of diverse type OPE, we can calculate average values for the common failure behaviour of two- and three-fold redundant systems with respect to 1-out-of-2 resp. 2-out-of-3 votes, obtaining the following tables 2 and 3:

	OO	PP	EE	OP OE PE
a	10,5%	2,3%	6,9%	5,5%
	average:		6,6%	
b	13,3%	6,6%	9,7%	9,6%
	average:		9,9%	

Table 2: Failure probability of 1-out-of-2 systems

	OOO	PPP	EEE	OPE
a	17,5%	4,7%	17,2%	13,2%
	average:		13,1%	
b	23,6%	15%	23%	20,8%
	average:		20,5%	

Table 3: Failure probability of 2-out-of-3 systems

These figures fulfill the relations stated in subsection 2.1, thus confirming the theoretical results of Littlewood and Miller: achievement of behaviour improvement by forcing diversity is expected in case of two-fold systems, but not in case of three-fold systems.

In this particular situation, however, even the positive influence exerted by two diverse methodologies is too low to justify the additional effort required for producing the same specification in two languages.

The reason for this lack of profitableness lies in the distribution of related errors shown in tables 4, 5, 6 for three different error types, where

- related errors are errors "related by symptoms", including both identical-cause errors and distinct-cause errors that produce acceptably similar symptoms

- specification errors are errors made at the specification phase of software development, including language inadequacies

- implementation errors are errors caused by a misinterpretation of the specification, rather than a mistake in the specification

- logic errors are errors caused by any other intrinsic inability to correctly develop a version.

Related Specification Errors

Error	Appears In		
	OBJ	PDL	ENG
S1	1,4,5,7		
S2			4,5,6
S3			4,5,6
S4			2,4
S5			6

Related Implementation Errors

Error	Appears In		
	OBJ	PDL	ENG
I1	2	1,2,3,4,5	
I2		1,4	
I3	1,2,5	4	1
I4	4	4	1
I5		2,4	5
I6	3	1	6
I7		1	4
I8	6		
I9		2	

Related Logic Errors

Error	Appears In		
	OBJ	PDL	ENG
L1	1		
L2	3		5
L3	7		
L4		1	
L5		1	
L6			4
L7			4

<u>Tables 4,5,6:</u> Related errors (from /4/)

The first of the three tables shows that there are only related specification errors within methodology classes. This means that diversity
of specification language was successful in preventing forced diverse
systems from being affected by common specification faults.

Also most of the related logic errors appearing in table 6 are class-
specific, concerning (apart from one exception) only single classes.

Thus the real cause for the high common failure probability of (both
forced and unforced) diverse systems is represented by the high number
of related implementation errors illustrated in table 5 and spreading
over all language classes.

In fact, as reported in /4/, many programmers were unable to understand
the specifications relying, instead, on the examples and on their in-
tuition, so that many misunderstandings could have been prevented by
giving more examples. This special case clearly shows that methodoloy
diversity may be successful, but obviously only with regard to the particular type of errors concerning the diversified development stage.
Thus, before expecting too much from forcing diversity at a given level, we should rather accurately consider the different error classes
which are likely to affect the final problem solution, investing at the
most promising stage the additional effort required by introducing different development techniques.

In particular, if the requirements strongly depend on complex numerical
computations, diversity should be introduced at algorithm level, whereas in case of lack of specification understandability, common implementation errors could be reduced by providing different clarifying examples.

3.   <u>COMMON FAILURE BEHAVIOUR OF FORCED AND UNFORCED DIVERSE SYSTEMS</u>
     <u>WITH RESPECT TO THE VOTER GRANULARITY</u>

3.1 Theoretical results

In case of outputs of complex data type, the consensus determined by
the voter may be at basic type level, adjudicating separately each output variable, or at complex type level, considering all output variables as a single result that is either correct or incorrect.

For example, if the system is supposed to produce a pair of results
(A,B) and the parallel versions actually produce the outputs (A,B),

(A,X) and (Y,B), where A≠Y and B≠X, then a 2-out-of-3 voter at component level would find each time two identical correct results and an incorrect one: (A,A,Y) and (B,X,B). Thus in this case a consensus value (A,B) exists, consisting of both components A and B identified as majority at the basic type level. By treating the output as a single result, on the contrary, the system has no consensus despite the fact that there is a majority on each of the output values.

The voter characteristic determining which output components to be jointly adjudicated is called its granularity. Thus voting systems with coarser granularity will grant a higher system reliability, but under some circumstances they may be too unforgiving, possibly excluding the contribution of essentially correct versions with some wrong unimportant details.

On the other hand, adjudication mechanisms with finer granularity will increase the system availability, but if the components are intended to be semantically connected, e.g. by forming a bit pattern, then making a decision on each output variable separately may create a nonsensical display.

Apart from these impacts of granularity on reliability and availability aspects, which are beyond the scope of this paper, we intend to study in this context the influence of granularity on the common failure behaviour of successfully forced diverse systems.

To simplify the problem representation we consider the case of a specification S defining an output o consisting of two components o_1 and o_2, each of which can be interpreted as being specified by a part S_1 resp. S_2 of S, according to Fig. 1:

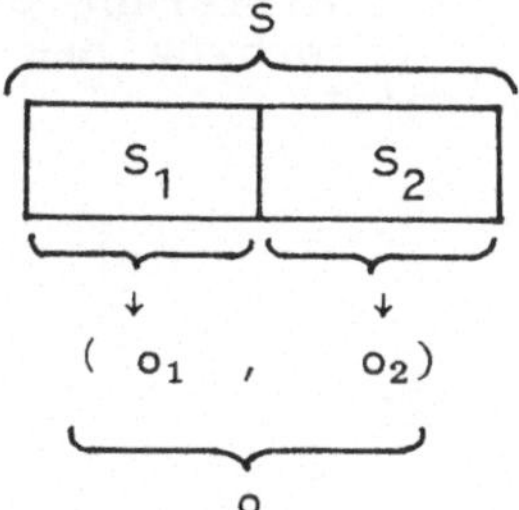

<u>Fig. 1:</u> Specification decomposition

The output variables may depend from one another in the sense that the calculation of o_2 may make use of values already obtained to determine o_1 (in the extreme case, both components might be identically defined and evaluated). Here we will, on the contrary, assume an independent decomposition, in that the development of the software specified by S_1 will not influence the development of the one specified by S_2.

Introducing analogous random variables to the ones adopted in section 2:

θ_A indicating the intensity function of S w.r.t. methodology A

θ_B indicating the intensity function of S w.r.t. methodology B

θ_{A1} indicating the intensity function of S_1 w.r.t. methodology A

θ_{B1} indicating the intensity function of S_1 w.r.t. methodology B

θ_{A2} indicating the intensity function of S_2 w.r.t. methodology A

θ_{B2} indicating the intensity function of S_2 w.r.t. methodology B, the development independence assumption yields then the following expressions:

$$\theta_A = \theta_{A1} + \theta_{A2} - \theta_{A1}\cdot\theta_{A2} \quad \text{and}$$
$$\theta_B = \theta_{B1} + \theta_{B2} - \theta_{B1}\cdot\theta_{B2}$$

The impacts of forced diversity on common failure probability will be represented for coarse and fine granularities by:

$$C := \theta_A^2 - \theta_A\cdot\theta_B$$

$$C_1 := \theta_{A1}^2 - \theta_{A1}\cdot\theta_{B1}$$

$$C_2 := \theta_{A2}^2 - \theta_{A2}\cdot\theta_{B2}$$

In case of indifference we expect according to the theoretical results of Littlewood and Miller reported in section 2 an improvement by forcing diversity:

$$E[C] \geqq 0$$

$$E[C_1] \geqq 0$$

$$E[C_2] \geqq 0$$

The following results hold for the particular case where each methodology will produce failures in a different component, (as expected for highly reliable versions, if diversity has been sucessfully forced), e.g. without restricting generality:

$$\theta_{A1} \equiv 0, \ \theta_{B2} \equiv 0$$

This yields:

$$C = \theta_A^2 - \theta_A\cdot\theta_B \leqq \theta_A^2 = \theta_{A2}^2$$

$$C_1 = \theta_{A1}^2 - \theta_{A1}\cdot\theta_{B1} = 0$$

$$C_2 = \theta_{A2}^2 - \theta_{A2}\cdot\theta_{B2} = \theta_{A2}^2$$

In particular:

$$C \geqq 0 \rightarrow C_1, C_2 \geqq 0$$

This means that in case of successful forced diversity separating failure occurrence of each version into disjoint output domains, if this actually improves the common failure behaviour (as expected) w.r.t. coarse granularity, it will do the same also w.r.t. fine granularity.

The other direction, however, does not always hold: comparison by a voter with fine granularity may show an improvement achieved by diversifying methodologies, which cannot be recognized by use of coarser granularity, i.e. it is possible that:

$$C_1, C_2 \geqq 0 \quad , \text{ but } C \leqq 0,$$

even if only methodology-dependent single faults may be assumed within each components. An intuitive explanation for this theoretical result can be found by observing the example presented in the next section.

3.2 Experimental Results of PODS and STEM

In this subsection we intend to analyze some data obtained by the PODS experiment in the light of the theoretical considerations previously presented.

The PODS project (s. /1/) developed 3 programs named CERL, HRP, VTT from the corresponding programming teams.

Diversity was enforced in a number of areas:

Halden and VTT used the formal specification language X, while CERL specifications were in free format.

CERL and VTT used a high level language, FORTRAN, while Halden programmed in Nord assembler.

Finally, CERL was constrained to use a 5th order polynomial as the main algorithm, while Halden and VTT had to use a table look-up algorithm, incorporating interpolation routines.

The diversity forced among the programming teams is summarized in Table 7.

Version	Spec. Language	Impl. Language	Algorithm
CERL	Free format	Fortran	Polynomial
Halden	Formal (X)	Assembler	Table
VTT	Formal (X)	Fortran	Table

<u>Table 7:</u> Summary of diversity among the programming teams (from /1/).

We can identify different degrees of forced diversity: HRP and VTT have been diversified with respect to the implementation language (one aspect).

CERL and VTT have been diversified with respect to the specification language and the main algorithm (2 aspects).

Finally, CERL and HRP have been diversified with respect to all aspects: specification and implementation languages and main algorithms (3 aspects).

The different degrees of forced diversity are summarized in Fig. 2.

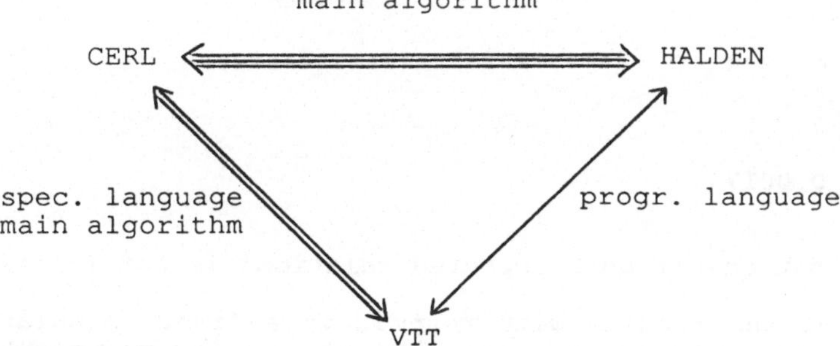

<u>Fig. 2:</u> Degrees of forced diversity

The results of a test consisting of 65.000 runs after a previous acceptance test of the three programs are shown in Table 8.

gol den	te st	number of faults detected in program:							
		C	H	V	CH	CV	HV	CHV	ANY
ANY		1152	32428	388	1134	0	108	0	32726
DL	<	0	0	33	0	0	0	0	33
DL	>	0	50	87	0	0	50	0	87
OS 0	1	0	0	0	0	0	0	0	0
OS 1	0	0	0	0	0	0	0	0	0
LS 0	1	0	28	28	0	0	28	0	28
LS 1	0	0	0	0	0	0	0	0	0
CB 0	1	0	0	0	0	0	0	0	0
C3 1	0	32	0	0	0	0	0	0	32
AL 0	1	0	0	0	0	0	0	0	0
AL 1	0	26	0	0	0	0	0	0	26
ST 0	1	0	0	0	0	0	0	0	0
ST 1	0	17	0	0	0	0	0	0	17
TA 0	1	0	0	0	0	0	0	0	0
TA 1	0	0	0	0	0	0	0	0	0
TF0 0	1	0	0	0	0	0	0	0	0
TF0 1	0	0	0	0	0	0	0	0	0
TF2 0	1	1115	0	0	0	0	0	0	1115
TF2 1	0	0	0	0	0	0	0	0	0
TF4 0	1	0	0	0	0	0	0	0	0
TF4 1	0	0	0	0	0	0	0	0	0
TF5 0	1	0	0	240	0	0	0	0	240
TF5 1	0	0	32388	0	0	0	0	0	32388

<u>Table 8:</u> PODS results with post-acceptance programs (from /2/)

We can observe that the HRP program has a very high failure probability, which is not particularly due to the language aspect distinguishing it from the other programs, but merely to a specification ambiguity. Before the testing phase we would have been indifferent between the diversified aspects and thus, according to the theoretical results of section 2 and to the following notation:

θ_C indicating the intensity function w.r.t. the methodol. used by CERL,

θ_H indicating the intensity function w.r.t. the methodol. used by HRP,

θ_V indicating the intensity function w.r.t. the methodol. used by VTT,

we would expect the following relations between the common failure probabilities of the three possible two-fold diverse systems:

$$E[\theta_C \cdot \theta_H] \leq E[\theta_C \cdot \theta_V] \leq E[\theta_H \cdot \theta_V]$$

The calculations performed for the experimental data, however, yield for the common failure probabilities P_{CH}, P_{CV}, P_{HV}:

$$P_{CH} = 0.0175$$

$$P_{CV} = 0$$

$$P_{HV} = 0.0017$$

This means that one of both expected relations is not fulfilled.

If we look at the results with respect to a finer granularity, we may distinguish two components consisting of the following output variables:

1. $\{DL,LS,TF5\}$ and

2. $\{CB,AL,ST,TF2\}$.

The remaining output variables will be neglected in the following, as they were always correct in the test results.

In this case we obtain for the common failure probabilities P_1 and P_2 of the respective components:

$$P_{1CH} = 0 \qquad P_{2CH} = 0$$

$$P_{1CV} = 0 \qquad P_{2CV} = 0$$

$$P_{1HV} = 0.0017 \qquad P_{2HV} = 0$$

We see that here the expected relations are fulfilled, confirming the theoretical considerations of the previous sub-section.

The intuitive explanation expressing in words the context represented by the formulae in section 3.1 is the following:

In spite of diverse methodologies successfully preventing both versions from containing errors in the same output component, their common failure probability might be nonetheless astonishingly high with respect to a coarse voter, if inputs cause contemporarily failures in different components (as for the pair CH).

The finer voter, on the other hand, only considers common failures occurring in the same component, which obviously represent only a sub-set of the previous ones.

4. CONCLUSION

This paper presented a study on two specific characteristics of the adjudicator in a diverse system - its majority and its granularity - in the light of the improvement expected to be achieved by forcing dissimilarity during development.

Known and original theoretical results were confirmed and explained by means of real-world examples.

On the whole from the observations analyzed it can be concluded that forced diversity may be an extremely powerful technique; its main restriction is represented by its strict relation to specific error classes, possibly resulting in an unjustified effort, if the fault categories chosen to be tolerated later show to be only partly representative for the application considered.

5. REFERENCES

/1/ M. Barnes, P.G. Bishop, B. Bjarland, G. Dahll, D. Esp, P. Humphreys, Y. Lahti, S. Yoshimura, A. Ball, O. Hatlevold
      PODS (the Project on Diverse Software)
      OECD Halden Reactor Project, HRP-323, June 1985

/2/ M. Barnes, P. Bishop, B. Bjarland, G. Dahll, D. Esp, Y. Lahti, H. Välisuo, P. Humphreys
      Software Testing and Evaluation Methods (the STEM Project)
      OECD Halden Reactor Project, HWR-210, May 1987

/3/   D.E. Eckhardt, L.D. Lee
      A Theoretical Basis for the Analysis of Multiversion Software Sub-
      ject to Coincident Errors
      IEEE Transactions on Software Engineering, Vol. SE-11, No. 12,
      December 1985

/4/   J.P.J. Kelly, A. Avizienis
      A Specification-Oriented Multi-Version Software Experiment
      13th International Symposium an Fault-Tolerant Computing, Milano,
      Italy, 1983
      IEEE Computer Society Press

/5/   B. Littlewood, D.R. Miller
      A Conceptual Model of Multi-Version Software
      17th International Symposium on Fault-Tolerant Computing, Pitts-
      burgh, USA, 1987
      IEEE Computer Society Press

/6/   B. Littlewood, D.R. Miller
      A Conceptual Model of the Effect of Diverse Methodologies on Coin-
      cident Failures in Multi-Version Software
      3rd International GI/ITG/GMA Conference on Fault-Tolerant Compu-
      ting Systems, Bremerhaven 1987
      Springer-Verlag, Informatik-Fachberichte Nr. 147

6. APPENDIX

Derivation of figures in 2.2, table 2 and table 3.

From [KEL83] we obtain the original data for the triplets OOO, PPP, EEE
and OPE with respect to each of the following failure types (regardless
of the order of the three outputs):

	OOO	PPP	EEE	OPE
V_1 : (G,G,G)	48.7%	44.8%	41.0%	44.5%
V_2 : (G,G,D)	3.7%	12.8%	8.3%	6.4%
V_3 : (G,G,U)	24.1%	27.4%	27.7%	28.3%
V_4 : (G,D,D)	1.4%	0.8%	1.6%	1.7%
V_5 : (G,D,U)	4.0%	8.8%	4.0%	5.0%
V_6 : (G,U,U)	7.5%	1.8%	10.3%	8.3%
V_7 : (G,U^*,U^*)	2.5%	1.2%	4.1%	1.8%
V_8 : (D,D,D)	0.1%	0.0%	0.0%	0.3%
V_9 : (D,D,U)	0.6%	0.7%	0.2%	0.6%
V_{10}: (D,U,U)	0.3%	0.6%	1.1%	1.3%
V_{11}: (D,U^*,U^*)	0.2%	0.0%	0.0%	0.1%
V_{12}: (U,U,U)	4.9%	0.1%	1.0%	1.3%
V_{13}: (U,U^*,U^*)	1.4%	0.0%	0.4%	0.3%
V_{14}: (U^*,U^*,U^*)	0.7%	1.0%	0.3%	0.1%

Where G represents a good point, D represents a detected error, U represents an undetected error that does not appear in one of the other two
versions, while U* is an undetected error that is related to both or
all three versions.

Table 3 easily results from these informations by adding the corresponding percentage values according to both error interpretations defined as a) resp. b), i.e.:

a) cases where at least 2 versions cannot detect an error:

$$V_6 + V_7 + V_{10} + V_{11} + V_{12} + V_{13} + V_{14}$$

b) cases where at least 2 versions cannot provide the correct answer:

$$V_4 + V_5 + V_6 + V_7 + V_8 + V_9 + V_{10} + V_{11} + V_{12} + V_{13} + V_{14}$$

Table 2, on the other hand, has been established by averaging the given data about triplets to represent homogeneous resp. inhomogeneous pairs.

For each pair OO, PP, EE the required estimations have been provided on the basis of the corresponding triplet OOO, PPP, EEE, taking into account the three possible pairs resulting from the triplet, e.g. (G,G,G) yields three good pairs (G,G), whereas (G,D,U) yields three pairs (G,D), (D,U) and (G,U).

Similarly, for the inhomogeneous pairs OP, OE, PE we obtain an average value by considering the inhomogeneous triplet OPE as including information on all of them.

On the whole we obtain for

a) cases where both versions cannot detect an error:

$$(V_6 + V_7 + V_{10} + V_{11})/3 + V_{12} + V_{13} + V_{14}$$

b) cases where both versions cannot provide the correct answer:

$$(V_4 + V_5 + V_6 + V_7)/3 + V_8 + V_9 + V_{10} + V_{11} + V_{12} + V_{13} + V_{14}$$

ACKNOWLEDGEMENT

The investigations for this paper have been sponsored in part by the Commission of the European Communities under the ESPRIT program (sub-item Software Technology, projects REQUEST and DARTS): the author expresses her thanks for the support.

Bausteintechnik
eine durchgängige Softwaretechnologie für Prozeßrechner- und Leitsysteme

Peter Fritz

SIEMENS AG, Bereich Energieerzeugung KWU

Abteilung L231 Prozeßleitrechner Softwarearchitektur

7500 Karlsruhe 21, Siemensallee 84

1. Einleitung

Der klassische Prozeßrechner der 60er und 70er Jahre stand als zentrale digitale Komponente isoliert in einer Leittechnik-Welt von analogen Meß-, Regel- und Anzeige-Einrichtungen. Er war ein Außenseiter und Sonderling – ein Buch mit sieben Siegeln für die meisten Leit- und Prozeßtechniker.

Heute ist der Rechner nur noch **eine** von vielen digitalen Komponenten- **ein** Spezialist unter anderen Spezialisten: dedizierte Geräte zum Messen, Erfassen, Steuern, Regeln, Überwachen, Anzeigen, Bedienen usw.

Alle diese Geräte arbeiten heute digital, sind mit einem Mikroprozessor bestückt und – im Rahmen ihrer Aufgabenstellung – programmierbar.

Die Programmierung und Hantierung dieser dedizierten Geräte ist in aller Regel so gestaltet, daß Leittechniker und Prozeßtechniker die Projektierung und Strukturierung dieser Geräte ohne besondere Spezialistenkenntnis vornehmen können (dies war eine wesentliche Grundvoraussetzung für die Akzeptanz digitaler Technik in der Prozeßtechnik!).

Doch obwohl der Prozeß- oder Leitrechner ein – aus technischer Sicht – gleichberechtigter Teilnehmer im Netz des Gesamtsystems darstellt, ist er in mancher Hinsicht ein Außenseiter geblieben; die Hantierung, Projektierung, Strukturierung der Rechnersysteme erfordert in der Regel Spezialisten. Die beim Rechner eingesetzten Softwarewerkzeuge sind strukturell, logisch und in der Oberfläche meist grundverschieden zu den Werkzeugen der dedizierten Automatisierungsgeräte.

Dieses liegt teilweise darin begründet, daß der Rechner oft als "Mädchen für alles" gesehen und eingesetzt wird, d.h. für Aufgaben, die mit den Automatisierungsgeräten nicht oder nur mit unverhältnismäßig hohem Aufwand zu lösen sind.

Und aus der dadurch entstehenden Vielfalt und Divergenz der Aufgaben erscheint eine einheitliche und durchgängige Hantierung und Strukturierung der Rechner-Software schwer erreichbar.

Zusätzlich wurde und wird der Prozeßrechner und die Prozeßrechnersoftware noch viel zu stark nach Gesichtspunkten und Konzepten der Rechnerfachleute, der Informatiker gestaltet, und nicht aus dem Blickwinkel der gesamten Leittechnik-Anlage oder des Prozesses.

Mit der im folgenden aufgezeigten Bausteintechnik wird der Versuch unternommen, eine Softwaretechnologie für Prozeßrechner zu entwickeln, welche einerseits die vielfältigen und durchaus divergenten Aufgaben und Anforderungen der Prozeßrechner lösen hilft und andererseits eine zu den dedizierten Automatisierungsgeräten strukturell und logisch durchgängige Hantierung, Projektierung und Strukturierung des Rechners erlaubt.

2. Was bedeutet " Durchgängigkeit "?

Durchgängigkeit von Systemen oder Systemteilen heißt, daß diese Systeme und ihre Funktionsweise sich dem Anwender und Benutzer in einer einheitlichen Form darstellen, wobei die technische Realisierung der einzelnen Systemkomponenten sehr unterschiedlich sein kann. Durchgängigkeit in diesem Sinne ist somit eine "Kompatibilität an der Benutzeroberfläche".

Im konkreten Fall bedeutet dies, daß innerhalb eines Prozeßleitsystems die Darstellung und Hantierung der Software im Prozeßrechner weitgehend identisch ist mit derjenigen in den Automatisierungsgeräten (AG). Das heißt, Projektierung, Parametierung, Strukturierung von Software im Rechner und im AG erfolgt mit dem gleichen Verfahren und mit den gleichen Werkzeugen, Dokumentation entsteht nach gleichen Richtlinien und in einheitlicher Form, Test und Diagnose ist mit gleichen Methoden und Tools durchführbar, Schnittstelle zu Visualisierungsstationen und Verhalten bei Bedieneingriffen sind identisch u.s.w.

Dadurch wird der Rechner aus seiner Außenseiterrolle innerhalb der Leittechnik befreit; die Hantierung und Strukturierung von Softwarefunktionen im Rechner kann in gleicher Weise vom Leittechniker und Prozeßtechniker vorgenommen werden, wie dieser es von den Automatisierungsgeräten her gewohnt ist – der Rechner wird durchschaubar und beherrschbar. So eröffnet sich dem Benutzer sogar die Möglichkeit, ohne Spezialistenkenntnisse die Softwarestruktur des Rechners zu gestalten und neue Funktionalität einzubringen.

Eine sich einheitlich darstellende Softwaretechnik reduziert nicht nur den Schulungs- und Personalaufwand beim Benutzer, sondern hilft auch Kosten bei der Projektabwicklung zu minimieren. Projekte zur Automatisierung von Anlagen (z.B. eines Kraftwerks oder einer Chemieanlage) können dadurch in allen Phasen – von Planung und Projektierung bis zu Inbetriebsetzung, Wartung, Pflege, Erweiterung im Betrieb u.s.w– mit einheitlichen Verfahren und einem einheitlichem Werkzeugsatz abgewickelt werden.

Vor allem auf den Gebieten der Projektierung und Inbetriebsetzung, welche den Hauptkostenanteil von Großprojekten ausmachen, wird ein spürbarer Ratioeffekt erwartet.

Nicht zuletzt bietet Durchgängigkeit dem Leittechnikhersteller die Möglichkeit, die Entwicklungskapazitäten auf die Erstellung und Pflege eines einheitlichen Werkzeugsatzes zu konzentrieren (statt getrennter Werkzeugsätze für die Software der Automatisierungsgeräte und der Rechner).

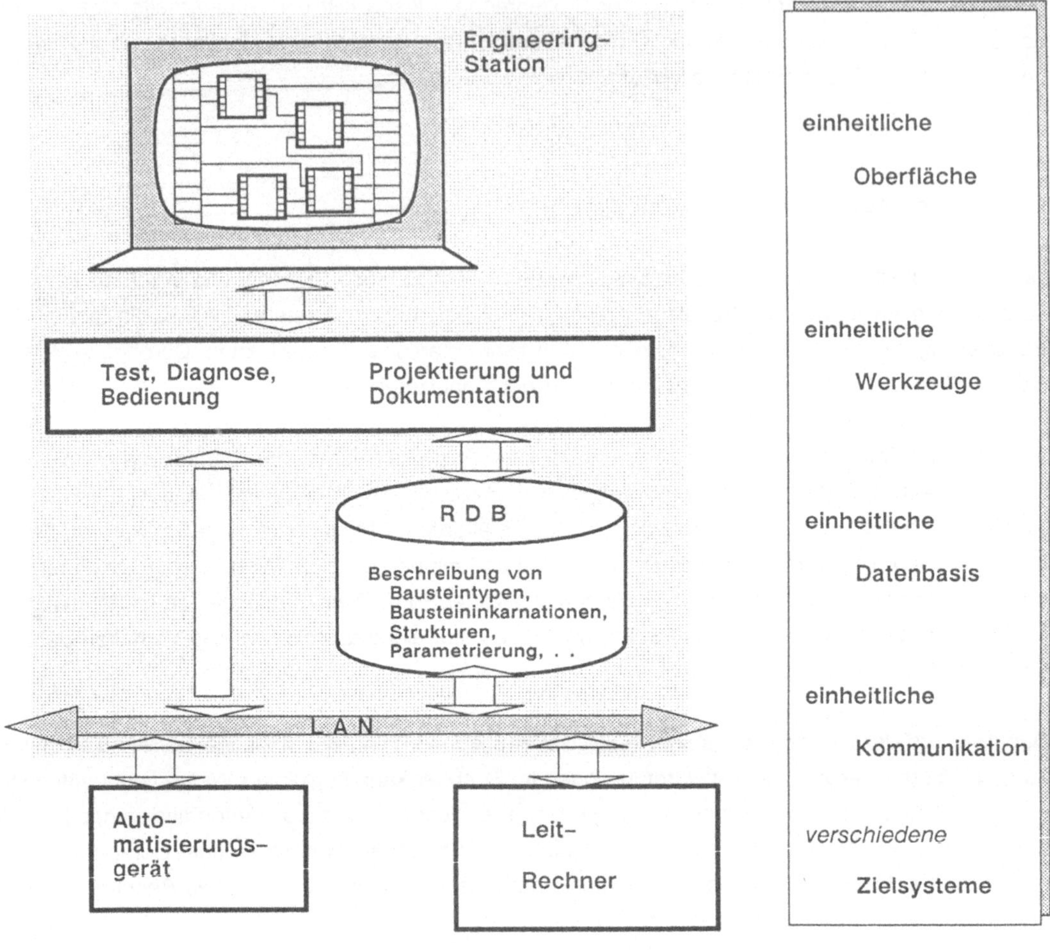

Bild 1: Durchgängige Softwaretechnik für Automatisierungsgeräte und Leitrechner

Mit welcher Technik läßt sich nun Durchgängigkeit in der Softwarewelt der Prozeßleittechnik erreichen?

Der Schlüssel zur Beantwortung dieser Frage liegt in der Bausteintechnik, der bewährten Softwaretechnik bei digitalen Automatisierungsgeräten, die seit über 10 Jahren zum Messen, Steuern, Regeln und Überwachen von Prozessen eingesetzt werden.

3. Die Bausteintechnik der Automatisierungsgeräte

Die bei Automatisierungsgeräten eingesetzte Bausteintechnik ist eine Softwaretechnologie, welche in direkter Ableitung aus der in Hardware realisierten, analogen und diskreten Mess-, Regel- und Steuerungstechnik entstand.

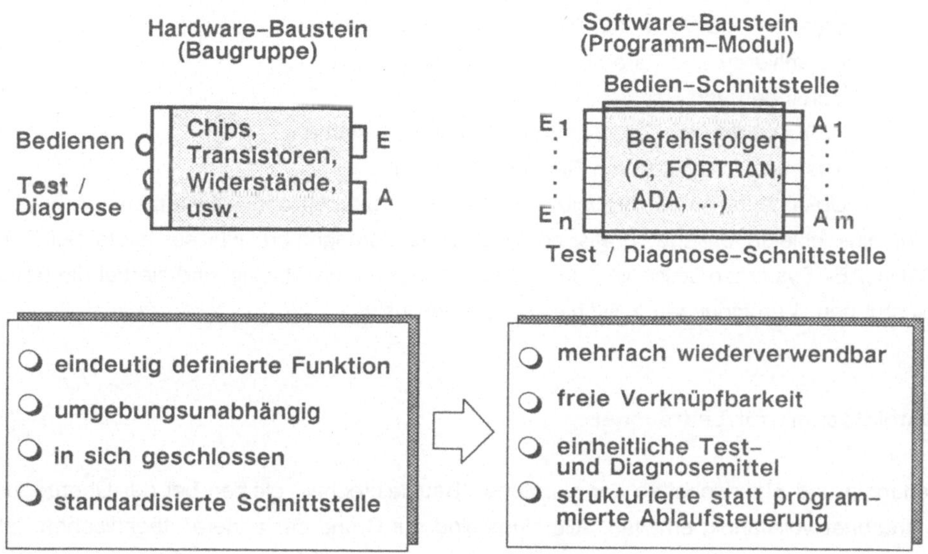

Bild 2: Von der Baugruppe zum Softwarebaustein – Merkmale von Bausteinen

Es wurden die Funktionen, die zuvor in Analogtechnik auf einer Baugruppe realisiert waren, einem entsprechenden Softwaremodul übertragen – dem **Funktionsbaustein.** Aus einer Reglerbaugruppe wurde ein Reglerbaustein, eine Motorsteuerbaugrupe ergab einen entsprechenden Motorsteuerbaustein u.s.w.
Neben ihrer speziellen **Funktion** sind diese Bausteine gekennzeichnet durch ihre **Eingänge** und **Ausgänge** (die Steckerleiste/-belegung der Baugruppe), über welche sie mit Daten versorgt werden und Ergebnisse abliefern können.

Zur Lösung einer bestimmten Aufgabenstellung (z.B. Regelung des Wasser-Dampfkreislaufs in einem Kessel) werden mehrere Funktionsbausteine zu einer Ablaufstruktur zusammengesetzt (Baugruppen in einen Rahmen stecken) und die Ein-/Ausgänge entsprechend dem gewünschten Datenfluss miteinander verbunden (Verdrahtung der Baugruppen bzw. der Rahmenrückwand).
Nach Aktivierung des Automatisierungssystems wird die Kette der Software-Funktionsbausteine zyklisch abgearbeitet, wobei durch relativ kurze Zyklen (0,1 Sek bis 0,5 Sek typisch) nach außen hin ein Verhalten erreicht wird, welches dem der traditionellen Funktionsbaugruppen in Analogtechnik entspricht.

Dabei bleibt dem Anwender, dem Benutzer des Bausteins verborgen, wie die Funktion eines Bausteine realisiert ist (z.B. welche Algorithmen, Programmiersprachen etc verwendet wurden). Ein Baustein ist vollständig – er benötigt zur Erfüllung seiner Funktion keine Nachbarbausteine – und abgeschlossen – er kann ausschließlich über seine spezifizierten Schnittstellen Daten erhalten (über Eingänge) oder weitergeben (über Ausgänge).

Die Hantierung mit vorgefertigten Software-Funktionsbausteinen
- Erstellen und Verwalten von Bausteinbibliotheken
- Instantiieren und Paramentieren von Bausteinen
- Verbinden von Bausteinen
- Strukturieren von Bausteingruppen und – systemen
- Test und Diagnose von Bausteinstrukturen
- Dokumentation von Bausteinen, Bausteinstrukturen und – systemen

wird über eine objektorientierte, grafische Oberfläche durchgeführt, wie sie heute bei Desktop- oder CAD/CAE- Systemen üblich ist. Die grafisch manipulierten Objekte sind hierbei die Bausteine, Bausteingruppen, Verbindungen zwischen Bausteinen usw.

4. Bausteintechnik für Leitrechner

Die Merkmale und Eigenschaften der skizzierten Bausteintechnik bleiben bei der Übertragung auf den Leitrechner vollständig erhalten. Allerdings sind auf Grund der anderen technischen Rahmenbedingungen und der vielfältigen Aufgaben des Leitrechners einige wesentliche Anpassungsarbeiten und auch strukturelle Erweiterungen nötig.

4.1 Anpassung der Bausteintechnik an die Rechnerwelt

Sowohl die Hardware als auch die Software der heutigen Automatisierungsgeräte ist direkt auf die speziellen Anforderungen der Grundautomatisierung zugeschnitten (Reaktionszeit, Ausfallsicherheit etc.): Externspeicher fehlen, Software (auch Bausteine) teilweise in PROMS, Programmierung der Bausteine in Spezialsprachen, einfache Ablaufsteuerung und Kommunikation durch Spezialbetriebssysteme, einfache Datentypen, kein Einbringen "fremder" Software möglich usw.

Beim Leitrechner werden ganz andere Bedingungen in der Hardware und im Betriebssystem vorgefunden. Der erste Schritt bei der Übertragung der Bausteintechnik auf den Rechner besteht daher darin, diese an die in der Rechnerwelt herrschenden Rahmenbedingungen anzupassen.

Der wesentliche technische Unterschied des Rechners zum AG liegt im Betriebssystem. Um die hohe Innovationsgeschwindigkeit auf dem Gebiet der Rechner – Hardware ohne Umstellung der Softwaresysteme nutzen zu können und unabhängig von Hardwareherstellern zu werden, werden in Zukunft beim Rechner hardware- und herstellerunabhängige Betriebssysteme wachsende Bedeutung gewinnen, auch im Markt der Prozeßautomatisierung und Prozeßtechnik. In den nächsten

Jahren wird hierbei das Betriebssystem UNIX eine zentrale Rolle spielen.

Übertragung auf den Leitrechner unter diesem Gesichtspunkt bedeutet daher nicht, Bausteintechnik auf eine bestimmte Rechnerhardware oder ein spezielles Prozeßrechner–Betriebssystem zu übertragen, sondern auf die von UNIX bereitgestellten und normierten Schnittstellen. Dadurch wird ein weites Spektrum verschiedener Rechnersysteme als Prozeßleitrechner einsetzbar – vom PC bis zum Supermini.

Allerdings bietet das Betriebssystem UNIX – auf Grund seiner Herkunft – wenig Unterstützung für eine prozeßorientierte Softwarearchitektur wie die Bausteintechnik. Es sind weder Bausteintypen noch Bausteininkarnationen als Verwaltungs– oder Ablaufobjekte bekannt. Gleiches gilt für Bausteinstrukturen und Verbindungen.

Es müssen daher Werkzeuge entwickelt werden, welche

1. die Abbildung der Objekte der Bausteintechnik auf die unter UNIX verfügbaren Ablaufobjekte (process) und Betriebsmittel (shared memory, message–queue, pipes, files..) vornehmen und

2. den Bausteinen beim Ablauf die erforderliche Ablaufumgebung bereitstellen.

Gleichzeitig muß der Tatsache Rechnung getragen werden, daß neben der bausteinorientierten Software noch andere Softwaresysteme wie z.B. Datenbanken etc. im Leitrechner zum Ablauf kommen. Hier muß nicht nur eine Koexistenz verschiedener Softwaresysteme, sondern auch wechselseitiger Datenaustausch und eine möglichst freie, projektierbare Verbindung und Verknüpfung solcher Softwaresysteme durch Adapterbausteine ermöglicht werden.

Neben dieser Offenheit des Leitrechners für von dritter Seite zugelieferte Software erwarten viele Anwender, daß zur Bildung von Anwendersystemen nicht nur (vom Lieferanten) vorgefertigte Bausteine eingesetzt werden können, sondern daß er selbst die Möglichkeit hat, mit einfachen Hantierungen und unter Verwendung von Standardprogrammiersprachen eigene Funktionen zu implementieren und ins System einzubringen.

Daher wird neben der Erstellung von Bausteinstrukturen auch die Erstellung von neuen Bausteintypen für den Rechner durch entsprechende Werkzeuge unterstützt. Der Bausteinalgorithmus wird dabei z.B. in ANSI–C programmiert (andere Programmiersprachen sind prinzipiell möglich und geplant); die eindeutige und vollständige Spezifikation der Bausteinschnittstellen (Ein– und Ausgänge) geschieht über eine in diesem Teil formalisierte (und damit von einem Praecompiler auswertbare) Baustein–Dokumentation.

4.2 Anpassung der Bausteintechnik an Leitrechneraufgaben

Automatisierungsgeräte haben einen klar beschriebenen Aufgabenumfang: Messen, Steuern, Regeln und Überwachen. Die Funktionen sind der Prozeßführung zugeordnet; eine Betrachtung von Vergangenheit (Auswertung etc) oder Zukunft (Prognose, Planung) oder des Umfeldes (Betrieb, Unternehmen) findet nicht statt.

Letztere Funktionen (Auswertung, Prognose,...) liegen im Aufgabenfeld des Leitrechners. Durch seine Fähigkeit, große Datenmengen über lange Zeiträume zu speichern, wegen der höheren und flexibler einsetzbaren Rechenkapazität und wegen der übergeordneten Stellung im Automatisierungsnetz wird er immer dann eingesetzt,

- wenn umfangreiche Datenmengen gespeichert und ausgewertet werden müssen
  (z.B. Störungsanalyse, Langzeitarchivierung von Prozeß- und Produktionsdaten,...),
- wo komplexe Berechnungen durchzuführen sind
  (z.B. Prognose, Simulation, Optimierung, Kenngrößenberechnungen, Korrelationsanalyse...),
- wo übergeordnete Aufgaben gelöst werden müssen
  (Betriebsführungsaufgaben wie Bilanzierung, Planung, Wartung) oder
- wenn Daten mit der Datenverarbeitung der Unternehmensleitebene ausgetauscht werden müssen.

Auf Grund dieser anders gelagerten und vielfältigeren Aufgaben des Leitrechners gegenüber den AG's muß die von dort vorliegende Bausteintechnik strukturell erweitert und ergänzt werden.

Bausteine der Automatisierungsgeräte verarbeiten charakteristischerweise Zustandsdaten: die Ausgänge eines Bausteins spiegeln einen Zustand wider; der Zustand ist – unabhängig von einer eventuell vorhandenen Weiterverarbeitung – immer vorhanden, er kann jederzeit von anderen Bausteinen über die projektierten Verbindungen gelesen werden (Hol – Prinzip; die Verbindungen sind meist als Verweis auf den entsprechenden Ausgang realisiert, auf welchen dann direkt zugegriffen wird). Die Weitergabe oder das Versenden von Nachrichten (Bring – Prinzip, z.B. Meldungen über Zustandswechsel) wird beim AG nicht unterstützt bzw. ist nur beim Verlassen der Bausteinstruktur erreichbar.

Um die möglichen Anwenderfunktionen im Rechner nicht auf solche zustandsverarbeitenden, nur die aktuelle Gegenwart betrachtende Architekturen einzuschränken, wurden den Verbindungen zwischen Bausteinausgängen und Bausteineingängen erweiterte Eigenschaften zugeordnet: Verbindungen zwischen Bausteinen im Leitrechner haben die Fähigkeit, Daten in bestimmbarem Umfang und mit verschiedener Zugriffscharakteristik zu speichern.

Drei Verbindungsarten werden unterschieden:
1. Die **Ringverbindung**. Sie ist eine Erweiterung der oben beschriebenen einfachen Verbindungen für Zustandsdaten; statt des letzten aktuellen Zustands sind jedoch die "n" letzten Zustände gespeichert und zugreifbar. Die Kapazität "n" des Ringspeicher wird bei der Strukturierung eines Bausteinsystems angegeben. Eine Ringverbindung ist einem **Ausgang** zugeordnet; beliebig viele Eingänge können die Daten aus dem Ring lesen.
2. Die **Warteschlangenverbindung**. Sie kann dynamisch auftretende Daten (Nachrichten, Meldungen, etc...) bis zu ihrer Verarbeitung zwischenspeichern. Die Warteschlange ist einem **Eingang** zugeordnet und kann Daten von beliebig vielen Ausgängen empfangen.
3. Die **Auftragsverbindung**. Sie ermöglicht den Aufbau einer Client-Sever-Beziehung zwischen zwei Bausteinen; die Auftragsverbindung puffert Aufträge bis zu ihrer Bearbeitung und liefert das Auftragsergebnis zurück an den auftraggebenden Baustein. Die Auftragsverbindung kennt den Ser-

ver und alle potentiellen Clients; diese kennen ihrerseits ihre jeweiligen Auftragspartner nicht, nur den Typ des Auftrags.

Die Art der Verbindung ist – wie der Typ der Daten selbst – ein festes Merkmal für jeden Ein – und Ausgang.

Diese Verallgemeinerung der Semantik von Verbindungen zwischen Bausteinen ist möglich, ohne die weiter oben beschriebenen Grundmerkmale der Bausteintechnik zu verletzen oder zu verändern. Auch an der Hantierungsoberfläche der Bausteinwerkzeuge treten diese erweiterten Eigenschaften zunächst nicht in Erscheinung; erst bei Anwahl der Attribute einer Verbindung (z.B., beim Anklicken derselben mit einem pointing – device) werden die unterschiedlichen Ausprägungen und Typen sichtbar und manipulierbar. Die Verbindungen zwischen Bausteinen in einem Automatisierungsgerät erscheint als Untermenge oder Spezialfall der aufgeführten Verbindungen, nämlich als Ringverbindungen der Kapazität 1.

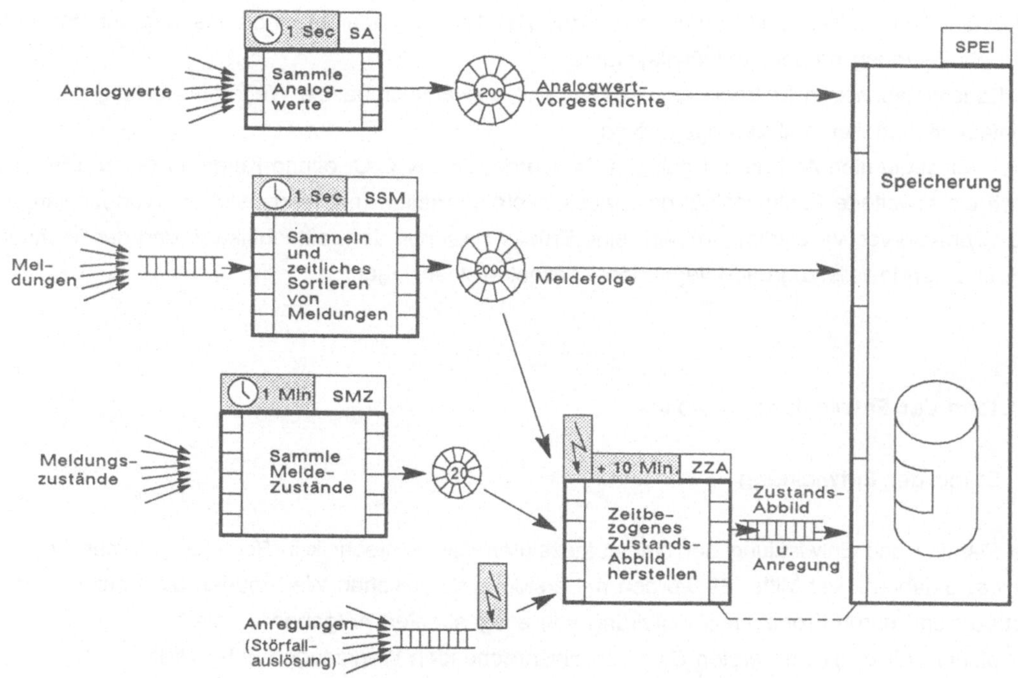

Bild 3: Beispiel einer Bausteinstruktur im Rechner:
Funktion "Störfallaufzeichnung"

Zu dieser erweiterten Möglichkeit im Datenfluss zwischen Bausteinen treten Verbesserungen in der Ablaufsteuerung von Bausteinstrukturen im Rechner. Neben zyklischen Aktivierungen von einzelnen Bausteinen bzw. Bausteinketten sind Aktivierungen auf Grund von Meldungen, Nachrichten, Aufträgen, Bedienungen u.s.w. oder auch Änderungen im Systemzustand wie Anlauf, shutdown etc. vorgesehen.

Durch diese Erweiterungen wird mit der Bausteintechnik im Rechner eine große Spanne von möglichen Anwendersystemarchitekturen erschlossen: zustandsorientierte, vorwiegend zyklisch arbeitende Anwendersysteme (wie in den AS) werden ebenso unterstützt wie messageorientierte, ereignisgesteuerte Systeme oder Client – Server – Strukturen. Es kann für jede Anwenderfunktion die jeweils geeignete Architektur ausgewählt werden, ohne daß auf eine einheitliche und durchgängige Hantierung und Dokumentation verzichtet zu werden braucht.

5. Zusammenfassung

Die Bausteintechnik bietet die Möglichkeit, die in Automatisierungsgeräten und in Leitrechnern eingesetzte Anwendersoftware in einer einheitlichen Softwaretechnologie zu gestalten und zu hantieren. Durchgängige Werkzeuge zur Projektierung und Parametrierung der Anwendersoftware, zur Dokumentation, zu Test und Diagnose und zur Visualisierung erleichtern die Planung, Inbetriebsetzung und Wartung großer Prozeßleitsysteme.

Die Bausteintechnik im Leitrechner ist dabei eine aufwärtskompatible Weiterentwicklung der Bausteintechnik von Automatisierungsgeräten.

Durch konsequenten Aufbau auf gültigen Standards wie UNIX, C, offene Kommunikation usw. und durch die erweiterte Funktionalität der Bausteinkommunikation (mit Ringspeicher, Warteschlangen und Client-Server-Verbindungen) wird eine Brücke zwischen der Leittechnikwelt und der Rechnerwelt und den jeweils zugrunde liegenden Konzepten geschlagen.

6. Stand der Entwicklung, Ausblick

6.1 Stand der Entwicklung

Die Planung und Entwicklung der hier aufgezeigten Bausteintechnik in Rechnersystemen begann vor ca. 3 Jahren. Seit Mitte '89 werden mit ersten prototypischen Werkzeugen die Konzepte konkretisiert und durch Probeimplementierung von ausgesuchten Aufgaben validiert.

Wir planen, Mitte '91 den ersten Satz von entsprechenden Werkzeugen unter UNIX verfügbar zu haben und in Kundenprojekten einzusetzen.

6.2 Ausblick

Im ersten Schritt der Entwicklung dieser Softwaretechnologie stand die Durchgängigkeit zur Leittechnikwelt und den dort eingesetzten Funktionsbausteinen im Vordergrund. Als Konsequenz daraus ergab sich ein weitgehend funktionsorientierter Ansatz (Baustein = Funktion).

Im Gegensatz hierzu wird heute bei vielen anderen Softwarekonzepten ein datenorientierter Ansatz verfolgt – basierend auf dem raschen Vordringen von relationalen Datenbanken mit normierten Schnittstellen und mächtigen Datenbank-Tools (wie Masken- und Reportgeneratoren, 4GL,usw.).

Es ist unser Ziel, in der Weiterentwicklung der Bausteintechnik diese Konzepte (Funktions- und Daten-Orientierung) weitestgehend miteinander zu verschmelzen. Als erster Schritt in diese Richtung ist geplant, die von den Bausteinen erzeugten Daten – Bausteinausgänge und Verbindungen – direkt in einer Datenbank abzulegen, so daß (ohne weitere Vorkehrungen) diese Daten mit den normalen DB-Tools zugreifbar und weiter verarbeitbar sind.

Literatur:

Eisenburger, L.: Basiskonzept für die Prozeßleittechnik,
       Interner Bericht SIEMENS AG, KWU L 231, Dez. 1988

Endres, A.: Software-Wiederverwendung: Ziele, Wege und Erfahrungen,
       Informatik-Spektrum 11 / 2, 1988

Fritz, P.; Praska, A.; Walz, H.: Bausteintechnik für Leitrechner,
       Interner Bericht SIEMENS AG, KWU L 231, Juni 1989

Klose, K.: Softwareentwicklung für Prozeßleitsysteme,
       atp Sonderheft NAMUR Statusbericht 1987

Stöckler, H.-P.: Fortschrittliche Strukturen von Leitsystemen,
       atp 7 / 1989

Strickling, A.: CAE in der Prozeßleittechnik,
       atp 2 / 1988

Ziemlich, K.: Rechnergestütztes Projektieren und Dokumentieren beim Einsatz von Leitsystemen
       atp 3 / 1988

. . .   Standard for programmable controllers – part 3: programming languages,
       IEC SC 65A / WG 6 / TF 3, march 1990

EIN WERKZEUG ZUR FUNKTIONSBAUSTEINORIENTIERTEN PROGRAMMENT-WICKLUNG FÜR DIE GERÄTE- UND KLEINAUTOMATISIERUNG

B. Rehwaldt
Humboldt-Universität zu Berlin
Sektion Elektronik
Bereich Automatisierungstechnik

Invalidenstraße 110, Berlin 1040

1. Einleitung

Gegenwärtig vollzieht sich in der Kleinautomatisierung ein Generationswechsel, der durch die Ablösung konventioneller Lösungen durch rechnerintegrierte Geräte und Systeme bestimmt ist. Auf der einen Seite stehen universelle Automatisierungs-gerätelösungen mit konfigurierbaren und programmierbaren Komponenten in Form von Mikrorechnerreglern und speicherprogrammierbare Steuerungen. Zum anderen finden Mikrorechnerkomponenten zunehmend zur Steuerung und Regelung von geräteinternen Prozessen ihre Anwendung. Letzteres wird als Geräteautomatisierung verstanden.

Das funktionsbestimmende Element einer jeden rechnerintegrierten Lösung ist die Software, die als Systemsoftware und Anwendersoftware in Erscheinung tritt. Die Systemsoftware umfaßt sowohl Betriebssystemkomponenten als auch hardware-spezifische Anpassungen und bildet den Laufzeitrahmen für die aufgabenabhängige Anwendersoftware. Die Effizienz der zur Anwendersoftwareerstellung verfügbaren Werkzeuge ist nicht zuletzt entscheidend für Qualität und Kosten der gesamten Automatisierungslösung. Gegenwärtig steht einer weitgehenden firmeninternen Unifizierung im Hardwarebereich oft noch eine Programmerzeugung auf assembler-nahem Niveau gegenüber, was nicht zuletzt als Ursache für sehr hohe Softwarekosten anzusehen ist.

Der Nutzung von universellen Hochsprachen stehen meist hardwarebedingte Ein-schränkungen und Echtzeitforderungen gegenüber. Oft sind spezielle Sprachimplemen-tationen für den eingesetzten Prozessortyp nicht verfügbar oder aber Speichermodelle und Laufzeiteigenschaften stehen im Widerspruch zu verfügbaren Ressourcen. Diese Überlegungen bilden den Ansatz für nachfolgend vorgestellte Sprachkonzeption. Wesentliche Prämissen sind dabei:
- Nutzbarkeit auch für Minimalsysteme (wenige kByte ROM/RAM)
- "programmierbare" Laufzeiteigenschaften
- aufgabenbezogener Funktionsumfang und
- minimaler Anpassungsaufwand für verschiedene Prozessoren.

2. Sprachkonzeption

Grundanliegen des Sprachentwurfs war es, dem Programmierer die Welt der proze-duralen Strukturen zu eröffnen, ohne ihn der Vorteile, die sich aus der enormen Codeeffizienz der Assemblerprogrammierung ergeben, zu berauben. Erreicht wird dies durch die Einführung eines Funktionsbausteinkonzeptes, bestehend aus Modulen und Modulblöcken.

Funktionsbausteine

Jeder Funktionsbaustein (vgl. Abb. 1) ist dabei ein logisches Objekt, das eine Zahl von Eingangssignalen unter dem Einfluß bestimmter Parameter in Ausgangssignale transformiert. Statussignale geben Auskunft über innere Zustände des Funktionsbausteins.

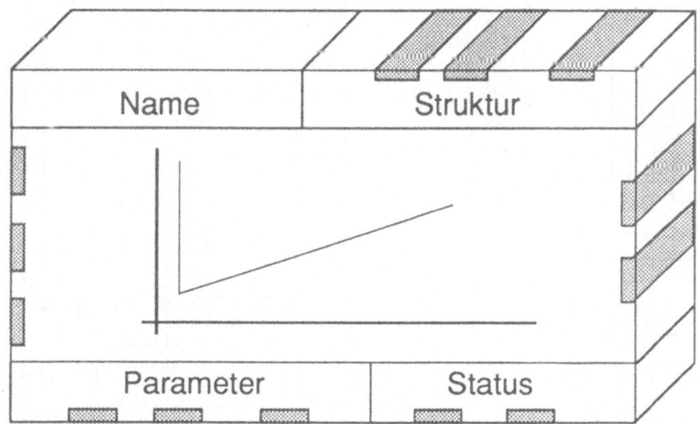

Abb. 1.: Funktionsbaustein

Die konkrete Funktion eines Bausteins kann dabei während des Programmablaufes durch Softwareschalter, gesteuert durch Struktursignale, spezifiziert werden. Module und Modulblöcke bilden zwei grundsätzliche Darstellungsarten zur Beschreibung eines logischen Objektes.

Im Interesse einer hohen Portabilität eines entsprechenden Sprachübersetzers wurden textsprachliche Darstellungsmittel als Basis für die Problemformulierung gewählt. Ein Programm besteht dabei aus einer Anzahl von Modulblöcken, also Funktionsbausteinen hochsprachlichen Ursprungs, woraus sich eine physische Programmstruktur ähnlich dem Funktionenkonzept der Programmiersprache "C" ergibt. Der Programmeintrittspunkt wird durch den ausgezeichneten Modulblock, "CONTROL", fixiert. In jedem Modul-block können wiederum beliebig viele andere Modulblöcke, aber auch Module aufge-rufen werden. Module haben ihren Ursprung auf nichthochsprachlichem Niveau. Hierbei handelt es sich also um Bausteine, die auf Zielsprachniveau (Assemblercode, FORTH,...) erzeugt und erst bei der Programmontage in das Anwenderprogramm integriert werden. Die gemeinsame Integration von Modulen und Modulblöcken in einem Programm ergibt eine logische Struktur wie sie Abb. 2. verdeutlicht.

Jeder Modulblock erscheint in der textsprachlichen Programmdarstellung als Prozedur, bestehend aus Deklarations- und Anweisungsteil. Im Deklarationsteil müssen alle innerhalb des Anweisungsteil verwendeten Datenobjekte mit einem bestimmten Datentyp eingeführt werden.

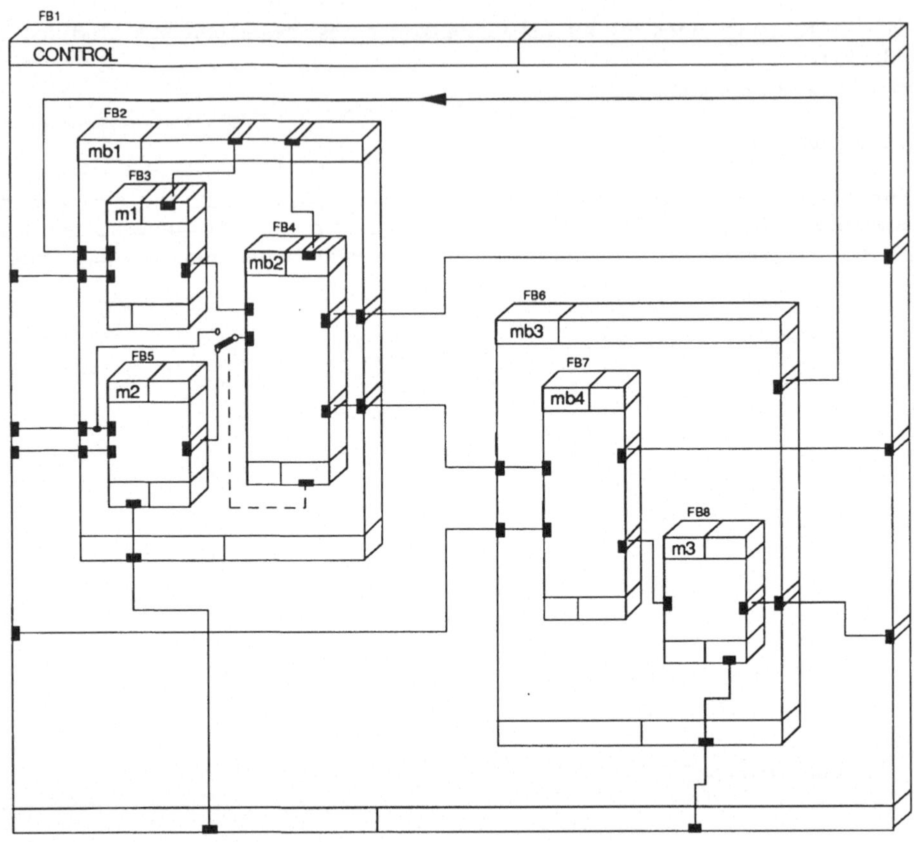

Abb. 2.: Funktionsbausteinverschachtelung
mb...Modulblock, m...Modul, FB...Funktionsbaustein

Datentypen

Die zulässigen Datentypen sind auf die Erfordernisse regelungs- und steuerungstechnischer Aufgabenstellungen, also die Umsetzung signalverarbeitender Strukturen zugeschnitten. Differenziert wird zwischen initialisierbaren Parametern und nichtinitialisierbaren Signalen (Daten). Darstellbar sind Datenobjekte "analogen" und "binären" Charakters. Zulässige Deklarationsdatentypen und mögliche Indizierungen zeigt Abb. 3. Datenobjekte mit gleicher Wertrepräsentation werden in Datenklassen zusammengefaßt, innerhalb dieser Typverträglichkeiten bestehen. Jede Typdeklaration wird durch ein Speicherklassenattribut ergänzt, das über Gültigkeit und Sichtbarkeit eines Datenobjektes entscheidet. Implizit sind alle Datenobjekte "LOCAL" deklariert, was ihre Gültigkeit auf den aktuellen Funktionsbausteinaufruf beschränkt. Den besonderen Erfordernissen, die sich aus dem möglichen Speicherverhalten eines Funktionsbausteins ergeben, wird durch die Speicherklasse "STATIC" Rechnung getragen. Ein so deklariertes Datenobjekt behält seinen Wert über die Funktionsbausteingrenze hinaus, bis zu eben diesem Funktionsbausteinaufruf im nächsten Programmzyklus. Die sich hieraus ergebende Vervielfachung eines Datenobjektes entsprechend der Aufrufe des Funktionsbausteins stellt eine Besonderheit gegenüber bekannten Universalsprachen dar. Ein drittes Attribut, "MONITOR", ermöglicht gleichzeitig einen zur Programmausführung asynchronen Daten- und Parameterzugriff zu Bedien- und Anzeigezwecken.

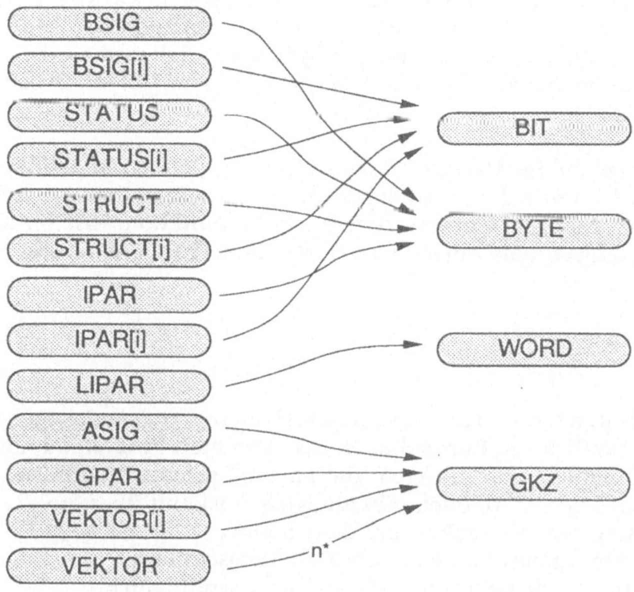

Abb. 3.: Datentypen und Datenklassen

Anweisungen

Der Anweisungsteil bestimmt den Funktionsinhalt des Funktionsbausteins. Die Syntax ist dabei stark an die der Universalsprache "C" angelehnt, was sich in Anweisungsstrukturen, dem Wert von Ausdrücken und ihrer bedingten Auswertung widerspiegelt. Einschränkungen gegenüber "C" sind vor allem in der Beschränkung auf die aufgabenspezifischen Datentypen und daraus resultierenden Konvertierungssinnfälligkeiten gegeben. Anstelle des Funktionsaufrufes in "C" tritt der Modulblock- oder syntaktisch äquivalente Modulaufruf. Neben einer einfachen Aufzählung der Aufrufargumente ist hier auch eine, die Programmlesbarkeit fördernde, Gruppierung der Signale und Parameter (vgl. Abb. 1.) zu Eingangssignalen, Ausgangssignalen, Parametern usw. zulässig. Zum Aufbau von Alternativen bzw. Auswahlstrukturen stehen Strukturanweisungen, wie IF..., IF...ELSE... und SWITCH... zur Verfügung, wobei über "MONITOR" deklarierte Signale auch Strukturbeeinflussungen durch Bedienereingriff möglich sind.

Module

Ein Modul ist ein Funktionsbausteinrahmen, der es gestattet, über den Modulaufruf einen Programmabschnitt in das Programm zu integrieren, der selbst nicht auf dem hochsprachlichen Niveau von "µpcl" entwickelt wurde. Der Modul besitzt eine zweigeteilte Struktur, bestehend aus Modulkopf und Modulkörper. Der Modulkörper ist der funktionsbestimmende, und somit der Teil des Moduls, der im Zuge der Programmmontage in das Anwenderprogramm integriert wird. Der Modulkopf beschreibt in einer fixen Struktur die Anforderungen des Modulkörpers, erwartete Übergabeparameter, deren Anzahl und Datenklasse. Die Köpfe der verfügbaren Module stehen dem Übersetzer in einer Bibliothek zur Verfügung und ermöglichen so die Integration dieser Funktionsbausteine in die semantische Überprüfung von Datentypverträglichkeiten und Aufrufmodalitäten.

Das Haupteinsatzgebiet der Module ist die Bedienung der Prozeßperipherie oder auch die Abwicklung von Kommunikationsdiensten mit der Bedien- und Anzeigetechnik sowie externen Geräten. Module werden aber auch in Funktionsabschnitten eingesetzt, in denen es auf die Zeitvorteile einer maschinennahen Programmierung ankommt.

3. Speichermodell

Das Speichermodell, als die dem Übersetzungsvorgang zugrundeliegende Speicherstruktur, bestimmt wesentlich die Kompaktheit des erzeugten Programms und ist damit maßgebend für die Minimalanforderungen, die an die Speicherressourcen eines Zielsystems gestellt werden. Das Speichermodell gibt zugleich Auskunft über die
- Speicherhaltung von Variablen und Konstanten,
- Art der Speicherorganisation (statisch o. dynamisch) und
- Parametervermittlung beim Modulblock- und Modulaufruf
Bei der Wahl des Speichermodells wurden die Prämissen auf eine zeiteffiziente Parametervermittlung gelegt. Erreicht wird dies durch eine weitgehende Vereinheitlichung in der Daten- und Konstantenvermittlung über die Adresse (call by reference). Eine hohe Effizienz des Zugriffs auf Datenobjekte wird durch ein statisches Speichermodell gewährleistet, wodurch aufwendige Adressrechnungen in den Übersetzungsvorgang verlagert werden. Abb. 4. veranschaulicht die gewählte logische Segmentierung des Speichers in Programm-, Daten- und Zeigersegment.

Deutlich werden hier der Bibliothekscharakter der Modulsammlung, Parametervermittlung durch Zeigerverschiebung, Variablenzugriff durch Zeigeradressierung und auch die Variablenvervielfachung durch Mehrfachaufrufe von Funktionsbausteinen.

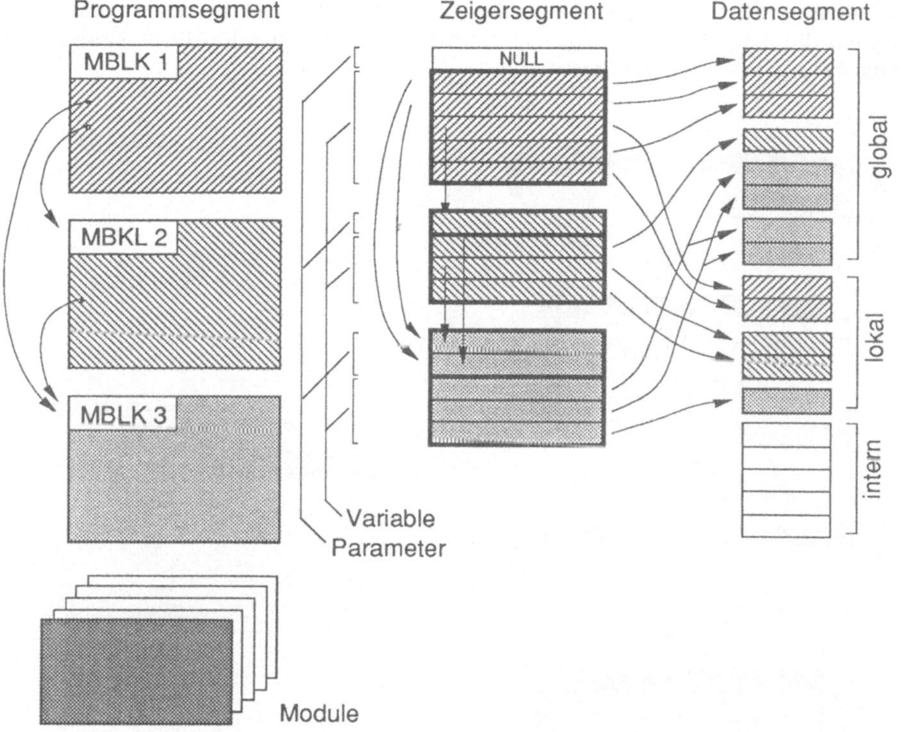

Abb. 4.: vereinfachtes Speichermodell

4. "µpcl"-Sprachübersetzer

Ein dominierender Ansatz für den Strukturentwurf des Sprachübersetzers ist die Vielfalt
potentieller Zielprozessoren. Hieraus leitet sich die Forderung nach einer aufwands-
minimalen Anpaßbarkeit des Entwicklungswerkzeuges an eine konkrete Zielhardware
ab. Dieser Aufwand wird im Falle des "µpcl"-Sprachübersetzers dadurch minimiert, daß
dieser in einen zielsystemneutralen und einen zielsystemspezifischen Teil zerlegt wird
(vgl. Abb. 5.).

Die informationelle Kopplung beider Teiltransformationen wird über einen virtuellen
Zwischencode realisiert. Dieser stellt eine Art Assemblercode für eine stackorientierte
virtuelle Maschine dar, die typische Eigenschaften eines realen Prozessors aufweist, ohne
aber dabei konkrete Besonderheiten zu berücksichtigen. Desweiteren basiert dieser
Zwischencode auf einem festen Speichermodell, das durch eine logische Segmentierung
die Anpassung an segmentierte als auch nichtsegmentierte physische Speichermodelle
zuläßt. Erreicht wird dadurch, daß der wesentliche Aufwand des Übersetzungsvorganges
im zielsystemneutralen Teil des Übersetzers steckt und demzufolge von einer Neuan-

passung nicht berührt wird. Der zielsystemspezifische Teil des Übersetzers, der Expander
hat lediglich die Aufgabe, die Befehle des virtuellen Zwischencodes durch Codesequenzen
des realen Prozessors zu expandieren.

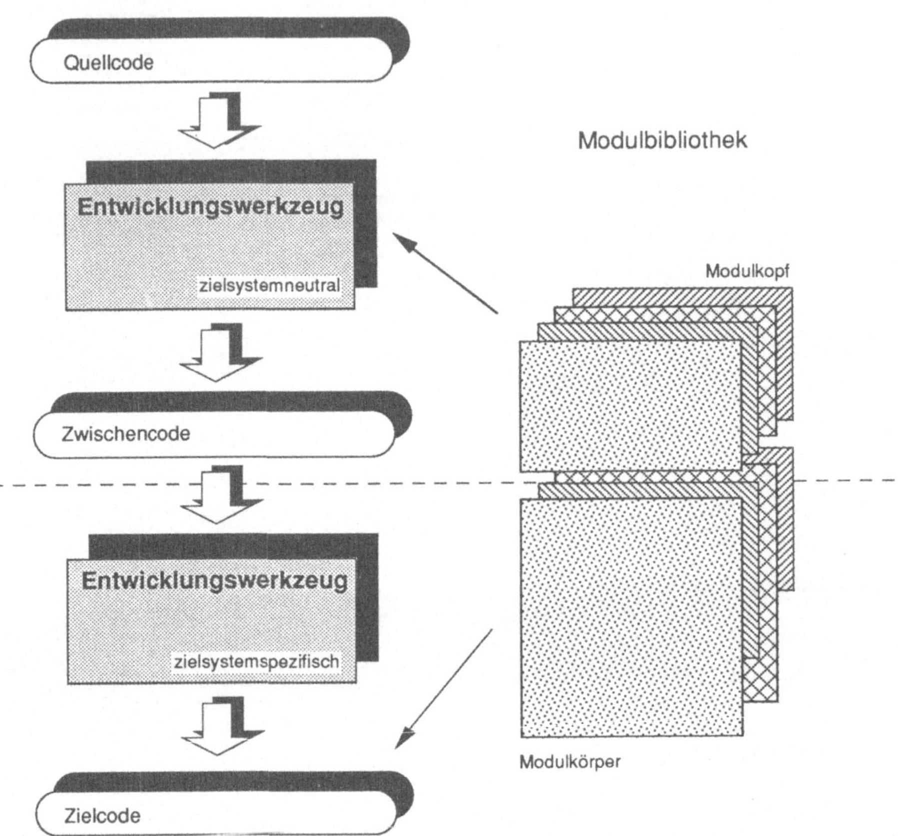

Abb. 5.: Übersetzerstruktur

Implementierung

Der "μpcl"-Sprachübersetzer ist in portablem "C" geschrieben und auf einem UNIX-
System implementiert. Der zielsystemneutrale Teil wurde als Zweipaßstruktur entworfen. Paß 1 vereint syntaktische und semantische Analyse sowie die Zwischencode-
generierung, wobei im Rahmen der Semantikkontrolle die Modulköpfe herangezogen
werden. Paß 2 schließt eine Zwischencodeoptimierung an.

Der zur sytaktischen Analyse konstruierte Programmverfolger wurde mit Unterstützung
der UNIX-Werkzeuge "YACC" und "LEX" entworfen.

5. Applikationen

Universeller Mikrorechnerregler

Das vorgestellte Programmierwerkzeug wurde mit besonderer Zielrichtung auf ein an der Humboldt-Universität/ Sektion Elektronik/ Bereich Automatisierungstechnik entwickeltes einchipmikrorechnerbasiertes prozeßnahes Automatisierungssystem mit feldbusgekoppelten Verarbeitungseinheiten entwickelt, um Firmwareentwickler, als auch Anwender ein effektives Werkzeug zur Anwendersoftwareentwicklung in die Hand zu geben.

Expansionsziel für den Zwischencode bei dieser Applikation ist eine einchipmikro-prozessor-(ZILOG Z8-Serie)basierte Mikrorechnerminimalkonfiguration mit wenigen kByte RAM/ROM. Als Laufzeitrahmen, des durch die Expansion zu erzeugenden Anwenderprogramms, steht ein Echtzeitsteuerprogramm mit Diensten zur feldbus-gestützten Intertaskkommunikation und Prozeßkopplung sowie ein Gleitkommapaket zur Verfügung. Für diesen Einsatz des "µpcl"-Werkzeuges wurde eine Modulbibliothek auf Assemblerniveau erzeugt, die Funktionsbausteine zu folgenden Komplexen bereit-stellt:
  - Analog- und Binärwerterfassung und Ausgabe
  - Intertaskkommunikation
  - analoge Grundbausteine
  - logische Grundbausteine.

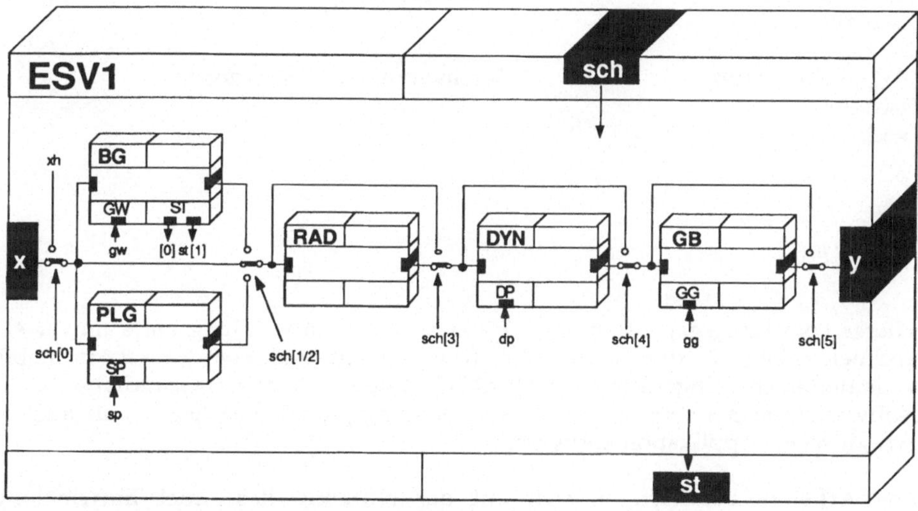

Abb. 6: Komplexer Funktionsbaustein ESV1

Abb. 6. zeigt einen durch Integration von Modulen erzeugten komplexen Funktions-baustein einer Eingangssignalverarbeitung. Das textsprachliche Äquivalent des Modul-blocks "ESV1" stellt sich formuliert in "µpcl" folgendermaßen dar:

```
ESV1 ([sch] E(x) A(y) S(st))
STRUCT  sch;          /* |_|_|1/GB|1/DYN|RAD/1|BG/PLG/1|H/A| */
STATUS  st;           /* |_|_|_|_|_|_|_|OG|UG| */
ASIG   x,y;


{
MONITOR GPAR   xh;              /* Handstellgröße */
LOCAL ASIG     x1,x2;
MONITOR VEKTOR  gw[2];          /* Grenzwert Begrenzer [OG,UG] */
MONITOR GPAR   gg;              /* Grenzwert Geschw.begrenzer */
MONITOR VEKTOR  dp[3];          /* Koeff.vektor Dynamikglied */
MONITOR VEKTOR  sp[10];         /* Stützstellenvektor für die */
                               /* Polygonapproximation */


IF(sch[0])
x = xh;
SWITCH(sch[1])
{
CASE sch[2]:  BREAK;                    /* direkt */
CASE     1:  PLG ((sp) E(x1) A(x1));
BREAK;                                  /* Polygonapproximation */
CASE     0:  BG ((gw) E(x1) A(x1) S(st[0],st[1]));      /* Begrenzer */
}
IF(sch[3])
RAD (E(x1) A(x1));                      /* Radizierer */
DYN ((dp) E(x1) A(x2));                 /* Dynamikglied */
IF(sch[4])
x2 = x1;
GB ((gg) E(x2) A(y));                   /* Geschwindigkeitsbegrenzer */
IF(sch[5])
y = x2;
}
```

Geräterechner

Ein weiterer Forschungsgegenstand o.g. Hauses ist die Entwicklung eines universellen Steuerrechnersystems für die Geräteautomatisierung mit konkretem Einsatz in mobilen Lichtwellenleiterschweißgeräten. Ein FORTH-System ist dabei Grundlage für die Gerätesoftwareimplementierung, sowohl der Betriebssystemkomponenten, als auch der einsatzspezifischen Applikationssoftware.

Laufende Arbeiten beschäftigen sich mit der Nutzung des "µpcl"-Entwicklungswerkzeuges zur Erzeugung letztgenannter Applikationssoftware. Hierbei wird durch die Erzeugung von entsprechenden FORTH-Secondaries der Zwischencode des "µpcl"-Sprachübersetzers direkt lesbar gemacht, wodurch im Entwicklungssystem auf einen Expander verzichtet werden kann. Weitere, der Peripherieanpassung dienende FORTH-Worte, werden als Module in eine "µpcl"-Bibliothek integriert und auf Quellsprachniveau verfügbar gemacht.

6. Zusammenfassung

"µpcl" ist ein Werkzeug für die funktionsbausteinorientierte Programmentwicklung zur Lösung steuerungs- und regelungstechnischer Aufgabenstellungen mittels universeller rechnerintegrierter Lösungen im Bereich der Kleinautomatisierung. Besonderer Wert wurde beim Sprachentwurf auf die Einführung funktionsbausteinorientierter Strukturen gelegt, die sowohl maschinennahe als auch hochsprachliche Programmiermöglichkeiten einschließen. Ein entsprechender Sprachübersetzer wurde dabei durch Trennung von zielsystemspezifischen und zielsystemneutralen Komponenten so zerlegt, daß er eine einfache Anpassung an verschiedene Zielsysteme gestattet.

Hier und in den sich daraus ergebenden Konsequenzen bezüglich der Universalität in der Anwendung für verschiedene Zielsysteme liegt auch der wesentliche Unterschied zu anderen vergleichbaren Systemen. Ein Vergleich ließe sich bezüglich des Anwendungsfalls "Mikrorechnerregler" mit Systemen zur Konfigurierung und Parametrierung von Prozeß- und Industriereglern oder auch kleineren Regelsystemen anstellen. Eine funktionsbausteinorientierte Programmierung hat sich hier generell durchgesetzt. Dabei wird jedoch durch die meisten Systeme eine Programmierumgebung mit abgeschlossenem Bausteinsortiment geboten. Auch was die Programmentwicklung angeht, wird in der Regel eine spezielle PC-Umgebung verlangt, wobei der XT/AT-Standard dominiert. Hierdurch begünstigt, wird der Trend hin zum grafischen Konfigurieren am Bildschirm auch im Bereich der Kleinautomatisierung durch erste Beispiellösungen belegt (Kompaktregler System 6000 von TSC Turnball Control Systems) [3].

Im vorgestellten System wurde der Schwerpunkt hingegen darauf gelegt, ein Entwicklungswerkzeug zu schaffen, das sich für die Programmierung einer Vielzahl von Systemen der Geräte- und Kleinautomatisierung eignet und dabei auch nicht an eine spezielle Entwicklungsumgebung gebunden ist. Dies wird durch eine konsequente Nutzung ausschließlich textsprachlicher Darstellungsmittel und die Programmierung in portablem "C" garantiert und dabei zu Gunsten einer hohen Portabilität auf grafische Elemente verzichtet.

Schrifttum

[1] *Rehwaldt B.:*
Konzeption und Entwurf eines Entwicklungswerkzeuges zur Erzeugung regelungstechnisch orientierter Anwenderprogramme für Mikrorechnerregler, speziell für das dezentralisierte prozeßnahe Automatisierungsgerätesystem DPA02.- Dissertation A.- Berlin: Humboldt-Universität, 1990.

[2] *Rehwaldt B.:*
Dokumentation zum "µpcl"-Sprachübersetzer.- Forschungsbericht.- Berlin: Humboldt-Universität, Sektion Elektronik, 1990.

[3] *Kuhn, U.:*
Neue Regler und Regelsysteme.- In: atp.- München: R. Oldenbourg GmbH, Sonderheft: Der Interkama Report `90, (1990).

Eine datenflußorientierte funktionale Programmiersprache für die Echtzeitdatenverarbeitung

A. Knoll, A. Schweikard und M. Freericks

Technische Universität Berlin
Institut für Technische Informatik, FR 2-2
D-1000 Berlin 10

I Einführung

Funktionale Programmiersprachen weisen gegenüber den bekannten imperativen Programmiersprachen, wie beispielsweise den Sprachen der ALGOL-Familie, eine Reihe von Vorteilen auf: Die Abstraktion von Maschineneigenheiten und algorithmischen Details ist wesentlich höher, deshalb kann die Spezifikation eines Programmes wesentlich näher am Problem orientiert sein. Als Folge davon sind in funktionalem Stil geschriebene Programme häufig wesentlich kürzer als ihre imperative Entsprechung; sie sind einfacher zu lesen, besser zu verstehen, einfacher zu verfassen und daher weniger fehleranfällig.

Es hat in letzter Zeit mehrere Ansätze zur Konzeption funktionaler Echtzeitprogrammiersprachen gegeben, die auf spezielle Einsatzfälle abzielen oder für theoretische Untersuchungen gedacht sind. Die Sprachen Lustre[1] und Signal[2] basieren auf der nichtprozeduralen Sprache Lucid[3] und eignen sich für die Programmierung synchroner Systeme. Silage[4] ist eine funktionale Programmiersprache für digitale Signalverarbeitung, die jedoch keine Rekursion zuläßt. Arctic[5] kommt aus dem Bereich der Klangerzeugung bzw. elektronische Musik, Ruth[6] ist datenflußorientiert und bietet die Möglichkeit, Zeiten über während der Laufzeit auszuwertende „Zeitbäume" einzubeziehen.

Die Programmiersprache ALDiSP[7] ist demgegenüber eine funktionale Programmiersprache für allgemeine Anwendungen in der Echtzeitdatenverarbeitung. Die Mächtigkeit von ALDiSP ist der von Scheme[8] oder ML[9] vergleichbar, ALDiSP kann auch für über den engeren Kreis der Echtzeitdatenverarbeitung hinausgehende Aufgabenstellung verwendet werden. Im folgenden werden wir uns jedoch auf die für die Echtzeitdatenverarbeitung erforderlichen Eigenschaften beschränken. Die in diesem Zusammenhang wichtigen Ziele bei der Entwicklung der Sprache waren:

+ Einfache Darstellung und Manipulation von Datenflüssen
+ Bereitstellung von Konstrukten zur zeitlichen Einplanung von Prozessen in Abhängigkeit von synchron und asynchron auftretenden Ereignissen
+ Verzicht auf Variablen
+ Einengung des Sprachumfangs auf solche Konstrukte, die in effizienten Code übersetzt werden können
+ Mechanismen für synchrone und asynchrone Ein-/Ausgabe
+ ·Reiche Auswahl vordefinierter parametrierter Datentypen samt mächtiger Operatoren
+ Möglichkeiten zur Behandlung von Ausnahmen
+ Leistungsfähiges Modulkonzept, Möglichkeiten zur Einbeziehung von in anderen Sprachen geschriebenen, ausgetesteten Code-Modulen

II Spracheigenschaften

Programm- und Kommunikationsstruktur

Ein ALDiSP-Programm besteht im allgemeinen aus verschiedenen Modulen; jedes dieser Module enthält Funktionen, die die auf die Datenobjekte anzuwendenden Algorithmen spezifizieren. Es gibt ein Hauptmodul, das Funktionen aus allen anderen Modulen importieren kann; dieses Hauptmodul enthält einen *Netzwerk-Teil,* in dem beschrieben wird, wie die einzelnen importierten Funktionen zusammenarbeiten, indem sie Daten von anderen Funktionen empfangen bzw. an sie senden.

Ein grundlegender Datentyp der Sprache ist der Datentyp *stream.* Ströme stellen potentiell unendlich lange Listen dar, deren Elemente erst bei Bedarf berechnet werden. In ALDiSP dienen sie zur Darstellung von zeitdiskreten (Prozeß-)Größen und zur Kommunikation zwischen Funktionen. Die Elemente des Stroms werden von einem Strom-Generator erzeugt, der bei jeder neuen Nachfrage durch eine stromkonsumierende Funktion jeweils ein neues Element erzeugt. Zu jedem beliebigen Zeitpunkt ist nur eine endliche Menge von Stromelementen (nämlich die bereits erzeugte) vorhanden, während die (unendliche) Menge der folgenden Elemente erst noch berechnet werden muß. Die den Strom-Generator realisierende Funktion bleibt inaktiv, bis eine stromkonsumierende Funktion ein neues Element anfordert (Call-by-Need-Semantik). Auf diese Art und Weise können potentiell unendlich lang andauernde Signale bequem manipuliert werden.

Das folgende ALDiSP-Programm illustriert die Benutzung von Funktionen und Strömen. Es erzeugt als Ausgabe einen Strom von Zahlen, die die Lösung des Problems von Hamming [10] darstellen: Zu produzieren ist die geordnete und wiederholungsfreie Folge natürlicher Zahlen, die durch keine anderen Primzahlen als 2, 3 und 5 teilbar sind. Das Problem kann in einer Sprache wie ALDiSP, die komfortable Stromverarbeitung samt automatischer Speicherverwaltung zur Verfügung stellt, gelöst werden, indem ein Strom generiert wird, der zunächst nur ein Initialelement 1 enthält. Dieser Strom wird sowohl ausgegeben, wie auch in drei Prozesse eingespeist, die die Stromelemente mit 2, 3 und 5 multiplizieren. Die von diesen drei Prozessen erzeugten Zahlen werden durch einen weiteren Prozeß (hier mit merge3Streams bezeichnet) sortiert, zu einem einzigen Strom verschmolzen und von mehrfachem Auftritt derselben Zahl befreit. Der vom Prozeß merge3Streams ausgegebene Strom stellt den Ausgabestrom dar, dessen Elemente wieder in die Multiplizierer eingespeist werden. Bild 1 zeigt graphisch, wie der Algorithmus arbeitet; Bild 2 zeigt das zugehörige ALDiSP-Programm.

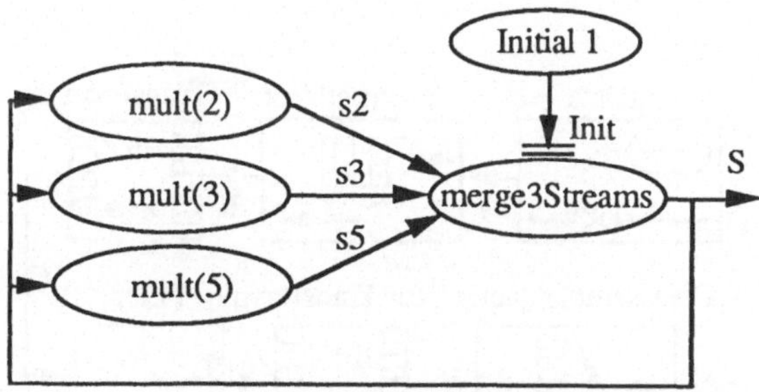

Bild 1: Netzwerk zur Lösung des Problems von Hamming

Die Topographie des Netzwerkes, d.h. die Angabe, wie der Datenfluß zwischen den Funktionen mult() und merge3Streams organisiert werden soll, spiegelt sich in dem mit net eingeleiteten Programmteil wider. Der Ausgabestrom S kann von anderen Funktionen weiterverwendet werden. Das dargestellte Programm spezifiziert den Algorithmus komplett; es sind keine zusätzliche Speicherverwaltungsmaßnahmen, Zuweisungen, etc. erforderlich bzw. möglich.

```
func mult (n) (aStream) = n*head(aStream) :: delay mult(n) (tail(aStream))


func mergeStreams (strA, strB) =
if   head(strA) < head(strB) then head(strA) :: delay mergeStreams(tail(strA), strB))
|    head(strA) > head(strB) then head(strB) :: delay mergeStreams(tail(strB), strA))
|    head(strA) = head(strB) then mergeStreams(tail(strA), strB) \ ---- suppress multiples
endif


func merge3Streams (strA, strB, strC) =
        mergeStreams (strA, mergeStreams (strB, StrC))


net
S = 1 :: delay merge3Streams(s2, s3, s5)
s2 = mult (2) (S)
s3 = mult (3) (S)
s5 = mult (5) (S)
in
  S
endnet
```

Bild 2: ALDiSP-Programm zur Realisierung des Netzwerkes von Bild 1

Die Funktionsresultate werden auf Ströme geschickt, indem die Funktion einem stream-Bezeichner zugewiesen wird. Sämtliche Maßnahmen, die erforderlich sind, um einen Strom zu erzeugen, ihm die Funktionsresultate zuzuordnen und auf ihm zu transportieren, verbergen sich hinter dem als Zuweisungssymbol verwendeten Gleichheitszeichen. Eine Funktion, die für eine Zahl definiert wurde, kann auch auf einen Strom angewendet werden. In diesem Fall wird automatisch dafür gesorgt, daß die Funktion auf jedes Element des Stromes angewendet wird. In ALDiSP funktioniert diese automatische Abbildung nicht nur bei Strömen, sondern auch bei homogenen Datentypen. Deren wichtigste Repräsentanten sind Felder (Arrays). Hier wird die Funktion auf alle Elemente des homogenen Datentyps angewendet. Die (zweistufige) Abbildung von Funktionen auf Ströme und Felder werde im folgenden Beispiel demonstriert:

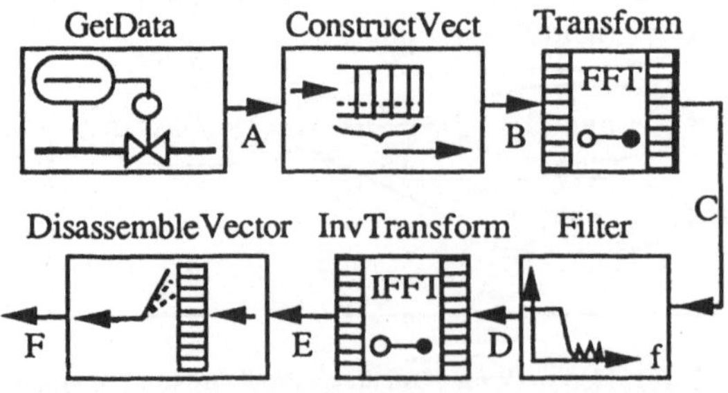

Bild 3: Netzwerk zur Auswertung einer Prozeßgröße

Das in Bild 3 dargestellte Netzwerk dient zur Analyse einer Prozeßgröße „Durchflußmenge". Dazu soll die in bestimmten Zeitabständen abgetastete Prozeßgröße zu Blöcken (Vektoren) gruppiert werden, eine blockweise Transformation in den Frequenzbereich erfolgen und im Frequenzbereich eine Tiefpaßfilterung der (nun komplexwertigen) Vektoren vorgenommen werden. Nach Rücktransformation wird wieder ein Strom bestehend aus skalaren Elementen erzeugt. Die Filterung im Frequenzbereich werde derart vorgenommen, daß die Amplitude der transformierten Vektoren mit der Filterfunktion multipliziert werde, während die Phasen unverändert bleiben: Jede Komponente des die Amplitudenwerte des Spektrums enthaltenden Vektors werde dazu mit dem korrespondierenden Dämpfungsfaktor der Filterfunktion multipliziert. Es werde angenommen, daß die Funktionen

```
GetData () = ...          \ Data acquisition
FFT (aVector) = ...       \ Input: Real vector, Output:Two real vectors
IFFT (twoVectors) = ...   \ Inverse FFT
AttnCoeffs (n) = ...      \ Generate n spectral coefficients (transfer function)
```

bereits innerhalb eines Moduls mit dem Namen FreqDomain definiert wurden. Dann nimmt das ALDiSP-Hauptmodul die in Bild 4 dargestellte Form an.

```
Imports GetData, FFT, IFFT, AttnCoeffs from FreqDomain
net
    A = GetData()
    B = VectorizeScalarStream (A, 256)
    (Ampl, Phase) = FFT (B)
    D = (Ampl * AttnCoeffs(256), Phase)    \ Apply " * " to all elements of both vectors
    E = IFFT (D)
    F = ListifyVector (E)                  \ Now we have a stream of scalars again
in
    F
endnet
```

Bild 4: ALDiSP - Programm für das Netzwerk nach Bild 3

Man beachte, daß FFT zwei Ströme erzeugt (strukturieres Ergebnis), nämlich für Amplitude und Phase. Wenn diese beiden Ströme getrennt gehalten werden, ist die Manipulation der Amplitude allein sehr einfach. Jede ALDiSP-Funktion kann multiple Ergebniswerte erzeugen, damit ist es nicht erforderlich, getrennt zu haltende Funktionsresultate auf einen Strom zu „multiplexieren" und vor Weiterbenutzung in der konsumierenden Funktion zu demultiplexieren. Von der automatischen Abbildung einer Funktion auf homogene Datenstrukturen wird in der Programmzeile in Bild 4 Gebrauch gemacht, in der die Filterung vorgenommen wird: Der Operator "*" wird auf alle Elemente der Blöcke/Vektoren Ampl und AttnCoeffs angewendet und hat als Resultat einen neuen Vektor mit skalaren Elementen. Die Sprache verfügt darüberhinaus über eine Reihe vordefinierter Funktionen zur Manipulation von Feldern (siehe [7]); es wird daher nur sehr selten erforderlich sein, Feldelemente einzeln zu behandeln.

Dieser Beispiel zeigt deutlich, daß der funktionale Ansatz ideal mit dem „Black-Box-Denken" des Systemanalysten und Regelungstechnikers harmoniert: Die Möglichkeiten zur Manipulationen von Strömen, die automatische Abbildung von Funktionen auf Ströme und insbesondere von Funktionen auf Felder macht die Transformation von

Block-Diagrammen in Programme extrem einfach. Der Programmierer muß komplexe Datenstrukturen nicht zunächst auseinandernehmen, ihre Komponenten behandeln und danach wieder zusammensetzen. Er kann diese Datenstrukturen stattdessen als Ganzes behandeln und sich darauf verlassen, daß der Abbildungsprozeß auf die einzelnen Elemente vom Compiler korrekt vorgenommen wird.

Wie viele moderne funktionale Programmiersprachen ist ALDiSP streng typisiert und polymorph. Das bedeutet, daß Funktionen mit Parametern unterschiedlichen Typs aufgerufen werden können (siehe dazu das folgende Beispiel). Der Übersetzer erkennt Typinkonsistenzen automatisch, indem er den Typ der Operation, die auf einen bestimmten Parameter angewendet wird, analysiert und feststellt, ob Operation und Operand zueinander typkompatibel sind. Sowohl Funktionen wie auch Operatoren können überladen werden, d.h. für vollständig unterschiedlich geartete Operanden definiert werden. Operatoren können in Post-, Prä- oder Infix-Notation verwendet werden, je nachdem, was besser lesbar oder leichter anzuwenden ist. Die Operator-Präzedenz ist in sechs verschiedenen Stufen einstellbar.

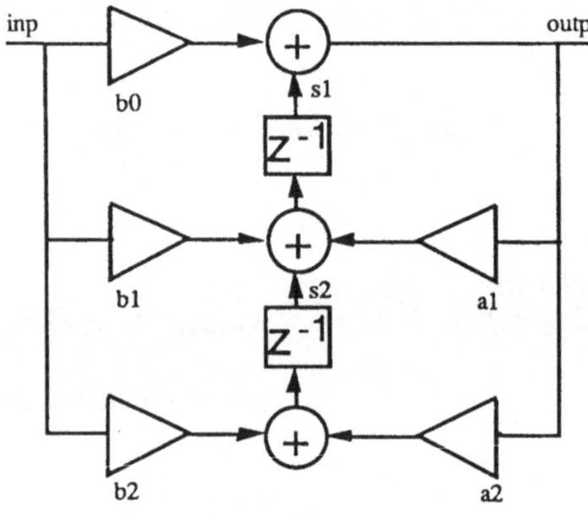

Bild 5: Filternetzwerk

```
func IIR (b0,b1,b2,a1,a2: number) (inp: stream) =
let
    outp = b0*inp + s1
    s1 = 0 :: delay  b1*inp + a1*outp + s2
    s2 = 0 :: delay  b2*inp + a2*outp
in
    outp
endlet
```

Bild 6: Implementierung des Filternetzwerks nach Bild 5

ALDiSP erlaubt ferner die Definition von Funktionen höherer Ordnung, d.h. Funktionen, die andere Funktionen als Eingabeparameter tragen oder Funktionen als Resultate abliefern. Ströme können rekursiv definiert werden und sie können um eine beliebige Anzahl von Elementen verschoben bzw. verzögert werden. Deshalb kann das Stromkonzept auch ideal in der Regelungstechnik zur Spezifikation von digitalen Reglern verwendet werden.

Als Beispiel werde das Filternetzwerk nach Bild 5 betrachtet (siehe z.B. [11]): Dieses Netzwerk kann durch die ALDiSP-Funktion nach Bild 6 direkt realisiert werden. Der Strom outp ist das Resultat der elementweisen Addition der Ströme inp (jedes Stromelement gewichtet mit b0) und s1. Der Strom s1 besteht aus der Konkatenation des Elementes "0" mit der gewichteten Summe aller anderen Ströme, die jeweils um ein Element verzögert wurden. Man beachte, daß die Funktion IIR Integer-Zahlen jedes Typs sowie auch Real- und Complex-Zahlen filtert, da die Operatoren "+", "*" und "::" für alle diese Datentypen definiert sind. Ein kürzere Spezifikation für das Filter als die in Bild 6 gezeigte ist wohl nur schwerlich vorstellbar.

Asynchrone Kommunikation und Ein-/Ausgabe

Ströme sind ein adäquates Mittel zur Repräsentation von Datenflüssen im synchronen Fall, in welchem die permanente Verfügbarkeit des jeweils nächsten angeforderten Stromelements garantiert werden kann. Für Situationen, in denen nicht angegeben werden kann, ob und wann ein nächstes Stromelement verfügbar wird, ist dieses Konzept weniger geeignet. Es ist zwar vorstellbar, daß ein Strom-Generator im Falle der Nichtverfügbarkeit eines Datenobjektes ständig Stromelemente erzeugt, die der konsumierenden Funktion die Nichtverfügbarkeit von Nutzinformation mitteilen, das aber wäre höchst ineffizient. Aus diesem Grunde wurden in ALDiSP die sogenannten *pipes* eingeführt. Syntaktisch sind pipes äquivalent zu streams, sie sind jedoch nicht nachfrage-, sondern datengetrieben. Eine pipe kann als Puffer angesehen werden, der Datenobjekte aufnimmt und speichert, sobald sie verfügbar werden. Eine konsumierende Funktion kann Daten aus pipes in genau der gleichen Weise wie bei streams benutzen. Wenn eine Funktion Daten aus der pipe entnimmt, wird diese elementweise geleert. Wenn ein Konsument auf eine leere pipe zugreifen will, so wird er *suspendiert*. Auf pipes ist ein Prädikat isAvailable definiert, welches den Wahrheitswert true annimt, sobald sich mindestens ein Element in der pipe befindet.

Sowohl streams wie pipes können direkt für Ein-/Ausgabe verwendet werden. Streams werden für synchrone Eingabe verwendet, bei der einzugebende Daten ständig verfügbar sind. Für asynchrone E/A werden pipes benutzt. Mit pipes und streams sind auf einfache Art und Weise Systeme realisierbar, die unterschiedliche Ankunftsraten aufweisen (multirate-systems). Logische Geräte, auf die ständig nichtblockierend zugegriffen werden kann werden als *register* bezeichnet, während blockierende Geräte *port* heißen. Normalerweise werden register mit streams assoziiert und ports mit pipes. Damit ist eine vollständige Verbindung des Programmsystems mit der Umwelt bzw. Peripherie möglich. Weil pipes und streams syntaktisch gleich und im Gebrauch ähnlich sind, ist es sehr einfach, eine mit einem E/A-Gerät verbundene pipe durch einen von einem Strom-Generator gespeisten Strom zu ersetzen. Durch die polymorphe Struktur der Sprache kann der Strom-Generator sogar in vielen Fällen Datenobjekte eines anderen Typs als die pipe liefern, ohne daß die konsumierende Funktion geändert werden muß. Diese Ersetzung kann in der Testphase eines Programms sehr nützlich sein, wenn nämlich zunächst Peripheriedaten vollständig reproduzierbar erzeugt werden sollen, bevor das Programmsystem mit der Umwelt verbunden wird.

Die Einbeziehung von Zeiten und das suspend-Konstrukt

In ALDiSP ist sowohl synchrone wie asynchrone Prozeßeinplanung möglich. Bei synchroner Einplanung wird eine Aktion zu bestimmten, äquidistanten Zeitpunkten ausgeführt, während bei asynchroner Einplanung ein Prozeß für eine gewisse Zeit suspendiert bleibt, bis sich entweder ein externer Zustand ändert oder ein Zeitüberwachungsmechanismus den Prozeß aktiviert. In ALDiSP gibt es nur ein einziges Konstrukt, das alle Möglichkeiten der Zeiteinplanung umfaßt, das suspend-Konstrukt:

suspend expr1 **until** expr2 **within** timeExpr1, timeExpr2

Wenn eine Funktion aufgerufen wird, die einen suspendierten Ausdruck enthält, wird ein anonymer Prozeß kreiert. Dieser Prozeß wird sofort suspendiert und bleibt inaktiv, bis expr2 wahr wird. Sobald dies der Fall ist, wird garantiert, daß der Prozeß *nach* Ablauf der Zeit timeExpr1 und *vor* Ablauf der Zeit timeExpr2 wieder aktiviert wird. Nach seiner Aktivierung wertet der Prozeß expr1 aus und terminiert. Ein einfaches Beispiel möge den Gebrauch von suspend veranschaulichen:

suspend OpenValve() **until** OverPressure **within** 0 ms, 0.5 ms

Hier wird ein Sicherheitsventil, das von einer Funktion OpenValve gesteuert wird, in einem Zeitintervall von 0,5 ms nach dem Auftreten von Überdruck geöffnet. Mit suspend sind Makros für alle anderen wichtigen Fälle zeitlicher Einplanung leicht zu schreiben, so z.B.:

```
\ ---- Synchrone Verzögerung
expr after time ≡ suspend expr until true within time, time


\ ---- Asynchron, Aktion wird sofort ausgeführt
expr1 when expr2 ≡ suspend expr1 until expr2 within 0 sec, 0 sec


\ ---- Asynchron mit Zeitüberwachung
when (expr1, expr2, timeoutPeriod) ≡
let
    StartTime = SystemClock()
in
    suspend expr1 until (expr2 or SystemClock - StartTime > timeoutPeriod) within 0 sec, 0
sec
endlet
```

Die timeExpressions dürfen zur Laufzeit auswertbare, dynamische Ausdrücke sein; sind sie jedoch bereits zur Übersetzungszeit bekannt, kann ein statischer Zeiteinteilungsplan bereits vom Übersetzer erstellt werden.

Mit suspend und dem isAvailable-Prädikat können asynchrone pipes sehr elegant manipuliert werden. So nimmt beispielsweise die wichtige Operation des Zusammenführens zweier pipes (von denen jede Daten verfügbar haben kann) die folgende Form an:

```
proc merge (pipe1, pipe2) =
suspend
    if isAvailable(p1) then head(p1) :: merge(p2,p1)
    else head(p2) :: merge(p1,p2)
    endif
until
    isAvailable(p1) or isAvailable(p2)
within 0 sec, 0 sec
```

Dies ist ein leichtverständliches Gegenstück zum ALT-Konstrukt aus Occam [8], das dazu benutzt wird, Daten, die durch unterschiedliche Kanäle bei verschiedenen Ankunftsraten ankommen, zusammenzuführen. Es ist offensichtlich, daß die Zusammenführung synchroner Ströme noch einfacher aufschreibbar ist. Mit Hilfe des suspend-Konstrukts sind auch Unterbrechungsbehandler leicht in ALDiSP zu formulieren.

Das Typsystem der Sprache
Das Typsystem von ALDiSP ist auf Prädikaten und nicht auf Termen aufgebaut: Eine beliebig gewählte Menge von Werten kann einen Typ darstellen. Auch Funktionen können zur Typvereinbarung herangezogen werden. Zum Beispiel kann ein Typ definiert werden, dessen Wertebereich alle Vielfache von 3 umfaßt:

type mult3(x) = **if** isInt (x) **then** x **mod** 3 = 0 **else** false **endif**

In den meisten Fällen sind solche komplexe Definitionen natürlich nicht erforderlich. Eine umfassende Menge von Grunddatentypen ist bereits vordefiniert, die meisten Grunddatentypen sind darüberhinaus parametriert. Vordefinierte Typklassen sind:

- Basistypen zum Aufbau des Typsystems
- Atomare Typen: Unstrukturiert, alle Arten von Zahlen
- Felder: n-dimensionale Reihungen von Objekten desselben Typs
- Produkttypen zur Darstellung von Listen
- Summentypen (Records)
- Maschinentypen: Register, Ports, Interrupts

Beipiele für atomare Typen sind Zahlen: nBitInteger (n), nBitCardinal (n), FixInt (n,m), ShortReal, Real, Longreal, etc. Alle gängigen Operationen auf diesen Typen sind vorhanden. Für Felder sind eine ganze Reihe weiterer Operationen vordefiniert, damit der Zugriff auf ein einzelnes Element praktisch nie erforderlich wird:

- Automatische Abbildung: [1,2,3] + 10 → [11,12,13]
- Selektive Änderung: UpdVector ([1,2,3], 1, 5) → [1,5,3]
- Reduktion: Reduce ("-", [1,2,3,4]) → ((1 - 2) - 3) - 4
- Erzeuge Untervektor: SubVector ([1,2,3], 0, 2) → [1,2]

Ähnliche Funktionen sind für zweidimensionale Felder verfügbar; ferner sind Funktionen vordefiniert, die Spalten und Zeilen ganzer Matrizen manipulieren, Diagonalen extrahieren, Determinaten bestimmen und Matrizen reduzieren.

Ausnahmebehandlung
Ausdrücke und Funktionen können bewacht werden, d.h. während ihrer Auswertung zu erwartende Fehler können spezifiziert und im Falle ihres Auftritts abgefangen werden. Falls ein Fehler auftritt, wird ein vordefinierter oder benutzerdefinierter Ausnahmebehandler aufgerufen. Prinzipiell hat ein bewachter Ausdruck die folgende Form:

guard expr
on   ExceptionDefinition
:

on   ExceptionDefinition
endguard

Wenn der Fehler auftritt, wird die entsprechende Behandlungsroutine aufgerufen. Nach der Auswertung des Behandlers wird die Auswertung von expr nicht fortgesetzt, sondern der im Behandler definierte Wert wird zurückgeliefert. Sollte es dennoch erforderlich sein, die Auswertung fortzusetzen, kann dies durch Aufruf der Funktion continue ebenfalls geschehen. Beispiel:

```
func reciprocalPlus3Vers1 (x) =    guard (1/x) + 3
                                   on DivisionByZero = 42
                                   endguard

func reciprocalPlus3Vers2 (x) =    guard (1/x) + 3
                                   on DivisionByZero = continue(42)
                                   endguard
```

Wird die erste Funktion mit dem Wert 0 aufgerufen, liefert sie als Resultat 42 ab. Die zweite Funktion liefert bei einem entsprechenden Aufruf den Wert 45, da mit der Auswertung hinter der Aufrufstelle des Ausnahmebehandlers fortgefahren wird.

Das Konzept der Ausnahmebehandlung wird auch verwendet für die Definition der Rundungsmodi im Falle von Zahlenüberläufen. Diese Rundungsmodi sind benutzerdefinierbar, die wichtigsten für Integer-Zahlen vordefinierten sind: Signaling, Ignoring, Realing, Complexing, Saturated, Wrapping. Vordefinierte Rundungsmodi für Real-Zahlen sind: RoundToZero, RoundToPlus, RoundToMinus, RoundToEven, RoundToOdd.

III Zusammenfassung und Ausblick

Es wurde eine mächtige polymorphe funktionale Programmiersprache für die Echtzeitdatenverarbeitung vorgestellt, die über Funktionen höherer Ordnung verfügt, die eine automatische zweistufige Abbildung von Funktionen auf unendliche und endliche Datenstrukturen vornimmt und die über universelle Möglichkeiten zur zeitlichen Einplanung von Aktionen verfügt. Ferner stellt sie die üblichen Konstrukte zur Modularisierung samt überladbarer Operatoren zur Verfügung. Sowohl synchrone wie asynchrone Datenströme können komfortabel gehandhabt werden, dies macht die Programmierung von Systemen mit unterschiedlichen Datenflußraten einfach. Direkter Maschinenzugriff und Hardware-Unterbrechungsbehandlung sind möglich. Ein sehr flexibles Typsystem, ein leistungsfähiger Ausnahmebehandlungsmechanismus und eine große Anzahl vordefinierter Funktionen ermöglichen die Bildung komplexer, der Problemstellung angemessenen Datenstrukturen und ihre Manipulation als ganzes, ohne sich in algorithmischen Details zu verlieren. Mit dieser Sprache können Probleme auf sehr hohem Niveau spezifiziert werden, oft als direkte Entsprechung ihrer mathematischen Beschreibung. Das bedeutet auf der anderen Seite, daß ein großer Teil des Aufwandes, den normalerweise der Programmierer übernimmt (beispielsweise für die Organisation der Speicherverwaltung) nunmehr vom Übersetzer übernommen werden muß. Demensprechend ist es wesentlich schwieriger, Code für Standard-Prozessoren zu erzeugen, als dies von imperativen Sprachen bekannt ist. Andererseits ist es zumindest prinzipiell wesentlich einfacher, aus einer funktionalen Beschreibung Code für Multiprozessorsysteme zu erzeugen, als sequentialisierten Code einer imperativen Sprache zu re-parallelisieren. Mit fortschreitender Compiler-Technologie wird der Unterschied in der Code-Qualität zwischen den beiden Sprachklassen zunehmend geringer werden, die Vorteile der höheren Abstraktion von der Maschine und der parallelen Ausführbarkeit funktionaler Programme werden jedoch bestehen bleiben.

IV Schrifttum

[1]   **P. Caspi, N. Halbwachs, D. Pilaud, J.A. Plaice**
Lustre, a Declarative Language for Programming Synchronous Systems
Proc. 14th ACM Symposium on Principles of Programming languages, München, 1987

[2]   **A. Benveniste, P. Bournai, T. Gautier, P. LeGuernic**
Signal: A Dataflow Oriented Language for Signal Processing
INRIA, Rapport No. 378, März 1985

[3]   **E. Ashcroft, W. Wadge**
Lucid, a Nonprocedural Language with Iteration
Comm. ACM, Vol. 20, No. 7, Juli 1977

[4]   **European Development Center**
Silage Compiler Reference Manual
Brüssel, 1989

[5]   **R. Dannenberg**
Arctic: A Functional Language for Real-Time Control
1984 ACM Symp. on Lisp and Functional Programming, Austin, Texas

[6]   **D. Harrison**
Ruth: A Functional Language for Real-Time Programming
LNCS 259, Parallel Arch. and Languages Europe, Springer-Verlag, 1987

[7]   **M. Freericks, A. Knoll**
ALDiSP - Eine applikative Programmiersprache für Anwendungen in der Echtzeitdatenverarbeitung und der digitalen Signalverarbeitung
Technischer Bericht, Techn. Univ. Berlin, FB 20 Nr. 90-9

[8]   **H. Abelson et. al.**
Revised^3 Report on the Algorithmic Language Scheme
MIT AI Memo 848a, Sept. 1986

[9]   **R. Milner**
A Proposal for Standard ML
1984 ACM Symp. on Lisp and Functional Programming, Austin, Texas

[10]   **P. Pepper, G. Egger**
The Opal Project
Interner Bericht, TU Berlin, 1987

[11]   **Leonhard, W.**
Digitale Signalverarbeitung in der Meß- und Regelungstechnik
Stuttgart: Teubner, 1989

[12]   **INMOS Ltd.**
Occam 2 Reference Manual
Prentice Hall, 1988

Architektur und Kommunikation

Untersuchungen zur Implementierung einer verkürzten Kommunikationsarchitektur (Mini-MAP-Konzept)

H. Rzehak, A.E. Elnakhal

Universität der Bundeswehr München
Werner-Heisenberg-Weg 39, 8014 Neubiberg

Zusammenfassung

Verschiedene Untersuchungen, darunter auch Arbeiten der Autoren, haben
gezeigt, daß der Kommunikationskontroller in lokalen Rechnernetzen mit
einer vollen Protokollhierarchie über 7 Schichten einen Engpaß darstellt. Um diesen Engpaß zu beseitigen wird die Verwendung einer verkürzten Kommunikationsarchitektur mit 3 Schichten (Mini-MAP-Konzept;
Process Communications Architecture der ISA) vorgeschlagen. Bei der
Implementierung in einem Tokenbus-Netz können verschiedene Alternativen gewählt werden. Eine erste Analyse ergibt, daß es von den Be-
triebsbedingungen abhängt, welche Alternative zu bevorzugen ist. Es
wird eine Pilotimplementierung entworfen, in der die verschiedenen
Alternativen und die volle Architektur (7 Schichten) in einem Netz
koexistieren können. Dies könnte auch im praktischen Einsatz verwendet
werden, um situationsgerecht die optimale Alternative auszuwählen. Abschließend wird die Übertragbarkeit einer der Alternativen auf
CSMA/CD-Netze diskutiert.

Abstract

Performance measurements reveal that the communication controller
implementing the full stack of ISO reference model is a bottlenack.
The paper discusses the benifits and applicability of a reduced communication architecture using the acknowledged services of the LLC
sublayer, which can be provided by two alternatives over a token bus
network (with and without immediate response). An implementation for
testing the performance of both variants is presented and first analytical results are discussed.

1. Motivation

Ausgangsbasis dieses Beitrages sind Leistungsuntersungen des Zeitbedarfes für die Kommunikation in MAP-Netzen. Die Ergebnisse der Untersuchungen zeigen, daß eine Minimierung des erforderlichen Zeitbedarfes
erfolgen kann, wenn man die zur Erbringung sicherer Datenkommunikation
benötigten Protokollsätze reduziert. Das Interesse gilt hierbei den
Zeitverhältnissen für zwei kommunizierende Prozesse, die sich im
allgemeinen Fall gemäß Bild 1 ergeben.

Im Einzelfall ist dabei die Zeit T_D (Durchlaufzeit) bis zum Eintreffen der Nachricht beim Empfänger oder die Zeit T_A (Antwortzeit) bis zum Eintreffen der Bestätigung bzw. der Ergebnisse beim Sender von besonderem Interesse.

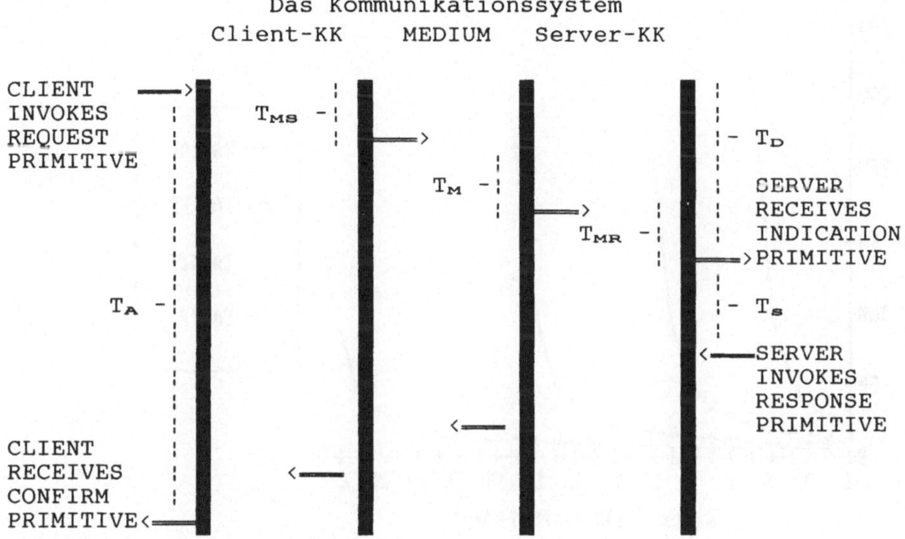

Bild 1   Das allgemeine Kommunikationsmodell zwischen zwei Anwenderprozessen

Um einen Überblick für die Größenordnung der Zeitverhältnisse zu geben, seien einige Ergebnisse aus Untersuchungen an einem MAP-Netz (vgl. [1], [2]) zusammengefaßt. Die Untersuchung wurde für eine Nachrichtenlänge von 1000 Oktett bei einer Übertragungsrate 5 Mbps (carrierband) vorgenommen. Zunächst sind in einer Tabelle die Durchlaufzeit und Antwortzeit für eine Bedienzeit $T_S = 0$ des Servers von Schicht zu Schicht, d.h. als Anwender gilt die jeweils höhere Schicht, in einem ansonst unbelasteten Netz zusammengestellt.

Schicht	TD,Schicht	TA,Schicht bei $T_S{=}0$
Mediumzugriff (MAC)	3 ms	6 ms
Verbindungskontrolle (LLC)	4,5 ms	9 ms
Netzwerk (NET)	6 ms	12 ms
Transport (TP)	9 ms	22 ms
Anwendungsprozeß (geschätzt) (z.B. lese einfache Variable)	20 ms	48 ms

Belastet man jedoch das Netz, so erhöht sich der Zeitbedarf erheblich. Meßergebnisse für die transportorientierten Schichten (Schicht 1 bis 4) unter einer gemäß Poisson Prozeß generierten Last sind in Bild 2 zusammengestellt.

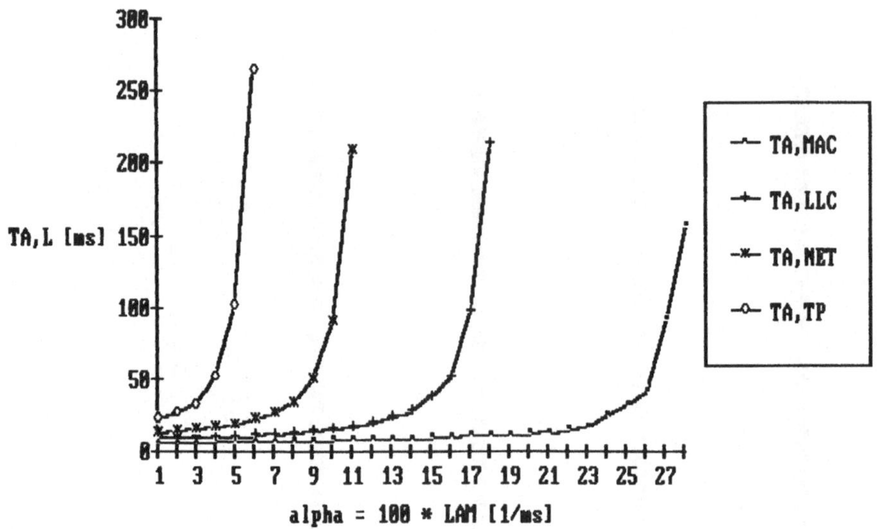

Bild 2   Antwortzeit der transportorientierten Schichten bei zunehmender Belastung

Zur meßtechnischen Erfassung der einzelnen Kurven wurde die Zwischenankunftsrate LAM schrittweise vergrößert bis LAM so groß ist, daß mehr Anforderungen eintreffen, als das System bedienen kann, was einen Pufferüberlauf zur Folge hat. Aus diesem Grund hören die aufgenommenen Kurven zu unterschiedlichen Werten des Angebotes auf. Damit verdeutlichen die Kurven nicht nur den Unterschied bei den Antwortzeiten, sondern zeigen auch quantitativ die Grenze der Belastbarkeit in den einzelnen Fällen.

Bild 3 zeigt den maximal erzielbaren Durchsatz auf der Transportebene für zwei unterschiedliche Quittierungsverfahren des Transportprotokolls, Nth Acknowledgement (NACK) und Explicit Acknowledgement (SACK).

Es ist zu bemerken, daß der maximal erzielbare Durchsatz bei NACK fast ein Mbps beträgt, während die physikalische Grenze des Übertragungsverfahrens den theoretischen Wert von fünf Mbps beträgt. Das bedeutet, daß man mit einer einzigen Transportverbindung im Sim-

plexverkehr nur ein Fünftel der zur Verfügung stehenden Medi-
umkapazität ausnutzen kann.

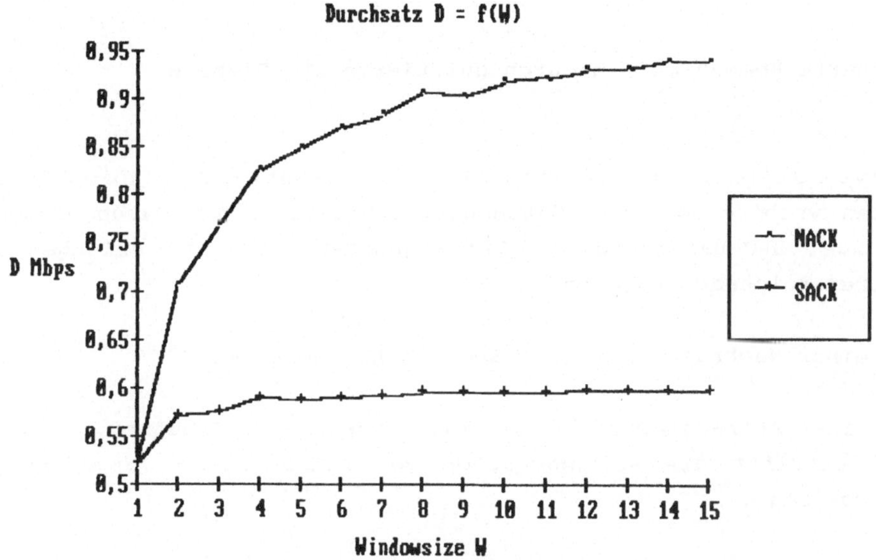

Bild 3   Maximaler Durchsatz einer mit Simplexverkehr betriebenen
          Transportverbindung als Funktion der Fenstergröße W für die
          Quittierungsstrategien SACK und NACK

2. Die verkürzte Architektur

Die oben aufgeführten Ergebnisse hängen natürlich von der Technologie
der Netzwerk-Kontroller ab. In dem Netz, in dem die Messungen durchgeführt wurden, ist jedoch für die Medienzugriffsschicht bereits ein
moderner VLSI-Baustein verwendet. Der Zeitbedarf läßt sich mit heuti-
ger Technologie nicht mehr um Größenordnungen reduzieren. Um den Zeitbedarf für Realzeitaufgaben zu verkürzen, wurde daher vorgeschlagen,
für spezielle Realzeitsegmente die Schicht sieben unmittelbar auf der
Schicht zwei zu implementieren (Mini-MAP-Konzept, vgl. [6]). Damit
sind einige Einschränkungen, z.B. durch Wegfall der Netzwerk-Schicht,
verbunden. Die Sicherung der Übertragung (Reihenfolge, Verlust oder
Duplizierung von Nachrichten und Datenflußkontrolle) muß an Stelle des
ISO-Transportprotokolls mit anderen Methoden erreicht werden. Hierfür

sind auch Lösungsvorschläge (z.B. das Alternating-Bit-Protokoll vgl. [9], und LLC-Type 3 vgl. [7]) bekannt.

2.1 Gesicherte Kommunikation durch quittierte LLC-Dienste

Der LLC-Type 3 (im folgenden kurz LLC3 bezeichnet) bietet Dienste zur quittierten verbindungslosen Datenübertragung bzw. zum Datenaustausch zwischen zwei LLC-Benutzern an. Die angebotenen Dienste erlauben dem LLC-Benutzer folgende Aktionen:

- Senden einer Nachricht zu einem anderen LLC-Benutzer.

- Senden einer Abfragenachricht an einen anderen LLC-Benutzer.
  Hierbei betrifft die Abfrage schon vom LLC-Benutzer vorgefertigte Daten (Polling).

Zur Erbringung dieser Dienste unterscheidet man zwischen zwei Mengen von Dienstprimitiven, die zu folgenden zwei Kommunikationsdiensten gehören.

A) Quittierte Datenübertragung "Acknowledged Connectionless Data Unit Transmission Service"

Dieser Dienst wird in der PROWAY-Terminologie mit "Send Data with Acknowledgement, SDA" bezeichnet. Hierbei wird eine Dateneinheit von LLC-Benutzer zu LLC-Benutzer übertragen, und die empfangende LLC-Instanz sendet eine Quittung an die sendende LLC-Instanz zurück, die ein Oktett als Statusmeldung enthält.

B) Quittierter Datenaustausch zwischen zwei LLC-Benutzern "Acknowledged Connectionless Data Unit Exchange Service"

Dieser Dienst entspricht dem RDR-Dienst in der PROWAY-Terminologie "Request Data with Reply". Es handelt sich um die Realisierung des Polling-Verfahrens auf LLC-Ebene. Dabei werden von einem LLC-Benutzer die Daten eines sich auf einem entfernten Knoten befindenden LLC-Benutzers abgefragt. Die empfangende LLC-Instanz quittiert die empfangene Abfragenachricht mit einer Antwort, die sowohl Statusmeldung als auch die abgefragten Daten enthält. Es wird in der Regel davon ausgegangen, daß die abzufragenden Daten schon aktualisiert bei der entfernten LLC-Instanz vorliegen. Naturgemäß unterscheidet man zwischen Dienst-Primitiven zur Erbringung der Kommunikation und Diensten zur Aktualisierung von Polling-Daten.

Die LLC-Protokollmaschine sichert die Nachrichten gegen Verlust, Miß-ordnung und Duplikation. Für jeden Kommunikationspfad ist ein Alter-nating-Bit-Protokoll vorgesehen, das neben der einfachen Form der Datenübertragung auch das Polling erlaubt. Ein Kommunikationspfad wird hierbei durch den Dienstzugangspunkt, Zieladresse und MAC-Priorität vollständig identifiziert (SSAP, DA, PRIO). Ändert sich in einer Nach-richt eine dieser drei Attribute, so gehört diese Nachricht einem anderen Kommunikationspfad und wird separat behandelt.

3. Implementierung von quittierten LLC-Diensten in einer Tokenbus-Umgebung

3.1 Alternativen im Token-Bus

Zur Erbringung der quittierten LLC-Dienste stehen der LLC-Instanz in einem Tokenbus-LAN folgende zwei Dienstklassen der MAC-Schicht zur Verfügung, vgl. [8]:

A) Request With No Response (RWNR)

Die Aufgabe der MAC-Schicht beschränkt sich hierbei auf das Senden der LLC-Nachrichten an den spezifizierten Knoten. Erhält der Knoten die Sendeberechtigungsmarke (Token), so sendet die MAC-Schicht die Nach-richten solange bis die Warteschlange leer oder die erlaubte Sendezeit abgelaufen ist. Die Sendeberechtigungsmarke wird dann an den benach-barten Knoten weitergegeben.

Die MAC-Schicht kann hierbei durch den CRC-Algorithmus (Cyclic Re-dundancy Check) die verfälschten Nachrichten entdecken und gegebe-nenfalls unterdrücken, sie unternimmt jedoch keine Aktionen um den Fehler durch Sendewiederholung zu beheben.

B) Request With Response (RWR)

Erhält der Knoten die Sendeberechtigung, so sendet die MAC-Schicht nur ein RWR-Frame an den Zielknoten und wartet auf eine gültige Antwort vom Zielknoten, dabei wird die Sendeberechtigungsmarke festgehalten. Bleibt die Antwort aus und läuft der Wiederholungstimer ab, so wird der RWR-Frame wiederholt gesendet und der Wiederholungszähler wird dekrementiert, die Sendeberechtigungsmarke wird nach wie vor festgehalten.

Diese Prozedur wird solange durchlaufen bis eine gültige Antwort empfangen wird, oder der Wiederholungszähler den Wert Null erreicht. Der MAC-Benutzer wird in jedem Fall über den Status der Sendung in Form einer Confirm-Primitive, die die Antwort (falls vorhanden) enthält, informiert. Empfängt eine MAC-Instanz einen RWR-Frame, so wird dieser an den MAC-Benutzer weitergereicht. Der MAC-Benutzer generiert eine Antwort und schickt sie an die MAC-Instanz, die diese Antwort unmittelbar (ohne auf die Sendeberechtigungsmarkesmarke zu warten) an den betroffenen Knoten sendet.

Der Wiederholungsmechanismus in der MAC-Schicht entlastet zwar die LLC-Schicht von der Sicherung der Datenübertragung gegen Verlust von Nachrichten, jedoch die Sicherung gegen Duplizierung von Nachrichten muß nach wie vor von der LLC-Schicht erbracht werden.

An dieser Stelle sei erwähnt, daß PROWAY zur Erbringung der LLC3-Dienste die unbedingte Ausnutzung der RWR-Dienste der MAC-Schicht vorschreibt. Der Grund dafür ist "das Garantieren von akzeptablen Zugriffszeiten für industrielle Applikationen". Diese Überlegung wurde offensichtlich von MAP-Gremien ohne Vorbehalt akzeptiert.

3.2 Vergleich der Alternativen

3.2.1 Betrachtung der Funktionalität

A) LLC3-Dienste mit RWNR-MAC-Diensten

Zu jedem Zeitpunkt darf für **jeden Kommunikationspfad** höchstens eine übertragene aber nicht quittierte Nachricht (Outstanding PDU) existieren.

Die Kommunikation entlang eines LLC-Pfades ähnelt also der Kommunikation entlang einer mit Fenstergröße W=1 operierenden Transportverbindung der Klasse vier des ISO-TPs.

B) LLC3-Dienste mit RWR-MAC-Diensten

Die Realisierung der LLC3-Dienste mit RWR-Diensten der MAC-Schicht
setzt aus der Sicht der LLC-Schicht folgendes voraus:

*In einem Tokenumlauf garantiert die MAC-Schicht, dass für die ge-
samte Zeit von der ersten Sendung einer Nachricht bis zum Ende der
notwendigen Wiederholungen (Retransmission) der empfangende Knoten
keine anderen Nachrichten ausser dieser empfängt.*

Das impliziert, daß zu einem gegebenen Zeitpunkt **für alle Kommunikationspfade** nur eine einzige Outstanding PDU beim Sender sein kann.
Man kann also alle Kommunikationspfade als die einzige mit Fenstergröße eins operierende Transportverbindung auffassen.

3.2.2 Betrachtung des Zeitbedarfes

Die Implementierung von LLC3-Diensten mit Request with Response führt
zur unvermeidbaren Blockierung des physikalischen Mediums, vgl.
Bild 4. Jeweils um die auftretende Blockierung erhöht sich netzweit
die Tokenumlaufzeit TRT.

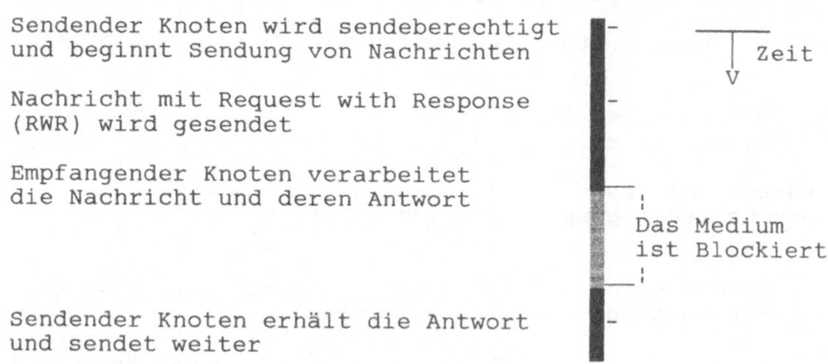

Bild 4   Blockierung des Mediums durch Request with Response

Betrachtet man nun den Zeitbedarf T_R für die <u>korrekte</u> Übertragung
einer Nachricht (Nachricht und Quittung) von LLC-Instanz zu LLC-
Instanz, so ergibt sich für beide Varianten das Zeitdiagramm gemäß

Bild 5. Hierbei sind die Zeiten zur Ausführung einzelner Aktivitäten in beiden Varianten als gleich angesehen.

Bei der RWNR-Variante muß zweimal auf das Medium zugegriffen werden. Bei jedem Zugriff ergibt sich die mittlere Wartezeit im Sendepuffer zu TRT/2, wobei TRT die mittlere Tokenumlaufzeit ist. Bei der RWR-Variante wird zwar auf das Medium nur einmal zugegriffen, allerdings hat man aufgrund der unvermeidlichen Blockierung (vgl. Bild 4) mit einer höheren Tokenumlaufzeit TRT˝ zu rechnen, somit ergibt sich die Wartezeit im Sendepuffer zu TRT˝/2.

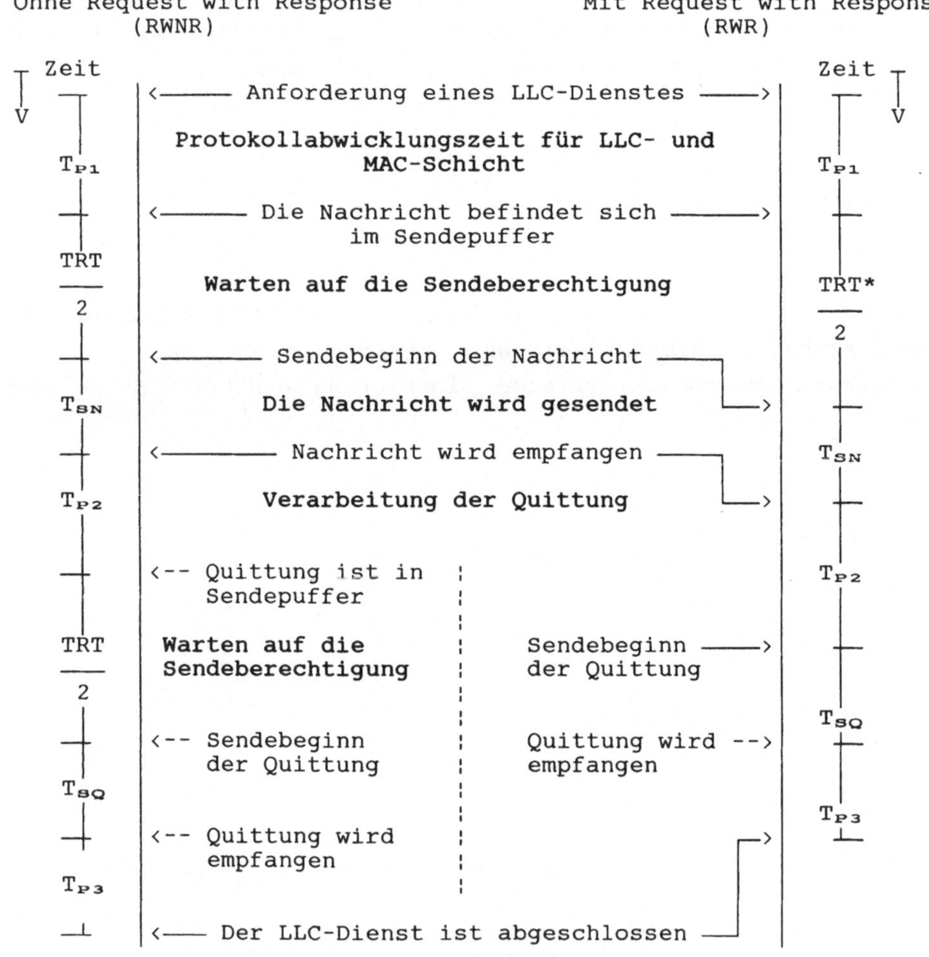

Bild 5   Zeitdiagramm zum Senden einer quittierten Nachricht auf LLC-Ebene mit den MAC-Diensten RWNR und RWR

Bildet man die Differenz zwischen den Zeiten beider Varianten gemäß

$$D = T_R(RWR) - T_R(RWNR) \tag{1}$$

so ergibt sich D zu

$$D = TRT - \frac{TRT^*}{2} \tag{2}$$

Gleichung (1) impliziert, daß die RWR-Variante nur bei $D > 0$ zu einem kleineren Zeitbedarf führt. Daher würde sich der Einsatz von Request with Response nur dann lohnen, wenn folgende Ungleichung erfüllt wird

$$TRT^* < 2\ TRT \tag{3}$$

Nimmt man nun an, daß die Zeit für die Tokenweitergabe viel kleiner als TRT ist und die Nachrichten eine konstante Länge haben, so kann die Bedingung (3) wie folgt umschrieben werden:

$$T_{SN} + T_{SQ} > T_{P2} \tag{4}$$

Der Einsatz von Request with Response scheint demnach für Nachrichten mit kurzer Länge ungeeignet zu sein, denn bei einem Übertragungsmedium mit der Kapazität 10 Mbps beträgt T_{SN} für eine Nachricht der Länge 100 Oktetts 80 μs. Überwiegend erwartet man im Bereich industrieller Fertigung Nachrichten mit kurzen Längen (kleiner als 100 Oktett), daher führt die obige Betrachtung zu der Annahme, daß die Erbringung von LLC3-Diensten mit Request with Response Diensten der MAC-Schicht fraglich ist.

4. Struktur einer Pilotimplementierung

Zur genauen Untersuchung der Varianten von verkürzten Kommunikationsarchitekturen wurde, ausgehend von einer MAP 2.1-Implementierung, eine Erweiterung zur Realisierung von quittierten Diensten der LLC-Schicht geplant. Dabei sollen die Anwenderprozesse die Möglichkeit haben zwischen den quittierten LLC-Diensten sowohl mit als auch ohne Request with Response zu wählen, vgl. Bild 6.

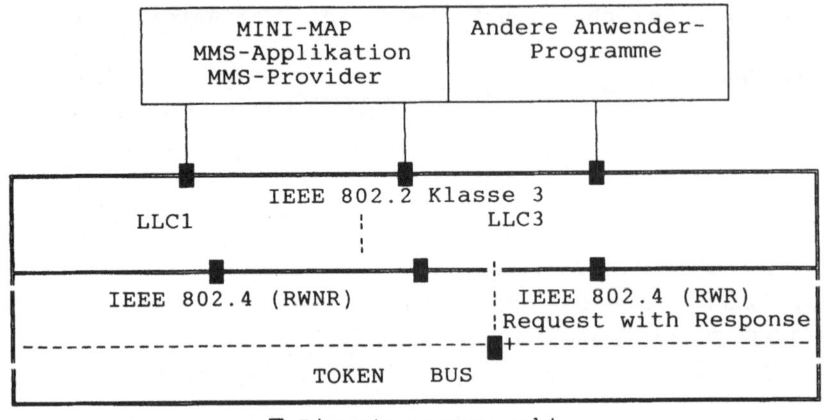

Bild 6   Struktur der Pilotimplementierung

Diese Architektur kann in einem MAP-Netz mit voller Architektur koexistieren, da das Zugriffsverfahren in keiner Weise verletzt wird. Die
MMS wird als Anwenderschnittstelle angeboten, so daß bei dem Anwenderprogramm gemäß voller MAP-Architektur keine Modifikation gemacht zu
werden braucht. Die Implementierungsumgebung besteht aus mehreren
Workstations mit VME-Bus-System, die jeweils über folgende drei Komponenten verfügen, vgl. Bild 7.

Kommunikation: Diese Komponente dient zur Abwicklung der Kommunikationsprotokolle in der Umgebung des Realzeitbetriebssystem-
kern VRTX32 (von Ready Systems). Das Zugriffsverfahren wird
hierbei von einem speziellen VLSI-Baustein abgewickelt.
VRTX32 erlaubt auch die Verbindung zur Applikation und Entwicklung.

Applikation: Diese Komponente verfügt auch über VRTX32 und realisiert
somit Anwenderprozesse in einer Realzeitumgebung. Es ist
unter anderem vorgesehen, die MMS-Provider auf diese Komponente zu verlagern, um somit eine Entlastung der Kommunika-
tionskomponente zu erreichen.

Entwicklung: Als Entwicklungsumgebung dient das Betriebssystem UNIX.
Von hier werden die entwickelten Programme in die Applika-
tions- und Kommunikationskarte geladen.

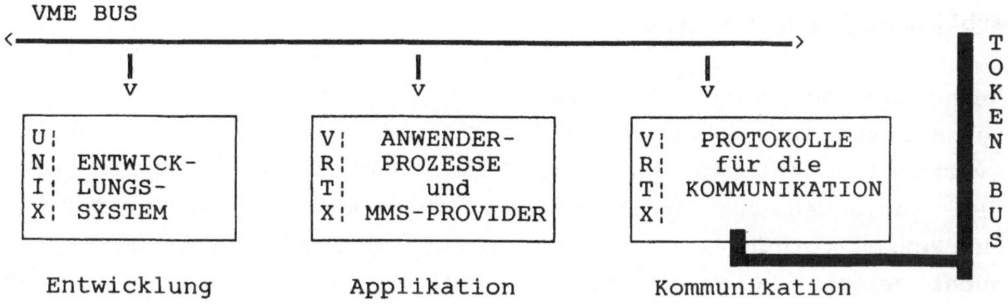

Bild 7   Komponenten eines Knotens im Labornetz

5. Quittierte LLC-Dienste in einem ETHERNET

Der LLC-Type 3 wird gewöhnlich in Zusammenhang mit dem Tokenbus diskutiert und es entsteht gelegentlich der Eindruck, daß dieser Dienst nur
implementierbar ist, wenn die MAC-Schicht einen RWR-Dienst (Request
with Response) anbietet. Die obigen Ausführungen zeigen, daß dies
nicht der Fall ist, und daß darüber hinaus die Benutzung der RWR-
Dienste nur in bestimmten nicht besonders häufigen Situationen Vorteile bringt. Die Ausführungen zeigen auch, daß quittierte LLC-Dienste
auf beliebigen Mediumzugriffsverfahren implementiert werden können.
Der entsprechende Teil der skizzierten Pilotimplementierung (Benutzung
von RWNR-Diensten) ist unmittelbar auf ein ETHERNET (bzw. ISO IS
8802.3) übertragbar (vgl. Bild 8).   Die Implementierung in einem
bestehenden ETHERNET kann z.B. ohne zusätzliche Eingriffe erfolgen,
wenn man das Alternating-Bit-Protokoll in einer Zwischenschicht
realisiert, die dann LLC1-Dienste benutzt und nach oben LLC3-Dienste
anbietet. Insofern bietet ein Tokenbus-Netz keine Vorteile gegenüber
einem ETHERNET.

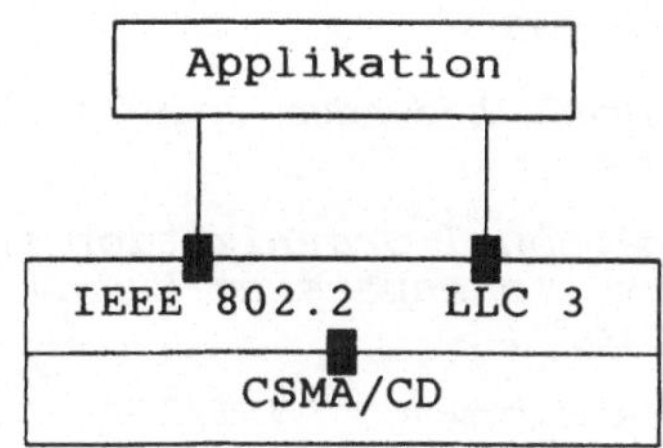

Bild 8   LLC3 in CSMA/CD Netzen

6. Überblick und weitere Arbeiten

Die Verwendung der verkürzten Kommunikationsarchitektur scheint die überlegene Lösung zur Reduzierung des Kommunikationsaufwandes und dem damit verbundenen Zeitbedarf zu sein. Im Rahmen dieses Beitrages wurden zwei Varianten zur Realisierung der verkürzten Architektur in Tokenbus-Netzen vorgestellt, die in einer Pilotimplementierung weiter untersucht werden. Die entstehende Implementierung soll zu einem späteren Zeitpunkt mit einer von einem US-Softwarehaus zur Verfügung gestellten MMS-Implementierung zu einem Mini-MAP Knoten ergänzt werden. Daher ergibt sich automatisch die Aufgabe, eine Zwischenschicht zu konzipieren, die die MMS-Dienste auf LLC3-Dienste abbildet und die MMS-Assoziationen verwaltet.

Literaturverzeichnis

[1] H. Rzehak; A.E. Elnakhal; R. Jäger: *Analysis of Real-Time Proper-ties and Rules for Setting Protocol Parameters of MAP Network; The Journal of Real-Time Systems, Vol. 1, No. 3, pp. 221-241, 1989*

[2] A.E. Elnakhal; H. Rzehak: *Messungen zur Beurteilung der Re-alzeiteigenschaften eines Kommunikationskontrollers für ein MAP-Netz; Messung; Modellierung und Bewertung von Rechensystemen und Netzen, GI/ITG-Fachtagung, Braunschweig, 26.-28.9.89; Informa-tik-Fachberichte Bd. 218, Springer-Verlag, S. 60-74*

[3] Marathe, M.V.; R.A. Smith: *Performance of a MAP Network Adapter; IEEE Network 2 (1988), No. 3*

[4] Ciminiera, L.; A. Valenzano: *Acknowledgement and Priority Mecha-nisms in the 802.4 Token Bus; IEEE Trans. on Industrial Electro-nics Vol. 35 (1988), No.2*

[5] Janetzky, D.; K.S. Watson: *Token Bus Performance in MAP and PRO-WAY; IFAC Workshop on Distributed Computer Control Systems. May-schoss (FRG); Sept., 30 Oct., 2, 1986*

[6] Manufacturing Automation Protocol Sepcification, *Version 3.0; European MAP Users Group, 1989*

[7] ISO 8802-2/PDAD2 *Proposed Draft Addendum 2: Acknowledged Connectionless-mode Service, Type 3 Operation*

[8] ISO IS 8802.4-1987 *International Organization for Standardiza-tion: "Local Area Networks, Tokenpassing Bus Access Method and Physical Layer Specification";*

[9] PROWAY-LAN *Industrial Data Highway; ANSI/ISA - S 72.01 - 1986; Instrument Society of America*

[10] Process Communications Architecture; *Draft Standard ISA-DS 72.03 - 1988; Instrument Society of America*

Kommunikationssicherheit in industriellen Netzen

Walter Fumy und Hans Peter Rieß
Siemens AG, Erlangen

Abstrakt: *In diesem Beitrag werden verschiedene Aspekte der Kommunikationssicherheit in industriellen Netzen herausgearbeitet. Dazu werden in diesem Zusammenhang wesentliche Charakteristika lokaler Netze zusammengestellt und daraus zu erwartende Bedrohungen und benötigte Sicherheitsfunktionen abgeleitet. Für die Kommunikationssicherheit grundlegende Mechanismen, sowie verschiedene Ansätze zur Einbettung dieser Mechanismen in die Kommunikationsarchitektur werden diskutiert. Konkrete Sicherheitsmaßnahmen für lokale Netze werden vorgestellt.*

1. Einführung

Die Rechnerkommunikation gewinnt auch im Bereich der industriellen Fertigung zunehmend an Bedeutung. Die anfänglich üblichen Automatisierungsbestrebungen auf Maschinenebene wurden inzwischen auf komplette Anlagen oder Fabriken ("Fabrik der Zukunft") ausgedehnt. Die industrielle Kommunikation wird vor allem über lokale Netze (LANs) abgewickelt, die sowohl zur Reduzierung des für die Übertragung von Prozeßsignalen notwendigen Verkabelungsaufwandes, wie auch zur Kopplung größerer Funktionseinheiten (Rechnerverbund) eingesetzt werden. Aus dieser Entwicklung erklärt sich, daß auch im industriellen Umfeld immer häufiger Forderungen nach verbesserter Kommunikationssicherheit erhoben werden.

Die Vernetzung von Rechnern, Steuerungen und Prozeßleitsystemen ist eine wesentliche Grundlage für das *Computer Integrated Manufacturing* (CIM). Damit kommt ihr eine strategische Bedeutung im Rahmen von Unternehmensplanungen zu. Zum Einsatz kommende Produkte müssen einerseits zukunftssicher und erweiterungsfähig, andererseits stabil und kompatibel sein. Diese Anforderungen können i.a. am ehesten mit Produkten erfüllt werden, die auf internationalen Standards basieren. Die Erprobung und Weiterentwicklung der offenen Kommunikation im Hinblick auf kostengünstige CIM-Realisierungen haben sich verschiedene internationale Initiativen zum Ziel gesetzt. Eines der bekanntesten Projekte ist das von General Motors initiierte *Manufacturing Automation Protocol* (MAP).

MAP beruht auf Normen, die die *International Organization for Standardization* (ISO) im Rahmen ihrer Standardisierung der *Open Systems Interconnection* (OSI) verabschiedet hat. Das CNMA-Projekt (*Communication Network for Manufacturing Applications*) ist eine MAP ergänzende europäische Initiative, die im Rahmen des ESPRIT-Programms gefördert wird. Von CNMA werden Kommunikationsstandards für den Fertigungsbereich spezifiziert, implementiert, validiert und propagiert. Dabei wird der Definition von Profilen für existierende bzw. in Entwicklung befindliche Standards große Bedeutung beigemessen. Der Schwerpunkt der Implementierungen liegt in der Schicht 7 des ISO-OSI-Modells, z.B. bei Automatisierungsprotokollen auf der Basis der *Manufacturing Message Specification* (MMS), *File Transfer, Access and Management* (FTAM), Netzmanagement und *Directory Services*.

Zwar werden Sicherheitsprobleme in vielen Bereichen der elektronischen Datenverarbeitung immer häufiger offenbar, doch sind ausgereifte Konzepte und Produkte zu deren Lösung insbesondere für Netze derzeit nicht in ausreichendem Maße verfügbar. Von einigen besonders sensiblen Anwendungen (z.B. im Bankbereich) abgesehen, befindet sich auch die Standardisierung von Sicherheitsfunktionen noch in einem relativ frühen Stadium. So spielt die sichere Kommunikation bis heute weder bei MAP noch bei CNMA eine größere Rolle.

Die Erarbeitung von Normen auf diesem Gebiet ist jedoch nicht zuletzt im Hinblick auf die zunehmende Verbreitung offener Kommunikationssysteme unumgänglich. Wesentliche Impulse für Sicherheitsstandards im Bereich Informationstechnik gehen von Gremien der ISO, sowie von CCITT und ECMA aus. Einen Schwerpunkt bilden die Aktivitäten der IEC/ISO/JTC1 Subkomitees SC27 (*Security Techniques*) und SC21 (*Open Systems Interconnection*). Das SC21 ist für die Sicherheitsarchitektur und allgemeine Sicherheitskonzepte (*Models, Frameworks*), SC27 u.a. für grundlegende Fragen zu Methoden, Anwendungen und Integration kryptographischer Verfahren zuständig.

Im folgenden Abschnitt werden zunächst einige Besonderheiten der Sicherheitsproblematik in industriellen Netzen herausgearbeitet. Dazu werden hierfür wesentliche Charakteristika lokaler Netze zusammengestellt und daraus zu erwartende Bedrohungen und benötigte Sicherheitsfunktionen abgeleitet. Abschnitt 3 behandelt für die Kommunikationssicherheit grundlegende Mechanismen und Möglichkeiten zu deren Einbettung in die Kommunikationsarchitektur. Im letzten Abschnitt schließlich werden spezifische Sicherheitsmaßnahmen für lokale Netze vorgestellt.

2. Bedrohungen und Sicherheitsfunktionen in lokalen Netzen

Der Begriff Informationssicherheit umfaßt den Schutz der Vertraulichkeit, Integrität, Verfügbarkeit und Authentizität von Informationen. Vertraulichkeit bedeutet, daß Informationen nur denjenigen zugänglich sind, für die sie bestimmt sind. Unter Integrität versteht man, daß zufällige und absichtliche Manipulationen (z.B. Einfügen, Fälschen, Verändern der Anordnung, Wiederholen oder Löschen) von Daten nicht unbemerkt bleiben können. Die Verfügbarkeit ist gegeben, wenn der Zugriff auf die Information bzw. die Speicher- und Kommunikationsmedien für Berechtigte stets unbehindert möglich ist. Authentizität schließlich stellt sicher, daß die Daten von der angegebenen Quelle stammen. Bedrohungen von Systemen sind neben einem unautorisierten Zugang vor allem Angriffe auf die Verfügbarkeit von Diensten. Zu den Bedrohungen von Daten gehören ein unautorisierter Zugriff (z.B. durch Abhören), unautorisierte Modifikationen und deren unautorisierte Vernichtung.

In der Regel muß für jedes konkrete System ein eigenes Sicherheitskonzept erstellt werden, in dem nach dem Studium der Qualität aller erdenklichen Angriffe (der Bedrohungsanalyse) und einer Abschätzung des Ausmaßes der zu erwartenden Schäden ein problemspezifischer Sicherheitsbegriff definiert wird. Die resultierenden Anforderungen im Hinblick auf verschiedene Sicherheitsaspekte sollten dann i.a. bereits bei der Konzeption eines Kommunikationsnetzes berücksichtigt werden. Für die Sicherheitsanforderungen bei der industriellen Kommunikation spielen einige spezielle Gegebenheiten eine Rolle:

- Als typische Kommunikationsmedien werden lokale Netze eingesetzt.
- Die Geheimhaltung der übertragenen Daten spielt häufig eine untergeordnete Rolle.
- Echtzeitanforderungen bedingen die Unterstützung vieler Sicherheitsdienste durch geeignete Hardware-Komponenten (z.B. Verschlüsselungschips).
- Im Automatisierungsbereich können die Kommunikationsbeziehungen zwischen Knoten in der Regel zentral projektiert werden.

Auch wenn in industriellen Netzen die Vertraulichkeit der übertragenen Daten nur selten zwingend erforderlich ist, so sollten doch Mechanismen zur Sicherung der Datenintegrität auch in diesem Bereich unverzichtbar sein. Diese Integritätsmechanismen werden u.a. zur Identifizierung der Kommunikationspartner, zur Authentifizierung von Nachrichten, oder für Zugriffskontrollen benötigt.

Aus einigen typischen Eigenschaften lokaler Netze ergeben sich unmittelbar spezifische Bedrohungen für die Informationssicherheit beim Betrieb eines solchen Netzes. Dazu zählen:

- die ungerichtete Kommunikation (*broadcast medium*);
- die Gleichberechtigung der Benutzer;
- der gemeinsame Adressraum aller Stationen in der ISO-Schicht 2;
- die hohe Bandbreite der Übertragungsmedien.

Die ungerichtete Kommunikation hat zur Folge, daß nicht nur legale Neuanschlüsse und autorisierte Veränderungen von LANs geringen Aufwand erfordern, sondern auch unbemerkte Manipulationen am Netz relativ einfach möglich sind. Daher stellen z.B. das Mithören fremder Nachrichten und das Vorgeben falscher Identitäten durch Mißbrauch bereits angeschlossener oder Hinzufügen zusätzlicher Stationen (z.B. von Netzdiagnose-Geräten) ernste Bedrohungen im LAN-Bereich dar.

Da eine beschränkte geographische Ausdehnung beim heutigen Stand der Technik nicht mehr als ein Charakteristikum von LANs angesehen werden kann, sind physische und organisatorische Schutzmaßnahmen, die in lokal beschränkten Systemen häufig sinnvoll und praktikabel sind, in LANs oft nicht ausreichend. Ein besonders extremer Fall sind mehrere über Bridges und *Wide Area Networks* (WANs) gekoppelte Teilnetze, die über einen gemeinsamen Adressraum in Schicht 2 verfügen können und dann im logischen Sinn als ein LAN betrachtet werden müssen. Die Ausdehnung industrieller lokaler Netze ist häufig allerdings bereits ohne die Berücksichtigung von Netzübergängen so groß, daß aus der lokalen Beschränkung keine effektive Schutzwirkung erwartet werden kann.

Dennoch wird, mit Ausnahme von Organisationen mit besonders stark ausgeprägtem Sicherheitsbewußtsein, beim Schutz der Daten auf lokalen Netzen meist auf den Einsatz effektiver technischer Maßnahmen verzichtet. Dadurch kann auch die Wirksamkeit anderer Maßnahmen der Informationssicherheit - etwa eines Paßwortschutzes für den Host-Zugang - empfindlich eingeschränkt werden. Werden nämlich diese Paßwörter im Klartext über ein ungesichertes lokales Netz übertragen, so können sie an jeder angeschlossenen Station mit nicht allzu großem Aufwand abgehört werden.

Sicherheitsmechanismen für lokale Netze unterscheiden sich prinzipiell nicht von solchen, die in andersgearteten Netzen zum Einsatz kommen. Unterschiede ergeben sich jedoch bei der Einbettung der Mechanismen in die Kommunikationsarchitektur. Sicherheitsmechanismen, die die Möglichkeiten für transparente Schutzmaßnahmen über Netzübergänge (Router, Gateways) hinweg nicht beschränken sollen, müssen oberhalb von Schicht 2 implementiert werden. In vielen Fällen erweisen sich jedoch LAN-spezifische Mechanismen, d.h. Mechanismen die unterhalb der Schicht 3 der OSI-Hierarchie eingreifen, als vorteilhaft.

In den höheren Schichten der OSI-Hierarchie sind unterschiedliche Protokolle (in den transportorientierten Ebenen z.B. ISO-8073/8473, TCP/IP; in den anwendungsorientierten Ebenen z.B. MAP/MMS, SINEC AP/TF) verbreitet. Durch die Einbettung von Sicherheitsfunktionen in die Schicht 2 kann die Kommunikation zwischen Stationen mit beliebigen höheren Protokollen und Anwendungen innerhalb eines LAN geschützt werden. Ferner können alle IEEE-LAN-Medien unterstützt werden, wobei die Sicherheitsfunktionen für MAC-Bridges transparent sind.

In LANs kann das Sicherheitsmanagement relativ einfach gestaltet werden. Es bietet sich an, alle zulässigen Verbindungen zentral zu projektieren und somit auch einen Zugriffsschutz zu realisieren. Die Schlüssel für eine Ende-zu-Ende-Verschlüsselung über Netzgrenzen hinweg können dagegen nicht vorab erzeugt und verteilt werden, sondern müssen z.B. beim Verbindungsaufbau angefordert werden. Unterliegen die kommunizierenden Stationen einer getrennten Verwaltung, so sind dazu relativ aufwendige Verfahren erforderlich.

3. Sicherheitsmechanismen und deren Integration in die Kommunikationsarchitektur

Zur Sicherung der Kommunikation gegen unbefugtes Mithören und unbemerkte Veränderung der übermittelten Daten durch Dritte kommen seit jeher kryptographische Verfahren zum Einsatz. Bis vor wenigen Jahren waren die Anwender dabei auf individuelle Lösungen angewiesen. Erst die stark zunehmende Vernetzung in der Datenverarbeitung und die damit verbundenen Anforderungen an die Kompatibilität von Kommunikations-

Sicherheitsdienst	Mechanismus	Verschlüsselung	Digitale Signatur	Zugriffskontrolle	Datenintegrität	Authentifikation	Verkehrs-Erzeugung	Leitweg-Kontrolle	Notariat
gegenseitige Partner-Authentifikation		+	+			+			
Authentifikation des Datenursprungs		+	+						
Zugriffskontrolle				+					
Verbindungsorientierte Geheimhaltung		+						+	
Verbindungslose Geheimhaltung		+						+	
Selektive Geheimhaltung		+							
Verkehrsfluss-Sicherheit		+					+	+	
Verbindungso. Integrität mit Wiederaufsetzen		+			+				
Verbindungso. Integrität ohne Wiederaufsetzen		+			+				
Selektive, verbindungsorientierte Integrität		+			+				
Verbindungslose Integrität		+	+		+				
Selektive, verbindungslose Integrität		+	+		+				
Sende-Nachweis			+		+				+
Empfangs-Nachweis			+		+				+

Abb. 1: Sicherheitsdienste und -mechanismen

systemen hat auch im Sicherheitsbereich dazu geführt, den Normungsgedanken aufzugreifen. Den Ausgangspunkt der Standardisierungsansätze für die sichere Kommunikation bilden das OSI-Referenzmodell (ISO 7498) und die OSI-Sicherheitsarchitektur (ISO 7498-2) [4].

Letztere identifiziert alle relevanten Sicherheitsdienste für die Kommunikation zwischen offenen Systemen, definiert eine Reihe von Sicherheitsmechanismen, die zur Bereitstellung dieser Dienste genutzt werden können und beschreibt Alternativen für die Einbettung in die Schichten des Modells. Sie unterscheidet 14 verschiedene Sicherheitsdienste, wobei eine Anwendung in der Regel nur einen Teil in Anspruch nehmen wird. In Abbildung 1 ist dargestellt, welche Sicherheitsmechanismen, einzeln oder kombiniert, für die Bereitstellung welcher Sicherheitsdienste als geeignet angesehen werden [4].

Der Einsatz von Chiffrierverfahren stellt die zentrale Technik zur Gewährleistung von Kommunikationssicherheit dar. Man unterscheidet symmetrische und asymmetrische kryptographische Mechanismen [2]. Erstere sind dadurch charakterisiert, daß für das Dechiffrieren einer Nachricht der gleiche Schlüssel benötigt wird, wie für ihre Chiffrierung. Dieser Schlüssel muß auf sicherem Wege zwischen den Teilnehmern ausgetauscht und generell geheimgehalten werden. Von W.Diffie und M.E.Hellman wurde 1976 das Grundkonzept einer asymmetrischen Verschlüsselung entwickelt, bei dem zum Dechiffrieren ein anderer Schlüssel als zum Chiffrieren verwendet wird [1]. Die entschei-

256

Sicherheitsdienst	Schicht						
	1	2	3	4	5	6	7
gegenseitige Partner-Authentifikation			+	+			+
Authentifikation des Datenursprungs			+	+			+
Zugriffskontrolle			+	+			+
Verbindungsorientierte Geheimhaltung	+	+	+	+			+
Verbindungslose Geheimhaltung		+	+	+			+
Selektive Geheimhaltung							+
Verkehrsfluss-Sicherheit	+		+				+
Verbindungso. Integrität mit Wiederaufsetzen				+			+
Verbindungso. Integrität ohne Wiederaufsetzen			+	+			+
Selektive, verbindungsorientierte Integrität							+
Verbindungslose Integrität			+	+			+
Selektive, verbindungslose Integrität							+
Sende-Nachweis							+
Empfangs-Nachweis							+

SILS SP3 SP4

SINEC Kryptobox

Abb. 2: OSI-Sicherheitsarchitektur

dende Eigenschaft eines asymmetrischen Kryptosystems besteht darin, daß es zwar
"leicht" ist, solche Schlüsselpaare zu generieren, es aber praktisch unmöglich ist, aus
einem der beiden zusammengehörigen Schlüssel den jeweils anderen zu berechnen bzw.
aus der Chiffrier- auf die Dechiffrieroperation zu schließen oder umgekehrt. Asymmetrische Verfahren werden vor allem für den Austausch von Schlüsseln in symmetrischen
Kryptosystemen (hybrides Schlüsselmanagement), sowie für solche Sicherheitsmechanismen eingesetzt, die mit symmetrischen Verfahren nicht erbracht werden können (z.B.
elektronische Unterschriften).

Ein breiter Einsatz von Sicherheitsmechanismen in offenen Netzen bedingt Interoperabilität, welche in der Regel durch Standardisierung angestrebt wird. Diese sollte neben den
Algorithmen auch geeignete Betriebsarten, damit realisierbare Sicherheitsmechanismen
und Sicherheitsdienste, deren Einbettung in Protokolle und Anwendungen, sowie zumindest Teile des Sicherheitsmanagements umfassen. Einen Überblick über Standardi-
sierungsaktivitäten auf dem Gebiet der Netzsicherheit findet man z.B. in [5].

In Abbildung 2 ist dargestellt, welche der sieben Schichten der OSI-Sicherheitsarchitektur zufolge für welchen Sicherheitsdienst als geeignet anzusehen sind (durch '+' gekennzeichnet). Durch eine Verschlüsselung in der Verbindungs-Schicht ist nach [4] nur eine
Geheimhaltung erreichbar (verbindungsorientiert oder verbindungslos). In der Vermittlungs-Schicht können dagegen mit Hilfe eines Verschlüsselungsmechanismus die folgenden Sicherheitsdienste angeboten werden: gegenseitige Partner-Authentifikation, Authentifikation des Datenursprungs, Datengeheimhaltung (verbindungsorientiert oder ver-

bindungslos) und Datenintegrität (ohne Wiederaufsetzen, verbindungsorientiert oder verbindungslos). In der Transport-Schicht ist zusätzlich ein Datenintegritäts-Dienst mit Wiederaufsetzen möglich. Die Sitzungs- und die Präsentations-Schicht werden für die Integration von Sicherheitsdiensten als ungeeignet angesehen; in der Anwendungs-Schicht schließlich kann jeder Sicherheitsdienst implementiert werden.

Für eine konkrete Implementierung ist es meist nicht zu empfehlen, die Sicherheitsfunktionen über die Schichten des Kommunikationssystems zu verteilen. Vielmehr sollte man je nach Anwendung, Umfeld und zu erwartenden Bedrohungen eine oder einige wenige Stellen auswählen, an denen die benötigten Sicherheitsdienste am wirkungsvollsten eingebettet werden können. Zum Beispiel ist in den meisten Fällen eine Verschlüsselung in einer der sieben Schichten ausreichend. Einige Kriterien für die Auswahl dieser Schicht sind im folgenden zusammengestellt (siehe hierzu auch [6] und [9]):

- Vollständige Verkehrsflußsicherheit kann durch Verschlüsselung nur in der physikalischen Schicht erreicht werden.

- Möchte man einen einfachen Schutz von Ende-zu-Ende-Verbindungen, so bietet sich eine Verschlüsselung in der Vermittlungs-Schicht an. Sie erlaubt sowohl die Geheimhaltung als auch die Prüfung der Integrität von Daten.

- Benötigt man Integritätsdienste mit der Möglichkeit des Wiederaufsetzens, so empfiehlt sich die Transport-Schicht.

Der nächste Schritt zur Konkretisierung der Sicherheitsarchitektur betrifft die Normung von Protokollen für die wichtigsten Sicherheitsdienste. Eine grobe Unterscheidung kann zwischen anwendungsorientierten und netzorientierten Diensten getroffen werden. Beide Arten sind sinnvoll und müssen sich in der Praxis nicht selten ergänzen. Wir beschränken uns im folgenden auf die Sicherheitsdienste in den unteren Schichten des OSI-Modells. Derzeit werden drei derartige Sicherheitsprotokolle von verschiedenen Gremien behandelt, die zum Teil über die in ISO 7498-2 entwickelten Vorstellungen hinausgehen (siehe auch Abbildung 2):

- Die IEEE Arbeitsgruppe 802.10 entwickelt einen *Standard for Interoperable LAN Security* (SILS) für die Verbindungsschicht [3];

- Vom Projekt *Secure Data Network Systems* (SDNS), das vom amerikanischen Verteidigungsministerium initiiert wurde, werden u.a. Sicherheitsprotokolle für die Vermittlungs- bzw. Transportschicht (SP3 und SP4) spezifiziert [7], [8].

SILS unterstützt die Authentifikation des Datenursprungs, die Datengeheimhaltung, die Datenintegrität und eine Zugriffskontrolle. Aus dem angestrebten Einsatz in IEEE-LANs ergibt sich zwangsläufig die Einbettung des Sicherheitsprotokolls in die Verbindungsschicht. Nach derzeitigem Stand wird das SILS-Protokoll zwischen *Medium Access Control* (MAC) und *Logical Link Control* (LLC) angesiedelt. Der Nachteil einer solchen Lösung ist der Verzicht auf die Möglichkeit, Sicherheitsdienste ohne Unterbrechung über Netzübergänge hinweg anbieten zu können (siehe Abbildung 3). Als Vorteile stehen dem die Einfachheit der Schnittstelle, die Transparenz für alle Protokolle ab Schicht 3 aufwärts, sowie ein weitgehender Schutz von Header-Informationen gegenüber.

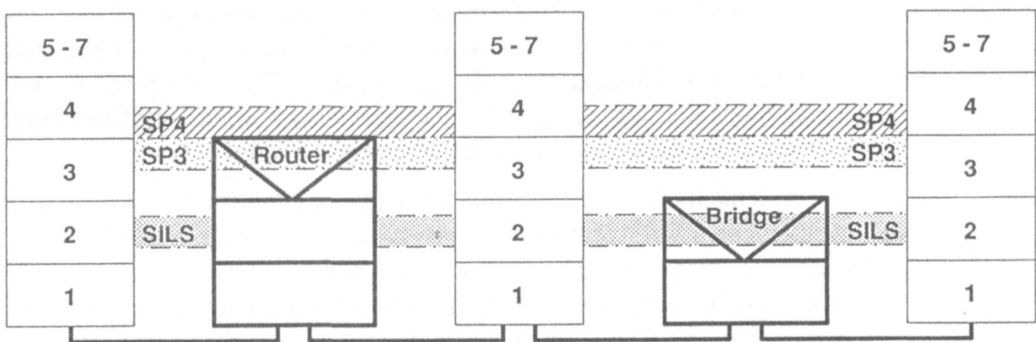

Abb. 3: Sicherheitsprotokolle und Netzübergänge

Für die Implementierung eines teilnetzunabhängigen Sicherheitsprotokolls bieten Einbettungen in die Schicht 3 bzw. 4 des OSI-Modells unterschiedliche Vorteile. Welcher Alternative der Vorzug zu geben ist, hängt vor allem von den individuellen Gewichtungen der verschiedenen Anforderungen an die Sicherheitsfunktionen ab. Das SDNS-Protokoll SP4 ist als Zusatz zum ISO-Transportprotokoll vorgesehen und funktional unterhalb der Transportschicht angeordnet; SP3 ist als eine Unterebene des ISO-Vermittlungsprotokolls definiert und direkt unterhalb des Transportprotokolls eingebettet. SP3 und SP4 operieren somit im wesentlichen auf denselben Datenstrukturen und sind daher in ihrer Grundfunktionalität kompatibel. Durch ihre Zugehörigkeit zu den jeweiligen Schichten haben sie jedoch auf unterschiedliche Zusatzinformationen Zugriff, mit deren Hilfe die Realisierung spezifischer Funktionen möglich ist. Die von SP3 und SP4 angebotenen Sicherheitsdienste sind Vertraulichkeit, Datenintegrität, Authentifikation des Datenursprungs und Zugriffsschutz (siehe auch Abbildung 2). Aufgrund der Einbettung zwischen den Schichten 3 und 4 können geschützte Datenpakete von nicht vertrauens würdigen Vermittlungsknoten verarbeitet werden, ohne daß der Schutz dort unterbrochen zu werden braucht (siehe Abbildung 3).

4. Sicherheitsmaßnahmen für lokale Netze

Das Hauptproblem bei der Sicherung der Geheimhaltung oder der Integrität der übertragenen Daten ergibt sich bei einem lokalen Netz durch die ungerichtete Kommunikation. Ein weitgehender Schutz gegen entsprechende Bedrohungen erfordert den Einsatz kryptographischer Verfahren. Wegen der Vielfalt der an industriellen Netzen betriebenen Stationen ist es sinnvoll, dabei ein Konzept zu verfolgen, das die Implementierung von Sicherheitsfunktionen ohne Eingriff in die Hard- und Software der Stationen gestattet. Im Hinblick auf eine hohe Flexibilität und damit breite Einsatzmöglichkeit von Sicherheitskomponenten können die folgenden Anforderungen formuliert werden:

- Unabhängigkeit von der Hard- und Software der abzusichernden Netzstationen;
- Unabhängigkeit von Transportprotokollen und Anwendungen;
- Einsetzbarkeit zur Absicherung einzelner Stationen und von Teilnetzen;

- Realisierung der grundlegenden Sicherheitsdienste Geheimhaltung, Datenintegrität, Sender-Authentifikation und Zugriffskontrolle;
- Transparenz für die Benutzer;
- Keine merkliche Beeinträchtigung der LAN-Leistungsdaten;
- Möglichkeit zu einem schrittweisen Ausbau des Systems.

Als Beispiel für ein System, das aufgrund derartiger Anforderungen konzipiert wurde, wird abschließend das SINEC Kryptobox-System kurz beschrieben. Dieses Produkt ermöglicht ein hohes Maß an Informationssicherheit in lokalen Netzen nach IEEE 802.3 (z.B. SINEC H1/H1FO). Seine vielfältigen Einsatzmöglichkeiten erstrecken sich sowohl auf die industrielle, wie auch auf die Bürokommunikation.

Die Sicherheitsmechanismen sind hier in von den Netzknoten unabhängigen in-line Boxen, sowie in einer Sicherheitsmanagement-Konsole implementiert. Dies gewährleistet eine maximale Hardware-Unabhängigkeit und Flexibilität bei der Installation. Der höhere Aufwand gegenüber einer etwa in die Stationsanschaltungen integrierten Lösung wird durch die Möglichkeit ausgeglichen, mit einer SINEC Kryptobox nicht nur einzelne Stationen, sondern ganze Teilnetze absichern zu können. Die Sicherheitsfunktionen werden dabei zentral administriert. Das Problem der Schlüsselverteilung, das bei kryptographischen Sicherheitssystemen häufig aufwendige Lösungen erfordert, wird durch eine zentrale Projektierung der Kommunikationsbeziehungen stark vereinfacht.

Durch die Plazierung der Sicherheitsmechanismen in der OSI-Schicht 2 wird die Unabhängigkeit von uneinheitlichen Protokollen der höheren Schichten und die leichte Anpassbarkeit an verschiedene Übertragungsmedien erreicht. Diese Einbettung erlaubt es, daß zwei Stationen auch dann sicher miteinander kommunizieren können, wenn sie an verschiedenen, durch Bridges (oder durch Halfbridges und ein WAN) verbundenen Netzsegmenten angeschlossen sind (siehe auch Abbildung 4). Damit die Leistungsfähigkeit des LAN nicht beeinträchtigt wird, kommt für die Realisierung vieler Sicherheitsdienste beim gegenwärtigen Stand der Technik nur ein symmetrisches Chiffrierverfahren in Betracht (in der SINEC Kryptobox wird ein DES-Chip eingesetzt). Die folgenden Sicherheitsdienste werden angeboten:

- Geheimhaltung der Daten durch Verschlüsselung;
- Integritätssicherung der Daten (und Sender-Authentifikation) durch eine kryptographische Prüfsumme;
- Die Kombination von Geheimhaltung und Integritätssicherung;
- Zugriffskontrolle durch das Sperren von Kommunikationsbeziehungen.

Nicht explizit eingetragene Kommunikationsbeziehungen werden entsprechend einer wählbaren Voreinstellung behandelt. Da in die Prüfsummenbildung eine Sender-Identifikation einbezogen wird, erhält man auch eine Sender-Authentifikation, deren Fehlen in heutigen lokalen Netzen eine wesentliche Schwachstelle darstellt. Die Projektierung der Sicherheitsparameter für die Kommunikationsbeziehungen wird vom Systemverantwortlichen an der Management-Konsole vorgenommen.

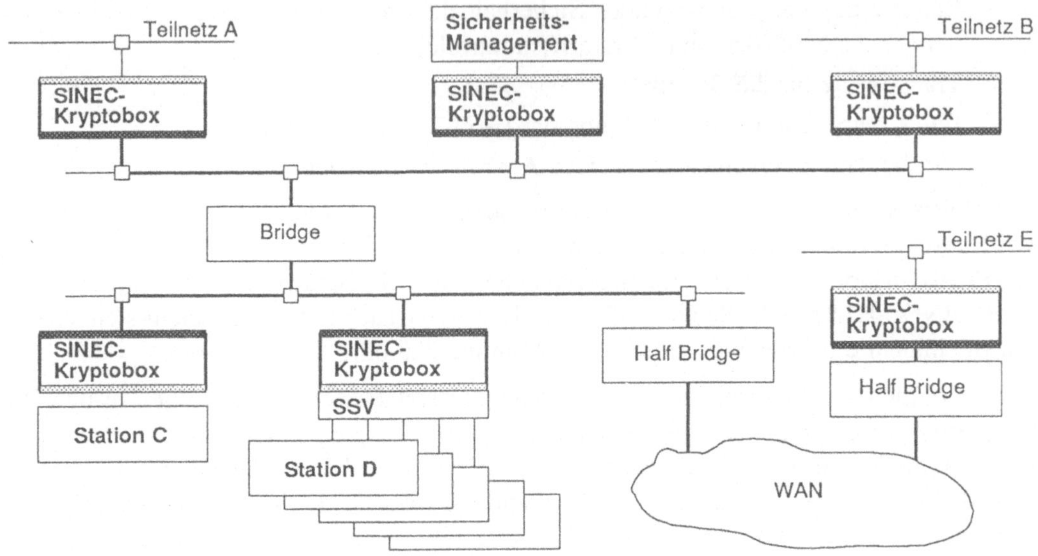

Abb. 4: Beispielkonfiguration

Im einfachsten Fall ist eine SINEC Kryptobox einer einzelnen Station bzw. der Management-Konsole zugeordnet. Sie kann jedoch auch Sicherheitsdienste für mehrere über ein Teilnetz oder einen Schnittstellenvervielfacher (SSV) angeschlossene Stationen erbringen. Falls in einem LAN Teilbereiche existieren, die keinen Bedrohungen ausgesetzt sind, ermöglicht ihr Einsatz als Bridge besonders kostengünstige Sicherheitslösungen. Verschiedene Konfigurationsmöglichkeiten sind in Abbildung 4 zusammengefaßt.

Literatur

[1]   Diffie, W.; Hellman, M.E.: "New Directions in Cryptography", *IEEE Trans. on Info. Theory*, **22** (1976), 644-654.

[2]   Fumy, W.; Rieß, H.P.: *Kryptographie: Entwurf und Analyse symmetrischer Kryptosysteme*, Oldenbourg, München, 1988.

[3]   IEEE 802.10: "Standard for Interoperable LAN Security", Draft 12/1989.

[4]   ISO International Standard 7498-2: *"Open Systems Interconnection Reference Model - Part 2: Security Architecture"*, 1989.

[5]   Kirkpatrick, K.E.: "Standards for Network Security", Proceedings 11th National Computer Security Conference (1988), 201-211.

[6]   Rutledge, L.S.; Hoffman, L.J.: "A Survey of Issues in Computer Network Security", *Computers & Security*, **5** (1986), 296-308.

[7]   Secure Data Network Systems: *"Security Protocol 3"*, 2/1989.

[8]   Secure Data Network Systems: *"Security Protocol 4"*, 7/1988.

[9]   Voydock, V.L.; Kent, S.T.: "Security Mechanisms in High-Level Network Protocols", *Computing Surveys*, **15** (1983), 135-171.

<u>MODELLIERUNG LEITUNGSGEBUNDENER STÖRBEEINFLUSSUNG</u>
<u>BEI OPTISCHEN EMPFÄNGERN</u>
J. Sippel
Fachhochschule für Technik Berlin-Lichtenberg

1. Einleitung

Die faseroptische Datenübertragung wird neben ihrer postalischen Nutzung auch immer mehr zur Datenübertragung im Sinne der Steuerung, Regelung und Automatisierung technologischer Prozesse eingesetzt. Oft ist dieser Einsatz verbunden mit der Notwendigkeit, auch in elektromagnetisch stark beanspruchter Umgebung eine zuverlässige Datenübertragung zu gewährleisten. Der Lichtwellenleiter (LWL) bietet hier eine gewisse Sicherheit, da er selbst unempfindlich gegenüber elektromagnetischen Störungen ist. Zu einer optischen Datenübertragungsstrecke gehören allerdings noch die elektrooptischen bzw. optoelektrischen Wandler. Besonders letztere reagieren hochempfindlich auf elektromagnetische Störungen und bestimmen u.U. maßgeblich die Zuverlässigkeit der Datenübertragung als Ganzes. Deshalb sollte an dieser Stelle großer Wert auf eine fundierte Arbeit zur Sicherung der elektromagnetischen Verträglichkeit (EMV) solcher Systeme gelegt werden.

2. Elektromagnetische Störungen in Automatisierungs- und Energieanlagen

Die Ursachen elektromagnetischer Störungen können sehr vielfältig sein (s. Abb. 1).
Neben den internen Störungen, die oft schaltungstechnisch oder topologisch bedingt sind wirken vor allem externe Netzstörungen, die, meist hervorgerufen durch Übergangsprozesse in leistungsstarken Verbrauchern sich über das öffentliche Niederspannungsnetz oder andere Versorgungsnetze ausbreiten. Nach den Untersuchungsergebnissen in /1/,/2/ haben die beobachteten transienten Störungen Amplituden nicht nur im Bereich von einigen Hundert, sondern bis einigen Tausend Volt.

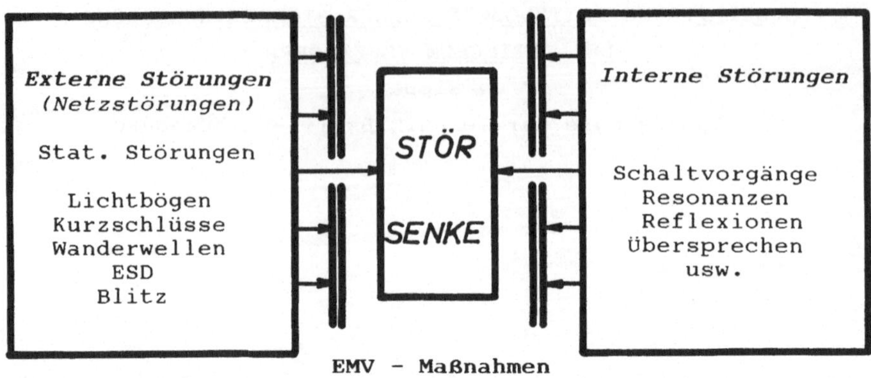

Abb. 1 Störbeeinflussungsarten

Die Entwicklung zeigt, daß die Niederspannungsnetze durch eine wachsende Anzahl von Oberwellenerzeugern wie Phasenanschnittsteuerungen, Motore, Schaltnetzteile oder auch Haushaltgeräte immer mehr gestört werden. Diese Störungen können auch durch Netzteile oft nur ungenügend abgeblockt werden. Vor allem über parasitäre Kapazitäten, die an jedem Netztransformator, Optokoppler, Relais u.s.w. existieren, werden die Störungen in die elektronische Schaltung geschleppt.
Die Störungen werden aber auch über andere Wege auf die betroffene elektronische Schaltung als Störsenke eingekoppelt (galvanisch, kapazitiv, induktiv, elektromagnetische Wellen bzw. Strahlung) und dabei durch bewußte (Erdung, Schirmung, Filterung u.s.w.) bzw. unbewußte EMV-Maßnahmen in ihrer Wirkung abgeschwächt.

Bei der Steuerung energetischer Prozesse, z.B. in Umspannwerken der Energieversorgung müssen Störquelle und Störsenke auf engstem Raum mit hoher Zuverlässigkeit funktionieren. Die energetischen Verhältnisse in solchen Fällen deutet Abb. 2 an.
Es zeigt sich, daß das Verhältnis von Störenergie zu gestörter Energie mehr als $10^{15}:1$ beträgt. Die Extreme entstehen hier einerseits durch die hohen geschalteten Leistungen in energetischen Objekten, die besonders im Fehlerfall weit in den MW-Bereich gehen, andererseits durch die extrem geringen Signalleistungen in optischen Empfängern von wenigen nW.

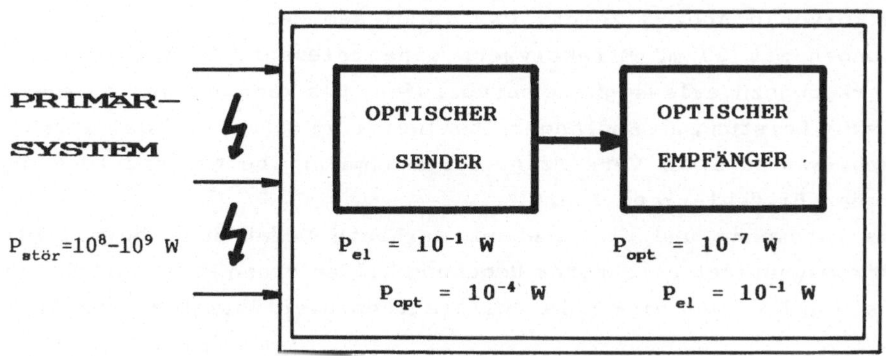

Abb. 2 Leistungsverhältnisse in Energieanlagen

3. Störbeeinflussungsmessungen an optischen Empfängern

Diese potentiellen Gefahren beim Einsatz von LWL-Übertragungsstrecken führten schon vor einigen Jahren zu der Erkenntnis, daß man sich mit dieser Problematik intensiver beschäftigen muß, um den erhöhten Zuverlässigkeitsansprüchen bei der Steuerung energetischer Objekte zu genügen /3/,/4/. In Laborversuchen wurde zunächst die Störfestigkeit verschiedener optischer Kurzstreckenempfänger in einem breiten Frequenzbereich untersucht /4/-/6/ (s. Abb. 3).

Es wurde festgestellt, daß die Störfestigkeit bezüglich leitungsgebundener Störeinkopplungen eine ausgesprochen spektrale Charakteristik besitzt. Dabei liegen die kritischen Frequenzen im Bereich der maximalen Verstärkung des linearen Traktes des optischen Empfängers. Die Bei-

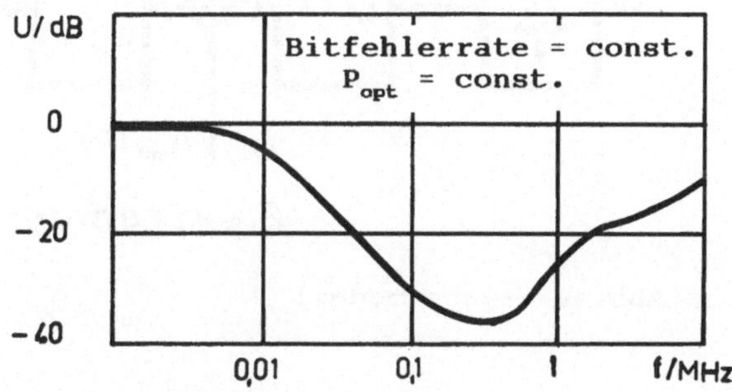

Abb. 3 Spektrale Abhängigkeit der zulässigen Störspannung für eine konstante Bitfehlerrate

spielkurve in Abb. 3 zeigt, daß in diesem Bereich bereits permanente
Störungen mit 20 mV Effektivwert eine relevante Verschlechterung der
Übertragungszuverlässigkeit hervorrufen. Das kann nur durch eine erhöhte
optische Leistung am Empfänger (teilweise um mehr als 5 dB) ausgeglichen
werden, was zu einer Verkürzung der maximalen Übertragungsfeldlänge bei
gleicher Bitfehlerrate führt.
Diese Tatsachen und die erwähnte erhöhte Gefährdung beim Einsatz in
elektromagnetisch belasteter Umgebung bilden einen Widerspruch, der ökonomisch gelöst und eine hohe Übertragungszuverlässigkeit zum Ziel haben
muß.

4. Zur Modellierung leitungsgebundener Störbeeinflussung bei optischen Empfängern

Um die elektromagnetische Verträglichkeit an solchen kritischen Stellen
zu planen und Optimierungsmöglichkeiten zu finden, ist es notwendig, den
Einwirkungsmechanismus theoretisch zu beschreiben. Man geht dabei in
Übereinstimmung mit der Natur transienter Störungen /7/ von folgendem
determinierten linearen System aus (s. Abb. 4).

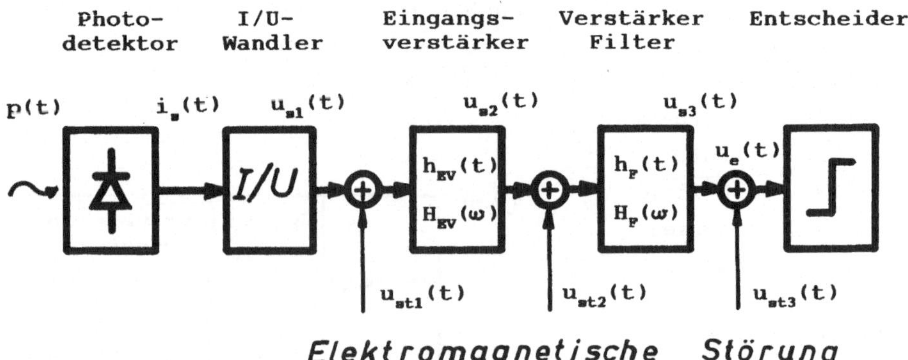

Abb. 4 Systemmodell

Die Systemelemente sind durch ihre Übertragungsfunktion H(ω) bzw. durch
ihre Pulsantwort h(t) charakterisiert. Die konkreten Zeitfunktionen für
Signal und Störungen werden durch die Elemente in entsprechender Art und

Weise gefiltert und an den Knoten additiv überlagert.

Die Spannung $u_{st1}(t)$ charakterisiert die Störungen, die über die Versorgungsspannung und die Sperrschichtkapazität der Photodiode direkt an den Eingang des ersten Verstärkers eingekoppelt werden. U_{st1} kann meist leicht über eine einfache RLC-Ersatzschaltung berechnet werden. Die folgende Konfiguration erweist sich als repräsentativ für die Mehrzahl der optischen Empfänger /8/-/10/ (s. Abb. 5).

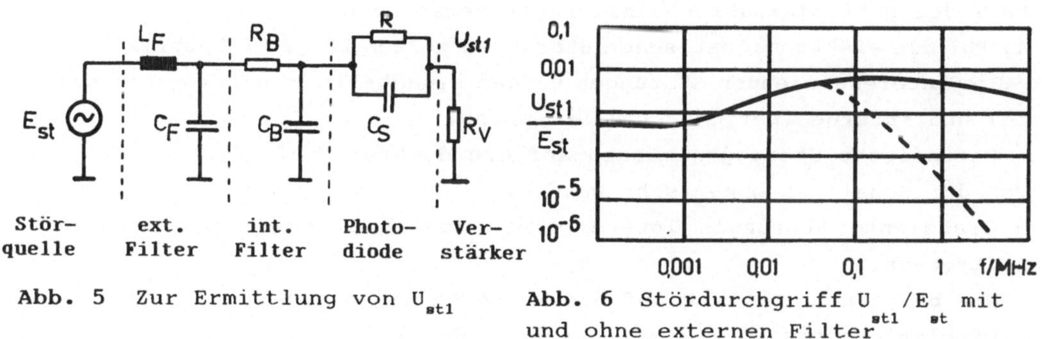

Abb. 5 Zur Ermittlung von U_{st1} **Abb. 6** Stördurchgriff U_{st1}/E_{st} mit und ohne externen Filter

Dabei ist die Drosselspule L_F eine von möglichen Schutzmaßnahmen, die vom Anwender im Sinne der Störfestigkeit getroffen werden sollte. Der Stördurchgriff U_{st1}/E_{st} mit (gestrichelt) und ohne (durchgehend) externen Filter ist in Abb. 6 dargestellt.

Die Störung wird über den gesamten Frequenzbereich um mehrere Größenordnungen abgeschwächt. Da aber die verarbeiteten Signalspannungen am Eingang des Eingangsverstärkers ebenfalls nur im Millivoltbereich liegen, kommt es an dieser Stelle bei Betriebsspannungsstörungen im Voltbereich bereits zur Beeinflussung.

Noch direkter betroffen sind u_{st2} bzw. u_{st3}, die intern oft überhaupt nicht gestützt sind und nur über ein externes Filter abgeblockt werden können. Geschieht das nicht, liegen die Betriebsspannungsstörungen einschließlich der durch die digitale Endstufe erzeugten Störungen in voller Höhe an u_{st2} und u_{st3} und damit am Eingang von Filter bzw. Entscheider.

Für das Systemmodell selbst werden die folgenden Einschränkungen festgelegt:

1. Das Modell ist linear. Diese Forderung sollte nicht im Widerspruch zur Praxis stehen, da der kritische Bereich für die elektromagnetische

Störbeeinflussung immer bei kleinen Aussteuerungen des Verstärkers und somit im linearen Bereich liegt. Diese Annahme wurde durch eine umfangreiche netzwerkanalytische Arbeit an einem konkreten Empfänger umfassend bestätigt.

2. Die Störungen werden additiv überlagert. Auch hier sollte zunächst kein Widerspruch entstehen. Kommt es bei größeren Störungen zu Nichtlinearitäten auf Grund des Sättigungsverhaltens des Verstärkers, hat man mit der additiven Einkopplung immer den worst-case erfaßt.

3. Als Zuverlässigkeitskriterium bei der Datenübertragung wird in jedem Fall die Bitfehlerwahrscheinlichkeit verstanden.

4. Für die elektromagnetischen Störungen gelten folgende Restriktionen:

- Die interessierenden Störungen werden eingeteilt in externe (transiente) und interne (periodische) Störungen.

- Periodische Störungen müssen auf Grund ihrer Häufigkeit prinzipiell von der Schaltung beherrscht werden.

- Transiente Störungen interessieren zunächst nur im Zeitraum ihres Auftretens.

- Die berechnete Wirkung auf die Spannung am Entscheider und damit die Bitfehlerwahrscheinlichkeit innerhalb des betrachteten Zeitintervalls wird dann einfach mit der Auftrittswahrscheinlichkeit der Störung multipliziert.

- Auf diese Art und Weise kann dann punktweise die Gesamtbitfehlerwahrscheinlichkeit z.B. durch Integration über der Störamplitude ermittelt werden.

Mathematisch hat man bei dieser Betrachtungsweise zwei Aufgaben zu lösen.

- die additive Überlagerung von Signal und Störungen in den entsprechenden Knotenpunkten

- die Signalfilterung durch die Übertragungsglieder mit Hilfe der Faltung bzw. der Multiplikation der Fouriertransformierten

Beispielsweise würde eine Lichtimpulsfolge der Form $p(t)$

$$p(t) = \Sigma \, b_k \, h_p(t-kT) \tag{1}$$

h_p.....Pulsform eines Einzelimpulses

T......Periode

b_k.....Information des k-ten Impulses

k......ganzzahlige Zählvariable $(-\infty < k < +\infty)$

im Photodetektor in folgenden Photostrom umgesetzt

$$i_s(t) = \frac{\eta \; e \; p(t)}{h \; f} \tag{2}$$

    e......Elementarladung

    η......Wirkungsgrad

Dieser Photostrom würde dann in Abhängigkeit von der Eingangskonfiguration des Strom-Spannungs-Wandlers in eine Signalspannung $u_{s1}(t)$ umgesetzt.

$$\underline{U}_{s1} = \underline{I}_s \; \underline{Z} \tag{3}$$

    $\underline{U}_{s1}, \underline{I}_s, \underline{Z}$......komplexe Werte von Eingangsspannung, Signalstrom und wirkendem Eingangswiderstand

Diese Spannung, in Addition mit der einwirkenden elektromagnetischen Störung $u_{st1}(t)$ ergibt die Eingangsspannung für den Eingangsverstärker (EV).

Dessen Ausgangsspannung ergibt sich dann über

$$u_{s2}(t) = (\; u_{s1}(t) \; + \; u_{st1}(t) \;) \; * \; h_{EV}(t) \tag{4}$$

bzw.

$$U_{s2}(\omega) = H_{EV}(\omega) \; F\left[\; u_{s1}(t) \; + \; u_{st1}(t) \;\right] \tag{5}$$

mit $h_{EV}(t)$........Impulsantwort des Eingangsverstärkers

       $H_{EV}(\omega)$........Übertragungsfunktion des EV

       F..............Fouriertransformation

Die weitere Berechnung von $u_{s3}(t)$ und $u_e(t)$ verläuft analog.

Um aus der berechneten Zeitfunktion $u_e(t)$ eine Bitfehlerwahrscheinlichkeit abzuleiten, geht man von folgenden Überlegungen aus:

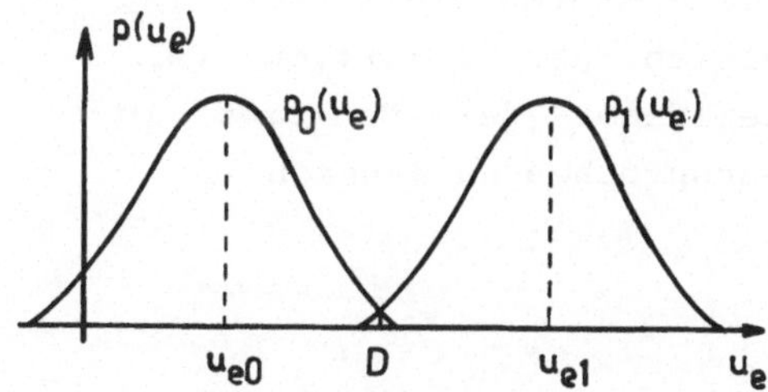

Abb. 7 Wahrscheinlichkeitsverteilungen der Spannung $u_e(t)$ bei 0- bzw. 1-Signal

1. Der Entscheider ist ideal. Das bedeutet, daß Werte von $u_e(t)$ zum Abtastzeitpunkt oberhalb der Schwelle D immer als 1, anderenfalls immer als 0 erkannt werden (s. Abb. 7).
Es gilt also zu ermitteln, mit welcher Wahrscheinlichkeit D bei 1-Signal unterschritten bzw. bei 0-Signal überschritten wird. Die Fehlerwahrscheinlichkeit im Störintervall T für eine gegebene Störamplitude U_{st} ergibt sich dann zu

$$P(U_{st}) = \int_{-\infty}^{D} p_1(u_e)\, du_e + \int_{D}^{+\infty} p_0(u_e)\, du_e \qquad (6)$$

$$P(U_{st}) = F_1(D) + (\ 1-F_0(D)\ ) \qquad (7)$$

mit $p_0(u_e), p_1(u_e)$ .....Wahrscheinlichkeitsdichtefunktionen der Spannung u_e im 0- bzw. 1-Zustand

F_0, F_1 ....Verteilungsfunktion der Spannung u_e im 0- bzw. 1-Zustand

2. Zur Ermittlung von $F_1(D)$ bzw. $F_0(D)$ im Falle einer deterministischen Zeitfunktion geht man von der Definition der Verteilungsfunktion F_x einer Größe x aus:

$$F_x(x_0) = P\ (\ x<x_0\ ) \qquad (8)$$

Dabei stellt $F_x(x_0)$ die Wahrscheinlichkeit dar, daß die Funktion x(t) zum Abtastzeitpunkt kleiner ist als x_0. Diese Wahrscheinlichkeit ist gleich der relativen Verweildauer der Funktion im entsprechenden Bereich

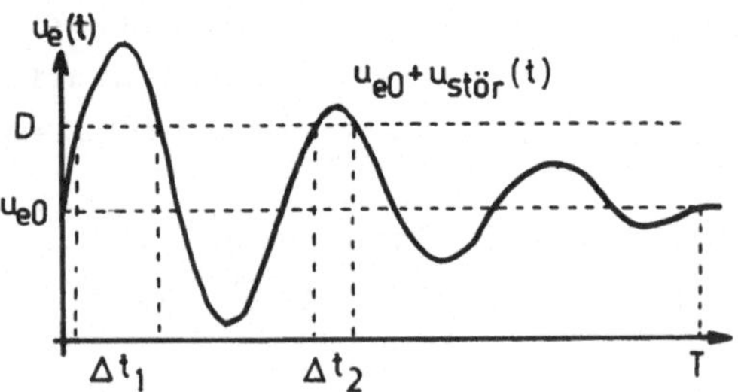

Abb. 8 Zur Berechnung der Fehlerwahrscheinlichkeit

$$F_x(x_0) = \frac{\Sigma\ \Delta t(x<x_0)}{T} \qquad (9)$$

Das bedeutet im konkreten Fall (s. Abb. 8)

$$F_0(D) = \frac{\Sigma \Delta t [(u_{e0} + u_{estör}) < D]}{T} \qquad (10)$$

Analog wäre

$$F_1(D) = \frac{\Sigma \Delta t [(u_{e1} + u_{estör}) < D]}{T} \qquad (11)$$

Damit kann dann über (6) und (7) die Fehlerwahrscheinlichkeit für die gegebene Störamplitude U_{st} während des Störintervalles T ermittelt werden.

3. Transiente Störungen treten statistisch, aber mit konkreten Amplitudenverteilungen $p(U_{st})$ auf. In /7/ und /11/ wird nachgewiesen, daß die großen Transienten, die vorwiegend die externen Störungen über das Stromversorgungsnetz verursachen, einem anderen, aber ähnlich stabilem Gesetz wie z.B. das Rauschen unterliegen.
Um die Gesamtfehlerwahrscheinlichkeit zu ermitteln, muß als letzter Schritt noch über die auftretenden Störamplituden integriert werden.

$$P_e = \int_0^\infty P(U_{st}) \; p(U_{st}) \; dU_{st} \qquad (12)$$

5. Schlußfolgerungen

1. Auf diese Weise ist es möglich, die elektromagnetische Beeinflussung beim Einsatz von LWL-Datenübertragungsstrecken in elektromagnetisch stark beanspruchter Umgebung, d.h. an einer äußerst empfindlichen Stelle theoretisch zu beschreiben und Zuverlässigkeitsbetrachtungen anzustellen. Diese Betrachtungen, deren Bedeutung mit der zukünftigen Verschärfung der EMV-Problematik immer mehr zunimmt, können auf beliebige ähnlich aufgebaute Systeme ausgedehnt werden.
2. Es ist möglich, die optischen Empfänger sowohl hersteller- als auch anwenderseitig einer EMV-mäßigen Optimierung zu unterziehen. Das

betrifft sowohl die Peripheriebeschaltung als auch die notwendigen
Maßnahmen zur externen Beschaltung wie Filterung der Versorgungsspannung
u.s.w. Damit sind dann auch ökonomische Abschätzungen verbunden.
3. Verschiedene Datenübertragungssysteme können miteinander verglichen
werden unter dem Gesichtspunkt der optimalen elektromagnetischen
Verträglichkeit in der gegebenen Umgebung.

/1/   Rehder,H.: Störspannungen in Niederspannungsnetzen.
         etz 100 (1979) 5   S.216-220

/2/   Meissen,W.: Transiente Netzüberspannungen.
         etz   107 (1986) 2   S. 50-55

/3/   Engelage,D.: Tendenzen der Anwendung von Lichtleitern
         für Energie- und Automatisierungsanlagen.
         Wiss. Ber. der IH Zittau 567 (1985) S.40-41

/4/   Tischer,K.: Untersuchungen zum Mikroelektronik-,
         Mikrorechner- und Lichtleitereinsatz in Kernkraftwerken.
         Forschungsbericht E 1305/85, IH Zittau, Sektion Elektro-
         energieversorgung, 1985

/5/   Hohlbein,R.:Beeinflussung der LWL-Technik beim Einsatz in
         Industrieanlagen. DA, IH Zittau, Sektion Elektroenergie-
         versorgung, 1987

/6/   Kollert,J.: Der Einfluß leitungsgebundener EMV-Störungen
         auf die Empfänger in optischen Datenübertragungskanälen.
         Ingenieurarbeit, IH Berlin, Sektion Elektrotechnik, 1990

/7/   Müller,W.F.;Lubenow,H.: Störimpulse in Nachrichtenkanälen.
         Nachrichtentechnik Elektronik 38 (1988) 1   S.8-9

/8/   Grimm,E.;Nowak,W.: Lichtwellenleitertechnik.
         Berlin: VEB Verlag Technik 1988

/9/   Hullet,J.L.; Moustakas,S.: Optimum transimpedance
         broadband optical amplifier design. Opt. and
         Quantum Electronics 13 (1981) S.65-69

/10/   Unwin,R.T.: A high speed optical receiver. Opt.
         and Quantum Electronics 14 (1982) S.61-66

/11/   Müller,W.F.: Ausbreitung von Störimpulsen.
         Nachrichtentechnik Elektronik 35 (1985) 1 S.2-4

Dipl.-Ing. Jürgen Sippel, Fachhochschule für Technik Berlin-Lichtenberg,
Fachbereich Elektrotechnik, Marktstr. 9-11, 1134 Berlin

EIN OSI-KOMMUNIKATIONSWERK AUF TRANSPUTER-BASIS FÜR DEN EINSATZ IN DER DEZENTRALEN PROZESSRECHENTECHNIK

Dietsch,H.,Ulrich,R.:
Lehrstuhl für Technische Elektronik - Informatik-Forschungsgruppe E
(Prof.Dr.-Ing.D.Seitzer)
Universität Erlangen-Nürnberg

Martensstr.3
D-8520 Erlangen, FRG

ZUSAMMENFASSUNG

Der Beitrag beschreibt ein Konzept, die (Teil-)Protokollschichten des ISO-OSI-Kommunikationsmodells zwischen den MAC- und APPLICATION-Dienstschnittstellen mittels eines Transputer-Netzwerks zu verwirklichen. Geringer Durchsatzverlust auf dem Weg durch die Schichtenhierarchie, ermöglicht durch flexible Anpassung der Verarbeitungskapazitäten an den Bedarf der OSI-Instanzen, ist das wichtigste Merkmal des vorgestellten Ansatzes: Die Transputer des "Kommunikationswerks" koordinieren ihre Parallelarbeit in hybrider Weise sowohl über ihre genuinen "Links" als auch über einen gemeinsamen Mehrtorspeicher. Die Leistungsdaten einer FDDI-"Dual Attachment Station" (TRANSPORT-Durchsatz: 40 Mbit/s) belegen die Brauchbarkeit des Konzepts. Durchgängigkeit bis zur APPLICATION-Schnittstelle wird am Beispiel der MAP-MMS-("Manufacturing Message Specification")-Dienste demonstriert.

1. EINLEITUNG: ZUM STAND DER HARDWARE-UNTERSTÜTZUNG FÜR OSI-IMPLEMENTIERUNGEN

Mit dem Aufkommen ISO-OSI-verträglicher lokaler Hochgeschwindigkeitsnetze ("High Speed LAN", HSLAN) wie z.B. FDDI ("Fiber Distributed Data Interface", (/1/)) verschärfen sich die Probleme bei Entwurf und Implementierung offener schichtenstrukturierter Kommunikationssysteme für den Einsatz in der dezentralen Prozeßrechentechnik:

(1) Wie können - ohne Minderung um Größenordnungen - die hohen Übertragungsraten der physischen Medien an der OSI-Anwendungsschnittstelle nutzbar gemacht werden?

(2) Wie können Teilnetze im prozeßnahen Bereich (basierend auf Feldbussen, ausgestattet mit schichtenreduzierten Architekturen) unter Erhaltung sowohl ihrer besonderen Eigenschaften als auch des Prinzips "Offenheit" in die OSI-Welt (CIM, MAP, MMS("Manufacturing Message Specification"), ...) eingebunden werden?

Zwar zielt dieser Beitrag auf das erstgenannte Problem, doch bietet die vorgestellte Lösung neben hohem Durchsatz auf hoher OSI-Ebene ausreichende Flexibilität für die Gestaltung von Übergängen zwischen "Full MAP"- und "Enhanced Performance"-Architekturen (EPA).

Für den Bereich zwischen standardisiertem physischen Medium und MAC-("Media Access Control")-Schnittstelle werden leistungsfähige VLSI-Bausteine angeboten, die mit den marktgängigen Mikrorechnerfamilien kompatibel sind. Herkömmliche 802.X-LAN-Anschaltungen für PCs und Workstations, ausgelegt als periphere Einheiten mit Universal(-Mikro)-Prozessor, ROM, lokalem Zweitorspeicher und netzwerkspezifischem VLSI-Steuerbaustein, erbringen an der TRANSPORT-Schnittstelle nur Durchsatzraten, die bis zum Zehnfachen unter denen der physischen Medien liegen. Als größtes Durchsatzhemmnis erweisen sich Kopiervorgänge sowohl innerhalb der Protokollschichten als auch beim Übergang zwischen den Protokollschichten.

Eine Zusammenfassung der grundlegenden Probleme und Lösungsansätze bei der Implementierung von OSI-Systemen wird in /2/ gegeben. Ein Überblick über aktuelle Versuche, VLSI-Technologie für die Implementierung von Protokollen oberhalb der MAC-Schnittstelle einzusetzen, findet sich in /3/. Obwohl einige der vorgestellten Lösungen zum Zweck der Adaption an unterschiedliche Protokolle programmierbar sind, ist doch allen gemeinsam, daß sie den Bereich jenseits der TRANSPORT-Schnittstelle nicht abdecken. Dies wird verständlich, wenn man als Beispiel für die auf der APPLICATION-Ebene üblicherweise vorherrschende Komplexität die ca. 90 MMS-Dienste mit ihren Protokollautomaten und ihren aufwendig zu kodierenden und zu dekodierenden Protokollnachrichten heranzieht: Um eine repräsentative Teilmenge dieses Dienstangebots mit akzeptablen Leistungsdaten zu verwirklichen, würden VLSI-Lösungen benötigt, die hinsichtlich Durchsatz und Flexibilität (Hochsprachen-Programmierung von Echtzeitanwendungen, Schutzkonzepte, Beobachtungshilfen, ...) modernen Standard-(Multi-)Mikrorechnern äquivalent sind.

Deshalb wird im folgenden der Vorschlag präsentiert, ISO-OSI-Funktionalität zwischen MAC- und APPLICATION-Schnittstelle mittels eines "hybrid gekoppelten" Transputer-Netzwerks, das zusätzlich zu den genuinen "Links" einen gemeinsamen Mehrtorspeicher aufweist, zu verwirklichen. Eine OSI-Implementierung auf Transputer-Basis stellt einen Kompromiß dar zwischen den bisher üblichen reinen Software-Lösungen und den diversen experimentellen VLSI-Ansätzen. Sie bietet Durchgängigkeit in einem weiten Bereich und fördert damit ein einheitliches Implementierungskonzept.

Das Funktionsprinzip für die vorgeschlagene Architektur sieht vor,

- daß im Mehrtorspeicher, auf den die schichtspezifischen Transputer zugreifen, die zu bearbeitenden und zwischen den Schichten (ohne Kopieroperationen) zu "transportierenden" Datenpakete residieren, und daß
- Signalisierungs- und Synchronisationsnachrichten über die "Links" übertragen werden.

In /4/ wird - ebenfalls mit Blick auf die Durchgängigkeit - vorgeschlagen, den höheren Protokollschichten Rechenelemente mit identischer Architektur zuzuordnen, die auf einen gemeinsamen Speicher zugreifen. Zusätzliche Transputer-"Link"-ähnliche serielle Verbindungen zwischen den schichtenspezifischen Rechenelementen werden in Betracht gezogen. Die flexible Anpassung der Rechenleistung an die Gegebenheiten der höheren Protokollschichten wird nicht behandelt. Die in /5/ und /6/ vorgestellten, von vornherein auf den Transputer-Einsatz ausgerichteten Ansätze beschränken sich auf den Bereich unterhalb der TRANSPORT-Schnittstelle.

Der Beitrag geht in der folgenden Weise vor:

(Auf die Grundzüge von ISO-OSI-Referenzmodell und Transputer-Technik wird nicht mehr eingegangen.)

- Zunächst werden die dem OSI-Modell innewohnenden Möglichkeiten für parallele Verarbeitung herausgearbeitet.
- Meßergebnisse für Durchsatzraten an den LLC- und TRANSPORT(TP4)-Schnittstellen werden präsentiert und diskutiert für die eingangs skizzierte "herkömmliche LAN-Anschaltung" und für zwei Transputer-Kopplungsvarianten,
  - die genuine Kopplung ausschließlich über "Links" und
  - die vorgeschlagene "hybride Kopplung" über "Links" und Mehrtorspeicher.
- Ausgehend von den Resultaten wird die Architektur eines OSI-Kommunikationswerks auf Transputer-Basis für den MULTIBUS II vorgestellt.
- Um die Tauglichkeit des Konzepts für Anwendungen in der Prozeßrechentechnik zu belegen, werden
  - eine implementierte FDDI-Anschaltung und
  - Emulationsergebnisse von MMS-Diensten auf einem Transputer-Cluster dargeboten.

2. PARALLELISIERUNGSANSÄTZE FÜR OSI-IMPLEMENTIERUNGEN

Bausteine schichtenstrukturierter Kommunikationssysteme sind zum einen die Automaten, die die Regeln der (Teil-) Protokolle implementieren, zum andern die Funktionseinheiten für Erzeugung und Dekodierung der Protokollnachrichten. Es gibt zwei Möglichkeiten, das Konzept nebenläufiger Prozesse anzuwenden (/2/):

- Jede Protokollinstanz wird als Prozeß oder als Prozeßsystem verwirklicht. Der Nachrichtenaustausch zwischen benachbarten Instanzen geschieht über Botschaften ("Server Model").

- Prozesse sind den Protokollnachrichten als Begleiter auf ihrem Weg durch die Schichtenhierarchie zugeordnet ("Activity-thread Model"). Nachrichten werden zwischen den Nachbarinstanzen mittels Prozeduraufrufen übertragen.

Offensichtlich bietet sich das "Server Model" an für OSI-Implementierungen auf Basis parallelverarbeitender Hardware.

Die diversen OSI-Vorkehrungen für die wechselseitige Überführung von Dienst- und Protokollnachrichten stellen Anforderungen an die Speicherorganisation: Um zeitraubendes Kopieren zu vermeiden, sollten die von den schichtspezifischen Protokollinstanzen zu bearbeitenden Datenpakete in einem gemeinsamen Speicher liegen.

Hinweise auf Parallelisierungsmglichkeiten, die dem OSI-Modell innewohnen, werden mit dem Bild 1 gegeben:

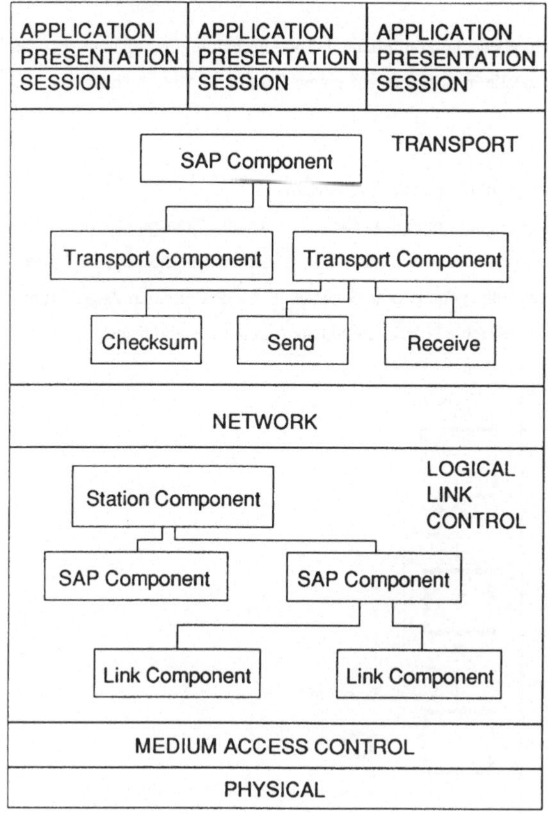

Bild 1: Zur Parallelisierung von OSI-Strukturen

- Die Protokollinstanzen bilden eine "Pipeline". Wenn die Bearbeitung der Protokolldaten länger dauert als deren Weiterleitung, ist die konstante Zuordnung von Rechenkapazität zu einer Protokollinstanz gerechtfertigt. Weil der Durchsatz einer Pipeline vom langsamsten Element bestimmt wird, muß auf die balancierte Ausstattung der Protokollschichten mit Hardware-Resourcen Wert gelegt werden.

- Für Empfangs- und Senderichtung können (z.B. in der TRANSPORT-Schicht) separate Resourcen eingesetzt werden, wenn der Aufwand für deren Koordination vernachlässigt werden kann.

- In Schichten, die verbindungsorientierte Dienste anbieten, existiert für jede Verbindung ein Protokoll-Automat. Bei stark belasteten Systemen können die Protokoll-Automaten auf mehrere HW-Prozessoren verteilt sein.

- Eigenständige Aufgaben einer Protokollinstanz, wie z.B. die Berechnung von Prüfsummen, können dedizierten Prozessoren zugeordnet werden.

- Die OSI-Schichten APPLICATION, PRESENTATION und SESSION weisen auf Grund innewohnender Vielfalt, Unabhängigkeit und struktureller Replikation viele Wege zur parallelen Verarbeitung.

3. EIN OSI-KOMMUNIKATIONSWERK AUF TRANSPUTER-BASIS

3.1 Vergleichende Messungen an Transputer-Kopplungsvarianten

Im folgenden werden für drei Architekturen und variable Paketlängen die Durchsatzraten an LLC- und TRANSPORT-Schnittstelle wiedergegeben und diskutiert:

- Unter dem Begriff "Standardarchitektur" wird die beschriebene herkömmliche LAN-Anschaltung verstanden, bei der jedwede Protokollfunktionalität oberhalb der MAC-Schnittstelle in Software auf einem Peripherie-Mikrorechner bzw. auf dem "Host"-Rechner realisiert ist.

- Zwei Transputer-Architekturvarianten werden untersucht:

  - Bei der genuinen "Link-Kopplung" (Bild 2a) ist kein gemeinsamer Speicher vorgesehen.

  - Der hier eingeführte Begriff der "hybriden Kopplung" bezeichnet eine Topologie, bei der die Transputer sowohl über "Links" als auch über einen gemeinsamen Mehrtorspeicher kommunizieren (Bild 2b). Dabei werden die seriellen "Link"-Verbindungen für Synchronisationszwecke genutzt, während der gemeinsame Speicher schnellen direkten Zugriff (unter Vermeidung von Kopiervorgängen) zu den in der Schichtenhierarchie bearbeiteten Datenpaketen gewährleistet.

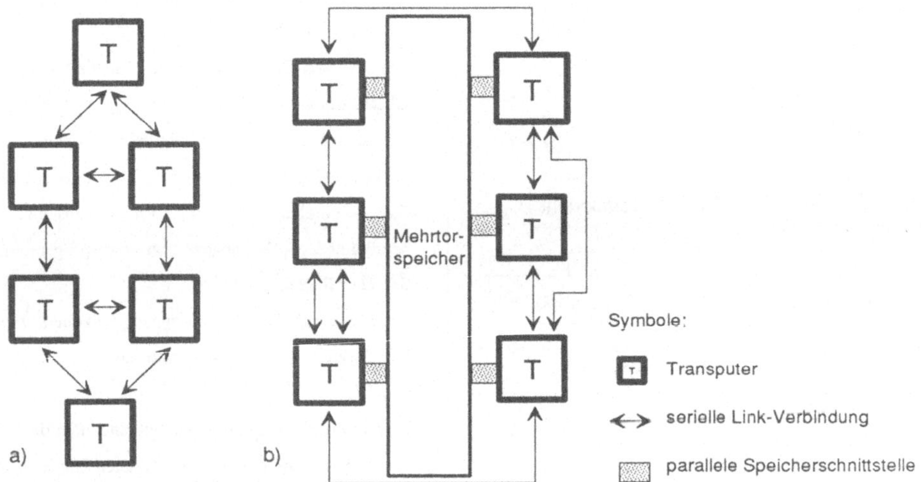

Bild 2: "Link-Kopplung" vs. "hybride Kopplung"

Für die Transputer-Architekturvarianten ist darauf hinzuweisen, daß jede Protokollinstanz auf genau einem Transputer implementiert wurde. In beiden Fällen beträgt die Bruttoübertragungsrate der "Links" 20 Mbit/s. Das in der hybriden Variante verwendete dynamische RAM weist eine "burst access"-Zykluszeit von 75 ns auf. Beide Meßobjekte wurden in OCCAM programmiert.

Die Zusammenschau der Ergebnisse zeigen Bild 3a für die LLC-Schnittstelle (/7/) und Bild 3b für die TRANSPORT-TP4-Schnittstelle (/8/). Bei der Darstellung der aus der "hybriden Kopplung" gewonnenen Ergebnisse symbolisieren die schraffierten Flächen Freiräume für Entwurfsentscheidungen (wie z.B. Art und Organisation des gemeinsamen Speichers).

Allen gezeigten Ergebnissen liegen zugrunde eine "leere" NETWORK-Schicht und ausgewogener bidirektionaler Verkehr.

Der lineare Zusammenhang zwischen Paketlänge und Durchsatz, der sich an beiden Schnittstellen der "Standardarchitektur" ergibt, macht deutlich, daß der Transportaufwand gegenüber dem Aufwand für die Bearbeitung der Pakete zu vernachlässigen ist.

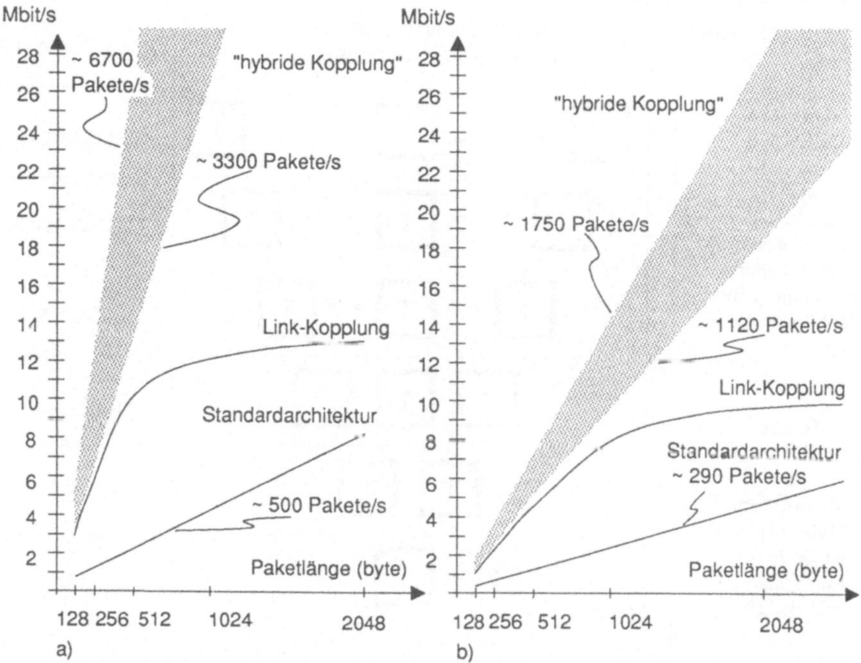

Bild 3: LLC- und TRANSPORT- Durchsätze im Vergleich

Bei der reinen "Link-Kopplung" müssen jeweils zwei Abschnitte der gewonnenen Kurven unterschieden werden: Bei kurzen Paketen wird der Durchsatz von der Prozessorkapazität bestimmt; mit wachsender Paketlänge gewinnt die relativ geringe Übertragungsrate der "Links" immer mehr Einfluß und führt schließlich zu einem Zustand der Sättigung, in dem Paketverlängerung, die in manchen Anwendungen (wie z.B. bei einem File Server) erstrebenswert wäre, den Durchsatz nicht erhöhte.

Völlig andere Ergebnisse zeigt die Variante der "hybriden Kopplung". Wiederum wird die Unabhängigkeit des Durchsatzes von der Paketlänge erreicht, jedoch mit wesentlich höheren Durchsätzen als bei der "Standardarchitektur". Es ist anzumerken, daß der Leistungszuwachs gegenüber den Konkurrenten in der Domäne liegt, in der lange Pakete vorherrschen (z.B. File Transfer, Program Download). Die Beobachtung zeigt, daß - bei leerer NETWORK-Schicht - der von einem LLC-Transputer bewältigte Durchsatz ungefähr dreimal höher ist als der eines TRANSPORT-Transputers.

3.2 Die Architektur eines OSI-Kommunikationswerks mit "hybrid gekoppelten" Transputer-Bausteinen für den MULTIBUS-II

In Bild 4 sind Verbindungsregeln und Topologieprinzip eines OSI-"Kommunikationswerks" auf Transputer-Basis skizziert. (Der Begriff "Kommunikationswerk" steht für eine aus dem traditionellen "E/A-Werk" abgeleitete neue Bedeutung.) Das annähernd dreieckige Anordnungsprinzip gründet sich auf der Beobachtung, daß auf dem Weg von unten nach oben durch die OSI-Schichtenhierarchie - mit zunehmendem Funktionsumfang - jede Schicht mehr Prozessorkapazität als die unterlagerte Schicht benötigt, wenn der Durchsatz an deren Dienstschnittstelle nach oben hin aufrechterhalten werden soll.

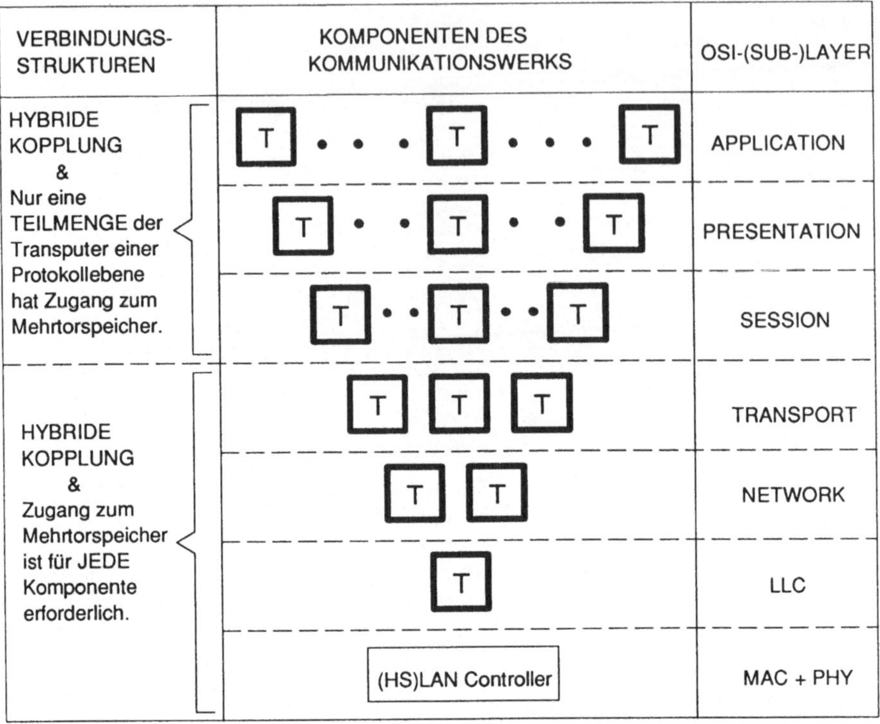

Bild 4: Prinzip eines OSI-Kommunikationswerks auf Transputerbasis

Aus der Prinzipdarstellung des Bildes 4 sollte nicht gefolgert werden, daß für das LLC-Protokoll genau ein Transputer und für das TRANSPORT-Protokoll drei Transputer vorzusehen sind: Die in Abschnitt 3.1 dargebotenen Meßergebnisse implizieren dieses Zahlenverhältnis, wobei zu betonen ist, daß das Verfahren der "hybriden Kopplung" in ähnlichem Maße wie die "Link-Kopplung" die flexible Einbindung zusätzlicher Komponenten unterstützt.

Hinsichtlich der Verbindungsregeln besteht ein Unterschied in der Behandlung der unteren und der oberen Schichten jenseits der TRANSPORT-Schnittstelle. In beiden Bereichen gilt das Prinzip der "hybriden Kopplung", doch hat in den unteren Schichten jeder Transputer (und natürlich die (HS-)LAN-Steuerung) direkten Zugriff auf den gemeinsamen Mehrtorspeicher, während in den höheren Schichten nur eine Teilmenge der Transputer dieses Privileg genießt. Für die höheren Schichten gilt: Auf Grund

- reduzierten Durchsatzangebots der unterlagerten Schichten,

- struktureller Vielfalt und

- von Besonderheiten, die die SESSION-, PRESENTATION- und APPLICATION-Schichten bei der Verbindungs-verwaltung aufweisen (Bild 1),

scheint für die überwiegende Zahl schichteninternen Transputer die reine "Link-Kopplung" ohne Zugriff auf den Mehrtorspeicher auszureichen, um die Durchsatzraten an den Dienstzugangspunkten zu optimieren.

Weitere Untersuchungen werden nötig sein, um Richtlinien für die optimale Zuweisung z.B. einer repräsentativen Teilmenge der MMS-Dienste zu einem APPLICATION-Transputer-Cluster mit adäquater Verbindungsstruktur zu formulieren.

Die Pilotimplementierungen lieferten grundsätzliche Erkenntnisse: Bei der Realisierung der TRANSPORT-Schicht wurden gute Ergebnisse mit dem Vorgehen erzielt, jeden Transputer mit identischem Funktionsvorrat auszustatten. Dabei stellt ein

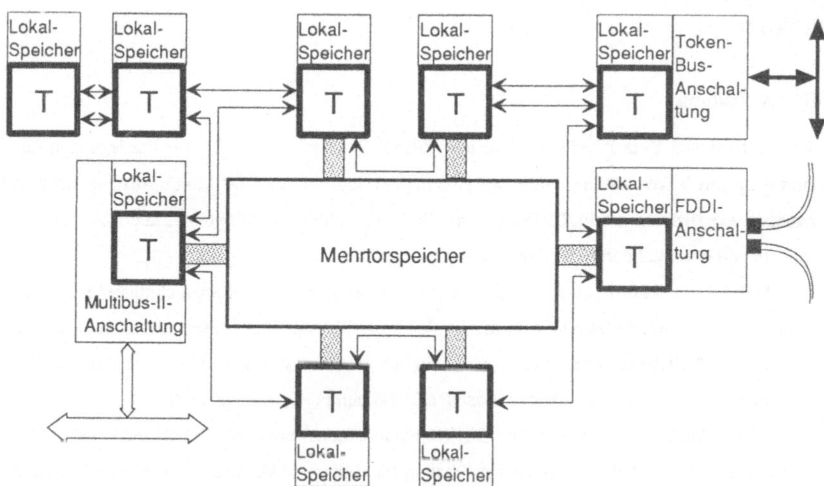

Bild 5: Architektur des Kommunikationswerks für einen Übergang zwischen FDDI und Token Bus

ausgewählter Transputer, der die Betriebsmittel seiner Partner zum Zweck des Auf- und Abbaus von Verbindungen verwaltet, die Ausnahme dar.

Bild 5 demonstriert die Architektur eines OSI-Kommunikationswerks mit hybrid gekoppelten Transputer-Modulen am Beispiel eines Übergangs auf höherer OSI-Ebene zwischen einem noch näher zu beschreibenden FDDI-"Backbone"-LAN und einem Token-Bus-LAN. An einen Mehrtorspeicher können bis zu sechs Transputer-Module angeschlossen werden. Andere Transputer-Module können per "Link-Kopplung" hinzugefügt werden. Im Beispiel stellt ein "hybrid gekoppelter" Modul die Verbindung zum MULTIBUS-II-Host-System her, ein anderer realisiert die FDDI-Anschaltung. Der Modul, der die Token-Bus-Anschaltung realisiert, benötigt wegen der relativ geringen Übertragungsrate keinen Zugriff auf den Mehrtorspeicher.

Im folgenden werden die Entscheidungen beim Entwurf des Mehrtorspeichers hinsichtlich Speicherkapazität, Speichertechnologie und Arbitrierungsverfahren skizziert: Um auf Stationen mit vielen gleichzeitig aktiven Verbindungen genügend Paketpuffer zur Verfügung zu haben, wurde die Speichergröße zu 4 MByte gewählt. Aus Kosten- und Platzgründen und, um die Übertragbarkeit der Ergebnisse zu gewährleisten, wurden dynamische RAM-Bausteine eingesetzt.

Speicherzugriff und Arbitrierung für den nächsten Zugriff laufen parallel und erbringen eine Zykluszeit von 125 ns (und eine Speicherbandbreite von 256 Mbit/s). Angesichts der Nettodatenrate des FDDI-Rings von 80 Mbit/s bleiben somit bei voller Last ca.80 Mbit/s für den Datenaustausch mit den Anwendungsprogrammen auf dem "Host"-Rechner und ca. 90 Mbit/s für die Zugriffe der "hybrid gekoppelten" Protokoll-Transputer. Für Module, die mittels DMA-Controller sequentiell auf größere Speicherbereiche zugreifen, wie z.B. der MULTIBUS-II-Modul, bietet der Mehrtorspeicher "Burst"-Zugriffe, die die Speicherbandbreite auf ca. 500 Mbit/s verdoppeln. Die Zahl der Tore wurde nach Messung des Zugriffsverhaltens eines Transputer-Moduls der verfügbaren Speicherbandbreite angepaßt.

4 ANWENDUNGEN

4.1 Eine FDDI-Anschaltung

Mit FDDI ("Fiber Distributed Data Interface") wird ein HSLAN-Standard angestrebt, der Festlegungen für einen doppelten Lichtwellenleiter-Ring mit Token-Passing und einer Übertragungsrate von ca. 100 Mbit/s trifft. In Erfüllung der ISO-OSI-Rahmenbedingungen spezifizieren die FDDI-Dokumente die (Teil-)Schichten PHYSICAL und MAC und eröffnen damit die Möglichkeit, höhere OSI-Protokolle in FDDI-Netzwerken zu installieren.

In /1/ wird ein Überblick über Prinzip und Leistungsdaten von FDDI gegeben; Entwicklungstendenzen hin zu integrierter Sprachübertragung (FDDI-II) werden skizziert. In dezentralen Prozeßrechensystemen wird FDDI vorwiegend als "Backbone"-Netzwerk auf der Prozeßleitebene eingesetzt werden, mithin in einer Position in der physischen Hierarchie, die die Problematik leistungsfähiger Übergänge zu konventionellen LANs und Feldbussen aufwirft.

Der FDDI-Modul als Bestandteil des vorgestellten OSI-Kommunikationswerks verwirklicht eine "Dual Attachment Station", die an beide LWL-Ringe angekoppelt ist und die von "Station Management"-Funktionen gesteuerte Fähigkeit zur Tolerierung diverser Ausfallarten besitzt. Er enthält

- die "SUPERNET"-VLSI-Chips der Fa. AMD,
- einen schnellen 256 KByte großen Zweitorspeicher, der von einem im "SUPERNET"-Chipsatz integrierten Arbiter verwaltet wird, sowie
- einen Transputer mit lokalem RAM und Schnittstelle zum Mehrtorspeicher.

Transputer und Zweitorspeicher stellen einen hohen Zusatzaufwand dar, der notwendig wird, weil der im FDDI-Chipsatz integrierte Arbiter nicht mit dem Mehrtorspeicher des Kommunikationswerks zusammenwirkt. Andererseits gereicht es zum Vorteil, daß dem lokalen Transputer des FDDI-Moduls LLC- und "Station Management"-Funktionen übertragen werden. Bei leerer NETWORK-Schicht können somit drei Transputer für die TRANSPORT-Aufgaben zur Verfügung gestellt werden. Den verbleibenden zwei Transputermodulen mit direktem Zugriff auf den Mehrtorspeicher werden die höheren Protokollschichten und die Steuerung der MULTIBUS-II-Schnittstelle zugewiesen. Bild 5 illustriert die Möglichkeit, zusätzliche Protokoll-Transputer für höhere Schichten per "Link-Kopplung" anzuschließen. Gateways und Konzentratoren können aufgebaut werden, indem netzwerkspezifische Transputermodule angefügt oder gegen Protokollmodule ausgetauscht werden. Hervorzuheben ist, daß das vorgestellte OSI-Kommunikationswerk mit geringem Aufwand um Hardware-Komponenten erweitert werden kann, die als Basis für die Realisierung von Anwendungsfunktionen ("File Server", "MMS-Client", ...) dienen.

4.2 Zur Implementierung von MMS-Diensten

Wie wichtig die Ausstattung der Schicht 7 mit Verarbeitungskapazität ist, zeigen die folgenden Untersuchungen: Während die Kodierungs- und Dekodierungszeiten für einfache MMS-Protokollnachrichten (z.B. bei den Diensten READ und WRITE) für einen 8086-Rechner bei 8 - 10 ms liegen, können sich diese Zeiten im Fall komplex strukturierter Pakete selbst auf einem 80286-Rechner in die Größenordnung von 25 ms erstrecken, was einem Durchsatz von deutlich unter 40 Paketen pro Sekunde entspricht. Angesichts der Komplexität, die die ASN.1-Notation bei MMS-Protokollnachrichten zuläßt, gelten für Ansätze, dedizierte VLSI-Schaltwerke für Kodierung und Dekodierung einzusetzen, die in der Einleitung formulierten Vorbehalte.

Die Realisierung eines MMS-Dienstes auf 3 über "Links" gekoppelte Transputer zeigt Bild 6: Je ein Transputer ist mit der Kodierung bzw. Dekodierung von MMS-Paketen befaßt; der dazwischen angeordnete Transputer verwirklicht den

Protokollautomaten und überprüft die Zulässigkeit der Dienstanforderungen. Um die Transporte über die "Links" und die Protokolldurchführung parallel ablaufen zu lassen, sind hochpriorisierte Sender- und Empfängerprozesse nötig. Die Reject-Prozesse gewährleisten Deadlock-Freiheit im Fehlerfall. Es ist anzumerken, daß je nach implementierter MMS-Dienstanzahl die Schicht 7 aus mehreren solcher Transputer-Dreiergruppen aufgebaut sein kann (Bild 1).

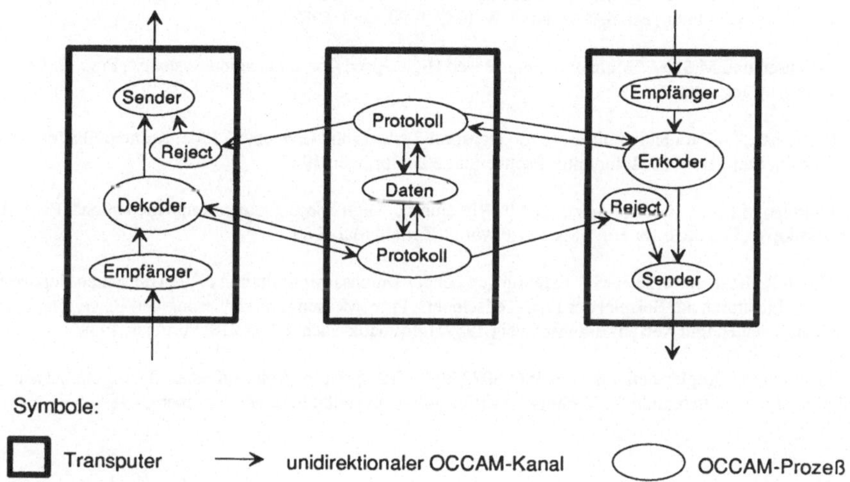

Bild 6: Struktur einer MMS-Protokolleinheit

Bild 7 faßt die gewonnenen Meßergebnisse für den Durchsatz in Abhängigkeit von der Paketlänge bei variiertem Paketbearbeitungsaufwand zusammen. Nur bei sehr einfach zu bearbeitenden Paketen, die zugleich länger als 2 KByte sind, begrenzt die "Link"-Übertragungsrate den Durchsatz. Typische Pakete im prozeßnahen Bereich der Fertigungsautomatisierung sind wesentlich kürzer, so daß die "Links" zu keiner Beschränkung führen und das vorgeschlagene Prinzip, nur eine Teilmenge der Transputer der höheren Schichten an den Mehrtorspeicher zu koppeln, seine Bestätigung findet.

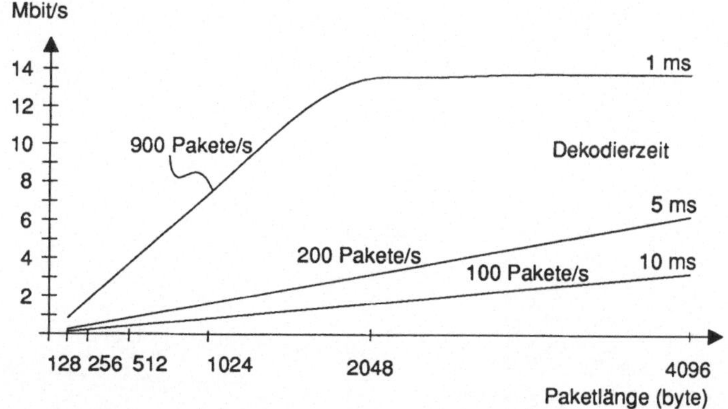

Bild 7: Durchsatzergebnisse der MMS-Emulation

Die Arbeiten, die diesem Beitrag zugrundeliegen, wurden im Rahmen des DFG-Sonderforschungsbereichs 182 "Multiprozessor- und Netzwerkkonfigurationen" an der Universität Erlangen-Nürnberg durchgeführt. Die Autoren danken für die ihnen zuteil gewordene Unterstützung.

LITERATUR

[1]   W.E.Burr und L.Zuquiu, "An Overview of FDDI", Proc. EFOC/LAN '88, Amsterdam, Juli 1988.

[2]   L.Svobodova, "Implementing OSI Systems", IEEE J. Select. Areas Commun., Bd. SAC-7, Nr. 7, S. 1115-1130.

[3]   A.S.Krishnakumar und K.Sabnani, "VLSI Implementations of Communication Protocols - A Survey", IEEE J. Select. Areas Commun., Bd. SAC-7, Nr. 7, S. 1082-1090, Sept. 1989.

[4]   M.N.Jensen und M.Skov, "Multi-Processor Based High-Speed Communication Systems", Proc. EFOC/LAN '88, Amsterdam, Juli 1988.

[5]   M.Zitterbart, "A Parallel Architecture for Transport Systems and Gateways", Proc. Kommunikation in verteilten Systemen, Stuttgart 1989, Informatik-Fachbericht 205, Springer, 1989.

[6]   D.Giarizzo,M.Kaiserswerth,T.Wicki und R.Williamson, "High Speed Protocol Implementation", Proc. IFIP Workshop on Protocols for High-Speed Networks, Zurich, Mai 1989.

[7]   R.Ulrich,R.Hinze und H.Dietsch, "Erfahrungen bei der Durchsatzoptimierung eines Transputer-Netzwerks für ISO-OSI-Architekturen am Beispiel der LLC-Teilschicht", Proc. Messung, Modellierung und Bewertung von Rechensystemen und Netzen, Braunschweig 1989, Informatik-Fachbericht 218, Springer, 1989.

[8]   I.Hagemeister, "Implementierung des ISO-8072/8073-Transportprotokolls auf einem Transputersystem", Diplomarbeit am Lehrstuhl für Technische Elektronik, Universität Erlangen-Nürnberg, 1989.

Verteiltes Monitoring- und Diagnosesystem für Ethernet

Jürgen M. Schröder
Institut für Nachrichtenvermittlung und Datenverarbeitung
Universität Stuttgart
Seidenstraße 36
7000 Stuttgart 1

Zusammenfassung: Lokale Netze bilden das Rückgrat von verteilten Systemen, wie sie in der Fertigungs- und Büroautomatisierung eingesetzt werden. Da alle Einheiten des verteilten Systems durch das lokale Netz verbunden werden, ist dessen Funktionalität ein entscheidender Faktor für die Funktion des Gesamtsystems. Um den Anwender bei der Unterhaltung eines lokalen Netzes zu unterstützen, wird, in Zusammenarbeit mit der Siemens AG, am Institut für Nachrichtenvermittlung und Datenverarbeitung (IND) ein Diagnosesystem entworfen. Das System besteht aus 2 Teilen: Einem verteilten Meßsystem für SINEC H1 (Ethernet) und einem zentral angeordneten Monitoring- und Diagnosesystem. Die Komponenten des verteilten Meßsystems ermöglichen eine automatische Beobachtung des lokalen Netzes während des Netzbetriebs. Sie werden vom zentralen Monitoringsystem gesteuert. Prototypen des verteilten Meßsystems befinden sich im Hause Siemens bereits im Einsatz.

1. Einleitung

Lokale Rechnernetze sind die Basis für verteilte Systeme, wie sie in der Fertigungsautomatisierung Anwendung finden. Die Funktionalität derartiger Systeme hängt wesentlich von der Verfügbarkeit des Rechnernetzes ab. Rechnernetze für die Fertigungsautomatisierung bestehen aus sehr vielen unterschiedlichen Komponenten, die zudem oft auch noch von verschiedenen Herstellern stammen. Aus diesem Grund erfordert die Wartung und Unterhaltung eines solchen Netzes sehr viel Erfahrung bezüglich des Gesamtsystems und bezüglich der verwendeten Komponenten.

Das benötigte Wissen steht allenfalls bei der Planung und der Inbetriebnahme eines Rechnernetzes zur Verfügung. Meist wird es auch nicht durch eine einzige Person verkörpert, sondern durch ein Expertenteam, das sich aus Spezialisten für die einzelnen Systemkomponenten zusammensetzt. Für den Betrieb eines Rechnernetzes steht dann das Expertenteam, meist aus Kostengründen, nicht mehr zur Verfügung.

Allgemein treten bei Netzen nach IEEE 802.3 (Ethernet) im Betrieb selten Fehler auf, Ethernet Netze gelten als "gutmütig". Bei dennoch aufretenden Fehlern benötigt der Anwender geeignete Werkzeuge zur Fehlerdiagnose, insbesondere wenn es sich um eine Netz handelt, das in der Fertigungsautomatisierung eingesetzt wird. In diesem Anwendungsbereich verursacht ein Netzausfall sehr hohe Kosten.

Fehlerdiagnose und Fehlervermeidung werden vereinfacht durch fortlaufende Beobachtung des Rechnernetzes. In dem hier vorgestellten Monitoring- und Diagnosesystem werden daher im Rechnernetz spezifische Meßkomponenten verteilt, die, ausgerüstet mit einem eigenen Netzzugang und ferngesteuert von einer zentralen Monitoring- und Diagnosestation, Messungen durchführen können. Die Meßergebnisse werden in den Meßsystemen zum Teil vorverarbeitet und dann zur Monitoring- und Diagnosestation übertragen. Bild 1.1 zeigt den prinzipiellen Aufbau des Monitoring- und Diagnosesystems.

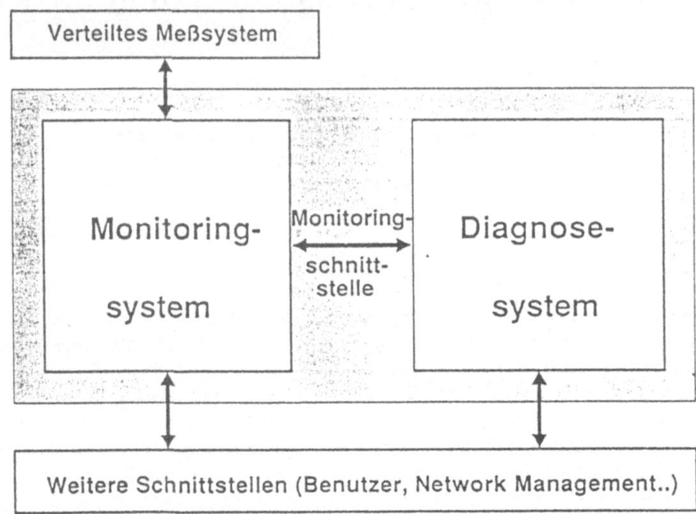

Bild 1.1   Aufbau des Monitoring- und Diagnosesystems

Die Steuerung des verteilten Meßsystems und die Auswertung der empfangenen Daten übernimmt das Monitoringsystem der Monitoring- und Diagnosestation. Dieses kann entweder vom Benutzer direkt bedient werden, oder es wird vom Diagnosesystem zur Aufnahme von fallspezifischem Wissen benützt.

In Abschnitt 2 des Aufsatzes wird das verteilte, modulare Meßsystem vorgestellt, das als Ergänzung weiterer Überwachungskomponenten wie LAN-Protokollanalysatoren verwendet wird. In den Unterpunkten des Abschnitts werden die Funktionen einzelner Meßteile erläutert. Im darauf folgenden Abschnitt wird das Monitoring- und Diagnosesystem vorgestellt, das mit dem verteilten Meßsystem zusammenarbeitet und dessen Kern aus einem Expertensystem besteht [4, 8, 9]. In den Unterpunkten dieses Abschnitts werden dessen einzelne Funktionsmodule näher erläutert. Abschließend wird auf den Stand der Implementierungen und die Erfahrungen mit Prototypen eingegangen.

2. Verteiltes, modulares Meßsystem

2.1 Allgemeines

Um das Rechnernetz im Betrieb beobachten zu können, werden im System Komponenten des Meßsystems verteilt. Bild 2.1 zeigt ein Beispiel für die Verteilung von Meßkomponenten im System. Die Segmentenden sind jeweils mit einem automatischen Reflexionsmeßteil abgeschlossen. Die Steuerung eines einzelnen Meßsystems kann zudem als sogenannte Echobox verwendet werden. Segmentkopplungen über Repeater werden mit Durchflußmessern überwacht. Einige Stationen werden mittels speziellen Meßfühlern direkt beobachtet.

2.2 Das Basissystem

Eine Komponente des verteilten Meßsystems setzt sich aus einem oder mehreren Modulen zusammen. Kern einer solchen Komponente ist eine CPU-Karte (Basiskarte), die auf dem Intel Prozessor i80186 basiert. Die Basiskarte verfügt über einen Ethernetzugang, der durch den Intel LAN-Koprozessor i82586 realisiert wurde. Die Karte verfügt ferner über mehrere parallele und serielle Ports. Über einen seriellen Port (RS232) kann sie mit einem Terminal oder PC lokal bedient werden. Bei Netzbetrieb wird sie jedoch über das Rechnernetz von der zentralen Monitoring- und Diagnosestation gesteuert. Neben 128 KByte Festwertspeicher und 128 KByte Schreib-/Lesespeicher verfügt die Karte noch über ein Bussystem, an das Meßerweiterungen angeschlossen werden können.

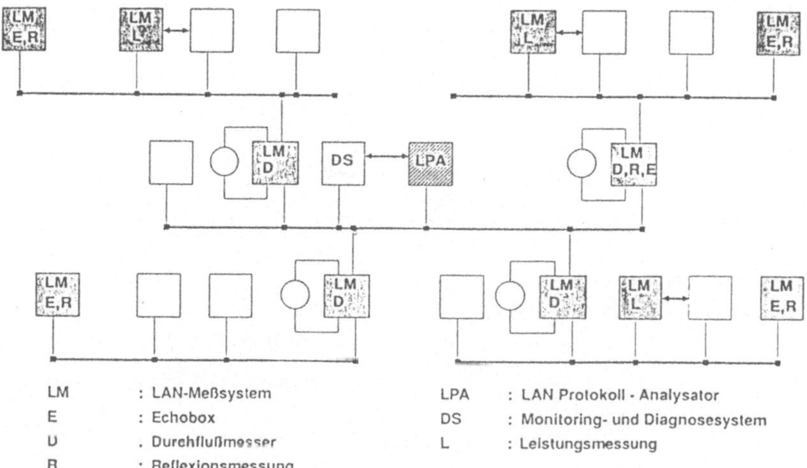

Bild 2.1   Beispiel eines lokalen Netzes mit verteiltem Meßsystem.

Die Basiskarte ohne Meßerweiterungen kann einfache Messungen und Funktionen durchführen. So kann z.B. die Erreichbarkeit aller Netzsegmente bei Netzbetrieb durch Abfragen von an den Segmenten installierten Meßkomponenten geprüft werden (Echobox). Weiter können einfache Statistiken über Netzlast, Anzahl der Kollisionen usw. erstellt werden.

2.2 Erweiterungen (Meßteile)

2.2.1 Durchflußmesser

Einzelne Segmente werden in lokalen Netzen nach IEEE 802.3 mit Repeatern verbunden. Die Verbindung von Segment und Repeater erfolgt über Transceiverkabel. Im Transceiverkabel werden die Datenströme von und zum Segmentanschluß getrennt geführt. Mit der Durchflußmessereinheit wird es möglich, den Verkehrsbeitrag eines einzelnen Netzsegmentes zum gesamten Verkehrsaufkommen zu bestimmen. Ferner ist die kurzzeitige Abtrennung des von der Monitoring- und Diagnosestation entfernt liegenden Netzsegmentes vom Gesamtnetz ferngesteuert möglich.

2.2.2 Automatische Reflexionsmessung

Entfernt man von einem installierten Netzsegment den Abschlußwiderstand und speist an derselben Stelle einen Impuls ein, so zeigt der Reflexionsverlauf eine charakteristische Funktion für das betreffende Netzsegment. Digitalisiert man den Reflexionsverlauf, so hat man die Möglichkeit, unterstützt durch ein entsprechendes Auswerteprogramm, Änderungen des Segmentkabels zu erkennen. Bild 2.2 zeigt das Prinzipschaltbild des automatischen Reflexionsmeßteils.

Der eingespeiste Meßimpuls hat dieselben elektrischen Eigenschaften wie der Impuls, mit dem ein Bit eines Datenpaketes übertragen wird. Bei einer maximalen Segmentlänge von 500m beträgt die einfache Signallaufzeit:

$$500 \text{ m} / (0.77 * c) = 2164.5 \text{ ns}$$

$$c = 30 \text{ cm/ns}$$

und somit die Meßzeit:

$$2164.5 \text{ ns} * 2 = 4329 \text{ ns}$$

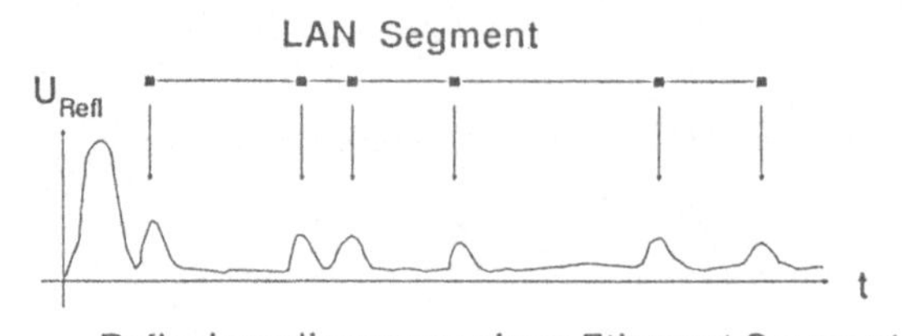

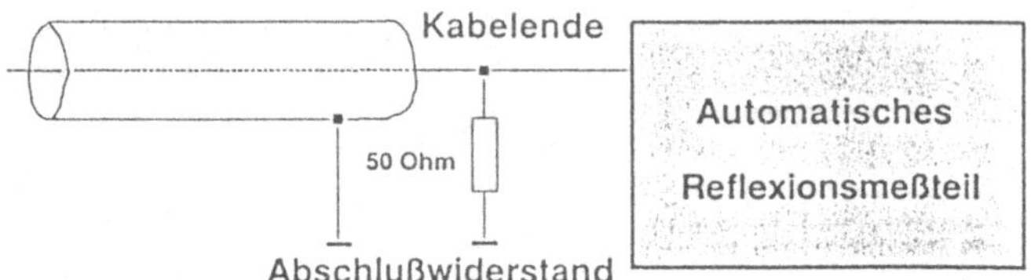

Bild 2.2   Prinzipschaltbild des automatischen Reflexionsmeßteils

Um die Reflexionen, hervorgerufen durch den Anschluß von Transceivern an das Netzsegment, noch erkennen zu können, müssen dem Reflexionssignal Amplitudenwerte im Abstand von 5 ns entnommen werden. Das realisierte Reflexionsmeßteil entnimmt dem Reflexionssignal 1024 Abtastwerte.

Das automatische Reflexionsmeßteil ersetzt den Abschlußwiderstand eines Segmentes und erlaubt die Durchführung einer Reflexionsmessung während des Netzbetriebs.

2.2.3 Weitere Meßteile

Für die Funktionalität des Gesamtsystems oder einzelner Teilbereiche kann es erforderlich sein, einzelne Stationen in ihrem Verhalten während des Betriebs genau zu beobachten. Hierfür sind besondere Meßteile vorgesehen und entwickelt worden.

In das verteilte Monitoringsystem lassen sich nicht nur Komponenten einbringen, die innerhalb des Projektes am IND entwickelt wurden, sondern alle fernsteuerbaren Meßgeräte. Als Beispiel hierfür seien Protokollanalysatoren genannt [1,2].

3. Monitoring- und Diagnosesystem

3.1 Allgemeine Betrachtungen

Die Auswertung der Daten, die das verteilte Meßsystem liefert, erfolgt in einer zentralen Monitoring- und Diagnosestation. Im normalen Netzbetrieb werden periodisch Meßdaten abgefragt und geprüft, ob sie innerhalb eines bestimmten Toleranzbereiches liegen. Liegen Sie außerhalb, dann werden entsprechende Einstiegspunkte für die Fehlerdiagnose generiert und die Fehlerdiagnose wird gestartet. Während der Fehlerdiagnose greift das System weiterhin auf das verteilte Meßsystem zurück, um fallspezifisches Wissen zu erwerben.

Für den Erwerb von fallspezifischem Wissen können auch Softwarekomponenten eingesetzt werden, die zum Test auf andere Stationen geladen, dort gestartet und beobachtet werden können.

Die Architektur des Diagnosesystems zeigt Bild 3.1. Die einzelnen Systeme, die den automatischen Wissenserwerb und das Monitoring des Netzes bei Netzbetrieb erlauben, werden über die Monitoringschnittstelle mit dem Diagnosesystem verbunden. Weiter ist für den automatischen Wissenserwerb eine Schnittstelle zum OSI-Netzmanagement geplant [5,6]. Für den Aufbau von hierarchischen Diagnosesystemen für große Netze ist eine Diagnosesystem-schnittstelle vorgesehen. Die Benutzerschnittstelle dient der Kommunikation mit dem Anwender.

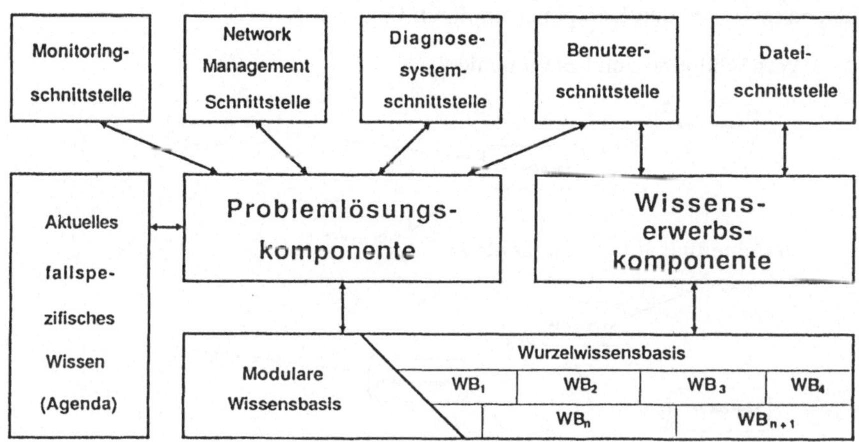

Bild 3.1   Architektur des Diagnosesystems

Für den Betrieb von lokalen Netzen ist sehr viel Wissen und Erfahrung nötig. Dies gilt erst recht, wenn in einem lokalen Netz Komponenten unterschiedlicher Hersteller, eventuell zudem mit verschiedenen Protokollarchitekturen, betrieben werden. Häufig ist das benötigte Wissen auf mehrere, zusammenarbeitende Personen verteilt. In der Wissensbasis des Diagnosesystems wurde das für die Diagnose benötigte Wissen deshalb in einzelne Module unterteilt, deren Wissensbereich in etwa dem der Bereichsexperten entspricht. Die Wissensbasis ist in eine sogenannte Wurzelwissensbasis und weitere der Wurzelwissensbasis untergeordnete Wissensbasismodule aufgeteilt [10]. In der Wurzelwissensbasis ist das Wissen über die Systemkonfiguration abgelegt. Die untergeordneten Wissensbasismodule enthalten Wissen über einzelne Komponenten, z.B. speziell über PCs. Eine Änderung in der Systemkonfiguration erfordert daher nur eine Änderung der Wurzelwissensbasis und eventuell den Einbau eines neuen Wissensbasismoduls.

Die Wissensbasismodule sind getrennt erstell- und wartbar. Für sich alleine sind sie als Wissensbasen für Diagnosesysteme der entsprechenden Komponenten einsetzbar. Durch diese Eigenschaften ist gewährleistet, daß Wissensbasen durch die Experten für die einzelnen Komponenten entworfen werden können.

Für den Wissenserwerb ist eine spezielle Wissenserwerbskomponente vorgesehen. Über eine Dateischnittstelle kann Wissen, daß nach einer entsprechenden Syntax in einem Textfile abgelegt wurde, in eine Wissensbasis eingelesen werden. Der Inhalt einer Wissensbasis kann entsprechend in einem Textfile ausgegeben werden. Interaktiv kann der Inhalt einer Wissensbasis durch den Benutzer verändert werden.

3.2 Wissensdarstellung

Das für die Diagnose benötige Wissen wird in zwei Kategorien unterteilt. Wissen über die Struktur des Rechnernetzes, die einzelnen Komponenten, die möglichen Diagnosen und deren Symptome wird in der Wissensbasis abgelegt. Vom Fehlerfall abhängiges Wissen, z.B. ob Symptome bestätigt werden konnten oder nicht, wird in der sogenannten Agenda gespeichert (Bild 3.1).

Die Wissensdarstellung für ein Netzwerk bedarf einer anderen Struktur, als die Wissensdarstellung für eine Komponente des Netzwerks. Aus diesem Grunde wurden zwei verschiedene Wissensbasistypen definiert. Mit der heterarchischen Wissensbasisstruktur läßt sich insbesondere die Netzarchitektur sehr gut nachbilden. Bild 3.2 zeigt das Prinzip der heterarchischen Wissensbasis. Mit der hierarchischen Wissensbasisstruktur können einzelne Komponenten sehr gut modelliert werden. Bild 3.3 zeigt das Prinzip der hierarchischen Struktur. Die Komponenten eines Netzwerkes werden durch Komponenten der Wissensbasis modelliert, die Verbindungen zwischen den Komponenten durch Beziehungen.

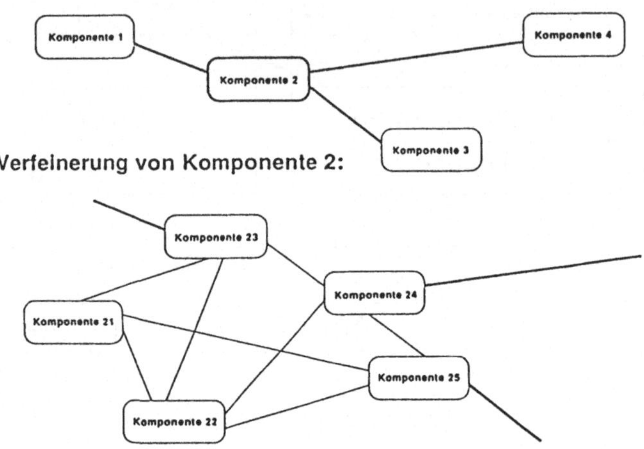

Bild 3.2   Komponentenverknüpfung in einer heterarchischen Wissensbasis. Prinzip der Verfeinerung.

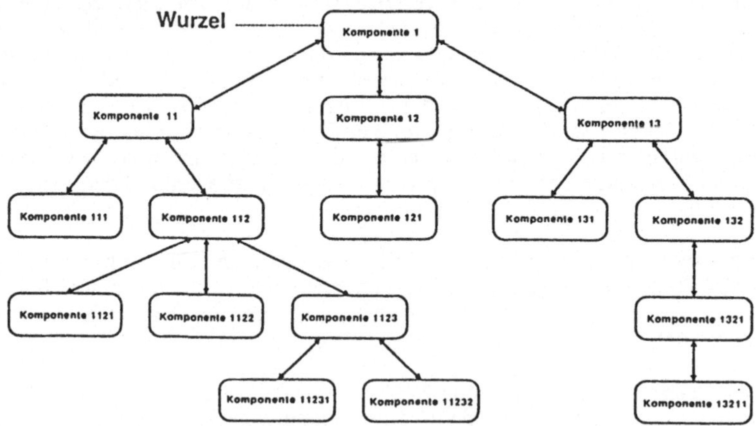

Bild 3.3   Komponentenverfeinerung in einer hierarchischen Wissensbasis.

Das für die Diagnose erforderliche Wissen wird mit Objekten dargestellt. Es sind folgende Objekte definiert:

- **Komponente**
  Eine Komponente ist ein physikalisch im Netzwerk vorhandenes Objekt. Hierzu zählen sowohl Hardware- als auch Softwarekomponenten. In einer heterarchischen Wissensbasis können Komponenten verfeinert werden, d.h. die betreffende Komponente wird durch ein Netz von Komponenten ersetzt, die die Funktionalität der verfeinerten Komponente genauer

beschreiben. Bild 3.2 erläutert den Vorgang der Verfeinerung. In einer hierarchischen Wissensbasis kann eine Komponente nur eine Vorgängerkomponente haben, aber beliebig viele Nachfolgerkomponenten. Die Struktur einer hierarchischen Wissensbasis ist baumförmig. Bild 3.3 zeigt die Komponentenstruktur einer hierarchischen Wissensbasis.

- **Diagnose**
Jeder Komponete ist mindestens eine Diagnose zugeordnet. Diagnosen werden unterteilt in Enddiagnosen und Diagnosen. Diagnosen führen, wenn eine Verfeinerung für die entsprechende Komponente angegeben ist, zur Fortsetzung der Diagnose, d.h. die Verfeinerung wird untersucht. Wird eine Enddiagnose bestätigt, dann muß die betreffende Komponente komplett ausgetauscht werden. Diagnosen werden durch Symptome bestätigt.

- **Symptome**
Erscheinungsbilder von Fehlern werden als Symptome bezeichnet. Symptome stehen in unmittelbarer Beziehung zu Diagnosen. Die Symptom - Diagnose - Beziehung wird durch eine Punktzahl bewertet, die angibt, mit welchem Beitrag ein Symptom zur Bestätigung einer Diagnose beiträgt. Ein Symptom kann mehreren Diagnosen zugeordnet sein.

- **Symptombewertungen**
Symptombewertungen verweisen auf die Methoden, mit denen ein Symptom bestätigt werden kann. Implizit kann das Zutreffen eines Symptoms immer vom Benutzer erfragt werden. Das System versucht jedoch, Symptome weitgehend automatisch zu bestätigen. Die entsprechenden Programme werden mit dem Objekt Symptombewertungen vereinbart.

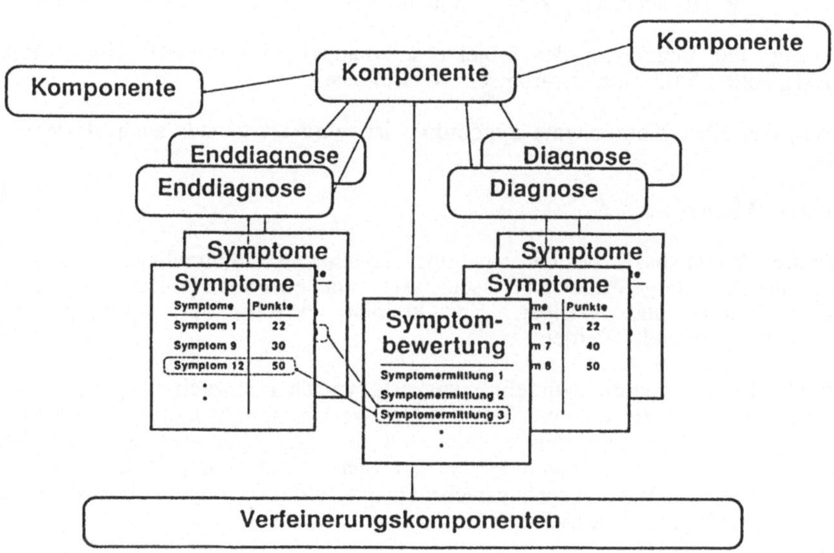

Bild 3.4   Relation der Objekte untereinander in einer heterarchischen Wissensbasis

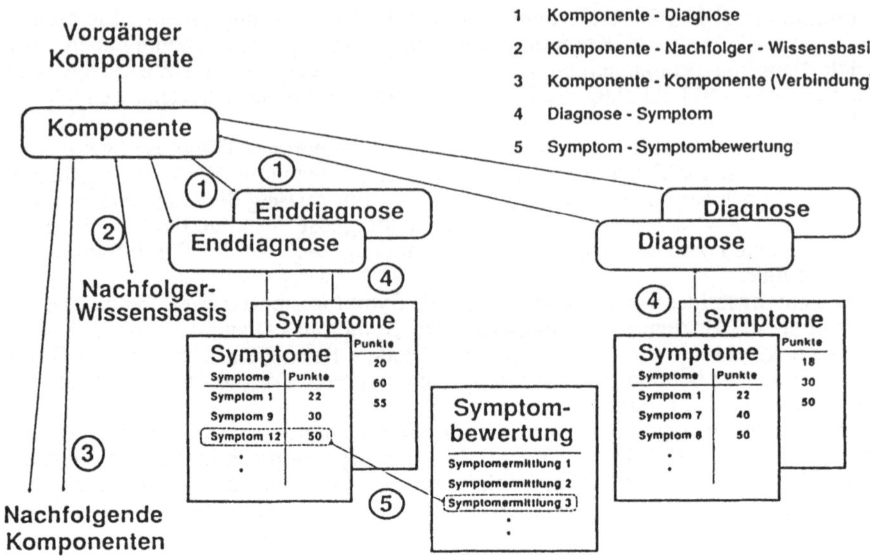

Bild 3.5   Relation der Objekte untereinander in einer hierarchischen Wissensbasis.

Bild 3.4 zeigt die Beziehung der Objekte zueinander für den Fall einer heterachischen Wissensbasis, Bild 3.5 für eine hierarchische Wissenbasis.

Das Wissen, das einer Komponente zugeordnet ist, wird als ganzes auch als Wissensrahmen bezeichnet.

3.3 Modulare Wissensbasis

Die modulare Wissensbasis besteht aus einer sogenannten Wurzelwissensbasis und dieser Wurzelwissenbasis untergeordneten Wissensbasen. Die einzelnen Wissenbasen sind getrennt voneinander erstell- und wartbar. Sie können einzeln für komponentenspezifische Diagnosesysteme verwendet werden.

Für ein lokales Netz besitzt die Wurzelwissensbasis typischerweise eine heterachische Struktur. Anhand dieser Wissensbasis kann eine defekte Netzwerkkomponente diagnostiziert werden, z.B. ein Transceiver. Für den Transceiver kann nun, je nach Typ, eine spezielle Wissensbasis nachgeladen werden, die den Anwender bei der Fehlersuche innerhalb des Wissensbereiches Transceiver unterstützt. Diese Vorgehensweise entspricht der Befragung eines Bereichsexperten. Dabei kann auch fallspezifisches Wissen ausgetauscht werden.

Da die Wissensbasen getrennt voneinander erstellt werden, muß, wenn fallspezifisches Wissen nach dem Laden einer neuen Wissensbasis weiterverwendet werden soll, eine Anpassung von Symptomen vereinbart werden. Dies geschieht durch das Globalisieren von Symptomen. Symptome, die von einer Wissensbasis in eine andere übergeben werden sollen, werden zu globalen Symptomen erklärt. Bild 3.6 erläutert den Vorgang des Ladens einer neuen Wissensbasis.

Das fallspezifische Wissen, das durch Bearbeitung der momentan geladenen Wissensbasis gewonnen wurde, ist in der aktuellen lokalen Agenda abgespeichert. Wird nun eine neue Wissensbasis geladen, dann wird zunächst die globale Agenda aktualisiert. Dies geschieht anhand einer Mapping Tabelle für die aktuelle Wissensbasis, in der verzeichnet ist, welche lokalen Symptome gleichbedeutend mit globalen Symptomen sind. Beim Laden der Nachfolger Wissensbasis wird eine neue lokale Agenda initialisiert. Anhand der Mapping Tabelle, die zur neuen Wissensbasis gehört, wird Information aus der globalen Agenda in die neue lokale

Agenda übertragen. Die Diagnose wird anhand der Information in der neuen lokalen Agenda fortgesetzt.

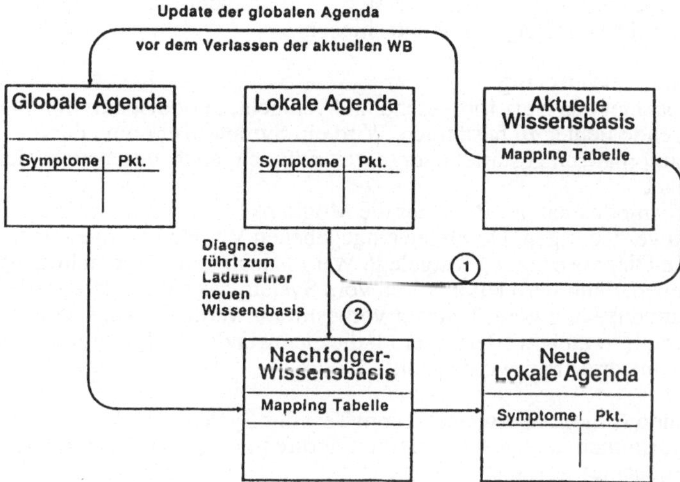

Bild 3.6   Verdachtsgenerierung beim Laden einer neuen Wissensbasis

3.4 Problemlösungskomponente

Die Problemlösungskomponente versucht anhand des in der Wissensbasis abgelegten Wissens, im Fehlerfall eine Diagnose zu stellen. Die Architektur der Problemlösungskomponente zeigt Bild 3.7. Die Ablaufsteuerung steuert mit Hilfe der Agendaoperationen die Diagnosesitzung. Die Agendaoperationen führen Operationen anhand der Informationen, die in der Agenda und in der aktuellen Wissensbasis enthaltenen sind, durch. Das Ergebnis der durchgeführten Operationen wird in die Agenda eingetragen. Zu den Agendaoperationen zählen die bekannten Diagnosealgorithmen wie Forward Chaining, Backward Chaining, Depth First, Breadth First, aber auch Algorithmen wie Verfeinerung, Laden einer neuen Wissensbasis usw., die speziell für diese Anwendung entworfen wurden.

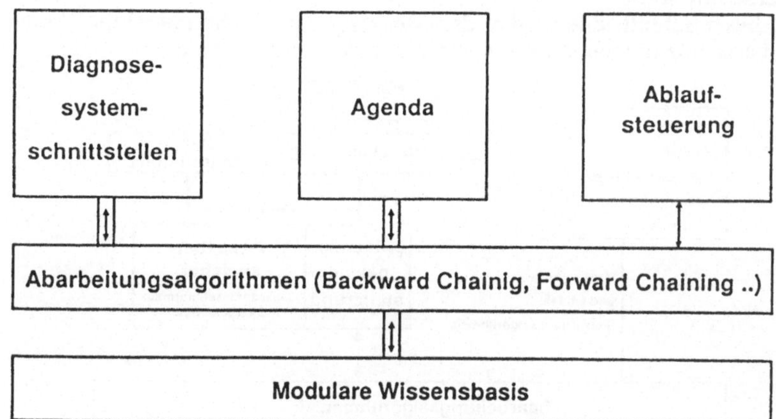

Bild 3.7   Architektur der Problemlösungskomponente

Die Regeln, wie die einzelnen Problemlösungsalgorithmen nacheinander angewandt werden, werden innerhalb der Ablaufsteuerung festgelegt. Diese Regeln können sich von Problemfall zu Problemfall unterscheiden. In der ersten Implementierung werden folgende Einstiegsmöglichkeiten für eine Diagnose unterschieden:

- **Automatisches Netzmonitoring**
  Das System beobachtet das Netz fortlaufend und versucht, periodisch Symptome für eventuell vorliegende Fehler zu bestätigen. Wird ein Symptom erkannt, dann wird die Fehlerdiagnose automatisch gestartet und, soweit möglich, auch automatisch durchgeführt.
- **Benutzergesteuert**
  In diesem Betriebsmodus hat der Benutzer die Möglichkeit, eine oder mehrere Komponenten zu verdächtigen. Durch die eingegebenen Verdächtigungen wird der Startpunkt für die Diagnosesitzung festgelegt. Wird jedoch keine Verdächtigung vom Benutzer angegeben, dann wird automatisch vom System das ganze Netz verdächtigt. In diesem Betriebsmodus kann vom Benutzer vorbestimmt werden, ob das System versucht, Symptome automatisch zu bewerten, oder ob die anzuwendende Symptombewertungsmethode vom Benutzer erfragt werden soll.
- **Expertenmodus**
  Dieser Betriebsmodus dient überwiegend dem Test des Systems. Es können die einzelnen Abarbeitungsalgorithmen benutzergesteuert aufgerufen und deren Auswirkungen auf den Agendainhalt beobachtet werden.

3.5 Wissenserwerbskomponente

Die Wissenserwerbskomponente unterstützt den Benutzer beim Eintragen von Wissen in die Wissensbasis. Bild 3.8 zeigt den Aufbau der Wissenserwerbskomponente. Sie verfügt über zwei Erwerbsschnittstellen:

- **Dateischnittstelle**
  Über diese Schnittstelle kann das Wissen, beschrieben in einer Sprache mit einer eigens definierten Syntax, über eine Textdatei in eine Wissensbasis eingelesen werden. Ebenso kann der Inhalt einer Wissensbasis in eine Textdatei ausgegeben werden. Über die Dateischnittstelle ist auch ein Anschluß des Configuration Managements an die Wissenserwerbskomponente möglich. Im Configuration Management ist die Konfiguration des Netzes abgespeichert. Durch Zugriffe auf die Datenbank des Configuration Managements ist die Wissenserwerbskomponente in der Lage, ein Gerüst für die Wurzelwissensbasis automatisch zu generieren.
- **Benutzerschnittstelle**
  Anhand dieser Schnittstelle wird es dem Benutzer ermöglicht, interaktiv Wissen in die Wissensbasis einzutragen, bzw. eingetragenes Wissen zu verändern.

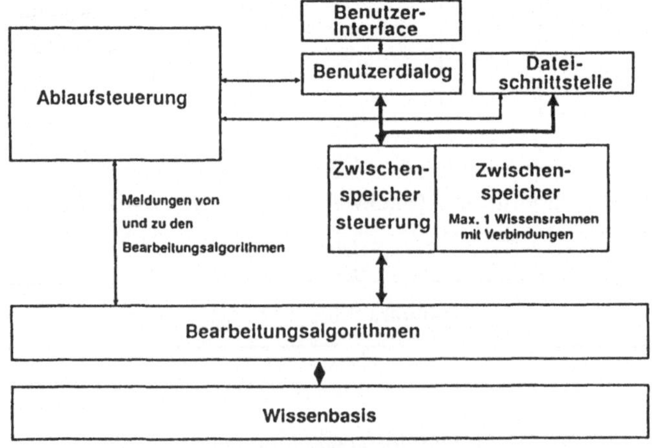

Bild 3.8   Architektur der Wissenserwerbskomponente

Das vom Benutzer eingetragene Wissen und das von einer Datei eingelesene Wissen werden intern auf eine Erwerbsschnittstelle abgebildet. Die Ablaufsteuerung steuert den gesamten Wissenserwerb mit Hilfe der Erwerbsfunktionen. Um die Widerspruchsfreiheit einer Wissensbasis zu gewährleisten, werden zwei unterschiedliche Konsistenzprüfungen durchgeführt. In einer ersten wird überprüft, ob ein eingetragener Wissensrahmen vollständig ist, d.h. zu einer Komponente gehört mindestens eine Diagnose, zur Diagnose mindestens ein Symptom. In der zweiten Konsistenzprüfung wird untersucht, ob alle eingetragenen Wissensrahmen mit der Netzstruktur der Wissensbasis verbunden sind.

4. Zusammenfassung und Ausblick

Das vorgestellte verteilte Monitoring- und Diagnosesystem erlaubt es, ein lokales Rechnernetz während des Betriebes umfassend zu beobachten. Im Fehlerfall werden die im Betrieb gesammelten Netzwerkdaten ausgewertet und automatisch, mittels der verteilten Meßkomponenten, fallspezifische Daten aufgenommen. Die Automatisierung der Aufnahme von fallspezifischem Wissen verkürzt die Zeit bis zur Diagnosestellung erheblich. Zudem ist das Monitoring- und Diagnosesystem durch laufendes Beobachten des Netzes im Betrieb in der Lage, Fehler, die nicht zum Ausfall des Netzes führen, dessen Funktionalität jedoch einschränken, zu erkennen und den Benutzer darauf hinzuweisen.

Das Diagnosesystem unterstützt den Anwender bei der Fehlersuche. Durch die Aufteilung der Wissensbasis in eigenständige Wissensbasismodule, die zudem getrennt voneinander erstellt werden können, ist das Diagnosesystem einfach auf jede Netzkonfiguration anpassbar.

Im Hause Siemens befinden sich mehrere Meßsysteme im Einsatz. Von den Meßteilen Durchfluß- und Reflexionsmesser existieren Laborimplementierungen, die am IND in Gebrauch sind.

Eine Prototypimplementierung des vorgestellten Diagnosesystems unter dem Betriebssystem MSDOS ist für das IND-Rechnernetz im Einsatz, wird aber nicht mehr weiter gepflegt. An einer Neuimplementierung des Diagnosesystems unter XENIX wird gerade gearbeitet, der Kern der Problemlösungskomponente und der Kern der Wissenserwerbskomponente sind bereits implementiert. Bei der Neuimplementierung wurde eine Vielzahl von Erfahrungen, die im Bereich der Wissensbearbeitung und der Wissensdarstellung mit dem Diagnosesystem unter MSDOS gewonnen wurden, berücksichtigt.

An dieser Stelle sei der Siemens AG, speziell der Abteilung AUT E 51 in Erlangen, für die technische Unterstützung und Vermittlung von Versuchspartnern für den Einsatz von Prototypen gedankt.

Schrifttum

[1]   *Bathelt, P; Pfeiffer, K.*: Functions und possible applications of the B5100 LAN protocol tester from Siemens. Proceedings EFOC/LAN 87, S.392-397, 1987.

[2]   *Bathelt P.*: Ein Werkzeug zur Analyse des Kommunikationsverhaltens lokaler Netze. Proceedings Prozeßrechensysteme '88, Informatik fachberichte, Vol. 167, S. 183-192, Springer Verlag, 1988.

[3]   *Dal Cin, Mario; Phillip, T.*: Expertensysteme für die Fehlerdiagnose. Informationstechnik it, 4/1988, S. 237-246.

[4]   *Haubner, H., J.; Kornmann, H.*: Fehlerdiagnose in verteilten Rechnersystemen mit dem Expertensystem REX. Tagungsband Prozeßrechensysteme '88, Informatik-Fachberichte, Vol. 167, S. 762-771, Springer Verlag, 1988.

[5]   *Heigert, J.*: OSI Netzwerk-Mamagement: Status und Perspektiven. DATACOM 1/90 , DATACOM-Verlag, 1990.

[6]   *Kiesel W.*: Kommunikationsnetze für verteilte Systeme Netzwerkmanagement in industrieellen lokalen Netzen. Tagungsband Prozeßrechensysteme '88, Informatik-Fachberichte, Vol. 167, Springer Verlag, 1988.

[7]   *Mescheder, B.; Westerhoff, T.*: Offene Architekturen in Expertensystem-Shells. Angewandte Informatik 9/88, S. 390-398, Friedr. Vieweg & Sohn Verlagsgesellschaft mbH.

[8]   *Puppe F.*: Diagnostik Experten Systeme. Informatik Spektrum 10/1987, S. 293-308, Springer Verlag.

[9]   *Puppe, F.*: Diagnostisches Problemlösen mit Expertensystemen. Informatik-Fachberichte, Vol. 148, Springer Verlag 1987.

[10]   *Reuter A.*: Kopplung von Datenbank und Expertensystemen. Informationstechnik it 3/1987, S.164-175, R. Oldenburg Verlag.

Durchgängige Anwenderschnittstelle für industrielle Kommunikationsnetze

W.A. Becherer
SIEMENS AG
AUT E 51
Günther-Scharowsky-Str. 1
8520 Erlangen

Zusammenfassung

Minicomputer werden in der Industrie für vielfältige Automatisierungsaufgaben eingesetzt. Dabei müssen sie in bestehende und neue Kommunikationsnetze integriert werden.

Kommunikationsnetze werden nach den Kriterien:

        a) Kosten
        b) Leistung
        c) Offenheit, d.h. Herstellerunabhängigkeit

ausgewählt.

Dadurch sind die Minicomputer gezwungen, sich an die unterschiedlichen industriellen Kommunikationsnetze anzupassen.
Um den Anpassungsaufwand für die Kommunikationsschnittstellen, auf Hardware-, Treiber- und Anwender-Ebene, zu minimieren, ist ein Architektur-Konzept für den Anschluß der verschiedenen Kommunikationsnetze notwendig.

In dem Vortrag soll, am Beispiel SINEC, ein Architektur-Konzept für die Schnittstellenanpassung im Minicomputer für die verschiedenen Kommunikationsnetze vorgestellt werden.

Kapitel 1 formuliert Anforderungen an Kommunikationsnetze und nennt notwendige Kommunikationsdienste in der Automatisierungstechnik. Dabei werden ebenso die Kriterien für eine Kommunikationsarchitektur und durchgängige Schnittstellen eines Minicomputers aufgestellt.
Ein modulares Hardware- und Software-Konzept für die Realisierung durchgängiger Anwender-Schnittstellen wird in Kapitel 2 aufgezeigt. Die Durchgängkeit wird anhand der MMS-kompatiblen Anwenderschnittstelle der SINEC-Technologischen Funktionen belegt. Eine Bewertung der Lösung, hinsichtlich Durchgängikeit, aber auch Performance und Wirtschaftlichkeit wird in Kapitel 3 vorgenommen.

1. Industrielle Kommunikation

1.1 Anforderungen an industrielle Kommunikationsnetze

Die Automatisierung in der Industrie erfolgt mit einem breiten Spektrum verschiedener Endsysteme. Sie reichen von mittleren DV-Anlagen und Minicomputers über PC's und Steuerungen wie SPS und CNC bis hin zu Kleinsteuerungen, Sensoren und Überwachungskomponenten.

Komplexe Automatisierungsaufgaben sind einerseits nur dadurch lösbar, daß die einzelnen Endsysteme via Kommunikationsnetz Daten austauschen, andererseits ist das Kommunikationsnetz die Vorraussetzung, um die vielen Endsysteme zentral zu programmieren, projektieren und zu diagnostizieren.

In der Automatisierungstechnik werden drei Klassen von industriellen Kommunikationsnetzen benötigt:

a)   Backbone Netzwerke wie Breitbandnetze siehe SINEC H2B oder FDDI
- o mit hohen Datenraten
- o für alle Netzdienste von der Bildübertragung bis zur Datenkommunikation
- o mit Endsystemen von unterschiedlichen Herstellern.

b)   Fabrik- und Zellennetze wie SINEC H1
- o mit hoher Flexibilität
- o großer Teilnehmerzahl
- o und einfacher Installation

c)   Low cost Netze wie SINEC L2
- o mit möglichst geringen Anschluß-und Leitungskosten.

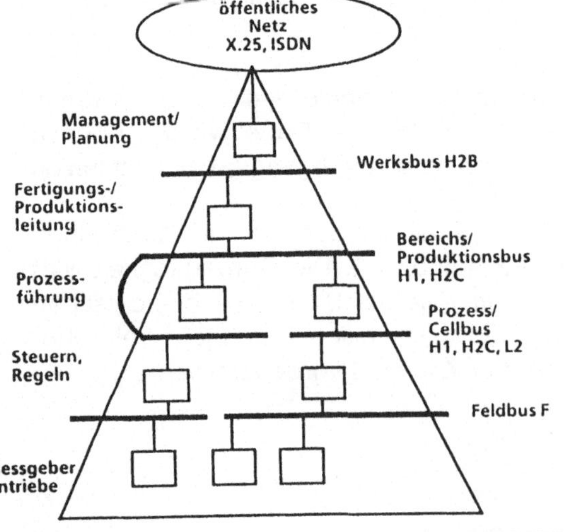

Protokolle		Leistung	
Typ	Übertragung	Durchsatz MBit/s	Entfernung km
High Performance			
H2B	Breitband Tokenbus	3x 10	> 10
H2C	Carrierband Tokenbus	5	< 2
H1	Basisband CSMA/CD	10	< 2,5
Low Cost			
L2	Tokenbus	0.5	< 0,5
Feldbus			
F	Tokenbus	0.01-0.1	1-2

Bild 1   **SINEC-Netze**

1.2 Kommunikationsdienste

Die Vielfältigkeit der Automatisierungsaufgaben erfordert unterschiedliche Kommunikationsdienste in einem Endsystem.

Das ISO-OSI 7 Schichten-Modell stellt einen Rahmen für die Einteilung der Kommunikationsprotokolle und der daraus resultierenden Dienste dar. Anhand dieses Modells lassen sich die Dienste im folgenden einfach definieren.

Eine industrielle Kommunikationsarchitektur muß mindestens drei Dienstschnittstellen bieten, um unterschiedliche Anwendungen aufsetzen zu können (siehe Bild 2).

a) Eine Schnittstelle zu den Manufacturing Messages Service (MMS ISO 9506), die in der Automatisierungstechnik besonders wichtig sind.

b) Eine Transportschnittstelle für Anwendungen, die die ISO-OSI Ebenen 5 bis 7 nicht benötien, wie z.B. der SINEC-Filetransfer.

c) Eine Ebene 2 Schnittstelle für größtmögliche Flexibilät als Zugangspunkt für nicht ISO-OSI-Dienste wie TCP/IP und Einfachst-Protokollen zu Sensor-und Überwachungskomponenten.

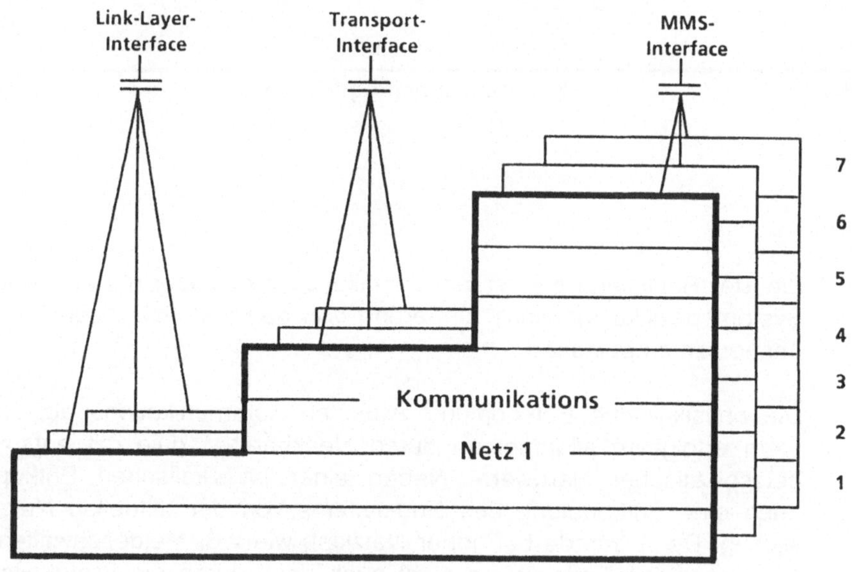

Bild 2   **Kommunikationsdienste in der Automatisierungstechnik**

1.3 Kriterien einer industriellen Kommunikationsarchitektur

Die Endsysteme unterscheiden sich sowohl hinsichtlich ihrer Hardwareeigenschaften wie Prozessorleistung, Speicheraustattung und Systembus als auch hinsichtlich ihrer Softwarestrukturen wie Betriebssystem und Ein-/Ausgabe-Schnittstelle (siehe Bild 3).

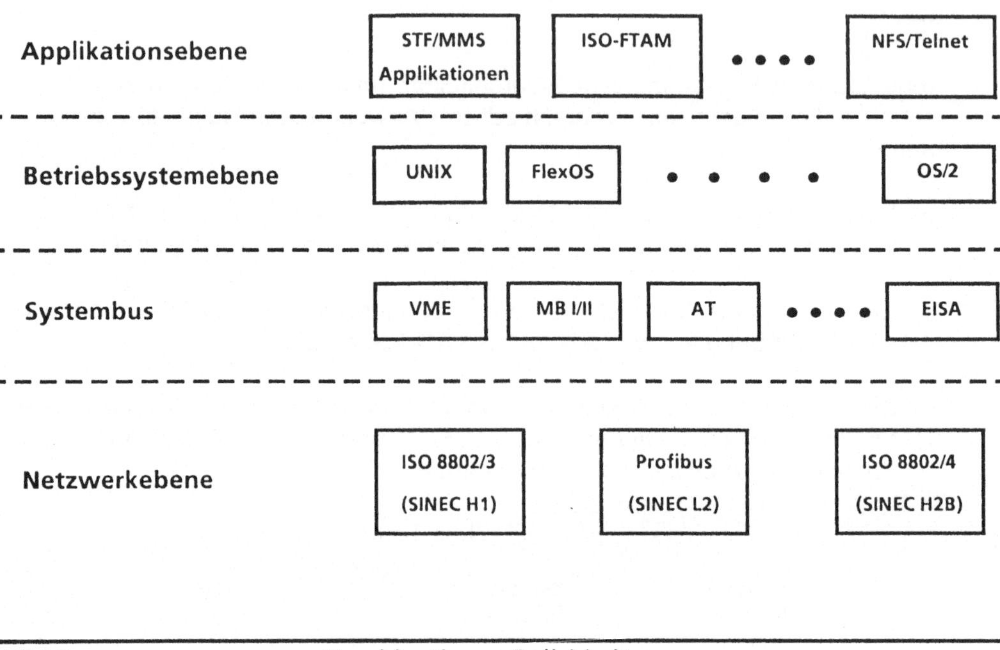

Bild 3   **Kombinationsmöglichkeiten**

Ziel der Realisierung einer Kommunikationsarchitektur muß es sein, die o.g. Systemspezifika möglichst unberührt von den verschiedenen Kommunikationsnetzen zu belassen.

Die physikalische Entkopplung zwischen Kommunikationsnetz und Endsystem erfolgt im allgemeinen durch steckbare Module mit entsprechender netzspezifischer Hardware. Neben einer physikalischen Entkopplung ist auch eine Entkopplung des Endsystems von der aktuellen Netzbelastung wichtig. Denn gerade bei hoher Netzlast, wie z.B. Meldeschwällen, darf der Prozessor des Endsystems nicht auch noch durch die Protokollabwicklung zusätzlich belastet werden.

Deshalb ist es sinnvoll, die Protokollabwicklung soweit wie möglich auf eine Spezial-Hardware, eine Kommunikationsanschaltung zu verlagern.
Auf einer Kommunikationsanschaltung sind die Voraussetzungen für eine effiziente Protokollabwicklung, sowie schnelle Interrupt-Verarbeitung durch ein entsprechendes Echtzeitbetriebssystem und hardwarenahe Programmierung vorhanden.

Eine Kommunikationsanschaltung besteht im wesentlichen aus drei Software-Komponenten,

a) dem Protokollabwickler für die o.g. Kommunikationsdienste,
b) einem Treiber zum Endsystem
c) einer Netzwerkmanagement-Komponente für Projektierung, Überwachung,Test und Diagnose.

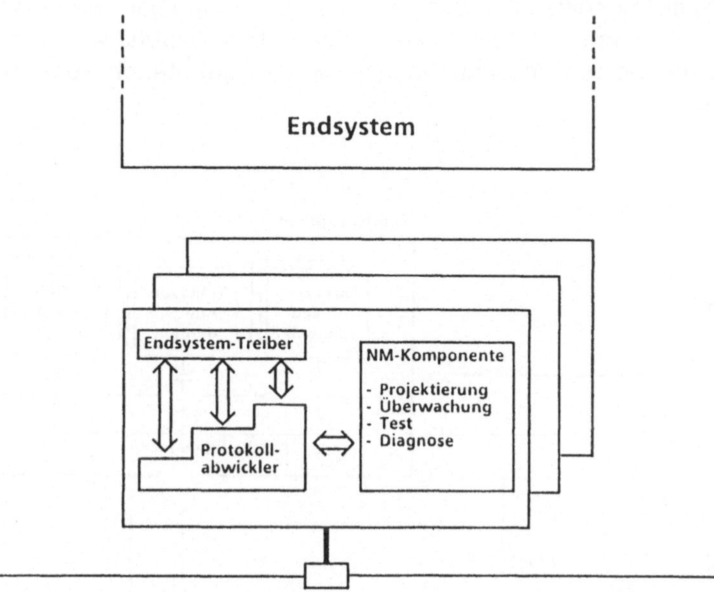

Bild 4   **Software-Komponenten einer Kommunikationsanschaltung**

1.4 Durchgängige Schnittstellen eines Minicomputers

Minicomputer haben eines der breitesten Einsatzgebiete in der Automatisierungstechnik. Die Anwender wollen ihre einmal realisierte Software auch für verschiedene Netze einsetzen, deshalb sind **durchgängige Anwenderschnittstellen** für die Minicomputer von besonderer Bedeutung.

Falls diese Anforderungen nicht durch eine gezielte Schnittstellenarchitektur in einem Minicomputer abgefangen wird, führt das zu hohen Entwicklungs- und Pflegeaufwänden beim Kunden und letztlich zum Verlust von Marktsegmenten.

Bei dem Entwurf einer Schnittstellenarchitektur in einem Minicomputer sind drei Ebenen zu beachten:

a) die physikalische Ebene mit Aufbautechnik, Stecker und elektrischen Eigenschaften
b) die Treiberebene mit Rückwandbus-Protokoll und Datenübergabemechanismen
c) die Anwenderebene, die sich an den bereits genannten Kommunikationsdiensten orientiert.

Die gewählte Schnittstellenarchitektur ist dann optimal, wenn sowohl die physikalische Ebene als auch die Treiberebene unabhängig von den o.g. Komunikationsnetzen und unabhängig von den gebotenen Kommunikationsdiensten ist.

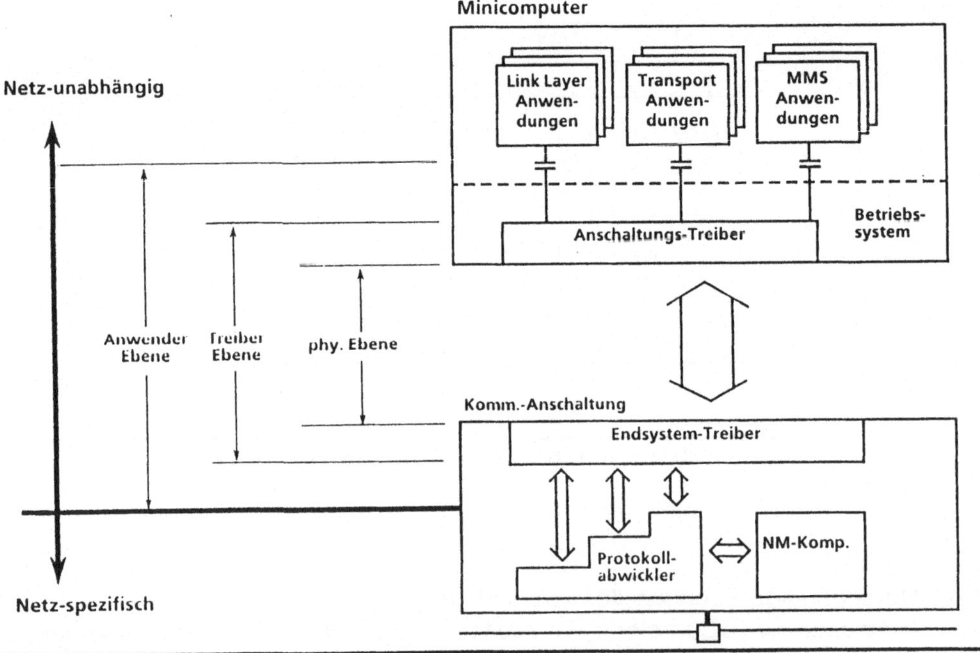

Bild 5   **Schnittstellenarchitektur eines Minicomputers**

2. Lösungsansatz für Minicomputer

Im folgenden wird ein Lösungsansatz, unter Berücksichtigung der bisherigen Anforderungen für Minicomputer vorgestellt.
Der Ansatz geht von netzspezifischen Kommunikationsanschaltungen aus, die mit einem modularen Hardware-und Software-Konzept dem Minicomputer durchgängige Anwenderschnittstellen ermöglichen und ihn dabei von der Protokollabwicklung entlasten.

2.1 Hardwarekonzept

Die Hardware besteht aus einer Trägerbaugruppe und verschiedenen, netzspezifischen Aufsteckmodulen, den Kommuniationsanschaltungen, die wahlweise verwendet werden können (siehe Bild 6).

Die Aufsteckmodule werden "face to face" auf die Trägerbaugruppe gesteckt und mit ihr verschraubt. "Face to face"bedeutet, daß die Bauteilseite des Aufsteckmoduls zur Trägerbaugruppe zeigt.
Die Aufsteckmodule haben die Höhe einer Doppeleuropakarte (233,4 mm) und etwa die halbe Tiefe (114mm).
Auf der nicht überdeckten Fläche der Trägerbaugruppe können Anpaßkomponenten wie Stecker, Spannungsversorgung und die wenigen Logikbausteine angeordnet werden.

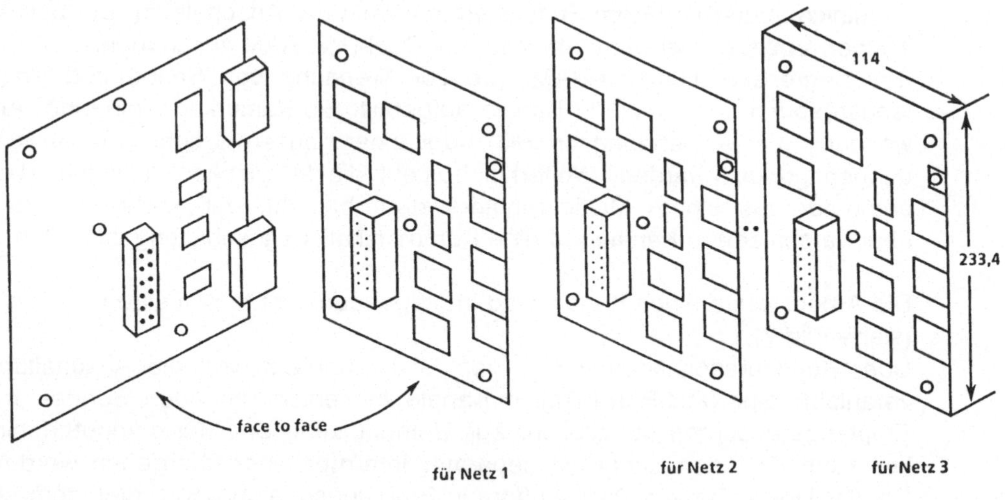

Bild 6   **modulares Hardwarekonzept**

Durch diese Aufbautechnik können Trägerbaugruppe und Aufsteckmodul in 1 1/3 Standardeinbauplätze untergebracht werden.

Die Aufsteckmodule enthalten immer ein Mikroprozessorsystem, ein Kommunikationsnetz-Interface mit Controllerbausteinen, einen genormten Stecker und für die Kopplung zum Minicomputer ein Dual-Port-RAM.

Das Dual-Port-RAM ist aus Minicomputersicht ein bis zu 256 KByte großer Speicher. Der Zugriff erfolgt über ein asynchrones Speicher-Interface mit Zugriffsbreiten von 8, 16 oder 32 Bit.

Zusätzlich zu der Speicherkopplung ist in den meisten Fällen eine Signalisierung über Interrupts sinnvoll. Die Interrupts an den Minicomputer werden von den Aufsteckmodulen generiert, können jedoch auf der Trägerbaugruppe auch ausgeblendet werden.

2.2 Treiberebene

Die Treiberebene ist für die Performance und das Verhalten des Gesamtsystems von entscheidender Bedeutung.

Einerseits müssen die Kommunikationsdienste der Anschaltung, wie flußgesteuerte Verbindungen und sichere Datenübetragung bis zur Anwenderschnittstelle erhalten bleiben, andererseits muß die Verwaltung der Treiberebene und die Anzahl der Interaktionen zwischen dem Minicomputer und der Kommunikationsanschaltung auf ein Minimum reduziert werden, um eine möglichst geringe Belastung des Minicomputers zu erreichen.

Die Treiberebene besteht aus einem Treiberpaar, dem Anschaltungstreiber im Minicomputer und dem Endsystemtreiber in der Anschaltung. Die beiden Treiber arbeiten über ein gemeinsames Dual-Port-RAM zusammen.

Das begrenzte Dual-Port-RAM wird zur Trennung von Sende und Empfangsresourcen in bis zu 32 Kanäle aufgeteilt. Ein Kanal kann von einer Anwendung (i.a. Anwendungsprozeß) oder einer ganzen Gruppe von Anwendungen genutzt werden. Weiterhin besteht die Möglichkeit, falls der Minicomputer aus einem Multicomputersystem besteht, daß mehrere CPU's über verschiedene Kanäle auf eine Kommunikationsanschaltung zugreifen.

Die Arbeitsweise eines Kanals wird im folgenden kurz beschreiben
(siehe Bild 7):
Über Konfigurationskommandos des Minicomputers wird die Anschaltung veranlaßt, das Dual-Port-RAM in Kanäle mit entsprechenden Sende- und Empfangsresourcen aufzuteilen. Zur Unterstützung der Multicomputerfähigkeit kann für jeden Kanal ein separater Interruptvektor übergeben werden. Bei der Übergabe von Datenpuffern in Empfangsrichtung, wird dies dem Minicomputer per Interrupt mit dem entsprechenden Vektor signalisiert.

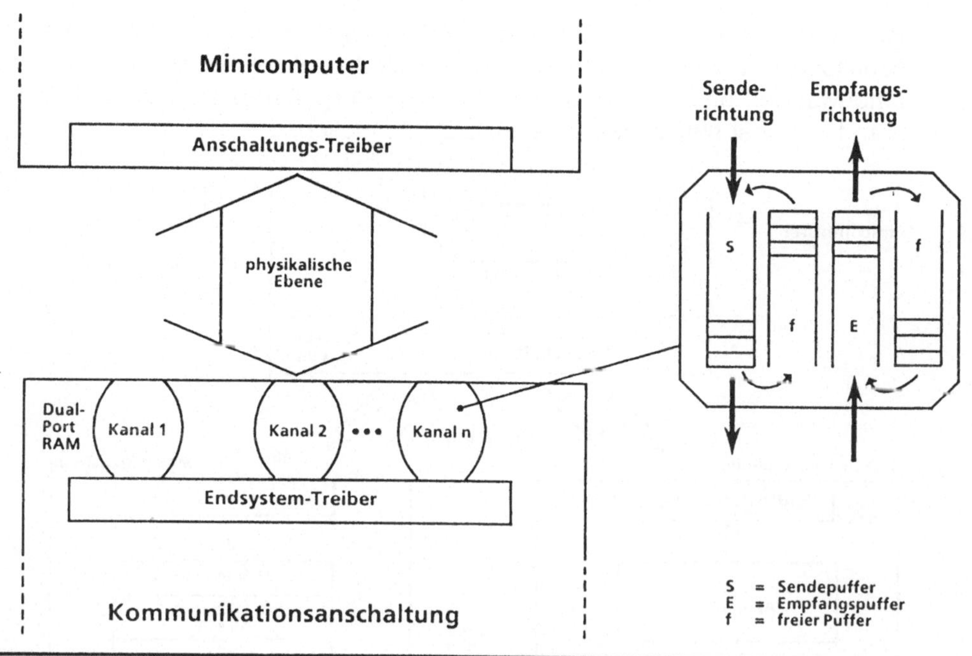

Bild 7 **Treiberebene**

2.3 Anwenderschnittstelle

Die Anwenderschnittstelle basiert
a) auf den Kommunikationsdiensten der Anschaltung und
b) auf der Einbindung der Schnittstelle in das jeweilige Betriebssystem.

Es würde den Rahmen des Vortrages sprengen, wenn für alle o.g. Kommunikationsdienste die Durchgängigkeit bewiesen würde.
Als Beispiel dient das MMS-kompatible Interface von SIEMENS, die Anwenderschnittstelle der SINEC Technologischen Funktionen (STF-Anwenderschnittstelle).

Die STF-Anwenderschnittstelle ist eine Bibliothek die zu den jeweiligen Anwenderprogrammen gebunden wird. Sie bietet dem Anwenderprogramm Aufrufe, mit denen MMS-kompatible Dienste angestoßen oder empfangen werden können. Aufgabe der Bibliothek ist es, die entsprechenden STF-Nachrichten zu erzeugen oder ankommende STF-Nachrichten zu interpretieren und geforderten Dienste abzuwickeln.

Ist das unterlagerte Netz wie im Beispiel (siehe Bild 8) SINEC H1, werden die STF-Nachrichten direkt über das SINEC-AP-Protokoll transferiert.

Ist ein MAP-Netzwerk unterlagert, werden die STF-Nachrichten auf der Anschaltung entsprechend dem ASN. 1 Encoding Rules in MMS-Nachrichten umgewandelt und vice versa. Die analoge Vorgehensweise ist auch bei einem Low Cost Netzwerken wie SINEC L2 möglich.

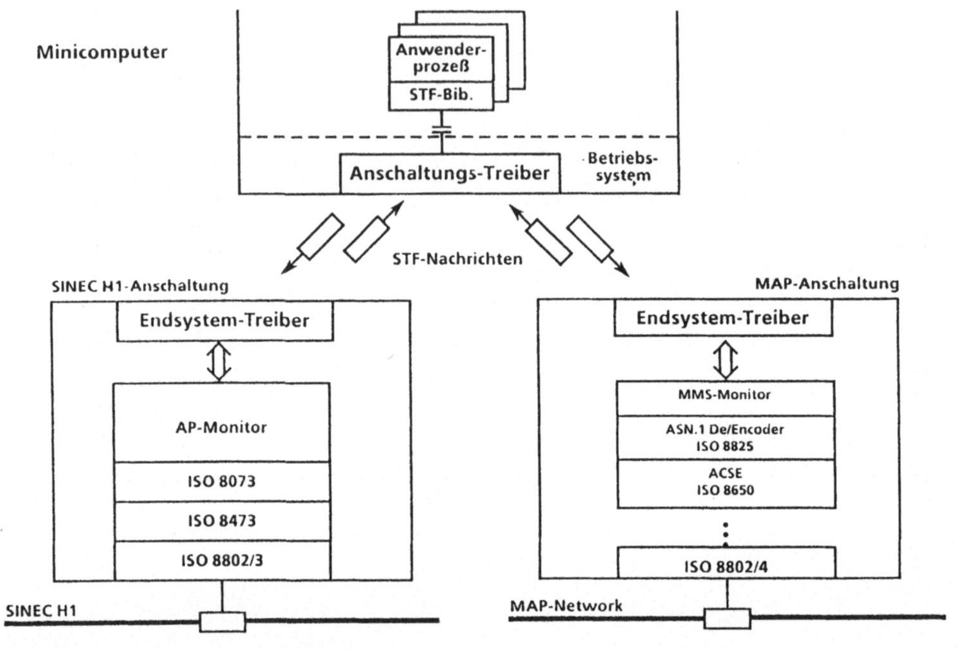

Bild 8	Durchgängige STF-Anwenderschnittstelle

3. Bewertung des Lösungsansatzes

Den Anwender von Kommunikationsdiensten in Minicomputersystemen interessiert weniger die Realisierung einer Kommunikationsarchitektur, vielmehr die geforderte **Durchgängigkeit** und ausreichende Performance zu akzeptablen Kosten.

3.1 Bewertung der Durchgängigkeit

Wie angedeutet wurde, läßt sich die Durchgängigkeit auch für sehr umfangreiche Anwenderschnittstellen erreichen.
Für die Transportschnittstelle wurde durch die Definition des X-Transport-Interfaces (XT-I) von der X-Open Group belegt, daß es möglich ist, durchgängige Schnittstellen für den gleichen Kommunikationsdienst auf Basis unterschiedlicher Netze zu definieren.

Für das Link-Layer-Interface ist eine durchgängige Schnittstelle am einfachsten festzulegen, da dort die Netzeigenschaften, wie z.B. maximale Nachrichtenlänge, Bestandteile der Dienstschnittstelle sind.

3.2 Bewertung der Performance

Messungen zeigen, daß die Realisierung der Kommunikationsarchitektur auf Basis von Kommunikationsanschaltungen erhebliche Performancevorteile ergeben.

Als Vergleichsmaßstab soll hier die für die Automatisierungstechnik wichtige Nachrichtenrate, d.h. die Anzahl der übertragbaren Nachrichten (ca. 100 Byte) pro Sekunde, auf ISO-OSI Ebene 7 dienen.

Bei dem SINEC H1-Anschluß von SICOMP M hat sich gezeigt, daß durch die Auslagerung der ISO-OSI-Ebenen 1 bis 4 auf eine Kommunikationsanschaltung eine Steigerung der Nachrichtenrate um den Faktor 4 möglich war. Durch die zusätzliche Auslagerung der ISO-OSI-Ebenen 5 bis 7 kann nochmals eine Steigerung um den Faktor 2 erreicht werden, so daß zwischen der Protokollabwicklung im Minicomputer und auf einer Kommunikationsanschaltung ein Performance-Unterschied bis zum Faktor 8 liegen kann.

3.3 Allgemeine wirtschaftliche Bewertung

Die Realisierung einer Kommunikationsarchitektur auf Basis der Anschaltungstechnik ermöglicht durchgängige Anwenderschnittstellen und gute Performance.

Falls diese Realisierungsmethode nur für **ein** Netz und nur für **einen** Protokollbwickler angewendet, ist sie aufwendiger als eine Implementierung im Minicomputer selbst mit einer reinen HW-Anschaltung für ISO-OSI-Ebene 1 und 2a.

Das beschriebene modulare Anschaltungskonzept kann für viele Endsysteme und damit in großen Stückzahlen eingesetzt werden. Dadurch sind Kostenvorteile für den Minicomputer möglich, die mit netz- und systemspezifischen Lösungen kaum zu erreichen sind. Dies gilt insbesondere dann, wenn auch noch vorhandene Standardträgerbaugruppen, z.B. für VME - Bus eingesetzt werden können.

4. Abkürzungen

AP	Automation Protocoll
CNC	Computer Numerical Control
ISO	International Organization for Standardization
OSI	Open System Interconnection
MAP	Manufacturing Automation Protocol
MMS	Manufacturing Message Services
SINEC	SIEMENS Network Architecture for Automation and Engineering
SINEC L2	SINEC Low cost Network
SINEC H1	SINEC Zellennetzwerk
SPS	Speicherprogrammierbare Steurung
STF	SINEC Technologische Funktionen

Implementierungsmethodik von Programmen und Entwicklungs-
umgebungen für modulare Mehrrechner-Prozeßautomatisierungssysteme

Wolfgang Fengler

Ulrike Sörgel

Almuth Wendt

Fakultät für Automatik und Informatik

Technische Hochschule Ilmenau

PSF 327   Ilmenau DDR-6300

ABSTRACT: In many cases decentralized process control systems have
an own processing unit for each partial problem, which can be progam-
med in special languages. For these special languages there are needed
design tools like specification systems, syntax-controlled editors,
compilers, recompilers, on-line-debuggers and knowledge- based design
aids. A Petri-Net-based development system is suitable for the design
of parallel processes.

KEY WORDS: process control, special languages, design tools, Petri
Nets

1. Einordnung der Ziel-Prozeßautomatisierungssysteme und der Entwicklungsmethodik

Mikroprozeßrechner kommen als kompakte Geräte, als modulare Systeme
oder geräteintegriert zum Einsatz. Modulare Systeme bilden dabei einen
Einsatzschwerpunkt der Mikrorechentechnik. Der Einsatz eines modularen
Automatisierungssystems ist dort technisch und ökonomisch von Vorteil,
wo funktionell und konstruktiv eine hohe Flexibilität durch Auf- und
Abrüstung bei kleinen Fertigungsloßgrößen erforderlich ist. Dadurch
wird eine hohe Anpassung an den jeweilig zu automatisierenden Prozeß
erreicht bei hoher Systemzuverlässigkeit und Verfügbarkeit /FENGLER
88,90B/.

Der Einsatz von dezentralen Prozeßcontrollern mit eigener Verarbei-
tungseinheit hat mehrere Gründe /FENGLER 86/:

- Der Prozeßrechner (Wartenrechner) wird von den unmittelbaren Ein-
  /Ausgabe-Operationen entlastet. Er erhält nur die für das Gesamtsystem benötigten und bereits aufbereiteten und verdichteten Daten.

- Durch die direkte Zuordnung eines Controllers zu einer Ein-/Ausgabe-Aufgabe wird eine wesentlich schnellere und unkompliziertere Reaktion auf Prozeßanforderungen möglich.

- Die Modularität und damit die Flexibilität des Systems wird erhöht.

Typische Anwendungen sind

- einzelne Steuerungs- oder Regelungsaufgaben
- Maschinen- oder Aggregate-Steuerung bzw. Regelung
- Einsatz als geräteintegrierter Rechner zur Meßwertverarbeitung und Lösung von Automatisierungsaufgaben in der untersten Ebene in Zusammenhang mit einem hierarchischen System.

Ein derartiges System kann durchgängig nach einem Firmware-Interpreterprinzip gestaltet werden /FENGLER 88,90B/. Jeder Modul enthält für die in ihm benötigten und zulässigen Operationen einen Fachsprachinterpreter. Dabei richtet sich die Komplexität und der Inhalt der Anweisungen nach dem Typ und damit dem Einsatzzweck des Moduls. Typischerweise kommen zum Einsatz:

- Zur Lösung von Regelungs-, Meßwerterfassungs- und Vorverarbeitungsaufgaben wird eine reglungstechnisch orientierte Sprache eingesetzt.

- Für die Programmierung von binären Steuerungsaufgaben kommen Steuerungsfachsprachen, wie z. B. Anweisungsliste und Steuernetz zur Anwendung.

Ein FORTH-Target-Interpreter ermöglicht als Zwischensprachinterpreter auch die Anwendung von Fachsprachen. Dabei wird in FORTH ein spezielles Wörterbuch geschaffen, dessen Anweisungen den Befehlssatz der Sprache bilden. Effektiv ist dieser Weg bei Anwendung eines extrem schnellen ASIC-FORTH-Prozessors /SCHNABEL 90/.

Im weiteren soll der Weg des dialogorientierten rechnergestützten Entwurfs verfolgt werden. Dabei geht es um die Entlastung des Ingenieurs von vielfältigen formalen Tätigkeiten (z. B. Dokumentieren) und damit vor allem um die weitgehende Unterstützung seiner heuristischen Fähigkeiten /FENGLER 90D/.

Dieses Ziel kann durch folgende Mittel erreicht werden:

- Verwendung fachgerechter, problemangepaßter Beschreibungsmittel.
- Verwendung von dialogunterstützten Entwicklungswerkzeugen.
- Verwendung von problem- und beschreibungsadäquaten Testmethoden.

2. Entwicklungswerkzeuge für Implemetierung von Automatisierungsproblemen über Fachsprachen und deren Implementierung

Die Leistungsfähigkeit objektorientierter Automatisierungssysteme wird entscheidend durch die Verfügbarkeit angepaßter oben genannter Fachsprachen bestimmt. Mit der Schaffung der Automatisierungssysteme ist die Implementierung der Entwicklungsumgebung für die Fachsprachen verbunden /FENGLER 90C/. Folgende Tools ermöglichen den Entwurf, die Implementierung, den Test und die Inbetriebnahme von Fachsprachsoftware:

- Spezifikationssysteme
- syntaxgesteuerte Editoren
- Compiler
- Recompiler
- On-line-Inbetriebnahmehilfen und
- wissensbasierte Inbetriebnahmeunterstützung.

UNIX-kompatible Betriebssysteme bieten mit der Verfügbarkeit von Scanner- und Parsergeneratoren (LEX und YACC) die Möglichkeit, effektiv einige der oben genannten Software-Tools zu erzeugen. Die lexikale und syntaktische Beschreibung der Fachsprache erfolgt mit entsprechenden Regeln in der LEX- bzw. YACC-Spezifikation, die semantische Analyse und Kodegenerierung mit C-Routinen. Der von YACC erzeugte Parser überprüft die Grammatik der Sprache auf Vollständigkeit, Folgerichtigkeit und Eindeutigkeit. Bei Erkennung einer bekannten Syntaxkonstruktion werden entsprechend notwendige Aktionen eingeleitet /SÖRGEL 89/.

Die von YACC benötigte Tokenfolge wird von dem mit LEX erzeugten Scanner produziert. Der Scanner liest das Quellprogramm und analysiert es hinsichtlich seiner Sprachelemente. Die lexikalische Analyse gliedert sich dabei in einen sprachabhängigen (Tabelle der reservierten Wörter, Bildung regulärer Ausdrücke) und in einen sprachunabhängigen Teil (Symbolklassen).

Als Datenstruktur zum Informationsaustausch zwischen den einzelnen Phasen der Implementierung der Tools Editor, Compiler und Recompiler wird eine Symboltabelle verwendet. Die Routinen zur Verwaltung der Symboltabelle sind ebenfalls weitestgehend sprachunabhängig.

Die Behandlung semantischer Probleme wie Markenverarbeitung und Überprüfung von Gültigkeitsbereichen sowie die Kodegenerierung der bei Fachsprachen vom Interpreter benötigten Zwischenkodedatei erfolgt mit Routinen der Programmiersprache C.

C als Zwischensprache ermöglicht die Portabilität des gesamten Systems zum Entwurf von Fachsprach-Tools. Aufgrund des Vorhandenseins einer Zwischensprache wird der Übersetzungsprozeß in zwei Stufen unterteilt, die unabhängig voneinander realisiert werden können:

- Übersetzung der Spezifikationen in C-Quellkode (lex, yacc) und
- Übersetzung C-Quellkode in ausführbares Programm (cc).

Die Fehlerbehandlung kann größtenteils in die Formulierung der Grammatiken einbezogen werden. Für kontextabhängige Fehler müssen entsprechende Routinen zur Fehlerüberprüfung und -behandlung geschaffen werden.

Im Rahmen der Schaffung einer Entwicklungsumgebung für ein Automatisierungssystem wurden unter UNIX folgende Fachsprach-Tools implementiert /LIPFERT 90/, /KRATOCHVIL 90/, /TENEV 89/, /SIEMENS 86/ (AWL: Anweisungliste, PROMAR: regelungstechnische Fachsprache, SS86: Spezifikationssprache):

- STEP5-/AWL-Compiler
- STEP5-/AWL-Recompiler
- PROMAR-Editor
- PROMAR-Compiler
- SS86-Editor
- SS86-Syntaxtester
- SS86-Analysesystem und
- Einchip-Emulator-Reassembler.

Die folgende in /SÖRGEL 89/ bereitgestellte Software ermöglichte eine effiziente Implementierung der Werkzeuge:
- Symboltabellenfunktionen

- Spezielle Funktionen zum dynamischen Reservieren von Speicherplätzen
- Organisation der Tabelle der reservierten Wörter und Funktionen zur binären Suche
- Erzeugen von Fehlermeldungen mit Hinweisen auf erwartete Eingaben
- Operandenmarkenbehandlung in einem zweiten Durchlauf
- Vereinbarung von Symbolklassen in der LEX-Spezifikation.

Bei syntaxgesteuerten Editoren ist die Verwendung der gleichen LEX- und einer ähnlichen YACC-Spezifikation möglich.

Zur Unterstützung der Inbetriebnahme von Fachsprachsoftware wurden Untersuchungen zur Implementierung einer wissensbasierten Debughilfe durchgeführt. Dabei hat es sich als sinnvoll erwiesen, das Wissen in folgende Bereiche zu strukturieren:

- Projektwissen (Informationen über das Programm, das getestet wird)
- syntaktisches Sprachwissen
- semantisches Sprachwissen
- Wissen über Debuggerfunktionen.

3. Ein Petri-Netz-basiertes Entwurfssystem

Das Petri-Netz-Entwurfssystem PENECA (Petri Net Computer Aidded Software Engineering) /FENGLER 90A/ dient zum Entwurf von Steuerstrukturen auf Basis von Petri-Netz-Beschreibungen (Platz-Transitions-Netze, PN). Für verschiedene Anwendungsfälle wurde das PN-Grundmodell $PN = (P, T, F, K, V, m_0)$ /PETRI 61/ erweitert um:

- spezielle Kantentypen (Test-, Inhibitor-, Setz- und Rücksetzkante)
- Bewertungsfunktion w_x der Transitionen und Ausgabefunktion w_y der Plätze ("Steuernetz")
- Aktionsfunktion w_{ap} der Plätze bzw. w_{at} der Transitionen ("Softwarenetz"
- Bewertung der Transitionen mit Zeitbedingungen und Wahrscheinlichkeiten
- Einführung von Makroplätzen und Makrotransitionen für die komplexe Beschreibung von hierarchisch untergeordneten Teilnetzen.

Bild 1 zeigt die wesentlichen Systemkomponenten. Die Benutzenoberfläche dient zur graphischen Editierung von Netzen, zur simulativen

Abwicklung entworfener Steuerstrukturen einschließlich Steuerung und
Protokollierung der Simulation und ermöglicht den Aufruf von benötigen-
ten Dienstfunktionen. Sie enthält eine umfangreiche Help-Unterstüt-
zung.

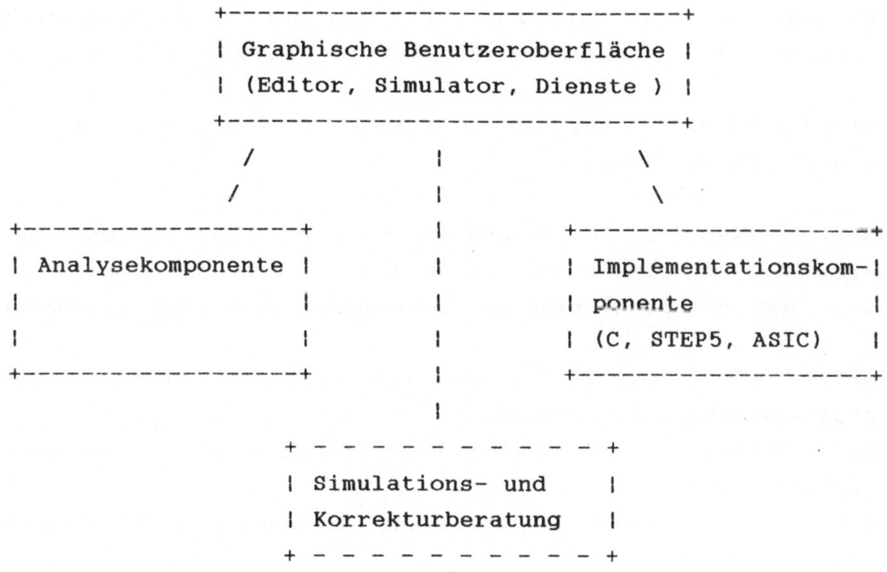

Bild 1: PENECA-Systemübersicht

Als Analysekomponete kann das System Petri-Netz-Maschine /STARKE 89/
genutzt werden. Es ermöglicht die Anwendung der dort entwickelten
formalen Analysen (z. B. Lebendigkeit, Konfliktfreiheit).

Die Implementationskomponete ermöglicht die Compilation von PN-Struk-
turen in die höhere Programmiersprache C bzw. die Steuerungsfachspra-
che STEP5 /SIEMENS 86/. In beiden Varianten ist der Aufruf von in der
Zielsprache geschriebenen Prozeduren an Plätzen (w_{ap}) bzw. Transitionen (W_{at}) möglich. Eine weitere Leistung der Implementationskomponete
ist die Übersetzung von PN-Beschreibungen in die Ausgangsnotierung
eines ASIC-Schaltkreis-Entwurfssystems, von der aus eine geschlossene
Überführung in Kundenwunschschaltkreise möglich ist.

Die Komponente Simulations- und Korrekturberatung unterstützt im Sinne
eines Expertensystems den Entwurf von PN-Steuerstrukturen durch Bera-
tungshinweise zur Steuerung der Simulationsläufe, zur Auswertung der
Simulationsergebnisse und zur Netz-Korrektur, sofern gewünschtes und
erzeugtes Verhalten differieren. Dabei werden auch Entwurfshinweise
gegeben.

Die Benutzeroberfläche und das Analysesystem in der beschriebenen Form stellen ein bereits arbeitsfähiges Teilsystem dar. Als Erweiterung ist der Übergang zu Gefärbten Netzen /JENSEN 81/ und Prädikat-Transitions-Netzen /GENRICH 81/ vorgesehen. Für die weiteren Komponenten (Implementation über Software, Hardware, Beratung) existieren Experimentalsysteme /FENGLER 82/, /UNGER 90/, /BUDACH 90/, /SÖRGEL 90/.

Als wichtigste bisherige Einsatzschwerpunte von PENECA werden gesehen (ergänzt nach /FENGLER 88/):

- Entwurf, Analyse, Simulation und Implementation von Protokollen
- Entwurf, Analyse, Simulation von technologischen Prozessen
- Entwurf, Analyse, Simulation und Implementaion von Parallelprogrammen
- Entwurf, Analyse, Simulation und Implementation Steuerprogrammen für programmierbare Steuerungen
- Entwurf, Analyse, Simulation und Implementation von komplexeren ASIC-Schaltkreisen.
- Lernsystem für studentische Ausbildung und Ingenieur-Weiterbildung.

4. Zusammenfassung und Ausblick

Die vorgestellte Implementierungsmethodik für Automatisierungsprobleme über Fachsprachen und Netze stellt einen effektiven Weg zum rechnergestützten Entwurf dar. Eine leistungsfähige Entwicklungsumgebung mit verschiedenen Software-Werkzeugen läßt sich problemadäquat mit einer ergänzten Sprachimplementierungumgebung unter UNIX schaffen. Zunehmend gewinnen die Methoden der Wissensverarbeitung bei der Software-Systeminbtriebnahme, der Korrektur und dem Entwurf an Bedeutung.

Literatur

/FENGLER 90A/
Fengler, W.; Ptacek, L.; Sörgel, U.:
Kurzbeschreibung des Petri-Netz-Entwurfssystem PENECA.
TH Ilmenau, Sektion TBKI, WB Computertechnik: 1990

/FENGLER 90B/
Fengler, W.; Philippow, I.: Entwurfsmethoden für industrielle Computersysteme.
Carl-Hanser-Verlag, München: 1990

/FENGLER 82/

Fengler, W.: Systematischer rechnergestützter Entwurf von Mikroprozes-
sorgeräten der Prozeßdatenverarbeitung.
Dissertation A an der TH Ilmenau: 1982

/PETRI 61/

Petri, C. A.: Kommunikation mit Automaten.
Darmstadt: 1961, Technische Hochschule, Dissertation

/STARKE 89/

Starke, P. H.: Der NET-Analysator PAN.
Berlin: 1989, PSI GmbH, Geschäftsbereich Software Engineering und
Simulation

/SIEMENS 86/

Berger, H.: Steuerungen programmieren in STEP 5.
München: 1986, Siemens AG, Abt. Verlag

/JENSEN 81/

Jensen, K.: Coloured Petri-Nets and the Invariant-Method.
1981: Ther. Comp. Sc. 14, S. 317-336

/GENRICH 81/

Genrich, H. J.; Lautenbach, K.: System Modelling with High Level Petri
Nets.
1981: Ther. Comp. Sc. 13, S. 109-136

/FENGLER 88/

Fengler W.: Entwurf und Technik von Mikroprozeßrechnern.
Ilmenau: 1988, Technische Hochschule, Dissertation B

/SCHNABEL 90/

Schnabel, O.: Methoden des Entwurfs von Prozessoren für die Prozeßda-
tenverarbeitung.
Ilmenau: 1990, Technische Hochschule, WB Computertechnik, Dissertation
A

/SÖRGEL 89/

Sörgel, U.; Fengler, W.: Untersuchungen zur Compilation von PROMAR
unter WEGA.
Ilmenau: 1989, Technische Hochschule, WB Computertechnik, TB 89/7.7.3

/FENGLER 90C/

Fengler,W.; Sörgel,U.: Implementierung einer Fachsprachentwicklungsumgebung unter einem UNIX-kompatiblen Betriebssystem.
Dresden: 22./23. 2. 1990, 1. Wiss. Arbeitskonferenz Informatik des Informatikzentrums der TU Dresden

/UNGER 90/

Unger, H.: Die PENECA-Implementationskomponente PNI.
Ilmenau: 1990, Technische Hochschule, WB Computertechnik, Forschungsbericht

/BUDACH 90/

Budach, A.: ASIC-Implementation für das Petri-Netz-Entwurfssystem PENECA.
Ilmenau: 1990, Technische Hochschule, WB Computertechnik, Diplomarbeit

/SÖRGEL 90/

Sörgel, U.: Wissensbasierte Debugsysteme.
Ilmenau: 1990, Technische Hochschule, WB Computertechnik, Diplomarbeit

/LIPFERT 90/

Lipfert, J.-U.: Compilation und Recompilation von MPSS-Programmen.
Ilmenau: 1990, Technische Hochschule, WB Computertechnik, Diplomarbeit

/KRATOCHVIL 90/

Kratochvil, L.: Untersuchungen zur Erstellung syntaxgesteuerter Editoren für PROMAR, SS86 und MPSS.
Ilmenau: 1990, Technische Hochschule, WB Computeretechnik, Diplomarbeit

/TENEV 89/

Tenev, P.: Spezifikationssystem unter UNIX.
Ilmenau: 1989, Technische Hochschule, WB Comuptertechnik, Diplomarbeit

/FENGLER 90D/

Fengler, W.; Sörgel, U.; Wendt, A.; Ptacek, L.: Entwurfsmethoden für industrielle Computersysteme.
Magdeburg: 27./28. 3. 90, Technische Universität, Tagung Mikrorechner-Interfacetechnik

<u>PROZESSVERTEILUNG IM VERLÄSSLICHEN ECHTZEITSYSTEM DREAM</u>

D. Neumann, H. Pietsch
Institut für Informatik und Rechentechnik der
Akademie der Wissenschaften der DDR
1199 Berlin, Rudower Chaussee 5

1. Einführung

Das Rechnersystem DREAM (Dependable REAltime Multicomputer) wurde als zuverlässiges Echtzeitsystem für lange Missionszeiten entworfen.

Hardwarearchitektur

Realisiert wird eine 3-Ebenenarchitektur [AHR 88]. Auf der obersten Ebene besteht das System aus bis zu 16 Multiprozessorsystemen MPS, die über 2 unabhängige 10 MBit Tokenbus LAN's miteinander verbunden sind. Die beiden LAN werden im Leistungs- und Zuverlässigkeitsverbund betrieben. Jedes MPS kann sich im Fehlerfall von den LAN isolieren. Jedes FAME besteht aus bis zu 16 Prozessormoduln PM, die über einen 32 Bit breiten Systembus miteinander verbunden sind. An jedem PM können am Lokalbus über ein Dual-Port RAM weitere intelligente Peripheriekarten angeschlossen werden. Jeder PM kann sich vom Systembus isolieren.

Entwurfskriterien für das Betriebssystem FEE

Das Betriebssystem FEE (Fault tolerant rEaltime Executive) ist ein Echtzeitkern, ergänzt um Komponenten zur Tolerierung von Hardwarefehlern.
Programmiert wird die Hardwarearchitektur mit prozeduralen Programmiersprachen (MODULA-2, CHILL). Die Granularität der Parallelität der Anwendersoftware liegt auf Prozeßniveau. Die Kommunikation erfolgt über einen Prozessormodul hinaus durch ein message-passing-Konzept. Innerhalb von Prozessormodulen können Prozesse auch über Regionen kommunizieren.
Als kleinste rekonfigurierbare Einheiten werden bezüglich der Hardware Prozessormodule und bezüglich der Anwendersoftware Prozesse verstanden. Als Fehlertoleranzprinzip wird der sanfte Leistungsabfall (graceful degradation) eingesetzt. Der Anwender kann zwischen mehreren Stufen der Absicherung bei Prozessorausfällen wählen:

- einfacher Prozeß ohne Absicherung,
- passives Prozeßreplikat auf Externspeicher (Restartprozeßtechnik),
- aktives Prozeßreplikat (Schattenprozeßtechnik).

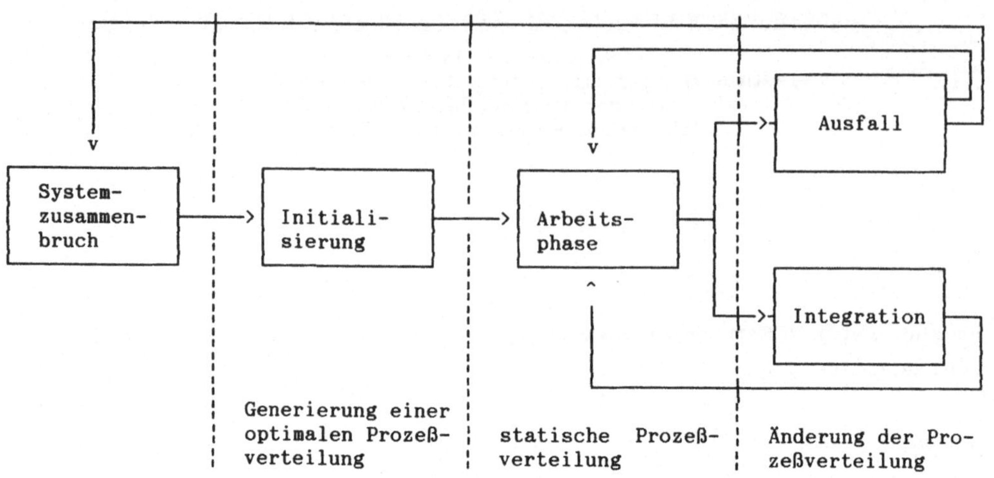

Bild 1   Globale Systemzustände und Prozeßverteilung

Ausgehend von der statischen Programmausrüstung, die für Echtzeitsysteme typisch ist, wird eine quasistatische Prozeßverteilung angewendet. In Bild 1 sind die Hauptzustände des Systems im Zusammenhang mit der Prozeßverteilung angegeben. Es wird deutlich, daß in der Arbeitsphase keine Prozeßverteilung stattfindet. Durch das Herauslösen dieser zeitaufwendigen Funktion aus dem Hauptzustand des Systems wird die Realisierung von Echtzeitbedingungen unterstützt.

Um den Organisationsoverhead einzuschränken, wurde die Sicht von FEE über das Gesamtsystem eingeschränkt und verteilt. Keine Komponente von FEE hat einen Überblick über das gesamte Multicomputersystem. Ein vollständiger Überblick existiert jedoch innerhalb jedes Multiprozessorsystems.

Programmiert und getestet wird das System von Hostrechnern, die über eine serielle Schnittstelle an beliebige Prozessormodule anschließbar sind.

2. Programmierung, Konfigurierung und Verteilung

In jedem verteilten Echtzeitsystem müssen für gleichartige Hardwareressourcen, die an unterschiedlichen Prozessoren angeschlossen sind, auch gleichartige Prozesse auf diesen installiert werden. Gleiches trifft auf Serviceprozesse zu, die den die Hardware steuernden Prozessen übergeordnet sind. Dabei kann ein Serviceprozeß durchaus

Anforderungen von mehreren Prozessen erhalten, die nicht unbedingt auf dem gleichen Prozessormodul liegen müssen. Beliebig viele Serviceebenen sind möglich.

Fallen Prozessoren aus, so müssen die mit ihnen ausgefallenen Serviceprozesse falls möglich (abhängig von noch verfügbaren Ressourcen), auf andere Prozessormodule verteilt werden (Abwärtskonfigurierung). Für die an spezielle Hardwareressourcen des ausgefallenen Prozessormoduls gebundene Prozesse ist das nicht notwendig.
Wird die verfügbare Hardwarekonfiguration um reparierte Prozessormodule bzw. zur Leistungssteigung um vollkommen neue Prozessormodule ergänzt, so müssen diese mit entsprechenden Prozessen (sowohl Hardwareabhängige, als auch Serviceprozesse) versorgt werden (Aufwärtskonfigurierung).

Der dem System DREAM zugrunde liegende Lösungsansatz geht von einer strengen Trennung zwischen Programmierung, Konfigurierung und Verteilung aus, ein Konzept, dem in der letzten Zeit zunehmende Aufmerksamkeit gewidmet wird [HEI 87], [SCH 90].
Programmiert werden Prozeßdefinitionen PD, aus denen durch die Konfigurierung die notwendige Anzahl von Prozeßinstanzen PI generiert werden. Nach dem Verteilen und Laden der PI werden sie durch den Start über das Betriebssystem zu Prozessen P. Die Programmierung der Prozeßdefinitionen erfolgt unabhängig zu einer konkret unterliegenden Hardwarekonfiguration. Alle zur automatischen Konfigurierung und Verteilung notwendigen Informationen erscheinen nicht im Quellcode der Prozeßdefinitionen, sondern müssen in einer getrennten Konfigurationsbeschreibung angegeben werden.

Aufgabe des Konfigurationsbehandlers ist es, die Konfigurationsbeschreibung auf ihre Konsistenz zu prüfen, in ihrer Darstellung zu verdichten und in Abhängigkeit von der aktuell verfügbaren Hardwarekonfiguration die notwendigen Prozeßinstanzen zu bestimmen (logische Prozeßinstanziierung). Dazu besteht er aus zwei Teilen. Der off-line-Teil ist dem Programmiersystem zugeordnet und liefert nach Konsistenzprüfung die verdichteten Basisdaten. Der on-line-Teil läuft als Systemprozeß von FEE und realisiert die logische Instanziierung sowie eine Initialverteilung.
Für alle an spezielle Hardwareressourcen gebundenen Instanzen steht mit der Initialverteilung die Zuordnung zu ihrem Prozessormodul (Verteilung) fest. Alle anderen Instanzen werden durch den Allokierer unter Beachtung von Kommunikationskosten und weiteren Nebenbedingungen verteilt [KRA 89]. Der Lader hat die Aufgabe, die seinem Prozessormodul zugeordneten Prozeßinstanzen zu laden (physische Prozeßinstanziierung) und zu starten.
Im Falle einer Änderung der Hardwarekonfiguration (Ausfall bzw. Integration von Prozessoren) wird der on-line-Teil des Konfigurationsbehandlers aktiviert.
Die am Konfigurieren, Verteilen und Laden beteiligten Komponenten von FEE sind in Bild 2 in ihrem Zusammenwirken dargestellt [HAM 90].

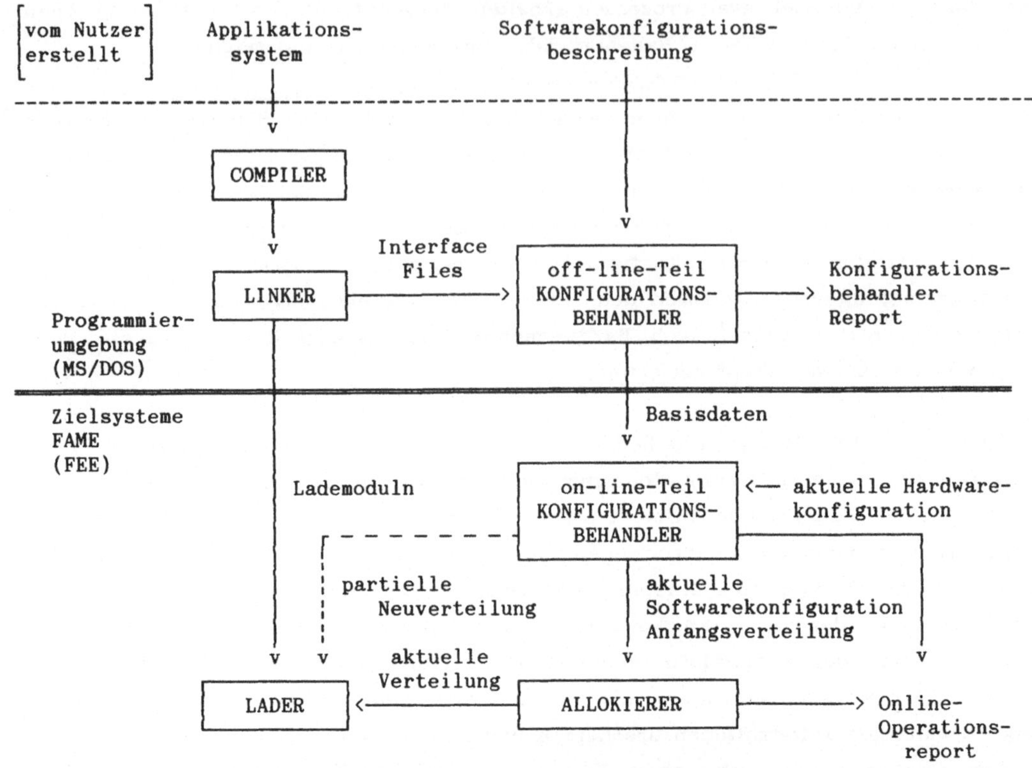

Bild 2   Einordnung des Konfigurationsbehandlers in das System der DREAM-Software

3. Konfigurationssprache

Die Konfigurationssprache ist eine Fachsprache, die in ihren sprachlichen Mitteln dem Verwendungszweck angepaßt ist. Folgende allgemeinen Zielstellungen wurden mit dem Sprachentwurf verfolgt:

(1) Die Konfigurationssprache umfaßt alle diejenigen sprachlichen Mittel, mit denen die Erzeugung, die Verteilung und die Verwaltung von Prozeßinstanzen beschrieben werden kann.

(2) Die Konfigurationssprache muß weitgehend unabhängig von der verwendeten Hardware-Architektur formuliert sein. Hardwareabhängigkeiten sollen sich höchstens in Parametern zu Konfigurationsanweisungen auswirken.

(3) Die Konfigurationssprache soll leicht zu handhaben sein und eine übersichtliche,

selbstdokumentierende Notation der Konfigurationsbeschreibung gestatten.
(4) Der Sprachentwurf soll für spätere Erweiterungen offen sein.

Diese allgemeinen Zielstellungen werden für den ersten Sprachentwurf durch die im Rechnersystem DREAM realisierten Mittel zur Unterstützung von Konfigurierungsaufgaben präzisiert und eingeschränkt. Angestrebte Erweiterungen werden auf Sprachkonstrukte zur Berücksichtigung von Systemeigenschaften und zur Festlegung der Granularität der Parallelität orientiert sein, wie sie z.B. in Mehrrechner-PEARL [MAN 89] anzutreffen sind. Zur Zeit werden folgende Gruppen von Anweisungen unterschieden:

(A) Anweisungen zur Strukturierung der Konfigurationsbeschreibung und zur Versionsverwaltung, z.B.

SCD (version);   (A1)

UsedPD (name_list);   (A2)

PDCharacteristics (name_list);   (A3)

ENDCharacteristics;   (A4)

SCDEND (version1);   (A5)

(B) Anweisungen zur Erzeugung von Prozeßinstanzen aus Prozeßdefinitionen, z.B.

NumberOfInstances (range_type, min_number, max_number);   (B1)

RatioToResource (res_type, number, number);   (B2)

RatioToPD (name, number, number);   (B3)

(C) Anweisungen zur Verteilung von Prozeßinstanzen, z.B.

DistanceToPD (name, alloc_type);   (C1)

DistanceToResource (res_type, alloc_type);   (C2)

UsesResource (res_type, number);   (C3)

UsesMemorySegment (access_type, seg_size, number);   (C4)

(D) Anweisungen zur Verwaltung von Prozessen, z.B.

StartsAt (time_kind, '12:00:00');   (D1)

RunsEvery (time_kind, '00:50:00');   (D2)

RunsUntil (time_kind, '22:30:00');   (D3)

Priority (number);   (D4)

Reliability (fault_tolerance_type);   (D5)

Displacement (displ_type);   (D6)

Die zugehörigen Parameter lassen sich in mehrere Parametersätze einteilen, die in ihrer semantischen Bedeutung weitgehend systemunabhängig sind. Ihr Wertespektrum wird durch die Hardware-Architektur und die Leistungen des Betriebssystems bestimmt. Sie beschreiben zum Beispiel:

(a) die im Rechnersystem verfügbaren Ressourcentypen
    (z.B. res_type: Harddisk, LAN, CPU-Zeit, Speicher),

(b) die Art und Weise des Zugriffs auf eine Ressource einer bestimmten Art

   (z.B. access_type: privat / geteilt, lokal / entfernt),

(c) die vom System unterstützten Fehlertoleranztechniken

   (z.B. fault_tolerance_type: Restart, Schatten),

(d) die Reichweite eines über einen Systemdienst vermittelten Zugriffs auf eine Ressource bzw. einer Konfigurationsanweisung innerhalb des Multicomputersystems

   (z.B. range_type: PM-lokal, MPS-weit, Multi-MPS-weit),

(e) den zu erwartenden Kommunikationsaufwand zwischen zwei Prozessen bzw. beim Zugriff auf eine Ressource

   (z.B. alloc_type: gleicher / verschiedener PM / MPS),

(f) gewünschte Anzahlen und Anzahlrelationen von Prozeßinstanzen,

(g) Prioritäten für Prozeßabarbeitung und Rekonfigurierung u.a.

Diese Parameter können sowohl als Schlüsselworte vorgegeben sein (z.B.(a)-(e)) als auch Zahlenangaben enthalten (z.B. (f),(g)). Durch Festlegung des Wertespektrums für jeden Parametersatz erfolgt die Anpassung der allgemeinen Methode zur Konfigurationsbehandlung an eine gegebene Hardwarestruktur und die Möglichkeiten des Betriebssystems. Diese Anpassung ist in der Regel einmalig zu Projektbeginn durch einen Systemverantwortlichen vorzunehmen.

Die Lexik der Konfigurationssprache wird vervollständigt durch die zum jeweiligen Applikationssystem einzuführenden Namen von Prozeßdefinitionen.

Eine erste Version der Konfigurationssprache ist als natürliche Erweiterung der Programmiersprache MODULA-2 implementiert. Trotz der damit verbundenen Beschränkung hinsichtlich der Wahl der Syntax wurde eine benutzerfreundliche Notation sowohl für die einzelnen Konfigurierungsanweisungen als auch für die Konfigurationsbeschreibung insgesamt erreicht. Nachfolgend einige Erläuterungen zu dem im Bild 3 gezeigten Beispiel für eine vollständige Konfigurationsbeschreibung.

Die Anweisungen (S1) und (S24) eröffnen bzw. schließen eine Konfigurationsbeschreibung mit dem Versionsnamen "FAME1". Mit den Anweisungen (S2) und (S3) werden die Namen der Prozeßdefinitionen in der Konfigurationsbeschreibung eingeführt. Die Anweisungen (S7) und (S11), (S12) und (S17) sowie (S18) und (S23) eröffnen bzw. schließen jeweils einen Anweisungsblock, innerhalb dessen sich alle Anweisungen auf die in der Parameterliste angegebenen Prozeßdefinitionen beziehen. Die Anweisungen (S4), (S5) und (S6) stellen die gewünschte Abarbeitungspriorität, die zu verwendende Fehlertoleranztechnik und das Maß der Verdrängungsfähigkeit von Prozessen bei Abwärtskonfigurierung (graceful degradation) ein. Zwischen Fehlertoleranztechnik und Verdrängungsfähigkeit besteht ein enger Zusammenhang, der im Abschnitt 5 näher beschrieben wird. Die Anweisungen (S4), (S5) und (S6) könnten auch ohne Namenliste innerhalb eines Anweisungsblockes verwendet werden. Mit (S9) und (S21) werden Zahlen zu erzeugender Instanzen, bezogen auf eine bestimmte Hardwareeinheit, mit unterer und oberer Grenze festgelegt. Die Anweisung (S16) zeigt die Möglichkeit, die Instanzenanzahl im Verhältnis zur Instanzenanzahl einer anderen Prozeßdefinition anzugeben. Mit den Anweisungen (S10) und (S22) wurde ein gewünschter Abstand zwischen den Instanzen verschiedener Prozeßdefinitionen bzw. zu einer Ressource als Maß für den Aufwand bei der Prozeßkommunikation bzw. beim Ressourcenzugriff angewiesen. Dabei wird durch die Parameter "samePm" bzw. "diffPm" die Lokalisierung auf ein und demselben bzw. auf verschiedenen Prozessormodulen ausgedrückt. Diese Art von Verteilungsanweisungen geht auf eine Studie von Schwan [SWA 86], [SWA 88] zurück. Mit (S8), (S13) und (S14) wird die Benutzung einer bestimmten Ressourcenart erklärt. Mit den Anweisungen (S19) und (S20) werden Angaben zum Scheduling-Regime an die Prozeßverwaltung gegeben.

```
SCD                   ('FAME2');                            (S1)

UsedPD                ('AA, AB, AC, AD, AE');               (S2)
UsedPD                ('AX, AY, AZ');                       (S3)

Priority              ('AA, AC, AD, AX', 5);                (S4)
Reliability           ('AB, AC, AY, AZ', restart);          (S5)
Displacement          ('AA, AX', likeRestart);             (S6)

PDCharacteristics     ('AA');                               (S7)
UsesResource          (cputime, 40);                        (S8)
NumberOfInstances     (fame, 1, 2);                         (S9)
DistanceToPD          ('AE', diffPm);                       (S10)
ENDCharacteristics;                                         (S11)

PDCharacteristics     ('AE');                               (S12)
UsesResource          (cputime, 20);                        (S13)
UsesResource          (lan, 2);                             (S14)
UsesMemorySegment     (global, 128, 2);                     (S15)
RatioToPD             ('AA', 2, 1);                          (S16)
ENDCharacteristics;                                         (S17)

PDCharacteristics     ('AB, AC, AD, AX, AY, AZ');           (S18)
StartsAt              (clock, '12:00:00');                  (S19)
RunsEvery             (duration, '00:30:00');               (S20)
NumberOfInstances     (pm, 1, 4);                           (S21)
DistanceToResource    ('disk', samePm);                     (S22)
ENDCharacteristics;                                         (S23)

SCDEND                ('FAME2');                            (S24)
```

Bild 3 Beispiel für eine vollständige Konfigurationsbeschreibung

4. Konfigurierung

Der Konfigurationsbehandler sichert eine einheitliche Behandlung von Systemanlauf und Rekonfigurierung. Dazu wurde die Konfigurationssprache so definiert, daß mit der Konfigurationsbeschreibung eine generische Beschreibung des Applikationssystems in dem Sinne entsteht, daß sie alle möglichen Softwarekonfigurationen in Abhängigkeit von allen möglichen Hardwarekonfigurationen in kompakter Form zusammenfaßt. Als wesentliche Informationen treten dabei die in Anweisungen der Gruppe (C) aus Abschnitt 3. zusammengefaßten Abhängigkeiten auf. Die Bezugnahmen auf Hardwareressourcen (vgl. Anweisung (C3)) können als die Basis der Softwarekonfiguration (BSK) aufgefaßt und durch folgende Abbildung beschrieben werden:

$$\text{BSK: MPD} \times \text{MR} \longrightarrow N_0$$

mit MPD $= \{PD_1, PD_2, \ldots, PD_{ad}\}$
    Menge der gegebenen Prozeßdefinitionen des
    Applikationssystems
  ad $=$ card (MPD)
    Anzahl der gegebenen Prozeßdefinitionen
  MR $= \{R_1, R_2, \ldots, R_{ar}\}$
    Menge der relevanten Ressourcentypen
  ar $=$ card (MR)
    Anzahl der relevanten Ressourcentypen

Damit wird eine Prozeßdefinition PD $\in$ MPD durch ein ar-Tupel beschrieben, wobei der i-te Wert die Anzahl der benötigten Ressourcen vom Typ R_i angibt.
Andererseits kann eine konkrete Hardwarekonfiguration (HK) durch folgende Abbildung allgemein ausgedrückt werden:

$$\text{HK: MR} \times \text{MPM} \longrightarrow N_0$$

mit MPM $= \{PM_1, PM_2, \ldots, PM_{am}\}$
    Menge der verfügbaren Prozessormoduln
  am $=$ card (MPM)
    Anzahl der verfügbaren Prozessormoduln
  MR siehe oben

Damit wird ein Prozessormodul PM $\in$ MPM durch ein ar-Tupel beschrieben, wobei der i-te Wert die Anzahl der verfügbaren Ressourcen vom Typ R_i angibt.
Die allgemeine Methode der Konfigurierung im System DREAM geht davon aus, daß durch

eine geeignete Matrizenmultiplikation

$$A_0 = HK * BSK$$

eine Abbildung

$$A_0: MPD \times MPM \longrightarrow \{0, 1\}$$

hergestellt werden kann, wo jede Prozeßdefinition $PD \in MPD$ durch ein am-Tupel beschrieben wird, wobei der i-te Wert die Möglichkeit oder Unmöglichkeit der Allokierung einer Instanz dieser PD auf dem Prozessormodul PM_i angibt. Nimmt man an, daß zu jeder Prozeßdefinition genau eine Instanz erzeugt wird, dann ist die Softwarekonfiguration nur dann mit der Hardwarekonfiguration verträglich, wenn es zu jeder Prozeßdefinition PD_j mindestens einen Prozessormodul PM_i mit $A_0(j,i) = 1$ gibt.

Konfigurierung unter realen Bedingungen

Die Konfigurationssprache enthält eine Reihe von Anweisungsformen, die Nebenbedingungen für das allgemeine Konfigurierungsproblem formulieren. Die einschneidenste Veränderung gegenüber dem allgemeinen Prinzip resultiert daraus, daß die Anzahl von Instanzen zu einer Prozeßdefinition verschieden von Eins sein kann und in der Regel von der Hardwarekonfiguration selbst abhängig ist. (Die minimale Instanzenanzahl kann auch Null sein, so daß eine Konfigurierung sogar dann lösbar sein kann, wenn nicht zu jeder Prozeßdefinition ein möglicher Prozessormodul gefunden wird.)
Bei der realen Konfigurierung werden zunächst die Instanzenanzahlen berechnet, die sich ohne weitere Bezugnahmen ermitteln lassen (vgl. Anweisung (B1)). Danach werden aus dem aktuellen Hardwareangebot die Instanzenanzahlen derjenigen Prozeßdefinitionen ermittelt, für die Relationen zu Ressourcen (nach Anweisung (B2)) vereinbart sind. Erst dann können die Instanzenanzahlen aus Relationen laut Anweisung (B3) unter Bezugnahme auf bereits berechnete Instanzenanzahlen ermittelt werden.
Dieser als **logische Instanziierung** bezeichnete Komplex kann mehrfach durchlaufen werden, solange noch freie Ressourcen vorhanden sind und nicht alle maximalen Instanzenzahlen (vgl. Anweisung (B1)) befriedigt wurden.
Allgemein ausgedrückt, wird in jedem Zyklus der Instanziierung eine Softwarekonfiguration SK erzeugt, die analog zur Basis der Softwarekonfiguration BSK als Abbildung

$$SK: MPI \times MR \longrightarrow N_0$$

mit $MPI = \{PI_1, PI_2, ..., PI_{ai}\}$
        Menge aller zu erzeugenden Instanzen

aufgefaßt werden kann. Analog zu A_0 kann nun eine Initialverteilung der Instanzen

$$AI_0 = HK * SK : MPI \times MPM \longrightarrow \{0, 1\}$$

gebildet werden. Die Softwarekonfiguration SK ist dabei nur solange mit der Hardwarekonfiguration verträglich, wie es zu jeder Instanz PI_j mindestens einen Prozessormodul PM_i mit $AI_0(j,i) = 1$ gibt. Dieses Abbruchkriterium sichert die Endlichkeit des Instanziierungszyklus' unter der Voraussetzung, daß jede Prozeßdefinition mit implizit angegebener Instanzenanzahl mindestens eine Ressource mit begrenztem Angebot benötigt. Durch den statischen Speicherbedarf der Prozesse ist diese Voraussetzung stets erfüllt.

Aufgrund unterschiedlicher Eigenschaften der in der MK erfaßten Ressourcen, ist die Matrixmultiplikation HK * SK nicht als einheitliche Operation beschreibbar. Für private Ressourcen ist ein Vergleich der angebotenen mit der geforderten Ressourcenkapazität erforderlich, während bei geteilten Ressourcen ein Test auf das Vorhandensein genügt.

Die nach dem ersten Instanziierungszyklus erzeugte SK wird auch als Minimalkonfiguration bezeichnet, da sie die vom Nutzer vorgesehenen minimalen Instanzenanzahlen realisiert. Sollte diese SK nicht mit der Hardwarekonfiguration verträglich sein, dann ist das System nicht arbeitsfähig. In diesem Fall muß entweder die Hardwarebasis erweitert oder die Konfigurationsbeschreibung geändert werden.

Zur Beseitigung der in der Verteilungsmatrix AI_0 im allgemeinen noch vorhandenen Mehrdeutigkeiten wird durch den Allokierer eine weitere wichtige Nebenbedingung genutzt, die im Beispiel der Anweisung (C1) angegebene Abstandsrelation als Maß für den Kommunikationsaufwand zwischen Prozessen. Die zuvor vom KH durchgeführten Konsistenztests garantieren jedoch nicht die Lösbarkeit des Verteilungsproblems. Für den Fall, daß der Prozeßallokierer keine Lösung findet, kann der Konfigurationsbehandler die Instanziierung schrittweise "zurückfahren" und den Allokierer jeweils mit verringerter Anzahl von Instanzen wieder aktivieren, bis entweder eine Lösung gefunden oder aber die Minimalkonfiguration unterschritten wird.

Die eindeutige Zuordnung der Prozesse zu Prozessoren unter den genannten Restriktionen läuft auf ein quadratisches Zuordnungsproblem hinaus. Die Lösung erfolgt im Allokierer mittels eines heuristischen Verfahrens, das schrittweise über die Vergabe von Straftermen eine Zuordnung vornimmt.

5. Rekonfigurierung

Dynamische Konfigurierung ist im System DREAM nur im Falle einer Änderung der aktuellen Hardwarekonfiguration (siehe Abschnitt 1) vorgesehen. Dabei sind zwei Fälle zu unterscheiden.

Abwärtskonfigurierung (graceful degradation)

Bei Ausfall von Prozessormodulen ermittelt der Konfigurationsbehandler die nicht mehr verfügbaren Prozesse und markiert sie als zu rekonfigurierende Prozesse, falls sie der Fehlertoleranztechnik **restart** oder **shadow** zugewiesen waren.

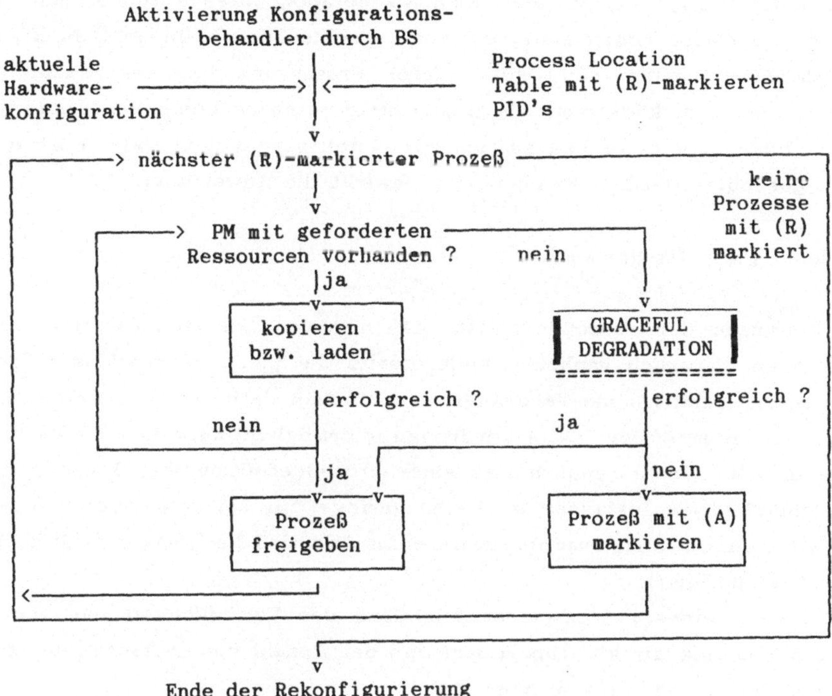

(R)-Markierung: Prozess ist zu rekonfigurieren und zeitweilig nicht verfügbar
(A)-Markierung: Prozess ist ausgefallen (nicht rekonfigurierbar)

Bild 4   Schema der Abwärtskonfigurierung (nach Ausfall eines PM)

Für jeden zu rekonfigurierenden Prozeß sucht der Konfigurationsbehandler einen noch arbeitsfähigen Prozessormodul, der über die notwendigen Ressourcen einschließlich ausreichend freiem Speicher verfügt. Findet der Konfigurationsbehandler einen solchen Prozessormodul, dann veranlaßt er sofort das Laden des jeweiligen Prozesses bzw. im Falle eines Schattenprozesses das Kopieren des ensprechenden Originals (vgl. Bild 4).

Findet der Konfigurationsbehandler keinen Modul mit ausreichendem Speicherangebot, so versucht er auf einem Prozessormodul, der die sonstigen Ressourcenanforderungen des zu konfigurierenden Prozesses befriedigt, die dort laufende Software abzurüsten.

Dabei werden nur solche Prozesse in Betracht gezogen, die eine niedrigere Verdrängung als der zu rekonfigurierende Prozeß haben. Die Einführung dieses speziellen Faktors zur Prioritätssteuerung bei der Rekonfigurierung ist deshalb erforderlich, weil mit den Fehlertoleranzklassen allein keine "gerechte" Bewertung aller Prozesse möglich ist. Einfache Prozesse wären in jedem Fall gegenüber Restart- und Schattenprozessen benachteiligt. Das ist aber nicht akzeptabel, da es durchaus sehr wichtige Prozesse dieser Klasse geben kann, die aus verschiedenen Gründen (z.B. Bindung an spezielle Hardware) selbst nicht auf anderen Prozessormoduln reproduziert werden können. Diese Prozesse können mittels Angabe einer entsprechenden Verdrängungsklasse durch den Nutzer vor einer Degradation zugunsten von Restart- oder auch Schattenprozessen geschützt werden. Beispielsweise bewirkt die Anweisung:

Displacement   ('AA', likeShadow),

daß die Instanzen der Prozeßdefinition 'AA' nur von Schattenprozessen verdrängt werden können. Natürlich impliziert andererseits die Fehlertoleranzklasse "Schatten" automatisch eine relativ hohe Verdrängung, ohne daß dafür eine Verdrängungsklasse angegeben werden muß. Der Faktor Verdrängung ermöglicht außerdem die Berücksichtigung der aktuellen Instanzenanzahl zu einer Prozeßdefinition. Für Prozeßdefinitionen mit ausreichend vielen Instanzen wird eine geringe, für solche mit wenigen Instanzen eine hohe Verdrängung angenommen. Zusammenfassend wird die dynamische Größe Verdrängung wie folgt definiert:
Die Verdrängung eines Prozesses ist dem Grad der Verläßlichkeit der verwendeten Fehlertoleranztechnik direkt proportional und der Anzahl von Instanzen der zugehörigen Prozeßdefinition indirekt proportional.

Führt eine Degradation nicht zum Erfolg bzw. gibt es keinen Prozessormodul, der dem sonstigen Ressourcenbedarf des Prozesses genügt, dann wird dieser Prozeß als nicht rekonfigurierbar markiert und zum nächsten zu rekonfigurierenden Prozeß übergegangen, solange, bis keine Prozesse mehr zu rekonfigurieren bzw. rekonfigurierbar sind. Sind im Ergebnis der Abwärtskonfigurierung weniger Instanzen zu einer Prozeßdefinition vorhanden als vorher, dann wird deren Verdrängung automatisch erhöht, damit bei nachfolgenden Abwärtskonfigurierungen nicht wieder Instanzen derselben Prozeßdefinition einer Degradation zum Opfer fallen.

Aufwärtskonfigurierung

Bei Integration von Prozessormodulen prüft der Konfigurationsbehandler zunächst, ob für die neue Hardwarekonfiguration noch eine entsprechende Softwarekonfiguration auf dem Externspeicher abgelegt ist. Das ist dann der Fall, wenn die letzte Abwärtskonfigurierung aus ebendieser Hardwarekonfiguration heraus erfolgt war. Trifft dieser

Fall zu, dann stellt der Konfigurationsbehandler durch Neuladen bzw. Umladen die der gespeicherten Softwarekonfiguration entsprechende Prozessorbelegung wieder her. War die der neuen Hardwarekonfiguration entsprechende Softwarekonfiguration nicht verfügbar, so muß der Konfigurationsbehandler die Softwarekonfiguration aus den Basisdaten neu berechnen. Auf diese Weise wird die Aufbewahrung aller möglichen Softwarekonfigurationen auf Externspeicher vermieden und andererseits gesichert, daß sich die Prozeßverteilung im Laufe mehrerer Ab- und Aufwärtskonfigurierungen nicht immer mehr vom Optimum entfernt. Die neu berechnete Prozeßverteilung wird mit der vorhandenen Prozessorbelegung verglichen und die neue Prozessorbelegung mit dem geringsten Aufwand an Lade- bzw. Umladeoperationen eingestellt. Nach der Aufwärts-konfigurierung werden in der Regel mehr Prozesse vorhanden sein als vorher. Für alle Instanzen der betreffenden Prozeßdefinition muß dann die Verdrängung der Prozesse wieder verringert werden.

6. Ausblick

Konfigurationsbehandler und Prozeßallokierer liegen als autarke funktionsfähige Komponenten vor. Sie werden zur Zeit als Systemprozesse unter FEE eingeführt. Weiterführende Arbeiten werden sich mit der Definition einer geschlossenen Grammatik für die Konfigurationssprache und deren Implementierung (Compiler), des Prüfens der Leistungsfähigkeit des Konzeptes sowie der Erweiterung um dynamische Änderbarkeit der Konfigurationsbeschreibung befassen.

Literatur

[AHR 88]   R.Ahrens, U.Kucharzyk
           "Systematik Approch to the development of a fault tolerant multiprocessor
           system"; XI. FTSD, Suhl 1988, Proc. pp. 26-34

[HAM 90]   D.Hammer, M.Heiner, R.Krahl, H.Pietsch
           "Tools for Software Distribution in Fault Tolerant Distribution Systems";
           EWDC-2, Florenz 1990

[HEI 87]   M.Heiner, H.Koenig, H.Pietsch
           "Requirements for languages for distributed programming"; MMPS '87,
           Eisenach Sep. 1987, Proc. Vol II pp. 23-31

[KRA 98]   R.Krahl, M.Heiner
"Modelling and Optimization of Processor Allocation in a Multiprocessor
System"; MMPS '89, Stralsund Okt. 1989. Proc. Vol I pp. 37-50

[MAN 89]   K. Mangold
"Mehrrechner-PEARL: Ein Wegweiser zu Multiprozessor-ADA ?"; PARLE '89,
Eindhoven Jun. 1989, Proc. Vol II pp. 90-98

[MAR 87]   P. Martin
"DELTA-4:   Definition and Design of an open Dependeable Distributed computer system architecture"; ESPRIT '87, Proc., pp. 790-797

[SCH 90]   A. Schill
"Migrationssteuerung und Konfigurationsverwaltung für verteilte objektorientierte Anwendungen"; Informatik-Fachberichte Nr. 241, Springer Berlin
1990

[SWA 86]   K. Schwan, A.K.Jones
"Specifing Resource Allocation for the Cm* Multiprocessor"; IEEE Software
(86)5, pp. 60-70

[SWA 88]   K. Schwan, et. al.
"A Language an System for the Construction an Tuning of Parallel
Programs"; IEEE Trans. on Software engineering 14(88)4 pp. 455-471

Objektorientiertes Automatisierungssystem

F.Baldauf, AEG Frankfurt; T.Batz, A.Sassenhof, FhG-IITB Karlsruhe

Der Beitrag stellt die Vorteile der Verwendung objektorientierter Techniken für die Erstellung und den Betrieb von Automatisierungssystemen dar. Die Forderungen aus der Automatisierungsaufgabe und die Merkmale der jetzt bereitstehenden Technik werden analysiert und auf die neuen Lösungen abgebildet. Beispielhaft werden die objektorientierten Strukturen für das Datenmanagement als Kernthema der Projektierung und des Betriebs von Realzeitsystemen vertieft. Die verteilte Realzeitdatenbasis und die Tools für ihre Verwaltung sind von zentraler systemtechnischer Bedeutung, ihre Ausführung in der neuen Technik bewirkt einen grundsätzlichen Strukturwandel der Systeme.

1.   Einführung

Der Trend in der fortschrittlichen Automatisierung geht zu objektorientierten Techniken. Sie ermöglichen die Beherrschung der immer komplexer werdenden Automatisierungsaufgaben und die Maintenierung der Systeme. Bisher war ihr Einsatz verhindert durch Kapazitäts- und Performance- Engpässe der Systembasis, durch die fehlenden Tools und den Mangel an ausgebildeten Mitarbeitern für ihre Erstellung und Behandlung. Jetzt sind die technischen Möglichkeiten der objektorientierten Automatisierung vorhanden und sollten genutzt werden.

Als Szenario für die Vorstellung und Einführung dieser neuen Techniken haben wir ein auf objektorientierten Strukturen basierendes Automatisierungssystem ausgewählt. Seine Position in der Automatisierungswelt ist gekennzeichnet durch

- **Anwendungsbreite.** Einsatz in wesentlichen Branchen der Industrie und in Infrastrukturnetzen. Einordnung in der Automatisierungspyramide von der Feldebene bis zur Produktions- oder Verteilebene mit Anschluß an die Unternehmensleitebene.

- **Rahmenarchitektur.** Baukastenartige Systemplattform aus leistungsgestuften Automatisierungsprodukten und Infrastrukturfunktionen mit der Integrationsfähigkeit für fremde Leistungen. Zusammenführung der verteilten und heterogenen Strukturen von Visualisierungssystemen, Prozeßrechnern und Steuerungen in einer einheitlichen Systemarchitektur. Transparente Anpassung an unterschiedliche Prozeßtopologien und Mengengerüste ohne Änderung interner Strukturen und Schnittstellen [BB88], [BB89].

2.   Merkmale und Architektur fortschrittlicher Systeme

Da die Liste der Merkmale fortschrittlicher Automatisierungssysteme äußerst umfangreich ist [Stö88], würde ihre komplette Behandlung den Rahmen dieses Referats sprengen. Wir konzentrieren uns daher bei der Kurzbeschreibung des Systems auf die wesentlichen Merkmale und Architekturelemente, soweit diese Einfluß haben auf die gewählte Systemstruktur eines verteilten Systems und auf die Technologieentscheidung für einen objektorientierten Ansatz.

- **Übersichtliche Strukturen** sind erforderlich in den Automatisierungskomponenten selbst und in ihrer Verbindung zu Gesamtsystemen. Nur so kann Flexibilität in der Konfigurierung, in der Aufnahme neuer Leistungen und eine einfache Pflege der Systeme garantiert werden.

  Die funktionale Ordnung in Automatisierungssystemen ist geprägt durch die gewählten Programm- und Datenstrukturen. Hierbei ermöglichen objektorientierte Techniken die strukturelle und verhaltensmäßige Abbildung des zu beherrschenden Prozesses im Automatisierungssystem und die Entkopplung der unterschiedlichen Abstraktions- und Funktionsebenen.

Die flexiblen Grundstrukturen ermöglichen auch eine evolutionäre Systementwicklung, d.h. eine kontinuierliche Einführung innovativer Leistungen/Produkte und Ausdehnung des Leistungs- und Funktionsumfanges. Grundsätzlich sind Objektstrukturen offen für zukünftige Erweiterungen und Erneuerungen.

Aufgrund der Flexibilität und Erweiterbarkeit der objektorientierten Strukturen bei Kontinuität in den einheitlichen Methoden ihrer Behandlung eignet sich diese Technik besonders gut, wenn Rahmenarchitektur und Basisleistungen später durch anwendungsspezifische Add-ons komplettiert werden sollen.

- **Offenheit für Systemkomponenten anderer Hersteller** ist durch die konsequente Orientierung an den bestehenden oder sich abzeichnenden Normen und Standards zu erreichen.

Allerdings müssen diese Standards noch ergänzt werden, weil sie wichtige für einen Realzeit-Betrieb und für die Maintenierbarkeit komplexer Systeme unbedingt erforderliche Leistungen bisher nicht abdecken, wie beispielsweise das systemweite Datenmanagement.

Normung benötigt darüberhinaus lange Abstimmungs- und Konsolidierungszeiträume, d.h. modernste Techniken sind selten genormt.

Ein Standardisierungseffekt kann allerdings durch Einführung eines tragfähigen Konzeptes in Schlüsseltechnologie (z.B. objektorientierte Techniken) und Kernthemenfeld (z.B. Datenmanagement) erreicht werden.

- **Rationalisierung in den Engineering-Aktivitäten** heißt Vermeidung von redundanten und damit fehlerbehafteten Arbeiten für die Planung, Erstellung und Pflege der Systeme, die bisher bei verteilten Daten- und Funktionsbeständen noch unerläßlich waren.

Die Möglichkeit der Prozeßmodellierung anhand der in Struktur und Verhalten realitätsbezogenen Objekte befreit den Projektierer von den bisher notwendigen Transformationen der Umwelt in die abweichenden Strukturen der Systeme. Vor allem die Unabhängigkeit der Applikation von speziellen Konfigurationen des unterlagerten Systems bedeutet einen Fortschritt in der Entwicklung verteilter Systeme.

Während die Harmonisierung der operationellen Schnittstellen offener Systeme durch Standards und Normen fortschreitet, wird die Adaption der heterogenen Strukturen mehr und mehr in das Engineering verlagert. Die in offenen Systemen einsetzbaren Softwarepakete bringen in der Regel ihre eigenen integrierten Tools für die Projektierung mit. Ungelöst blieb bisher die Verbindung der Pakete auch in der Toolebene. Redundante, u.U. inkonsistente Projektierung war die Folge.

Durch die Definition von Toolschnittstellen, einer gemeinsame Datenablage für die globalen Beschreibungsdaten, einer einheitlichen Bedienumgebung kann die bestehende Problematik entschärft werden.

- **Objektorientierte Techniken** finden in einem fortschrittlichen Automatisierungssystem ihre Anwendung in Bedienoberflächen, Programmiermethodik und Modellierung von Prozessen, Produkten und Systemen.

Bezüglich des Programmierstils ist ein objektorientiertes System dadurch gekennzeichnet, daß in ihm nur über Nachrichten miteinander interagierende Objekte existieren. Diese Objekte sind Exemplare von (Typ-) Klassen, die die gekapselten Daten und die darauf operierenden (public-) Methoden (information hiding) definieren. Durch die hierarchische Anordnung der Klassendefinitionen wird die Vererbung von Daten und Methoden (inheritance) unterstützt und die Redefinition von Methoden ermöglicht, wobei die Auswahl der auszuführenden Methoden erst zur Laufzeit erfolgt (Polymorphismus, late binding).

Unter objektorientierter Modellierung wird die im Rahmen der Projektierung des Automatisierungssystems vorzunehmende Abbildung von Prozessen, Produkten und Systemen der realen Welt auf rechnerinterne Objekte verstanden. Dazu werden in der zu modellierenden Welt typische Prozeß-, Produkt- bzw. System-Klassen bzgl. ihrer Datenbestände und ihres Verhaltens identifiziert und anschließend in einer Hierarchie strukturiert (Vererbung), einerseits um Gemeinsamkeiten zum Ausdruck zu bringen und andererseits um diese Gemeinsamkeiten nur einmal bearbeiten zu müssen bzw. an zentraler Stelle modifizieren zu können.

3. Objektorientierte Projektierungsdatenbasis (PDB)

Die Projektierungsdatenbasis mit den in ihr integrierten Methoden spielt im Lebenszyklus eines Automatisierungssystems eine zentrale Rolle. Im folgenden werden zunächst die wichtigsten Anforderungen an die Funktionalität dieser Datenbasis dargestellt.

-   **Systembeschreibung.** Sie enthält eine übersichtliche und aktuelle Beschreibung des Automatisierungssystems. Dazu werden sämtliche Objekte gespeichert, die von globalem Interesse sind. Es müssen hierarchisch verfeinerbare funktionale und topologische Strukturen und die dazugehörenden textuellen Beschreibungen verwaltet werden.

-   **Integration unterschiedlicher Nutzer**[1]. Innerhalb eines Gesamtengineering sind unterschiedliche Teilaufgaben zu leisten, die jeweils einzelne Aspekte des Gesamtsystems behandeln und dazu dem Benutzer ihre spezielle "Sicht" (mit u.U. eigenem unterlagerten Datenmodell) auf das System bieten. Die Integration dieser individuellen Sichten und die Abbildung auf das Datenmodell der PDB muß unterstützt werden. Die Synchronisation und Koordination der Zugriffe auf Daten, die mehr als einen Nutzer betreffen, muß gesichert sein.

-   **Modellbildung des Prozesses im Automatisierungssystem.** Zusammengehörende Daten des Prozeßes dürfen nicht in eine Vielzahl von Einzeldaten innerhalb des Projektierungssystems zerfallen; das Projektierungssystem muß ein identisches Bild des zu automatisierenden Prozeßes beschreiben.

-   **Konsistenzsicherung.** Maßnahmen zur Erkennung von Inkonsistenzen, Unterstützung zum Wiedererreichen eines konsistenten Zustands und Beschreibungen des Projektierungsstandes müssen vorhanden sein.

-   **Komfortable Handhabung.** Die übersichtliche und einfache Verwaltung und Änderbarkeit einer Vielzahl ähnlicher Daten einschließlich der auf sie wirkenden Methoden ist zu unterstützen. Dazu sollen zentrale Bibliotheken, die bereits die wesentlichen Vorbelegungen und Muster enthalten, verwendet werden.

Neben der strukturellen Einordnung der PDB in die Systemarchitektur ist ihre Stellung im Systemlebenszyklus von Bedeutung. Im folgenden soll die Qualität der zeitlich versetzten Aufgaben der PDB und die dazu erforderlichen Interaktionen diskutiert werden. Der Lebenszyklus besteht entsprechend einem Phasenmodell [BS90] aus mehreren Phasen[2] (Analyse/Angebot, Konfigurierung, Projektierung, Programmierung, Test/Simulation, Inbetriebnahme, Wartung und Weiterentwicklung [**Abb.3.0-1**]) mit entsprechenden Rücksprungmöglichkeiten in frühere Phasen.

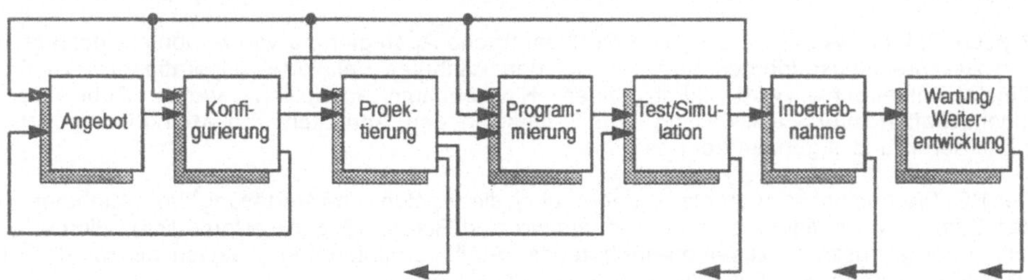

Abb. 3.0-1   Zusammenhänge der Engineering-Phasen

Diese einzelnen Phasen werden kurz beleuchtet im Hinblick auf die Erzeugung und Modifikation von Daten der Projektierungsdatenbasis.

---

[1] Nutzer sind hier die Tools für MMI-,SPS- und Leittechnik-Projektierung in Bezug auf Standards und die applikationsspezifischen Add-ons.

[2] Bei den hier aufgezählten Phasen kann es sich nur um eine für diesen Kontext repräsentative Auswahl handeln, jedes Unternehmen wird "sein" firmenspezifisches -- von dem hier vorgestellten abweichendes -- Phasenmodell haben.

- **Analyse (Angebot)/Konfigurierung.** Es werden die zu verwendenden Automatisierungsstationen mit Prozessoren, Speicherbausteinen-, Bildschirmgeräte, Betriebssysteme und Standardsoftware (z.B. Kommunikationssystem) ausgewählt und konfiguriert. Dies erfolgt am zweckmäßigsten mittels strukturierter Objekte [BS90][KSS89][BBK88][L+85], die in der Regel einer Standardbibliothek entnommen werden können.

- **Projektierung.** In der Projektierungsphase wird der konkrete Prozeß modelliert. Hierfür werden Prozeß-Objekt-Klassen definiert, aus den typisierten Datenstrukturen und Standardmethoden werden Exemplare abgeleitet[3], auf die Stationen des dezentralen Systems verteilt und an die Applikationsprogramme angebunden.

- **Programmierung.** Bei der Programmierung werden die anwendungspezifischen Teile und u.U. die bereits verfügbaren Standards ergänzt. Die Erweiterungen der Standards stehen anschliessend für alle weiteren Projekte als neue Bausteine in der Standardbibliothek zur Verfügung.

- **Test/Simulation und Inbetriebnahme.** Test/Simulation und Inbetriebnahme können, wenn sie keinen Rücksprung in die Phase Programmierung oder Projektierung bedingen, bei kleinen Veränderungen zur Modifikation einzelner Projektierungswerte (z.B. Sollwerte oder Ober-/Untergrenze) oder zur Modifikation der Verteilung einzelner Objekt-Attribute führen (z.B. ein Remote-Zugriff wird durch den schnelleren Zugriff auf eine Kopie ersetzt).

- **Weiterentwicklung.** Die Weiterentwicklung bedingt auf alle Fälle einen Rücksprung in eine der früheren Phasen (Konfigurierung, Projektierung, Programmierung oder Angebot) und eine damit verbundene Aktualisierung der PDB.

- **Online-Änderungen.** Veränderungen des Realzeitsystems im laufenden Betrieb, bedingen - wenn das Maintenierungssystem offline und damit nicht Teil des Realzeitsystems ist - Möglichkeiten zur (am besten automatischen) Aktualisierung der PDB, um die Konsistenz zwischen dem Realzeitsystem und seiner Beschreibung wieder herzustellen.

3.1 Projektierung von Prozeßobjekten

Bisher haben wir in diesem Kapitel die Engineering-Aspekte allgemeiner Objektklassen und ihrer Exemplare diskutiert. Am Beispiel der für die Prozeßdatenverarbeitung benötigten Modellbildung durch sog. Prozeßobjekte (PO) wollen wir den objektorientierten Lösungsansatz für eine Projektierungsdatenbank konkretisieren. Die Prozeßobjekte verfügen über eine zusätzliche Semantik, wie sie im folgenden geschildert wird. Die Realisierung erfolgt mittels der allgemeinen strukturierten Objekte **[Abb. 3.2-1]**.

- **Prozeß-Objekt-Typ.** Der PO Typ ist die thematische Aggregierung von Attributen, dazu gehörenden Beschreibungsattributen und den auf den Attributen definierten Operationen. Die Prozeß-Objekt-Typhierarchie wird durch einen Klassenbaum modelliert, wobei Subklassen die Eigenschaften ihrer Basisklassen erben, diese in den Attributen und Methoden jedoch auch modifizieren und ergänzen können.

  Der PO-Typ beinhaltet auch die Angaben über die topologische Verteilung[4] von Attributen (wo ist das Original, wo befinden sich Kopien und wo sind Remote-Zugriffe erforderlich). Hierbei werden aus Applikationssicht zusammengehörende Attributgruppierungen einschließlich der darauf wirkenden Methoden definiert (=Sichten auf PO-Typen)[5].

---

[3] Die Vorgänge für Klassendefinition und Exemplarbildung werden in Kap.3.1 beschrieben.

[4] Dadurch erfolgt ein Teil der globalen Kommunikationsfestlegung für die Aktualisierung von Kopien.

[5] Bei einer konkreten Konfiguration ergibt die Obermenge der Sichten auf einen PO-Typen, die einem lokalen Datenbasis-server zugeordnet sind, die Fragmentierung (s. auch Kap.4.0)

Bei der Erzeugung[6] eines PO-Typen legt man den Typnamen und die zu dem Typ gehörenden Attribute fest. Die Attribute besitzen ebenfalls wieder einen Namen, einen für dieses Attribut in allen Exemplaren der Klasse gültigen Wert, sowie projektierbare Beschreibungsattribute wie Defaultwert, Meldesperre oder Wert-ist-gültig-Flag. In bestimmten Fällen ist es wünschenswert, einen Attributwert nur einmal für Attribute in unterschiedlichen PO-Typen zu definieren, die zueinander in keiner Vererbungsbeziehung stehen. Dies ist durch ein Link-Attribut (Verweis auf ein Attribut eines POs eines anderen PO-Typs) möglich.

Dann werden die in einem PO neben den dedizierten Methoden (z.B. Lese-Sollwert, Setze-Sollwert, Lese-Istwert) gewünschten objektinternen Methoden (z.B. PID-Regelung oder Protokollierung) mittels der entsprechenden Funktionsblöcke definiert. Dazu werden die entspechenden Funktionsblöcke einer Bibliothek entnommen und als Succesors an den PO-Typ angehängt. Die Funktionsblöcke werden parametriert und an die PO-Attribute angebunden. Dazu werden die Parameter der Funktionsblöcke mit den entsprechenden Attributnamen des PO-Typs verknüpft. Zusätzlich existieron zu don dedizierten Methoden Before- bzw. After-Methoden, die vor bzw. nach Ausführung einer dedizierten Methode aktiviert werden (z.B. Plausibilitätsprüfungen, Aktivierung abhängiger Objekte etc.).

Nun muß noch die typspezifische Vertoilung der PO-Attribute und Funktionsblöcke auf Sichten[7] erfolgen. Da dabei Attribute gleichzeitig in verschiedenen Sichten benötigt werden können, erfolgt entsprechend ihrer Verwendung in den Funktionsblöcken die Entscheidung, in welcher Sicht das Orginal des Attributs, eine Kopie oder ein Verweis auf das Original liegt.

Als letzte dem Typ zugeordnete Information erfolgt die Zuordnung der Sicht zu der Anwender-Instanz, welche den betreffenden Ausschnitt aller POs dieses Typs verarbeitet, sowie die Zuordnung der Instanz zur entsprechende Station (z.B. erfolgt dadurch die Protokolierung sämtlicher Fehlermeldungen in der Station S_3).

- **Prozeß-Objekt.** Ein PO ist die Instanziierung eines PO-Typs, welche -soweit sie nicht redefiniert werden[8] - die Vorbelegungen des PO-Typs übernimmt. Im Allgemeinen sind von den Typen eine größere Anzahl Exemplare (ca. 10 bis mehrere hundert, z.B. bei Stellmotoren in einer größeren Fabrik) abgeleitet.

Bei der Projektierung der POs werden jetzt -im Gegensatz zur globaleren PO-Typ-Projektierung- weitere für das einzelne Exemplar spezifische Werte projektiert. So erfolgt z.B. eine Zuordnung zu den speziellen Meßwerten des Prozesses und dem entsprechendenden Wertesatz für die Visualisierung im Anlagenbild, die gerade diese eine PO abwickelt.

3.2 Vorteile durch objektorientierte Konzeption der Projektierung

Die objektorientierte Projektierung bietet eine ganze Reihe von Vorteilen gegenüber der herkömmlichen Projektierung. Wir werden die wesentlichen kurz zusammenfassen.

- **Effektivität.** Durch Ausnutzung der Klassenhierachie bzw. des Vererbungskonzepts erhält man eine starke Änderungsfreundlichkeit; leistungsfähige Änderungsoperationen (z.B. Hinzufügen oder Veränderungen von Attributwerten an zentraler Stelle der Klassenhierachie führt zur automatischen Aktualisierung in sämtlichen betroffenen Subklassen) ersetzen die mehrfache Ausführung fehleranfälliger, zeitraubender Primitivänderungsoperationen. Der Projektierer wird durch dieses Vorgehen von den Routinetätigkeiten entlastet (dem ständigen Eingeben gleicher Werte) und kann sich intensiver mit den Ausnahmen beschäftigen.

---

[6] Die hier vorgestellte Reihenfolge ist eine von mehreren möglichen, so könnten z.B. auch zuerst Funktionsblöcke definiert werden.

[7] Die Sichten stellen die Schnittstelle der Nutzer zu den Prozeßobjekten dar.

[8] Wie Speicherart und Wert der Attribute in der Sicht (Remote, Kopie oder Original) und die Verteilung der Sichten auf Komponenten.

Eine Veränderung der Konfiguration des Automatisierungsgeräts in der Projektbibliothek ist durch die Verwendung der flexiblen und doch typgeprüften Strukturen auf einfache Weise möglich.

Die Erzeugung neuer Typen durch Vererbung aus ähnlichen bereits vorhandenen unter Hinzufügung der noch fehlenden Teile (Programming by Difference) ermöglicht eine schnelle und kostengünstige Entwicklung **[Abb.3.2-1]**.

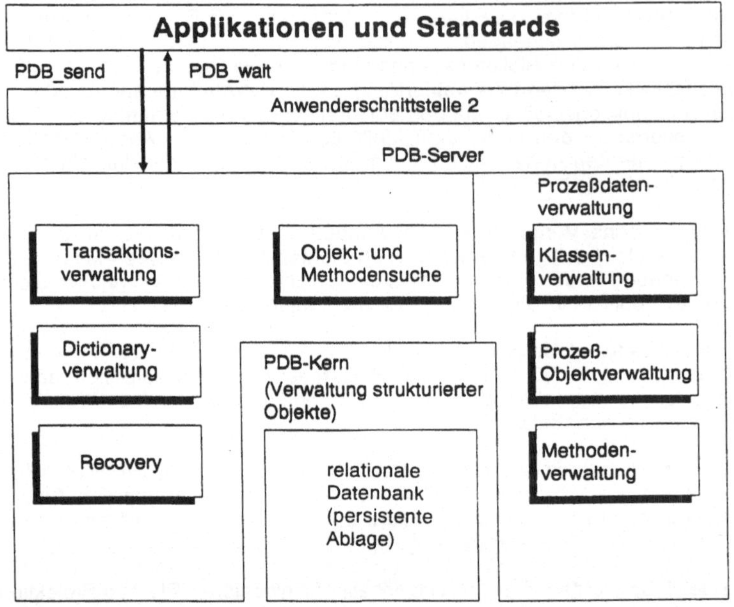

Abb.3.2-1   Architektur der Projektierungsdatenbank

Zur Objektidentifizierung werden eindeutige Objekt-Ids vergeben. Es handelt sich um Surrogate, d.h. systemweit eindeutige Identifikatoren, die keinerlei Hinweise auf die Art des Objekts, seinen Typ, seinen Verwendungszweck oder seine Lage beinhalten. Dadurch ist es möglich, die Objekte ohne Auswirkungen auf die Objekt-Id auf ein anderes Automatisierungssystem zu verlagern oder ihren Typ zu verändern. Derartige Änderungen bedingen dann keine Folgeänderungen in den Applikationsprogrammen.

- **Konsistenz.** Die Zusammenführung aller globalen und wichtigen Informationen in einer einzigen Datenbank stellt eine projektumfassende Dokumentation dar und gibt über Projektstände, noch durchzuführende Aktionen und frühere Systemzustände Auskunft. Die Konsistenzprüfung hat Zugriff auf alle global wichtigen Daten und kann neben Typunverträglichkeiten, nicht abgewickelten Bearbeitungsschritten auch nicht verwendete aber vorhandene Objekte oder Objektteile erkennen.

- **Integration.** Das zentrale Austauschdatenformat für die Kommunikation zwischen den Instanzen des operationellen Systems, die PO-Typen, werden in der PDB verwaltet. Dies dient der syntaktischen Integration der unterschiedlichen und verteilten Applikationen.

-   **Wiederverwendbarkeit.** Bereits entwickelte und getestete Objekte wie z.B. Funktionsblöcke oder sogar komplette strukturierte Typen bzw. Exemplare wie Automatisierungsteilanlagen oder PO-Typen (mit Attributwerten, -Verteilung und angebundenen Funktionsblöcken) können zur späteren Wiederverwendung (vor allen Dingen auch in anderen Projekten) in Bibliotheken gespeichert werden. Dies gilt ebenso für die Beschreibung der verwendeten Hardware und Hardwarekomponenten. Diese sollten mit ihren Konfigurierungsmöglichkeiten nur ein einziges Mal beschrieben werden[9].

4. Objektorientierte Realzeitdatenbasis (RDB)

Basierend auf den allgemeinen Anforderungen an ein innovatives Automatisierungssystem sowie aus Architekturaspekten läßt sich das folgende Anforderungsprofil an eine operationelle Datenbasis[10] für automatisierungstechnische Belange skizzieren:

-   **Dezentralisierung des Datenbestandes** aufgrund von Zeit-, Topologie-, Last- und Verfügbarkeitsanforderungen.
-   **Realitätsbezogene Modellierung der Prozeß-Objekte** wegen gestiegener Komplexität bzw. erhöhtem Mengengerüst mit flexibler Erweiterung der Modellierung.

-   **Offene Schnittstelle mit system- und datenstruktur-unabhängigem Zugriff** zur Adaption vorhandener/zukünftiger Applikationen sowie zur flexiblen Erweiterung der System- bzw. Datenstruktur.

-   **Integration von Standardverarbeitungen in die Datenbasis** zur Reduzierung der Komplexität von Applikationen.

Diesem Anforderungsprofil an eine Realzeitdatenbasis kann mit einem objektorientierten Lösungsansatz entsprochen werden:

-   Reale (Teil-)Prozesse werden durch entsprechende komplexe Prozeß-Objekte adäquat modelliert.

-   Erweiterungen/Modifikationen/Spezialisierungen von Prozeß-Modellen sowie deren Wiederverwendbarkeit werden durch Vererbungsmechanismen innerhalb eines Klassenbaums weitgehend unterstützt.

-   Durch die Methodenschnittstelle der modellierten Objekte (information hiding) bleibt die interne Datenrepräsentation in Struktur und Verteilung für die Applikationen unsichtbar, d.h. Modifikationen der Systemstruktur (z.B. Rekonfiguration, Lastausgleich, Erweiterung) haben keine Rückwirkungen auf die Applikationen.

-   Standardverarbeitungen wie Mittelwertbildung, Archivierung usw. können in Form entsprechend vererbbarer Klassen-Methoden in die Datenbasis integriert werden.

-   Die späte Bindung erleichtert die einheitliche Handhabung von Objekten unterschiedlichen Typs durch die Applikationen (z.B. getRefValueAsFloat in einer repeat-until-Schleife).

-   Der aus Laufzeit- bzw. Lastgründen notwendigen Verteilung von Datenbasisinhalten kann mit einer entsprechenden Objektverteilung entsprochen werden.

Durch einen objektorientierten Lösungsansatz werden die wesentlichen Anforderungen an eine Datenbasis für automatisierungstechnische Belange somit a priori erfüllt, wobei jedoch die folgenden für offene Realzeitsysteme interessierenden Belange noch einer besonderen Betrachtung bedürfen:

---

[9] Die Verwendung eines Expertensystems könnte dabei für die Konfigurierung komplexer Systeme erhebliche Hilfestellung leisten.

[10] Im Gegensatz zur Engineering Datenbasis.

- Die durch Realzeitbelange erforderliche Verteilung kann der Modellierung logisch zusammengehöriger Daten zu einem Prozeß-Objekt widersprechen, d.h. nicht nur vollständige Objekte müssen über die dezentrale Systemstruktur verteilbar sein, sondern auch Fragmente eines Objekts (z.B. attributsweise), wobei diese Verteilung für die Applikationen jedoch transparent sein muß.

- Die flexible Erweiterung vorhandener Modellierungen zur Laufzeit legt eine interpretative Lösung nahe, während die Realzeitanforderungen eine kompilative Lösung erzwingen. Es ist somit notwendig Typinformationen über den Compile/Link-Prozeß hinaus zu erhalten, insbesondere über die Klassenzugehörigkeit eines Objekts und über die Attribute und Methoden dieser Klasse.

- Die Applikationen werden nicht generell eng an die Datenbasis gekoppelt sein (gebunden), d.h. das Aktivieren von Methoden kann nicht in jedem Fall prozedural erfolgen. Für solche entkoppelten Applikationen muß eine Message-Schnittstelle bereitgestellt werden, die über Objekt-Id, Methoden-Id und Methoden-Parametern Zugriff auf den RDB-Service erlaubt.

4.1 Modellierung von Prozeß-Objekten

Das in der PDB in Form einer Prozeß-Objekt-Typhierarchie projektierte Prozeßmodell findet in der RDB seine Entsprechung in einem adäquaten Klassenbaum, der die Attribute und Methoden in einer geeigneten Erbhierarchie strukturiert, wobei insbesondere auch die Fragmentierung verteilter Objekte Berücksichtigung findet.

Neben diesen projektierten Prozeß-Objekt-Klassen (PO-Klassen) sind in einer intern vorhandenen, programmierten PO-Basisklasse allgemeine Attribute und Methoden enthalten, die alle PO-Klassen erben und die Verwaltung dieser PO-Klassen und ihrer Exemplare organisieren. Insbesondere sind hier zu nennen:

- Das Attribut "Prozeß-Objekt-Id" in der PO-Basisklasse, dessen Konstruktor gleichzeitig dafür sorgt, daß ein erzeugtes Exemplar sich bei der RDB-Objektverwaltung registriert (Eintrag in Objekt-Tabelle).

- Eine Methode zur Klassifizierung von Exemplaren, d.h. für ein identifiziertes Objekt kann zur Laufzeit seine Klassenzugehörigkeit festgestellt werden.

- Methoden, die Standard-Operationen auf Attributen ermöglichen (remote Zugriffe, lesen/schreiben/rücksetzen, aktivieren/deaktivieren, lock/unlock etc.).

- Eine Klasse, die den Aufruf von PO-Methoden über eine Message-Queue für entkoppelte Applikationen abwickelt.

- Eine Dispatcher-Klasse, bei der Objekte für einen Methoden-Aufruf zyklische oder einmalige Zeitaufträge anmelden können.

- Eine Klasse für Organisatorische-Objekte, in der dynamisch zur Laufzeit Aggregationen von Objekten erfolgen können (z.B. eine DefineList-Operation von MMS).

- Ein Name-Service und ID-Service zur Umsetzung (Translate) bzw. Anbindung von Applikations-Namensräumen an den projektierten RDB-Namensraum (Klassen, Methoden, Attribute, Objekte).

- Unterschiedliche Ausprägungen eines Link-Attributs mit zugehörigen Methoden, die eine Verkettung von Objekten ermöglichen, z.B. zur Triggerung von Methoden anderer Objekte bzw. zum Zugriff auf deren Attribute über Methoden.

4.2 Realisierungskonzepte und Generierung der Realzeitdatenbasis

Der RDB liegt eine Client-Server-Architektur zugrunde, wobei jede Komponente eines dezentralen Systems in der Regel besteht aus (**Abb.4.2-1**):

- einem RDB-Client zur Annahme eines Auftrags und zur Transaktionsverwaltung,
- einem RDB-Server zur Verwaltung der lokal gelegenen Objekte bzw. Objekt-Fragmente.

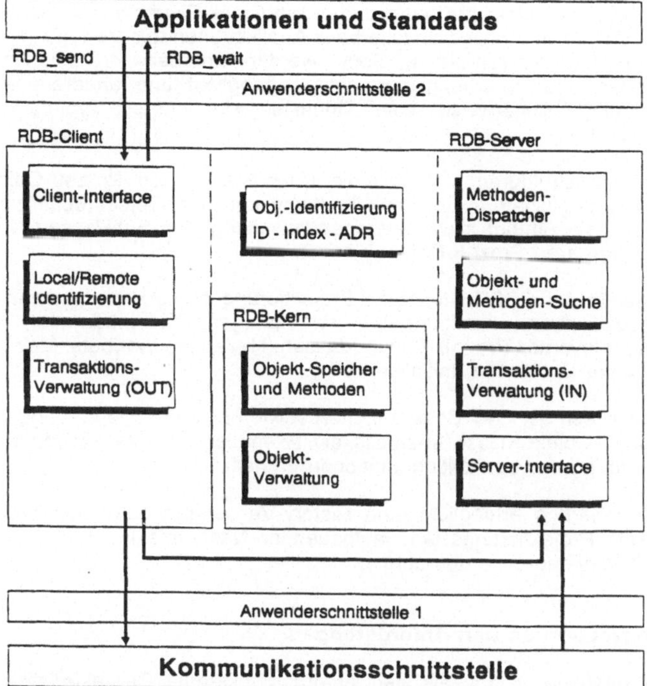

Abb.4.2.-1   Struktur der Realzeitdatenbasis

Die RDB ist somit konzipiert als über das gesamte Netzwerk verteilter Service, wobei die Lokalisierung des zuständigen Servers in mehreren Schritten erfolgt. Wie aus Bild 4.2.-1 ersichtlich, ist der dezentrale RDB-Service (ebenso wie der zentrale PDB-Service) über die Anwenderschnittstellen AS1 und AS2 mit den Applikationen bzw. dem Kommunikationssystem verbunden.

- **Beispiel.** An der AS2 wendet sich eine Applikation mit einem Remote Procedure Call (RPC) an den RDB-Service (Service-Id bzw. -Name).

In der AS2 wird der lokale RDB-Client als Adressat identifiziert (prozedural bei hinzugebundenen Applikationen, sonst über eine Message-Queue entkoppelt). Der RDB-Client lokalisiert das im RPC spezifizierte Objekt.

Ist dieses Objekt lokal vorhanden, wird der Auftrag an den lokalen RDB-Server weitergeleitet, im anderen Fall wird über die AS1 die logische Teilnehmer-Nummer in eine physikalische Komponentenadresse umgesetzt und der dort befindliche RDB-Server über die AS1-Schnittstelle mit dem Auftrag beaufschlagt.

Der RDB-Server identifiziert das Objekt und aktiviert die im RPC angegebene Methoden-Id mit den vorgegebenen Parametern.

Die RDB wird basierend auf den Betriebssystemen UNIX, OS/2 und iRMX implementiert in C++ [Str87, Lip89], einer objektorientierten Spracherweiterung von C. Eine SMALLTALK-ähnliche Klassenlibrary [GOP90] stellt in C++ realisierte Basisdatenstrukturen wie z.B. Dictionary, LinkedList usw. bereit.

Als wesentliches Merkmal dieses Ansatzes ist dabei die deskriptive Klasse Class hervorzuheben. Von dieser Klasse Class wird für jede Klasse des Klassenbaums ein Exemplar als statische member-class kreiert. Dieses Class-Exemplar spezifiziert die Klasse von der es ein Member-Attribut ist durch Angaben wie z.B. Klassen-Name, Klassen-Id, Verweis auf Basisklassen, Beschreibung der Attribute und Methoden etc. Durch dieses Konzept stehen für jedes Objekt zur Laufzeit über **ein** Class-Exemplar pro Klasse die notwendigen Typinformationen für verteilte und online erweiterbare Applikationen zur Verfügung, ohne die gesamte Implementierung interpretativ zu realisieren. Somit werden einerseits die zeiteffiziente kompilative Umsetzung der Methoden und Vererbungshierarchien ermöglicht und andererseits die Erhaltung zur Laufzeit notwendiger, üblicherweise vom Compiler eleminierter Typinformation gewährleistet.

In diesen Klassenbaum der Basisdatenstrukturen fügt sich die Basis-Klasse von Prozeß-Objekten harmonisch ein und erweitert die Typidentifizierung und -beschreibung um eine Objekt-Identifizierung bzw. -lokalisierung sowie um eine globale Attribut- bzw. Methodenverwaltung. Die Subklassen hiervon sind dann die vom Anwender projektierbaren Prozeß-Objekt-Klassen.

Basierend auf dieser vom Anwender in der PDB hinterlegten Projektierung erfolgt eine automatische Generierung einer komponentenspezifischen[11] konfigurierten, ablauffähigen RDB-Inkarnation durch Extraktion der für die Zielkomponente relevanten Prozeß-Objekt-Klassen (Hierarchie, Attribute, Methoden, Verteilung etc.) und -Exemplare (Name, ID, Klasse, Initialwert etc.).

Aus diesen Informationen wird automatisch der C++ Code generiert sowie ein Make-File erzeugt, wobei jede vom Generator erzeugte Prozeß-Objekt-Klasse ebenfalls ein Exemplar der deskriptiven Klasse Class als Member-Attribut zur Erzeugung der Laufzeitinformationen enthält.

In Ergänzung zu dieser offline Generierung ermöglicht eine History-Verwaltung dem Generator die Erzeugung additiver, auf dem letzten Projektierungsstand aufbauender Nachprojektierungen, die zur Modifikation einer RDB zur Laufzeit verwendet werden können.

4.3   Spiegelung der Lösungskonzeption an den Anforderungen

Die in ihren Lösungskonzepten vorgestellte Datenbasis stellt aufgrund folgender Charakteristika eine dem anfangs skizzierten Anforderungsprofil adäquate Lösung dar:

-   **Realzeitorientierung**
    -   optimale Verteilung von Objekten und Objektfragmenten,
    -   Ablage der Informationen im wesentlichen im Hauptspeicher,
    -   Reduktion von Projektierungsinformationen auf den zwingend notwendigen Umfang (z.B. deskriptive Objekte zur Beschreibung der Klassen und der Klassenzugehörigkeit von Objekten, aber compilierende Generierung der Algorithmik von Methoden, IDs statt Zeichenketten etc.)[12] .

-   **Objektorientierung**
    -   Zusammenfassung aller logisch zusammengehörenden Daten zu einer identifizierbaren Einheit unabhängig von der Verteilung,
    -   Integration von Verarbeitungs-Methoden in die Datenbasis bzw. deren Bindung an Objekt-Klassen.

-   **Offenheit**
    -   unterschiedliche Ankopplungsmechanismen (Interprozeßkommunikation oder prozedural),
    -   Umsetzung von applikationsspezifischen Namensräumen in den projektierten Namensraum,
    -   Umsetzung applikationsspezifischer Telegramm- und Prozedur- Schnittstellen in die o.g. Ankopplungsmechanismen (z.B. Umsetzung eines Signal-Kanals in ein PO-Attribut).

---

[11]Komponenten sind hier als funktional autonome Teilsysteme zu verstehen, z.Bsp. Station am LAN oder CPU einer Mehrprozessor-Station. Eine Komponente besitzt in jedem Fall eine Architektur, wie in Kap.4.1 dargestellt.

[12]Die RDB enthält also im Gegensatz zur Projektierungs-Datenbasis (PDB) im wesentlichen nur die hauptspeicherresidenten Prozess-Objekt-Exemplare mit ihren Daten und Methoden und nur rudimentäre Klasseninformation

5.   Zusammenfassung und Ausblick

Die objektorientierten Techniken ermöglichen eine den realen Strukturen angepaßte Automatisierungswelt. Ein Umdenken zwischen den bisher spezifischen Prozeß- und Automatisierungsstrukturen entfällt durch die wirklichkeitsnahen Modellierungsmöglichkeiten für Strukturen und Verhalten der Systeme.

Mit der Einführung der objektorientierten Strukturen schliessen wir die Lücke in einigen von den Normengremien bisher nicht behandelten Themenfeldern. Wir sind überzeugt, daß besonders unsere im Bereich des Datenmanagements verteilter Systeme liegenden Aktivitäten richtungsweisendend sind und damit standardisierenden Charakter für zukünftige Entwicklungen haben.

Wir geben durch die Ausführung von exponierten Teilen des Systemkerns in der neuen Technik auch unserer Überzeugung Ausdruck, daß wir die objektorientierten Techniken heute für ausgereift halten, den hohen Qualitäts- und Stabilitätsanforderungon eines vielfach verwendeten Systemkerns zu genügen.

Anhang   Literatur

[BB88]   Brandt, S.; Baldauf, F.: Leitlinien und Grundstrukturen der Automatisierungstechnik (Teil 1) In: K.-P. Fähnrich (Hrsg.) KOMMTECH 88, Band V, Seite 18.1.01-18.1.18. OnLine GmbH, 1988.

[BB89]   Brandt, S.; Baldauf, F.: Leitlinien und Grundstrukturen der Automatisierungstechnik (Teil 2) In: G.Schmidt/H.Steusloff (Hrsg.) INTERKAMA-Kongreß 89, Seite 303-312. Oldenbourg, 1989.

[BBK88]   Batz, T.; Baumann, B.; Köhler, D.: A Data Model Supporting System Engineering. In: Proceedings of the twelfth Annual International Computer Software & Application Conference COMPSAC '88. IEEE, 1988.

[BS90]   Batz, T.; Steusloff, H. (Hrsg.: PROSYT (Fachlicher Endbericht des Projektes PROSYT), in Vorbereitung. Springer-Verlag, Berlin/Heidelberg/New York/Tokio, 1990.

[GOP90]   Gorlen, K.E.; Orlow, S.M.; Plexico, P.S.: Data Abstraction and Object-Oriented Programming in C++. John Weley & Son, Chichester etc., 1990.

[KSS89]   Krömker, D.; Steusloff, H.; Subel, H.-P. (Hrsg.): PRODAT und PRODIA, Dialog-und Datenbankschnittstellen für Systementwurfswerkzeuge, Beiträge zur graphischen Datenverarbeitung, Springer-Verlag, Berlin/Heidelberg/New York/Tokio, 1989.

[L+85]   P.C. Lockemann Anforderungen technischer Anwendungen an Datenbanksysteme. In: Blaser, a., Pistor (Hrsg.), P., Proceedings der Tagung Datenbank-Systeme für Büro, Technik und Wissenschaft, Informatik Fachberichte, Band 94, Seite 1-26, Karlsruhe, 1985. Springer-Verlag, Berlin/Heidelberg/New York/Tokio.

[Lip89]   Lippman, S. B.: A C++ Primer. Addison Wesley, Reading etc., 1989.

[Stö88]   Stöckler, H.-P.: Fortschrittliche Strukturen von Leitsystemen. NAMUR-Statusbericht 88. Oldenbourg, 1988

[Str87]   Stroustrup, B.: The C++ Programming Language. Addison Wesley, Reading etc., 1987.

Anwendungen

COSMOS-2D -
Ein System zur vollautomatischen, optischen und geometrie-invarianten Vermessung von ebenen Strukturoberflächen

Bernhard Bürg, Helmut Guth, Andreas Hellmann
Kernforschungszentrum Karlsruhe GmbH, Institut für Datenverarbeitung in der Technik
Postfach 3640, D-7500 Karlsruhe

1 Einleitung

Die Qualitätskontrolle gewinnt in der Industrie zunehmend an Bedeutung. Ständig steigende Anforderungen an die Güte von Produkten zwingen zu umfangreichen Kontrollen während der Fertigung. Die technische Komplexität der herzustellenden Artikel bedingt immer aufwendigere und kostenintensivere Produktionsverfahren.

Eine Prüfung der Endprodukte verhindert allerdings nur die Freigabe eines fehlerhaften Artikels für den Markt und nicht auch unnötige Kosten nicht korrekter Produktionsprozesse. Um Ort, Zeitpunkt und Art von Fehlern exakt zu ermitteln, werden immer mehr auch die Zwischenprodukte überprüft (Abb. 1). Dabei genügt nur selten noch die Information, daß Fehler vorhanden sind. Vielmehr wird die Ursache von Fehlern gesucht, um den Produktionsprozeß korrigieren zu können.

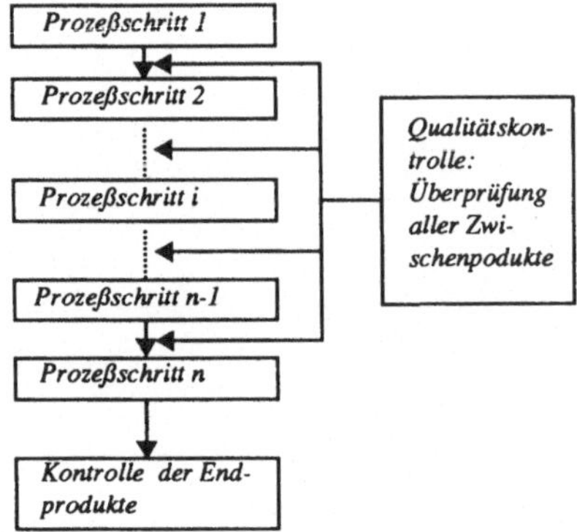

Bild 1: *Bei steigender Verfahrenskomplexität genügt eine Kontrolle der Endprodukte nicht mehr, denn jeder fehlerhafte Prozeßschritt i vererbt die Fehler an seinen Nachfolger i+1. Dadurch entstehen Kosten, die durch die Aufsummierung der Prozeßschrittkosten i bis n anfallen. Im Idealfall entstehen aber nur jeweils die "unvermeidbaren" Kosten des fehlerhaften Prozeßschrittes i. Dazu ist allerdings eine Kontrolle aller Zwischenprodukte ohne deren Zerstörung oder unzulässiger Veränderung erforderlich.*

Die Ergebnisse einer zeit- und kostenintensiven Sichtprüfung, die in der industriellen Qualitätskontrolle noch häufig angetroffen wird, sind oft subjektiv und für den Prüfer ermüdend. Zudem liefern sie nicht immer die gewünschte detaillierte Information über Maßhaltigkeit des Prüflings, da der Umfang der Messungen die Personalkapazität oft bei weitem übersteigt.

Diese Problematik zeigt sich insbesondere in der Mikrofertigung. Auf kleinstem Fertigungsnutzen werden tausende von Strukturen integriert. Die hohen Kosten der Produktionsprozesse zwingen hier in besonderem Maß zu ausgedehnten Zwischenprüfungen. Die Maßhaltigkeit so vieler Strukturen läßt sich jedoch manuell nicht mehr ermitteln.

Zu diesem Zweck wurde das Strukturvermessungssystem COSMOS-2D (Computer Based System for Measurement of Optical Structures in 2-Dimensional Space) entwickelt, das auf der Grundlage der Digitalen Bildanalyse vollautomatisch, geometrie-invariant und schnell vor allem mechanische Mikrostrukturen zwei-dimensional vermißt. Dieses System wird im vorliegenden Beitrag beschrieben.

2 Aufgabenstellung

Das Anwendungsspektrum des entwickelten Strukturvermessungssystems ist vielfältig und beschränkt sich nicht nur auf Messungen im mikroskopischen Bereich. Die Anforderungen gehen allerdings von einem Produktionsverfahren, dem sog. **LIGA**-Verfahren (**Lithographie, Galvanik, Abformung**), für meist mechanische Mikrostrukturen aus, das im Rahmen des Arbeitsschwerpunktes Mikrotechnik im Kernforschungszentrum Karlsruhe (KfK) entwickelt wurde. Die Anforderungen an das System sollen deshalb auf der Grundlage dieses Produktionsprozesses, der im folgenden kurz beschrieben wird, zusammengestellt werden.

Bild 2: Wabenförmige LIGA-Struktur, die beispielsweise als Filter eingesetzt werden kann (Stegbreite 8 µm). Zum Test von COSMOS-2D wurden definierte Fehler eingearbeitet.

Mit dem **LIGA**-Verfahren können Strukturen im Mikrometerbereich mit einem Aspektverhältnis bis zu 100 und Abweichungen in der Strukturwanddicke im Submikrometerbereich hergestellt werden. Das Aspektverhältnis ist das Verhältnis zwischen Strukturwandhöhe und Strukturwandbreite. Dabei sind beliebige Geometrien (mit konstanter Strukturhöhe) realisierbar [Becker85]. Beschleunigungssensoren, optische Multiplexer, hochpräzise Düsen, Mikrostecker oder Filter (Abb. 2) sind eine kleine Auswahl möglicher Kandidaten für die Fertigung mit dem LIGA-Verfahren.

Der Herstellungsprozeß beginnt mit dem Entwurf der Struktur in einem CAD-System. Die CAD-Daten dienen der Steuerung eines Elektronenstrahlschreibers, mit dem eine primäre Maske hergestellt wird. Über die aus der primären Maske gefertigte Röntgenmaske wird mit Synchrotronstrahlung ein Resist bestrahlt, dessen nichtresistente Teile weggeätzt werden. Durch verschiedene Galvanik- und Abformungsschritte entsteht ein "Werkzeug", mit dem durch weitere Abformung die Endprodukte in Serie hergestellt werden.

Sowohl alle Zwischenprodukte als auch die Endprodukte müssen einer zerstörungsfreien Vermessung der Strukturoberflächen unterzogen werden. Das Strukturvermessungssystem soll **beliebige geometrische Formen** zwei-dimensional vermessen können, um eine aussagekräftige Beschreibung der tatsächlich vorliegenden Strukturform zu geben.

Langfristiges Ziel der Fertigung von Mikrostrukturen mit dem LIGA-Verfahren ist die Serienherstellung in einer Produktionsstraße. Der Meßvorgang muß deshalb ab der Justierung des Substrats unter dem Mikroskop über die Vermessung der Strukturen bis zur Auswertung der ermittelten Formparameter **vollautomatisch** ablaufen, und darf dabei eine bestimmte Taktzeit des Prozesses nicht überschreiten.

Das **unterschiedliche Reflexionsverhalten** der durch die Vielfalt der Prozeßschritte bedingten Materialien darf die Meßergebnisse nicht verfälschen.

Weiterhin soll die Möglichkeit bestehen, sich ohne langwieriges Einlernen unter Berücksichtigung des vollautomatischen Prozeßablaufs auf wenige **kritische**, zu vermessende **Bereiche** zu beschränken.

Schließlich sollen die Systemkosten niedrig gehalten werden, möglicherweise durch den Verzicht auf einen hochpräzisen Tisch, ohne an Meßgenauigkeit zu verlieren.

3 Systemkonzept

Die Forderung nach dem vollautomatischen Ablauf des Vermessungsvorgangs ist eine Randbedingung, die das Einlernen von Referenzstrukturen und kritischen Bereichen verbietet. Da die CAD-Daten ohnehin vorhanden sind, soll das CAD-System die Information zur Steuerung des Vermessungsystems liefern.

Ausgangspunkt hierzu ist die Überlegung, daß jede Struktur durch ihre Randlinien, die Strukturkanten, beschrieben wird. Die Strukturgeometrie wird durch einen geschlossenen Kantenzug dargestellt, der sich in sog. Primitiven, nämlich Strecken und Kreisbögen zerlegen läßt. Durch die Definition von Ecken als ein weiterer Primitiven-Typ läßt sich eine hierarchische Form der Beschreibung einer beliebigen Struktur-Geometrie herleiten. Dabei bilden die Ecken als Kombination zweier Geraden oder einer Geraden und eines Kreisbogens oder zweier Kreisbögen die unterste Ebene der geometrischen Struktur-Beschreibung. Jeweils zwei Ecken, deren Ausprägung durch Attribute beschrieben wird, begrenzen einen Kreisbogen oder eine Gerade. Diese Linien stellen die zweite Beschreibungsebene dar. Auf der dritten Ebene schließlich erfolgt durch Kombination der auf der zweiten Ebene dargestellten Information die Beschreibung der kompletten Strukturgeometrie oder Teilen davon. Auf diese Weise ist sowohl die vollständige und exakte Darstellung als auch die rechnerinterne Handhabbarkeit jeder beliebigen Strukturgeometrie gewährleistet.

Der Primitiven-Satz läßt sich aufgrund des modularen Aufbaus der Beschreibungshierarchie und deren Analogon in der Befehlsstruktur nahezu beliebig um spezielle, komplizierte Geometrien erweitern. Obwohl sich z.B. eine spiralenförmige Linie durch eine Folge von Kreisbögen kontinuierlich ändernder Radien gut beschreiben ließe, ist es sinnvoll - weil in der Verifikation weniger zeitaufwendig - einen neuen Primitiven-Typ "Spirale" einzuführen.

Im CAD-System werden auf einer zusätzlichen Informationsebene (Layer) Hinweise auf kritische Bereiche und Dimensionen eingetragen, die örtlich mit den Geometrie-Daten in der Original-Layer korrespondieren (Abb. 3). Dem Entwurfsingenieur bietet sich damit schon beim Design der Struktur die Möglichkeit, Kontroll- und Meßwünsche, individuell auf jede Struktur abgestimmt, zu formulieren, die Strukturen in beliebige Bildabschnitte zu zerlegen und für jeden Bildabschnitt durch die Wahl seiner Größe, Auflösung und Genauigkeit frei zu bestimmen (Bürg89).

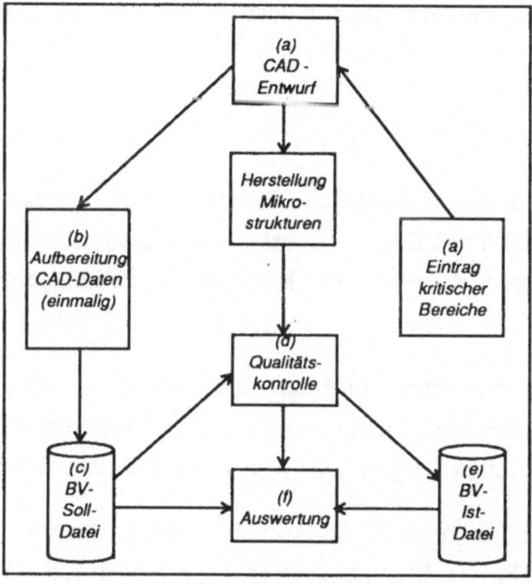

__Bild 3__: Konzept der Strukturkontrolle
- Eintrag kritischer Bereiche im CAD-System
  beim Strukturentwurf (a)
- Verknüpfung der Daten kritischer Bereiche mit
  den Original-CAD-Daten der Struktur (b)
- Formulierung von Bildverarbeitungs-(BV)-
  Befehlen aus den Verknüpfungsdaten (b)
- Abspeichern der BV-Befehle in einer BV-Soll-
  Datei (c)
- Verifikation und Vermessung der Strukturen
  durch Abarbeitung der BV-Befehle durch das
  BV-System (d)
- Abspeichern der Ist-Daten in einer BV-Ist-Da-
  tei (e)
- Auswertung der Messung durch Vergleich von
  Soll- und Ist-Datei (f)

Aus den korrespondierenden Informationen beider "Layer" des CAD-Systems werden in einer Datenaufbereitungsphase Bildverarbeitungs (BV) - und Steuer-Befehle formuliert und in eine Datei (Soll-Datei) geschrieben, die die Grundlage für die Steuerung des kompletten Meßvorgangs bildet. Die so aufgebaute Daten- und Informationsbasis zwischen CAD- und BV-System garantiert die einfache Erweiterbarkeit des Befehlssatzes.

Das Kontrollsystem fährt die in der Soll-Datei beschriebenen Bildabschnitte auf dem Substrat an, zieht ein Bild mit einer Video-Kamera ein und arbeitet sukzessiv die BV-Befehle ab, indem die Struktur-Primitiven in diesem Bild verifiziert und die Distanzen gemessen werden. Die Ergebnisdaten werden in einer zweiten Datei (Ist-Datei) gespeichert. Wie in der Soll-Datei die Soll-Geometrie der zu vermessenden Strukturen abgespeichert ist, so beschreibt die Ist-Datei jetzt die Ist-Geometrie der kontrollierten Strukturen. Aus einem Vergleich der Soll- mit der Ist-Datei lassen sich nicht nur Position, Art und Größe von Fehlern ermitteln, sondern auch eine mathematisch exakte Beschreibung der Ist-Geometrie.

Durch die Unabhängigkeit der Auswertung der Bildabschnitte untereinander lassen sich Strukturen **und** Strukturteile an beliebiger Position und ohne direkte Verbindung in Form von Kantenzügen zum Substratkoordinatensystem ausmessen. In einem frei wählbaren Rahmen können durch die Art der Detektion und Mustererkennung auch ungenau positionierte Strukturen vermessen werden.

Die notwendige Auflösung und Genauigkeit erfordern die Verarbeitung von Grauwertbildern der Größe 512 * 512 Pixeln mit 256 Graustufen pro Pixel.

4 Aufbau des Systems

4.1 Funktion

Für den Aufbau des Vermessungssystems COSMOS-2D wurde eine Multi-Prozessor-Architektur gewählt, bei der sich die Systemleistung auf einfache Weise durch Zuschaltung weiterer Prozessoren steigern läßt.

Die Aufgaben werden funktional in zwei Gebiete aufgeteilt, in die Verwaltung, die vom sog. Masterrechner übernommen wird, und in die Auswertung von Strukturbildern, die den übrigen Prozessoren (Slaves) zugeteilt wird.

4.2 Hardware

Für die Hardware wurden aus Kostengründen nur auf dem Markt erhältliche Standard-Komponenten eingesetzt. Basis des Systems ist der VME-Bus. Einschließlich des Bedien-Terminals und des Farbmonitors für die Bildwiedergabe finden alle Hardware-Komponenten in einem 19" Rack mit 33 Höheneinheiten (HE) Platz.

Als Master-Rechner, der gleichzeitig als Basis-CPU, Steuerrechner und Bus-Controller dient, wird der Prozessor Motorola 68020 mit 25 MHz verwendet. Neben dem mathematischen Co-Prozessor 68881 weist er 4 MB lokales, dynamisches RAM auf. Der CPU direkt zugeteilt ist eine 120 MB-Platte (Bild 4).

Bei dem Betriebssystem handelt es sich um das Real-Zeit-System OS-9. Trotz des Fehlens "harter Echtzeitbedingungen" im System COSMOS-2D hat sich der Einsatz dieses Betriebssystems bewährt, da es einfach handhabbar ist, wenig Speicher-Platz benötigt und sich auch zur Software-Entwicklung eignet.

Für die Kommunikation zur Außenwelt ist ein Ethernet-LAN-Controller vorhanden, der mittels OS-9-Internet das TCP/IP-Protokoll beherrscht.

Der Bildeinzug erfolgt mit einer Videokarte. Das Video-Signal einer SW-Kamera wird hier in ein Bildraster von 512 x 512-Pixel mit jeweils 256 Graustufen digitalisiert. Gleichzeitig enthält die Karte einen Bildspeicher für die Aufnahme von zwei 512x512-Bildern. Aus dem digitalisierten Bild kann mittels einer speziellen BV-Karte für die Bildvorverarbeitung [Massen88] in Video-Norm ein Linienbild mit 1-Pixel breiten Linien berechnet werden..

An die Digitalisierungskarte ist eine CCD-Kamera mit Sony-Sensor 2/3" und 756x581 Bildpunkten angeschlossen.

Das Multi-Prozessor-Konzept der Slaves ist mit Transputern realisiert. Dabei werden T800 von Inmos mit 25 MHz und jeweils 4MB dynamischem RAM eingesetzt. Durch die neu auf den Markt gekommenen 4 MBit-Chips lassen sich insgesamt 8 Prozessoren dieser Ausrüstung in einem einzigen VME-Bus-Slot unterbringen.

Die Transputer sind in Form eines trinären Baumes verschaltet, das bedeutet, daß von den jeweils 4 vorhandenen Kommunikationsleitungen (Links) einer mit dem übergeordneten Transputer (Vater-Transputer) verbunden ist, während mit den restlichen Links die Verbindungen zu drei anderen Transputern (Söhne) hergestellt werden. Ein Link des Root-Transputers ist physikalisch mit dem VME-Bus verbunden. Er läßt-

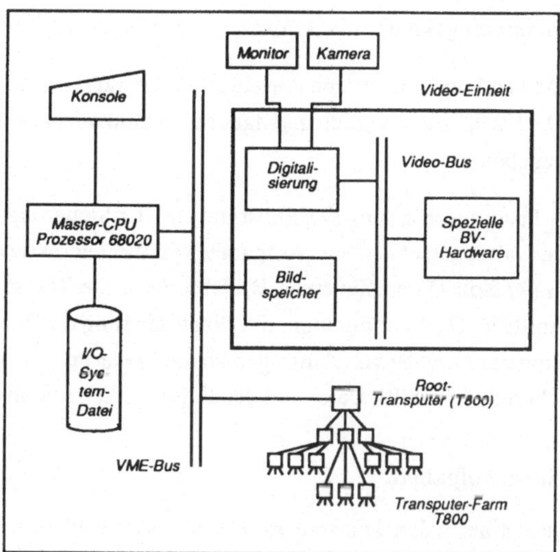

Bild 4: Die COSMOS-2D-Hardware ist in Form einer Multi-Prozessor-Architektur konzipiert. Die Master-CPU steuert die Auftragsverwaltung. Jeder Transputer der Transputer-Farm übernimmt die vollständige Abarbeitung eines Bildes. Die Digitalisierungseinheit ist um eine spezielle Bildverarbeitungs-Hardware erweitert, die in Real-Zeit die Bildvorverarbeitung durchführt.

zudem den direkten Zugriff auf seinen Speicher vom VME-Bus zu (dual-ported-RAM).

Die Transputer enthalten kein übliches Betriebssystem. Sie werden zu Beginn wellenartig mit identischen Programmen geladen. Der Lader erkennt automatisch die Konfiguration. Das Hinzuschalten weiterer Transputer erfordert keinerlei Software-Änderung. Auf jedem Transputer läuft zusätzlich zur Anwender-Software ein sog. Routing-Programm, das den Transport von Aufträgen und Ergebnissen steuert.

Der Routing-Algorithmus ist in Form eines "Farm"-Konzeptes realisiert: Der jeweilige "Knoten"-Transputer schickt einen Auftrag in den Ast des trinären Baumes, aus dem die Bereitschaft eines Transputers für die nächste Abarbeitung gemeldet worden ist.

4.3 Software

Für die Implementierung der Software wurde wegen der leichten Portabilität, der Möglichkeit des strukturierten Programmierens und der optimalen Übersetzung des Codes in Maschinenbefehle die Programmiersprache **C** gewählt. Die Software läßt sich einteilen in Programme, die auf dem Masterrechner ablaufen und Programme, die identisch von jedem Slave ausgeführt werden.

4.3.1 Master-Aufgaben

Zu Beginn einer Substratvermessung muß zunächst die Position des Substrats auf dem Mikroskoptisch festgestellt werden. Dazu werden automatisch die Schwerpunkte von drei Markierungskreuzen, die auf dem Substrat an vordefiniertem Ort aufgebracht sind und das Substratkoordinatensystem beschreiben, ermittelt und daraus eine dreidimensionale Transformationsmatrix berechnet, mit der sämtliche Substratkoordinaten in Tischkoordinaten übergeführt werden können.

Um den vollautomatischen Betrieb des Systems zu gewährleisten wurde für die Ermittlung der z-Koordinate der Markierungskreuze ein softwaregesteuerter Autofokus entwickelt.

Auf dem Mikroskoptisch ist eine Anschlagskante montiert, so daß eine Positionierung des Substrats manuell auf 0,1 bis 0,2 mm möglich und damit ein automatisches "Einfangen" der Markierungskreuze mit der Kamera gegeben ist.

Neben der Positionserfassung des Substrats und der Steuerung des Autofokus übernimmt der Masterrechner alle verwaltungstechnischen Aufgaben wie die interaktive Bestimmung der Prozeßablaufparameter, das Lesen der Soll-Daten für einen Bildabschnitt, die Transformation der Mittelpunktskoordinaten des Bildabschnitts in Tischkoordinaten, die Positionierung des Mikroskoptisches (x,y,z), den Bildeinzug, die Organisation der Vergabe von Aufträgen an die Slave-Prozessoren, die Rücknahme der Ergebnisdaten und ihre Speicherung in der Ist-Datei und das Führen einer Bildabschnitts- und einer Substrat-Statistik.

4.3.2 Slave-Aufgaben

Die Aufgabe eines jeden Transputers besteht in der Bildvorverarbeitung, der Strukturverifikation und der Vermessung der Strukturen eines kompletten Grauwert-Bildes der Größe 512 x 512 Bildpunkte (Pixel).

In der Bildvorverarbeitung werden mit einem Sobel-Operator die Gradientengröße und -richtung berechnet. Dadurch entstehen zwei neue Bilder, ein Bild **A**, in dem die Grauwert-Kanten des Originalbildes dargestellt sind, und ein Bild **B**, in dem zu jedem Pixel von **A** die berechnete Gradienten-Richtung in Grad gespeichert ist. Aus den Bildern **A** und **B** werden mit dem Verfahren nach Korn [Korn88] direkt 1-Pixel breite Linien ermittelt und in Bild **A** abgespeichert. Sie beschreiben nun die Strukturkanten des Originalbildes.

Das Linien-Bild **A** und das Gradienten-Richtungs-Bild **B** sind Eingabe für die Mustererkennung. Der dazu gehörende Soll-Daten-Abschnitt enthält alle Bildverarbeitungsbefehle, die für die Verifikation der Strukturkanten notwendig sind. Die Positionsdaten sind Koordinaten bzgl. des Substratkoordinatensystems und werden mit der Transformationsmatrix in Tischkoordinaten übergeführt. Nach Umrechnung in Bildkoordinaten werden alle Pixel-Linien, die sich in einer frei wählbaren Umgebung der vermuteten Position des gesuchten Strukturmerkmals, z. B. einer Ecke, in Bild **A** befinden, auf Pixel-Listen abgebildet, wobei nicht nur für jedes Pixel die Bildkoordinaten und die Gradientenrichtung aus **B** und deren 1. und 2. Ableitung gespeichert werden, sondern auch die globale Information über die Anzahl von geschlossenen und nicht geschlossenen Linien in dem betrachteten Fenster.

Der gesuchte Eckpunkt ist das Pixel, dessen beidseitige **k** Umgebungspixel in der Summe die geringsten Abweichungen zwischen ihren Richtungen und den berechneten Richtungen der korrespondierenden Pixel der gesuchten Ecke aufweisen, wobei **k** von der Glattheit der Kanten und damit vom Substratmaterial und der gewünschten Auflösung nebeneinanderliegender Strukturecken abhängt. Diese Methode hat den Vorteil, daß zum einen in jedem Fall eine mögliche Eck-Position gefunden wird, falls überhaupt Linien im Fenster vorhanden sind, und zum anderen vorhandene Ecken unabhängig davon, wie spitz oder abgerundet sie sind, identifiziert werden, wobei die berechnete minimale Differenzensumme ein Maß für die "Güte" der Eck-Form ist. So lassen sich Ist-Ecken über einen Schwellwert, der von dem Soll-Öffnungswinkel der Ecke abhängig ist, verwerfen.

Auf die gleiche Weise werden alle Eck-Verifikationsbefehle abgearbeitet und man erhält eine Ist-Ecken-

Liste. In einer globalen Geometriebetrachtung erfolgt eine exakte Zuordnung der Ist-Ecken zu den Soll-Ecken. Eine Überarbeitung der Zuordnung ist deshalb notwendig, weil in einem Fenster durchaus mehrere Ecken gleicher Soll-Parameter auftreten können und die beschriebene Suchmethodik positionsunabhängig greift.

Die weitere Mustererkennung setzt nun auf der Kenntnis der exakten Position der Ist-Ecken auf. Für die Verifikation von Linien wird im Bild ein entsprechendes Fenster ausgehend von den Linienendpunkten (=Ist-Eckpunkte) berechnet, in dem eine Pixel-Liste nach dem oben beschriebenen Schema erstellt wird. Darin wird die vorgeschriebene Geradheit oder Krümmung der Linie nachvollzogen, Ortsdifferenzen in Form von Standardabweichungen berechnet und mögliche Unterbrechungen registriert.

Die Vollständigkeit einer Struktur wird nachgewiesen durch die erfolgreiche Verifikation aller erforderlichen Soll-Linien. Abstände werden nun entsprechend den Bildverarbeitungsbefehlen zwischen detektierten Ecken oder aber zwischen beliebigen Kanten an den vorgegebenen Stellen gemessen. Neben Vollständigkeits- und Distanzprüfungen werden auch noch Flächenprüfungen und exakte Kreisberechnungen mit Mittelpunktsbestimmung und Ermittlung der Standardabweichung vom Radius durchgeführt.

Der Befehlssatz ist modular aufgebaut und kann unter Beachtung der hierarchischen Beschreibung der Struktur beliebig erweitert werden. Zur Beschleunigung der Verifikation von z.B. spiralenförmigen Strukturen wurden ein neuer Primitiven-Typ "Spirale" und ein neuer BV-Befehl "Verifiziere eine spiralenförmige Linie." eingeführt. Gleiches ließe sich auch bei anderen, mathematisch beschreibbaren Kurvenverläufen wie z.B. Hyperbeln oder Parabeln verwirklichen.

Die ermittelten Ergebnisse werden in einem Ist-Daten-Abschnitt, der vom Aufbau her dem Soll-Daten-Abschnitt ähnlich ist, gespeichert und an den Master-Rechner zurückübergeben. Dort fließen die wichtigsten Ergebnis-Daten in eine Gesamtstatistik, mit deren Hilfe sich der Anwender nach dem Ende des Meßvorgangs auf einen Blick informieren kann, ob das ganze Substrat in Ordnung ist, oder ob einzelne Abschnitte Fehler aufweisen. Über fehlerhafte Abschnitte wird eine Einzelstatistik geführt, die es erlaubt, das entsprechende fehlerhafte Bild sofort unter dem Mikroskop zu positionieren.

Eine detailliertere Auswertung erfolgt durch spezielle Auswerteprogramme, die die gemessenen Werte aus der Ist-Datei mit den Soll-Werten aus der Soll-Datei vergleichen.

5 Erfahrungen mit dem System und Ergebnisse

Langfristiges Ziel war die Entwicklung einer Zielmaschine, die ein Bild durchschnittlich in 1 Sekunde vollständig vermißt. Der Zeitbedarf für die Mustererkennung und Berechnung der Geometrie hängt natürlich direkt von der Komplexität der Strukuren und vom Umfang des Vermessungsauftrags ab. Es ist deshalb sinnvoll, eine obere Zeitbedarfsgrenze anhand eines "Worst-Case-Bildes" abzuschätzen. Dazu wurde ein Substrat gewählt, das auf einem Nutzen von 2 x 5 cm etwa 600.000 Strukturen in Wabenform aufweist. Davon sollten ca. 36.000 Waben vermessen werden. Unter Beachtung einer vernünftigen Auflösung ergaben sich etwa 1000 Bilder mit je 36 Waben. Die Strukturverifikation beinhaltete damit die Detektion von 216 Ecken je Bild, also insgesamt ca. 220.000 Ecken, ebenfalls je Bild die Berechnung von 216 Geradengleichungen sowie Standardabweichung und Maximalabweichung der Geraden als ihr Gütemaß. Der Vermessungsauftrag sah weiterhin die Berechnung von mehr als 55.000 Distanzen für das ganze Substrat vor, also etwa 55 je Bild.

Mit dem Einsatz der Spezial-Hardware für die Bildvorverarbeitung und 15 Transputern erreicht das System für solche "Worst-Case-Bilder" im Durchsatz einen Zeitbedarf von ca. 1 Sekunde. Eine Ecken- und eine Geradenverifikation erfolgt dabei durchschnittlich in etwa 5 Millisekunden. Die Ermittlung einer Distanz beschränkt sich bei bekannten Endpunktpositionen auf die Berechnung von Differenzen und läuft im Mikrosekundenbereich ab. Bei entsprechender Anforderung kann durch Verdoppelung der Transputerzahl die System-Performance bis zum zeitlichen "Nadelöhr" des Systems, die Auftragsverwaltung, von etwa 0.5 Sekunden pro Bild im Durchsatz, gesteigert werden.

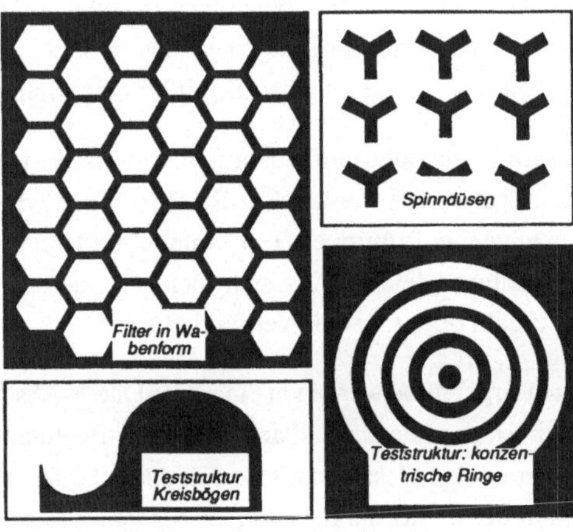

Abb. 5: Strukturen, die mit dem LIGA-Verfahren hergestellt wurden: Filter in Wabenform und Spinndüsen. Zum Test der Bildverarbeitungsroutinen wurden die Kreisbogen- und die Ring-Teststruktur gefertigt. Die Kreisbogen-Teststruktur weist in der oberen Hälfte vier Viertel-Kreisbögen in Folge auf, wobei der Radius jeweils von links nach rechts konstant zunimmt. Die Drehrichtung ändert sich dabei nach dem zweiten Kreisbogen. Die konzentrischen Ringe dienen dem Test von BV-Software für die eindeutige Verifikation von Kreisen bestimmter Radien in einer Szene mit vielen Kreisen ähnlicher oder auch stärker abweichender Radien.

In weiteren Testläufen mit speziell für die Bildverarbeitung entwickelten Strukturen, die entsprechende Schwierigkeitsgrade aufweisen (Abb. 5), konnte gezeigt werden, daß das System beliebige (zwei-dimensionale) Strukturgeometrien sowohl in Form von Positiv- als auch als Negativ-Strukturen unter Verwendung der am Prozeß beteiligten Materialien vermessen kann. Die Reproduzierbarkeit liegt bei gleicher Tischposition bei +- 1 Pixel, das bedeutet bei einer 50-fachen Vergrößerung (kleinstes verwendetes Objektiv) +- 1.1 μm, bei einer 1000-fachen Vergrößerung (größtes verwendetes Objektiv) +- 50 nm (rein rechnerisch).

6 Abschließende Betrachtung und Ausblick

Das System COSMOS-2D erfüllt alle gestellten Anforderungen: In einem vollautomatischen Prozeß vermißt es beliebige Mikro-Strukturen eines Substrats unabhängig von ihren geometrischen Formen und ihrer

Lage auf dem Substrat. Optimale Strukturen und kritische Bereiche werden schon beim CAD-Entwurf definiert. Dadurch entfällt ein langwieriges Suchen und Einlernen von Referenzstrukturen. Das System liefert eine detaillierte Beschreibung von Strukturfehlern, die eine abgestufte Beurteilung der Qualität des Produkts zuläßt. Trotz Verwendung eines kostengünstigen Tisches wird durch den Einsatz von Mustererkennungsmethoden eine hohe Präzision der Meßergebnisse erreicht.

COSMOS-2D wurde als Strukturvermessungssystem im Bereich der Mikrofertigung konzipiert und entwickelt. Dennoch ließe sich durch Austausch oder Modifikation der Systemkomponenten "Bildgeber = Kamera mit Mikroskop-Optik" und "Substratträger = Mikroskop-Tisch" ein Einsatz im makroskopischen Bereich vorstellen. Eine Anwendung ist überall dort denkbar, wo viele Strukturen mit komplexer Geometrie schnell und präzise zu vermessen sind und wo eine CAD-ähnliche Konstruktionsbeschreibung der Strukturgeometrie vorliegt.

7   Literatur

[Becker85]   E.W. Becker, W. Ehrfeld, P. Hagmann, A. Maner, D. Münchmeyer: *Herstellung von Mikrostrukturen mit einem großen Aspektverhältnis ...*, KfK-Bericht 3995, Kernforschungszentrum Karlsruhe 1985.

[Bürg89]   B. Bürg, H. Guth, A. Hellmann: *Bildanalytische Qualitätskontrolle in der Mikrofertigung*, Informatik-Fachberichte Nr. 219, S. 168 - 172, Springer-Verlag, 1989.

[Korn88]   A. Korn: *Toward a Symbolic Representation of Intensity Changes in Images*, IEEE Transactions on Pattern Analysis and Machine Intelligence, Vol. 10, No. 5, Sept. 1988.

[Massen88]   R. Massen: *Real-Time grey-level and colour image pre-processing for a vision guided biotechnology robot*, Proc. Int. Conf. Robot Vision and Sensory Controls, 115-122, IFS (Conferences) Ltd and authors, ISBN 0-948507-780, Feb. 1988.

INTEGRIERTE ANWENDUNG VON BILDVERARBEITUNG UND NEURONALEN NETZEN ZUR ZEICHENERKENNUNG

M. Preisenberger, F. Raupp, S. Schießl, M. Hoffmann
Elektronik-System-Gesellschaft mbH, München

1. EINLEITUNG

Die in den vergangenen Jahren erhebliche Leistungssteigerung in bezug auf Speicherkapazität und Rechengeschwindigkeit, ebenso wie die Einführung neuer Software-Technologien etwa aus der Künstlichen Intelligenz, haben für den Computer neue Anwendungen erschlossen. Ein Gebiet, in dem sowohl die verbesserten Möglichkeiten bei der Hardware als auch bei der Software wesentlich zum breiteren Einsatz führten, ist die digitale Bildverarbeitung. War bisher ihre Anwendung z. B. im industriellen Bereich auf einfache Szenen in einer abgeschlossenen, voll kontrollierbaren Umgebung mit einer meist geringen Anzahl von zu erkennenden Objekten beschränkt, so rückt heute auch ihr Einsatz für komplexere Aufgaben in weniger kontrollierbarer Umgebung und mit einer nicht von vorne herein vollständig beschriebenen Menge zu erkennender Objekte in den Bereich des Interesses.

2. ANWENDUNG UND PROBLEMSTELLUNG

Eine solche etwas komplexere Anwendung findet sich im Rahmen von Systemen zur Zufahrtskontrolle bei Hof- und Geländeeinfahrten oder an Straßenauffahrten. Die Zufahrt wird hierbei entweder durch eine Wachperson mittels einer Kennplakette oder dem Kfz-Kennzeichen oder z.B. durch Schranken und Magnetkartensysteme kontrolliert. Diese Überprüfung kann von einem Bildverarbeitungssystem, das Kfz-Kennzeichen lesen kann, auch automatisch durchgeführt werden.

Ein Bildverarbeitungssystem, das dies leistet, muß folgende Teilaufgaben lösen:

- Auffinden des Kfz-Kennzeichens im Bild
- Segmentierung der alphanumerischen Zeichen des Kfz-Kennzeichens
- Lesen der relevanten Information

Bei den für diese Anwendung vorliegenden, nicht vollständig kontrollierbaren Umgebungsbedingungen ergeben sich typisch die folgenden Probleme:

- Die Position und Winkellage des Schildes ist nicht genau definiert.
- Der Kontrast und Verschmutzung des Schildes können stark variieren.
- Die Anzahl der auf dem Schild befindlichen Zeichen ist unterschiedlich.
- Die verwendeten Zeichensätze sind nicht einheitlich.
- Der Aufbau der Kennzeichen ist unterschiedlich.

3.   LEISTUNGSANFORDERUNGEN UND ZIELSETZUNG

Zusätzlich zur Lösung der unter 2. aufgeführten Probleme gibt es für ein in der Praxis kostengünstiges, einsetzbares System weitere Anforderungen und Randbedingungen. Dies betrifft insbesondere:

- Einfache Anpaßbarkeit an wechselnde Beobachtungsgeometrie,
- Einfache Erweiterbarkeit bezüglich verschiedener Kfz-Kennzeichen-Typen,
- Hohe Zeichenklassifikationssicherheit (ca. 99 %),
- Kurze Auswertezeiten (ca. 3 sec pro Bild und Kennzeichen).

Um zu einer kostengünstigen Lösung zu kommen und um aufwendige Hardware-Entwicklungen zu vermeiden, sind verfügbare Hardware-Standardkomponenten einzusetzen.

Zielsetzung der im folgenden näher beschriebenen Realisierung des entsprechenden Bildverarbeitungssystems war der Nachweis der Erreichbarkeit der geforderten Leistungen im Laborversuch mit deutschen Kfz-Kennzeichen, wobei zwei Zeichensätze zugelassen waren.

4.   REALISIERUNG

4.1   Laborsystem

Der Versuchsaufbau besteht im wesentlichen aus einer Schwarz-/Weiß-CCD-Videokamera, einem Video-Monitor und einem Rechnersystem.
Bei der Konfiguration des Rechensystems waren die bereits in 3. erwähnten Anforderungen in bezug auf Rechenzeit, Verfügbarkeit und

Kosten zu berücksichtigen. Dies führte zu einer Rechneranordnung bestehend aus:

- IBM-kompatiblen PC mit Intel 80368 und 25 Mhz Taktfrequenz,
- Prozessorkarte mit Array-Prozessor speziell für Zeichenklassifikation unter Verwendung Neuronaler Netze (Netz-Prozessor als PC-Zusatzkarte),
- Bildverarbeitungssystem mit speziellen Karten für A/D-Wandlung, Bildspeicherung, nichtlineare Filter, Histogrammberechnung und Grauwertanalyse.

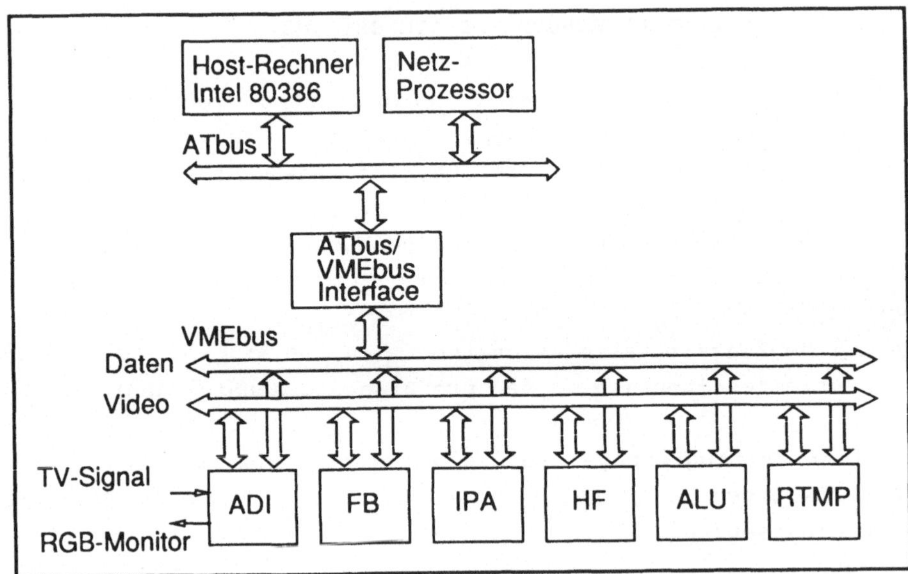

Abb. 1: Rechnerkonfiguration

ADI	Analog/Digital-Interface	HF	Histogramm-Prozessor
FB	Bildspeicher	ALU	Arithmetik-Prozessor
IPA	Array-Prozessor	RTMP	Filter-Prozessor

Abbildung 1 zeigt das Blockschaltbild der Rechnerkonfiguration. In dieser Konfiguration übernimmt der Intel-Prozessor die Host-Rechnerfunktion. Er steuert die Bildverarbeitungsprozessoren und den Netz-Prozessor.

Die auf diesem System implementierte Software besteht aus drei Komponenten:

- Hauptmodul zur Steuerung des Datenflusses und zur Integration von Teilergebnissen,
- Bildverarbeitungsmodul zur Ansteuerung der Bildverarbeitungsprozessoren,
- Klassifikationsmodul zur Zeichenklassifikation auf dem Netz-Prozessor.

Nach dem Start befindet sich das System in einem Wartezustand, in dem fortlaufend das von der Kamera kommende Videosignal digitalisiert wird, bis durch einen externen Impuls (z. B. Bedienereingabe, Induktionsschleife) das Videobild im Speicher des Bildverarbeitungssystems abgelegt wird. Das Hauptprogramm stößt nun den Bildverarbeitungsprozeß zur Auswertung des gespeicherten Bildes an. Dieser extrahiert aus dem Bild die Segmente, die als Zeichen des Kfz-Kennzeichens in betracht kommen. Diese Segmente werden in Form einer binären Bildpunktmatrix an den Netz-Prozessor zur Klassifikation weitergereicht.

Aus dem Klassifikationsergebnis für die Segmente und ihrer Position im Bild wird dann wieder vom Hauptprogramm das endgültige Ergebnis, das gelesene Kfz-Kennzeichen generiert.

Für eine einfache Anpassung bzw. Erweiterung des Zeichensatzes für die zu klassifizierenden Zeichen, die etwa dann erforderlich wird, wenn unterschiedliche Perspektiven zu berücksichtigen sind, gibt es neben den oben aufgeführten Moduln eine Trainingskomponente.

4.2   Verfahren

Als Verfahrensansatz wurde eine Kombination aus Standard-Bildverarbeitungs- und Neuronale-Netz-Algorithmen gewählt.

Durch die Vorschaltung eines Bildverarbeitungsprozesses vor die eigentliche Klassifikation mit einem Neuronalen Netz werden zwei Aufgaben gelöst:

- Reduktion der Eingangsdatenmenge
- Skalierung der Zeichen des Kfz-Kennzeichens

Das in den Bildspeicher eingelesene Bild besteht aus 512 x 512 Bildpunkten zu je 8 Bit. Alleine der Transport dieser Datenmenge über den AT-Bus vom Bildspeicher in einen verarbeitenden Prozessor kann zu Übertragungszeiten im Sekundenbereich führen. Damit wird ein Reduktionsprozeß notwendig, der nur Bildsegmente liefert, die einzelne Zeichen enthalten. Nach dieser Segmentierung hat man darüber hinaus die Möglichkeit, die Zeichen zu skalieren und erreicht somit eine Entfernungsinvarianz des Verfahrens innerhalb gewisser Grenzen.

Zur Klassifizierung der von der Bildverarbeitung extrahierten Zeichensegmente werden Neuronale-Netz-Algorithmen verwendet. Obwohl für eine solche Klassifikationsaufgabe normierter Zeichen auch konventionelle Verfahren bekannt sind, wurden Neuronale Netze für den Einsatz ausgewählt. Gründe dafür sind die Flexibilität und Offenheit solcher Systeme, die eine leichte Anpassung an verschiedene Situationen ermöglichen (veränderte Beobachtungsgeometrie, verschiedene Zeichensätze usw.), die leichte Implementierbarkeit, sowie eine hohe Fehlertoleranz, d. h. Sicherheit bei der Klassifikation (z. B. bei ungünstigen Lichtverhältnissen, Verschmutzung usw.). Ein weiterer Aspekt ist die Möglichkeit, Neuronale-Netz-Algorithmen maximal zu parallelisieren und damit bei Verwendung entsprechender Hardware eine erhebliche Einsparung an Rechenzeit zu erzielen.

4.2.1   Bildverarbeitung

Die Bildverarbeitung zur Segmentierung der Buchstaben und Ziffern des Kfz-Kennzeichens besteht aus den folgenden Verarbeitungsschritten:

- Bildvorverarbeitung,
- Histogrammanalyse,
- Konturextraktion,
- Bestimmung des jeweils kleinsten umfassenden Rechtecks,
- Segmentvorauswahl,
- Zeichenskalierung und -extraktion.

Durch die Bildvorverarbeitung werden auf der Basis nichtlinearer Filteroperationen die einzelnen Buchstaben und Ziffern des Kfz-Kennzeichens verstärkt. Die Abbildung 2a zeigt ein Ausgangsbild und 2b das

Abb. 2a: Orginalbild

Abb. 2b: Bild nach Vorverarbeitung

Abb. 2c: Bild mit gefundenen Konturen

Ergebnis nach der Vorverarbeitungsoperation. Im nachfolgenden Schritt wird unter Verwendung des Histogramm-Prozessors ein Grauwert-Histogramm berechnet. Aufgrund der Vorverarbeitung erhält man ein ausgeprägtes bimodales Histogramm.

Zur Trennung des Hintergrundes von den interessierenden Segmenten wird mit einem etwa in [1] beschriebenen Standardverfahren zur Behandlung bimodaler Histogramme ein Schwellwert zwischen den beiden Grauwerthäufungen errechnet. Mit diesem Schwellwert wird ein Binärbild erzeugt, aus dem die Konturen der verbliebenen Segmente in Form eines Konturkettenkodes extrahiert werden. Für die Extraktion der Konturkettenkodes wurde ein Algorithmus nach [2] verwendet.

Für jedes Segment wird aus dem Konturkettenkode das entsprechend kleinste umfassende Rechteck errechnet. Anhand der Höhe und Breite jedes Rechteckes werden in einer Vorauswahl solche Elemente zurückgewiesen, die wegen ihrer Größe keine alphanumerischen Zeichen enthalten können. Die verbleibenden Segmente werden jeweils bezüglich ihrer Höhe skaliert und in binäre Matrizen mit je 14 x 9 Bildpunkten abgebildet. Die Matrizen bilden die Eingabe für die Zeichenklassifikation durch das Neuronale Netz.

4.2.2 Neuronale Netze

Zur Klassifizierung der von der Bildverarbeitung extrahierten Zeichensegmente werden Neuronale-Netz-Algorithmen verwendet. Eine Reihe verschiedener Netztypen wurden auf ihre Eignung für den Einsatz im Erkennungssystem untersucht (ART: Adaptive Resonance Theory, Counterpropagation, Hamming und Backpropagation Netze [3],[4],[5],[6]).

Für das gestellte Problem erwiesen sich Backpropagtion Netze als am besten geeignet. Sie zeigten einerseits eine hohe Klassifikationssicherheit und erfüllen damit die gestellten Anforderungen an die Erkennungswahrscheinlichkeit. Andererseits ist mit diesem Netztyp eine Verarbeitungsgeschwindigkeit zu erreichen, die den harten Zeitrestriktionen des Gesamtsystems genügt.

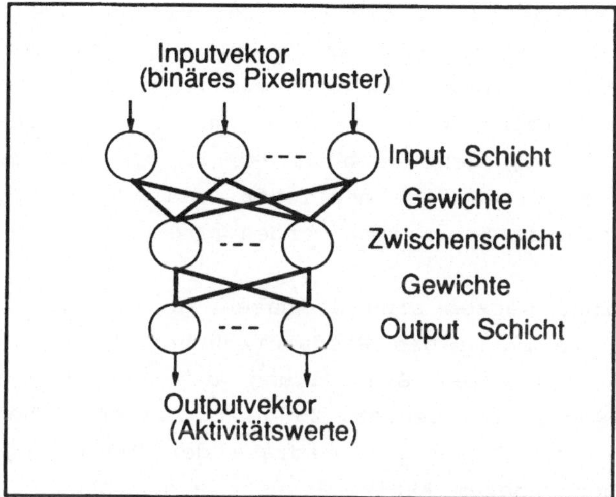

Abb. 3: Schema eines Backpropagation Netzes

Das Schema eines Backpropagation Netzes zeigt Abbildung 3. Der Algorithmus basiert auf dem Konzept eines Perceptrons (in Schichten angeordnete Knoten bzw. Neuronen, [7],[8],[9]) mit Zwischenschichten (Hidden Layers). Die Ausgangswerte der Knoten in einer Schicht werden an Knoten der nächsten Schicht weitergegeben, modifiziert mit Gewichtsfaktoren. Der Eingangswert jedes Knotens ist die Summe der Ausgangswerte der Knoten der vorhergehenden Schicht, die jeweils mit den Gewichten multipliziert wurden. Jeder Knoten wird angeregt durch den Eingangswert und eine Aktivitätsfunktion (in der vorliegenden Arbeit wurde die Sigmoid-Funktion als nichtlineare Anregungsfunktion verwendet).

Bei der Anwendung muß in einem ersten Schritt die Netztopologie festgelegt werden, d.h. die Anzahl und Dimension der Netzschichten. Für die Auslegung der Netztopologie gibt es keinen theoretisch fundierten Ansatz. Beim speziellen Problem ist man deshalb auf Erfahrungswerte angewiesen.

Beim Erkennungssystem für Kfz-Zeichen ist die Dimension der Input- und Outputschicht vorgegeben. Die Eingabeschicht ist durch die Größe der von der Bildverarbeitung gelieferten Pixelmatrix bestimmt. Dies sind für das betrachtete System 14 x 9 = 126 Knoten. Die Ausgabedimension ist gegeben durch die Zahl der zu unterscheidenden Zeichen. Im vorliegenden Fall waren es 38 Kategorien verschiedener Symbole.

Frei zu wählen blieben damit die Zahl der Hidden Layers und deren Dimensionen. Das Auswahlkriterium war hier, die Dimension und damit die Zahl der Verbindungen bzw. Gewichte möglichst klein zu halten. Eine große Zahl von Verbindungen erfordert nicht nur eine höhere Speicherkapazität, sondern geht auch direkt in die Rechengeschwindigkeit ein. Andererseits sollte eine möglichst optimale Performance gemessen an den Leistungsvorgaben erreicht werden.

Beim Einsatz eines Netzes sind 2 Phasen zu unterscheiden: die Trainingsphase und die Phase des Wiedererkennens (Recall). In der Lernphase wird zur optimalen Einstellung der Gewichte die "verallgemeinerte Delta Regel" angewendet. Zu jedem Musterzeichen wird das Fehlerquadrat, d.h. die Summe der Quadrate der Differenzen zwischen gewünschtem und berechnetem Aktivitätswert eines Knotens, bestimmt. Die Lernprozedur besteht darin, die Gewichte so zu verändern, daß die Fehler minimiert werden. Bei der Fehlerminimierung geht man von der Outputschicht rückwärts schichtweise bis zur Inputschicht vor (Backpropagation of error) [6].

In der Recallphase wird das Netz in Vorwärtsrichtung durchlaufen (Feedforward). Dabei werden die Gewichte aus der Lernphase zur Berechnung der Ausgangsaktivitäten verwendet. Sie garantieren eine optimale Abbildung der Zeichen auf einen zugeordneten Outputvektor.

Das Vorgehen beim Erkennen von Kfz-Kennzeichen ist in Abbildung 4 dargestellt. Der Prozeß des Trainings ist entkoppelt von der eigentlichen Zeichenzuordnung. Die beim Durchlaufen der Lernzyklen auf eine minimale Fehlerfunktion eingestellten Gewichtsfaktoren werden abgespeichert und beim Recallprozeß wieder geladen.

Beim Kfz-Lesesystem erfüllten Netze mit 2 Hidden Layers mit jeweils etwa 50 Knoten, d.h. etwa 10^4 Verbindungen, die beschriebenen Anforderungen.

Zur Beschleunigung des Klassifikationsprozesses wurde ein spezieller Array-Prozessor verwendet, auf dem die Recall Prozeduren ablaufen. Die Zeichenerkennung verläuft wie in Abbildung 4 dargestellt. Über eine Schnittstelle wird die von der Bildverarbeitung erzeugte binäre Pixelmatrix eines Zeichens auf den Netz-Prozessor geladen. Dieser berechnet den Ausgangsvektor des Netzes, aus dem die Zuordnung des entsprechenden Zeichens ermittelt wird. Diese Information wird an den Host-Rechner zurückgegeben.

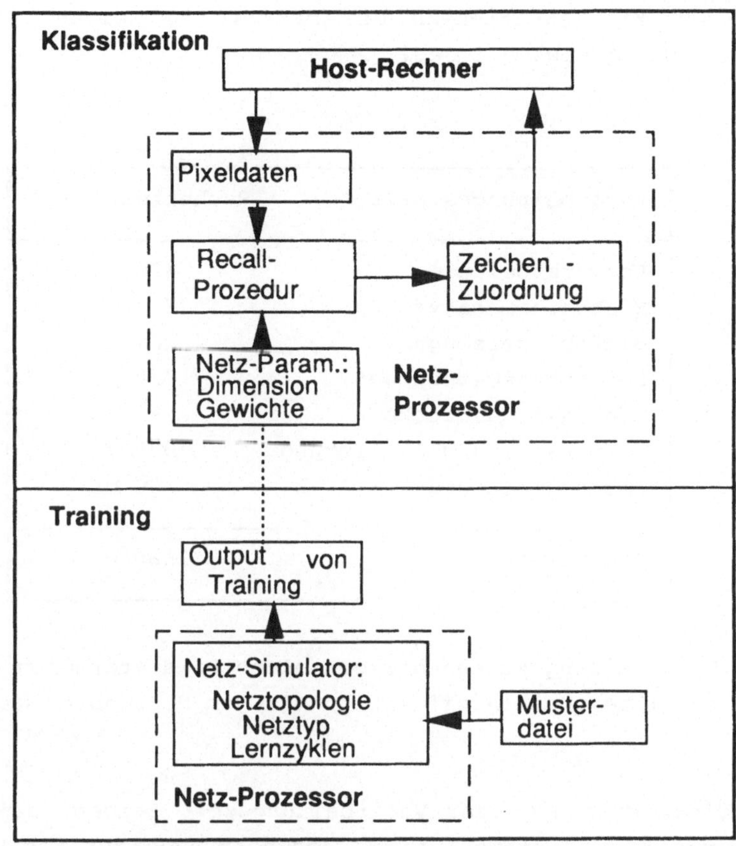

Abb. 4: Abblauf bei Klassifikation und Training

5.   Ergebnisse

5.1   Rechenzeiten

Ein wichtiger Gesichtspunkt bei der Systemauslegung war die Erzielung kurzer Verarbeitungszeiten. In Tabelle 1 sind die Verarbeitungszeiten für verschiedene Schritte der Bildverarbeitung und die Klassifikation mit dem Neuronalen Netz zusammengestellt.

Die Rechenzeiten für die Klassifikation durch das Neuronale Netz hängen von der Netztopologie und der Zahl der zu erkennenden Zeichen ab. Für ein Netz mit der oben beschriebenen Auslegung (4 Schichten: 126,50,50,38 Knoten pro Schicht) betrug die Zeit zur Wiedererkennung eines einzelnen Zeichens etwa 10 msec. Für die Klassifizierung aller

Zeichen eines Kfz-Kennzeichens mit maximal 10 Zeichen benötigt die Netzprozedur damit etwa 0,1 sec.

Verarbeitungsschritt	Rechenzeit '386
Vorverarbeitung	820 ms
Histogrammanalys	220 ms
Binärbilderzeugung	50 ms
Kettenkodeberechnung	720 ms
Minimale Rechtecke	50 ms
Vorauswahl und Skalierung	900 ms
Klassifizierung	100 ms
Gesamtzeit	2860 ms

Tab. 1:   Verarbeitungszeiten der einzelnen Bildverarbeitungs-
schritte und Klassifikation mit dem Neuronalen Netz

Die bisherigen Messungen der Verarbeitungzeit zeigen, daß eine Reaktionszeit von 3 sec erreicht wird. Abschätzungen zur Reduktion dieser Zeit durch weitere Soft- und Hardware-Optimierungen, z.B. durch Verwendung dem Konzept entsprechenden und verfügbaren Prozessorkarten, ergaben, daß eine Reaktionszeit von 2 sec erreichbar ist.

5.2   Klassifikationsgüte

Die mit 50 bis 200 Musterzeichen trainierten Netze wurden auf ihre Zuverlässigkeit bezüglich der Klassifikation der Zeichen untersucht. Zur Simulation von Verschmutzungen der Zeichen, Unschärfen durch schlechte Beleuchtungsverhältnisse und kleinen Änderungen der Perspektive wurden die Musterzeichen mit einem Zufallsverfahren verrauscht. Eine Reihe unterschiedlich stark modifizierter Zeichen wurden auf ihre korrekte Klassifikation getestet. Die Erfolgsquote mit dem oben beschriebenen Netz betrug dabei 99,5 %.

Untersucht wurden ebenfalls die Ergebnisse der Klassifikation bei Drehungen des Kfz-Schildes bzw. der Zeichen in der Bildebene. Das Netz war dabei nur mit ungedrehten Zeichen trainiert. Einige Ergebnisse für 15 Zeichen sind in der Tabelle 2 unten zusammengestellt. Bis zu Drehungen um etwa 3° – 4° werden die Zeichen in der Regel korrekt erkannt. Bei größeren Winkeln treten Fehler bei ähnlichen Zeichen auf. So wird bei den in der Tabelle aufgeführten Zeichen das R als P interpretiert. Verwechslungen treten auch auf zwischen U und V sowie S und 5.

Drehung	M	N	O	P	R	S	T	U	V	W	X	5	6	7	8
7,5°					x		x			x	x	x		x	
6°					x			x			x	x			
4,5°					x						x				
3°					x										
1,5°															
0°															
-1,5°															
-3°			x												
-4,5°					x			x							
-6°					x					x					
-7,5°								x							

Tab. 2: Klassifikationsergebnis in Abhängigkeit vom Drehwinkel
Die mit x gekennzeichneten Zeichen wurden nicht korrekt klassifziert.

Bei der Zeichenzuordnung sollen von der Bildverarbeitung fälschlich als Zeichen extrahierte Muster, wie z.B.Schmutzflecken u.ä., erkannt und ausgesondert werden. Dies konnte durch eine Analyse der Aktivitätswerte der Neuronen der Outputschicht erreicht werden. Es zeigte sich, daß Muster, die stark von den gelernten Zeichen abweichen, eine größere Fluktuation der Aktivitätswerte des Ausgabevektors um den Mittelwert aufweisen. Die Einführung eines Grenzwertes für die Varianz der Outputaktivitäten erlaubte so - bei geeigneter Wahl des Limits - eine einwandfreie Abtrennung von Zufallsmustern.

5.3   Anpaßbarkeit

Bei stark von der ursprünglichen Perspektive abweichenden Kamerawinkeln können die Zeichen nicht mehr eindeutig klassifiziert werden. Bei einer wesentlichen Änderung der Bildperspektive ist es darum notwendig, das Netz mit den Zeichen in der neuen Anordnung zu trainieren. Da die Dimension der Pixelmatrix und der Umfang des Zeichensatzes fix bleiben, kann auch die Netztopologie erhalten bleiben. Das unverändert dimensionierte Netz wird dann mit neuen Musterzeichen, die aus der neuen Bildperspektive aufgenommen wurden, trainiert. Die Dauer der Lernphase beläuft sich je nach Dimension des Netzes auf maximal einige Stunden. Andere Zeichensätze sind ebenso einfach durch das Training mit den entsprechenden Zeichenmustern implementierbar.

6.   ZUSAMMENFASSUNG

Die erarbeitete hybride Systemlösung - als Kombination von angepaßten Bildverarbeitungsalgorithmen und Neuronalem Netz - erfüllt die an sie gestellten Anforderungen und bestätigt die in sie gesetzten Erwartungen.

Insbesondere durch den Einsatz eines Neuronalen Netzes erzielt das System

- große Verarbeitungsgeschwindigkeit,
- hohe Klassifikationswahrscheinlichkeit,
- Robustheit gegenüber Störungen,
- leichte Anpaß- und Erweiterbarkeit,

ohne dabei auf problemspezifische Hardwarelösungen angewiesen zu sein.

Damit wird die Möglichkeit eines breiten Einsatzes in Verkehrsüberwachungs- und leitsystemen ermöglicht, ohne daß organisatorisch komplizierte oder kostenintensive Modifikationen an Fahrzeugen notwendig werden.

Literatur

[1]   Haberäcker, P. "Digitale Bildverarbeitung"
        Hanser Verlag, München (1985).

[2]   Pavlides, T. "Graphics and Image Processing".
        Springer Verlag, Berlin (1982).

[3]   Carpenter,G.A. und Grossberg,S. "Absolutely Stable
        Learning of Recognition Codes by a Self-Organizing Neural
        Pattern Machine". Computer Vision, Graphics, and Image
        Processing (1986).

[4]   Hecht-Nielson,R. "Counterpropagation Networks"
        Proceedings of the International Conference on Neural
        Networks". San Diego, (1987).

[5]   Hopfield,J.J. "Neural Networks and Physical Systems
        with Emergent Collective and Computational Abilities".
        Proceedings of the National Academy of Science", Vol. 79
        (1982).

[6]   Rumelhart,D.E., Hinton,G.E. und Williams,R.J. "Parallel
        Distributed Processing: Explorations in the Micro-
        structures of Cognition. Vol. 1: Foundations". MIT Press,
        Cambridge, MA (1986).

[7]   Minsky,M. und Papert,S. "Perceptron: An Introduction to
        Computational Geometry". MIT Press, Cambride, MA. (1969).

[8]   Nilsson,N.J. "Learning Machines: Foundations of Trainable
        Pattern Classifying Systems". McGraw Hill, New York (1965)

[9]   Rosenblatt,F. "Principles of Neurodynamics: Perceptrons
        and the Theory of Brain Mechanisms". Spartan, New York
        (1962).

EIN WISSENSBASIERTES FEHLERFRÜHDIAGNOSESYSTEM FÜR WERKZEUGMASCHINEN AUF DER BASIS EINES PC MIT SIGNALPROZESSORKARTE

Autoren: D. Neumann, R. Isermann

Technische Hochschule Darmstadt, Inst. f. Regelungstechnik, Landgraf-Georg-Str. 4, D-6100 Darmstadt

Kurzfassung:

In dieser Arbeit wird ein wissensbasiertes Fehlerdiagnosesystem vorgestellt, dessen Hardware aus einer Kombination von einer Floating-Point Signalprozessorkarte und einem handelsüblichen IBM-kompatiblen PC-AT besteht. Hierzu wird ein Fehlerdiagnose - Software - System entwickelt, welches mit mathematischen Prozeßmodellen und Signalmodellen arbeitet. Praktische Ergebnisse werden für eine Werkzeugmaschine gezeigt.

1. EINLEITUNG

Im Zuge fortschreitender Automatisierung fertigungstechnischer Prozesse kommt einer Verbesserung der Überwachung und Fehlerdiagnose von Werkzeugmaschinen eine besondere Bedeutung zu. Durch den Einsatz von geeigneten Überwachungssystemen ist zu erwarten, daß eine entscheidende Steigerung von Zuverlässigkeit, Verfügbarkeit und Sicherheit erzielt werden kann.

Zur Zeit beschränken sich die Verfahren der Prozeßüberwachung meist auf Grenzwertabfragen einiger meßbarer Signale. Auf der anderen Seite zeigen einige Arbeiten, daß Expertensysteme für den Bereich der Fertigungstechnik entwickelt werden /1/, /2/, /3/. Diese arbeiteten jedoch alle ohne direkte Prozeßkopplung im Benutzerdialog. Weitere Möglichkeiten ergeben sich durch Ankopplung von wissensbasierten Systemen an den zu überwachenden Prozeß, um auch direkt Meßsignale in die Überwachung einbeziehen zu können /4/, /5/.

Im Bereich der Signalauswertung kommen mehrere Methoden in Betracht. Hierbei sind besonders die Spektrum- und Cepstrumanalyse zu nennen. Die in den letzten Jahren entwickelten Methoden der modellgestützten Fehlerdiagnose erlauben eine tiefere Einsicht in den zu überwachenden Prozeß. Je nach zugrundeliegendem Prozeßmodell unterscheidet man Diagnosemethoden auf der Basis von *Parameterschätzung von dynamischen kontinuierlichen Prozeßmodellen*, wo sich Fehler als Änderungen physikalischer Parameter interpretieren lassen /6/, /7/ und *Methoden der Zustandsbeobachtung*, bei der Fehler sich in Änderungen der Signale bzw. Zustandsgrößen ausdrücken /8/.

Bei der Überwachung von Werkzeugmaschinen mit diesen modellgestützten Methoden fallen sowohl große Datenströme als auch rechenzeitintensive signalanalytische Auswertungen an. Aus diesem Grunde werden prozeßnah leistungsstarke Schnittstellenrechner benötigt, welche über ein entsprehendes Netzwerk in der Lage sind mit übergeordneten Leitrechnern zu kommunizieren (CIM).

2. ON-LINE FEHLERDIAGNOSE AN WERKZEUGMASCHINEN

Das im folgenden beschriebene Überwachungskonzept läßt sich in vier verschiedene Aufgaben untergliedern (vgl. Bild 1). Die einzelnen Komponenten und deren Zusammenwirken werden nachfolgend näher erläutert.

<table>
<tr><td>

Meßwerterfassung
- 4 A/D-Kanäle
- 6 Kanäle 16 Bit direkt digital I/O

</td></tr>
<tr><td>

Signalanalyse
- Filterung und mathem. Operationen
- Spektralanalyse und Korrelation
- Parameterschätzung

</td></tr>
<tr><td>

Klassifikation
- Bayes - Test
- Vergleich mit Referenzzustand

</td></tr>
<tr><td>

wissensbasierte Diagnose
- Suche in hierarchischem Fehlerbaum
- Hinzunahme statistischer Information

</td></tr>
</table>

Bild 1: Aufgabenbereiche des wissensbasierten Fehlerdiagnoseprogrammes

2.1 MESSWERTERFASSUNG

Die verwendete Signalprozessorkarte DSP32C wurde mit einer A/D-Wandlerplatine aufgerüstet, welche 4 Analogkanäle und eine digitale Schnittstelle zur Verfügung stellt. Damit ist es möglich analog anfallende Sensorsignale wie Strom, Beschleunigung oder Kräfte und auch digitalisierte Größen der inkrementalen Weg- oder Drehgeber oder Informationen der numerischen Steuerung gleichzeitig zu erfassen und in den Rechner einzulesen. Die in 16 Bit Integer Format vorliegenden Werte werden unter Berücksichtigung eines eventuellen Sensor-Offsets in ihre entsprechende physikalische Größe umgerechnet und im RAM oder auf Festplatte abgespeichert.

2.2 SIGNALANALYSE

Die in den Meßwerten enthaltene Information wird mit verschiedenen mathematischen Verfahren verdichtet um hieraus charakteristische und überwachbare Kenngrößen zu gewinnen. Als mögliche Routinen stehen zur Verfügung:

- Integration, Differentiation und Trenddetektion
- digitale Filterung (Butterworth) und Effektivwertbildung
- Auto- und Kreuzkorrelation
- Spektralanalyse (FFT und parametrische Spektralschätzung)
- Parameterschätzung mit kontinuierlichen dynamischen Prozeßmodellen
- Mustererkennung

2.3 KLASSIFIKATION

Hier können Abweichungen der in der Signalanalyse berechneten charakteristischen Kenngrößen von Referenz- oder Nominalwerten aus dem störungsfreien Betrieb erkannt werden. Die dazu verwendeten statistischen Verfahren leiten sich vom Bayes-Test /ll/ ab. Bei Erkennung von Änderungen werden der entsprechenden Kenngröße zugeordnete Symptome aktiviert, deren Auswertung im wissensbasierten Diagnosesystem erfolgt.

2.4 WISSENSBASIERTES DIAGNOSESYSTEM

Nach Auswertung aller Kenngrößen und Aktivierung der entsprechenden Symptome entsteht eine Symptomkombination, welche dem momentanen Fehlerzustand der Maschine zugeordnet ist. Die Verknüpfung der Symptome zu einer Symptomkombination geschieht mit Hilfe eines zweistufigen hierarchischen Fehlerbaumes (Bild 2). Am Ende jeder Kante des Fehlerbaumes steht ein Zeiger, der auf eine entsprechende Fehlermeldung deutet. Durch Hinzunahme von statistischem Erfahrungswissen kann die getroffene Diagnoseentscheidung zusätzlich bewertet werden.

Das eigentliche Expertensystem zur Fehlersuche besteht aus den Elementen *Wissensbasis, Problemlösung und Wissensakquisition.* Der genaue Aufbau und das Zusammenwirken dieser Komponenten zur Fehlersuche findet man in /9/, /10/.

Die Wissensbasis enthält zwei Formen von Information:

1) *analytisches Wissen*

- analytische Prozeßmodelle aus der theoretischen Modellbildung
- Berechnungs- und Schätzverfahren zur Bestimmung der Kenngrößen
- Normales Prozeßverhalten
- Prozeßgeschichte und Fehlerstatistik, soweit quantifizierbar

2) *heuristisches Wissen*

- Auf Erfahrungswissen des Wartungsfachpersonals basierende Zusammenhänge zwi-

schen Symptomen und Fehlern in Form von Fehlerbäumen
- Prozeßgeschichte und Fehlerstatistik, falls nur qualitativ bekannt

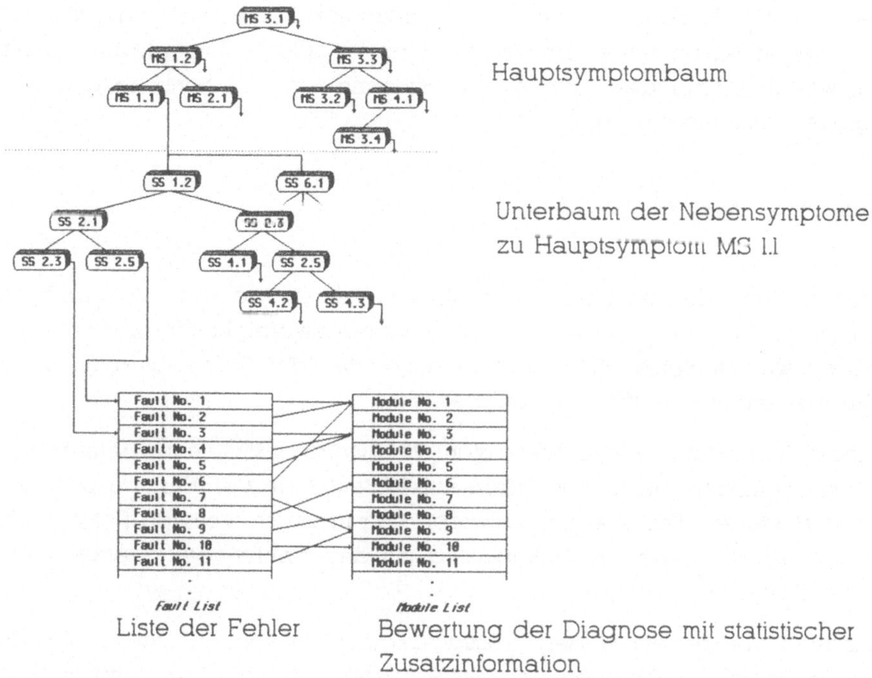

Bild 2: Beispiel für einen hierarchischer Fehlerbaum (MS - Hauptsymptom, SS - Neben-
symptom)

Analytisches Wissen gewinnt man aus einer theoretischen Modellbildung auf der Basis
physikalischer Gesetzmäßigkeiten, welche in praktischen Untersuchungen verifiziert
wurden. Es stellt sich in Form von prozeßbeschreibenden Differentialgleichungen,
parametrischen Signal- oder Spektralmodellen, etc. dar.

Heuristisches Wissen ergibt sich aus dem Erfahrungsschatz des Bedienungs- und War-
tungspersonals und läßt sich in Form von WENN ... DANN - Beziehungen einbringen.
Diese ergeben entsprechend umgeformt die auf UND/ODER - Verknüpfungen aufge-
bauten Fehlerbäume zur Symptomkombinationssuche.

Das hier vorgestellte Diagnosesystem bedient sich eines zweistufigen, hierarchischen
Fehlerbaumes (Bild 3). In der ersten Stufe werden speziell ausgewählte Hauptsympto-
me gesucht, welche charakteristisch für eine Fehlerklasse oder ein Maschinenteilsystem
sind. Bei Auffinden eines solchen Hauptsymptomes verzweigt sich das System in einen
zugeordneten Unterbaum um das Vorhandensein aller zur fehlerbestimmenden Symptomkombination gehörenden Symptome abzuprüfen. Dieser Vorgang geschieht im
"Depth-First-Search" auf der Basis des "modus ponens". Der dazugehörige Fehlerbaum

wurde in Form von doppeltverketteten Listen abgelegt, um ein einfaches Einfügen von Listenelementen und ein Rückwärtssuchen bei negativen Entscheidungen zu ermöglichen.

Zur Bewertung der Diagnose werden die sogenannten Vertrauenswürdigkeiten s_{ji} aller an einer Symptomkombination beteiligten Einzelsymptome multiplikativ miteinander verknüpft, was innerhalb der Zuverlässigkeitstheorie einer Reihenschaltung (UND-Verknüpfung) der Elemente entspricht:

$$f_i = \prod_{j=1}^{n} s_{ji} , \quad 0 \le f_i, s_{ji} \le 1 \tag{1}$$

Der daraus entstehende Fehlerindex f_i ergibt ein Maß für die Eintretenswahrscheinlichkeit des Fehlers Nummer i. In den Symptomvertrauenswürdigkeiten s_{ji} sind statistische Einflußgrößen wie Meßfehler, Rechenungenauigkeiten der Signalauswertung, etc. enthalten. Sie werden im Klassifikatormodul bestimmt.

Die Diagnose kann durch Hinzunahme von statistischer Information (Betriebserfahrung, A-priori Ausfallwahrscheinlichkeiten durch Ausfallstatistik, letzter Wartungszeitpunkt, etc.) weiter erhärtet werden. Diese Angaben lassen sich in einem weiteren Faktor v_i berücksichtigen, dessen multiplikative Verknüpfung mit dem Fehlerindex f_i eine weitere Bewertung der Diagnose zur Folge hat.

Zur Gewinnung von neuem Wissen (*Wissensakquisition*) können sowohl der Benutzer als auch die Betriebserfahrungen des Systems selbst dienen. Der Benutzer kann entweder als "Supervisor" Veränderungen an der Wissensbasis oder dem Regelwerk vornehmen oder als reiner "User" zur Eingabe von Beobachtungen aufgefordert werden. Eine selbstlernende Komponente ergibt sich durch interne Protokollierung aller aufgetretenen Fehler und Einarbeitung in Ausfallwahrscheinlichkeitsdichten für alle Maschinenkomponenten.

3. IMPLEMENTATION AUF VORLIEGENDES RECHNERKONZEPT

Die obengenannte Struktur läßt sich in der folgenden Form auf den Diagnoserechner übertragen (vgl. Bild 3).

Der Signalprozessor besitzt vier 40 Bit breite Akkumulatoren und unterstützt mit seiner Multiplizierer-/Addiererkaskade für Floating-Point Operationen genau die für arithmetische und signalanalytische Berechnungen nötige Struktur. Weiterhin ist eine sehr schnelle Kommunikation mit Peripherie (A/D-Wandler, digitale Schnittstelle) über memory-mapped I/O und Host-Rechner durch einfache Befehle möglich. Es lassen sich Datentransferraten bis zu 20 MByte/s erreichen. Somit war es sinnvoll, die Aufgaben der Meßwerterfassung und Signalanalyse im Signalprozessor anzusiedeln. Die Programmierung erfolgt in einem der Sprache C sehr ähnelndem Assemblercode, desweiteren existieren Treiberbibliotheken insbesondere zur Kommunikation mit dem Host-AT für die Programmiersprache Turbo-Pascal 5.5. Damit wurde erreicht, daß alle zeitkritischen Routinen laufzeitgünstig auf der Signalprozessorkarte implementiert werden konnten. Im Vergleich mit einem Standard-AT mit einer Taktfrequenz von 12 MHz und numerischem Koprozessor 80287 ergab sich eine Steigerung der Rechengeschwindigkeit für die Signalanalyse um etwa den Faktor 100. Eine Abtastung aller vier analogen und sechs 16 Bit breiter digitaler Kanäle ist damit mit einer Abtastzeit von 50 kHz möglich, welche im wesentlichen durch die eingesetzten Sample-/Hold-Verstärker und A/D-Wandler begrenzt wird.

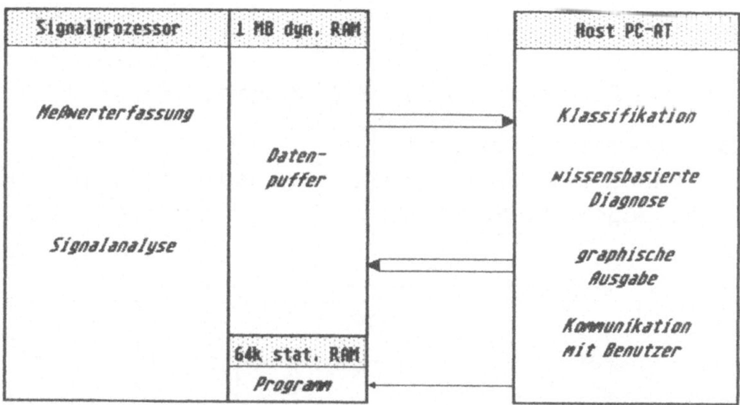

<u>Bild 3:</u> Verteilung der Aufgabenbereiche auf die vorliegende Rechnerstruktur

Die Signalprozessorkarte verfügt über eigene RAM-Bereiche von 64 KByte schnellem statischen und 1 MByte dynamischem RAM. Ein Zugriff darauf kann sowohl von dem Signalprozessor als auch vom PC aus erfolgen. Dementsprechend wird das statische RAM als Programmspeicher für die Signalprozessorkarte, das dynamische als Datenpufferbereich und zur Kommunikation mit dem Host-AT benutzt.

Die auf dem Host-AT parallel zur Datenerfassung und Signalanalyse laufenden Aufgaben der Klassifikation und wissensbasierten Diagnose wurden in der Programmiersprache Turbo-C 2.0 implementiert. Falls der Klassifikator nach Auswertung aller Kenngrössen Symptome in den Symptompufferbereich übertragen hat, wird das Diagnosesystem aufgerufen und es erfolgt eine Fehlersuche. Desweiteren dient der PC zur graphischen Ausgabe auf dem Bildschirm und zur Benutzerkommunikation.

4. EXPERIMENTELLE ERGEBNISSE

Das System wurde an einer Bügelsägemaschine erprobt. In diesem Kapitel werden beispielhaft für die Fehlerdiagnose mittels Spektralanalyse und Parameterschätzung einige Ergebnisse gezeigt. Die an der Maschine angebrachte Sensorik ergibt sich aus Bild 4:

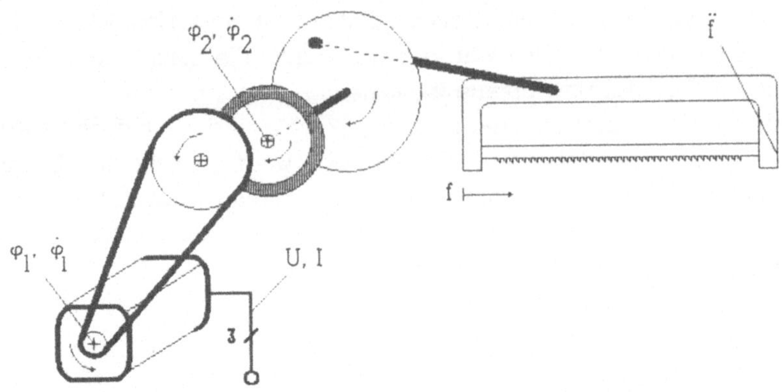

Bild 4: Aufbau und gemessene Signale an der Bügelsägemaschine

Dementsprechend standen als Signale zur Verfügung:
- Der Motorstrom I und Spannung U einer Phase der antreibenden Maschine
- Die Beschleunigung $\ddot{f}$ des Bügels in Schnittrichtung

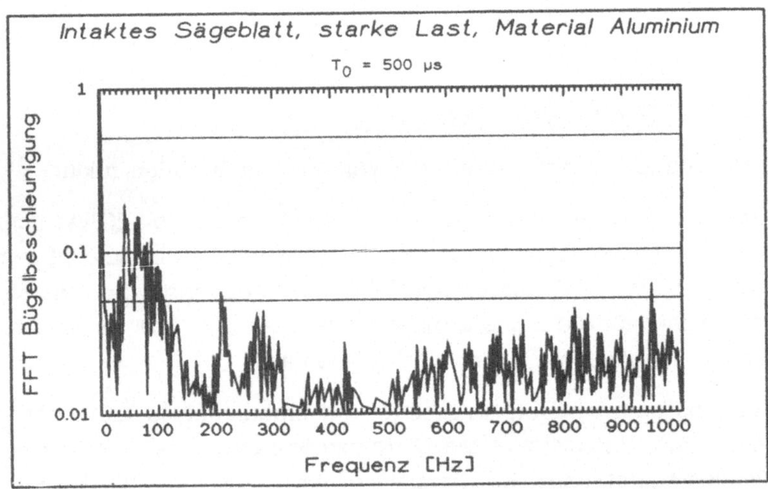

Bild 5a: Frequenzspektrum der Bügelbeschleunigung bei starker Normalkraft und Schnitteingriff, intaktes Sägeblatt

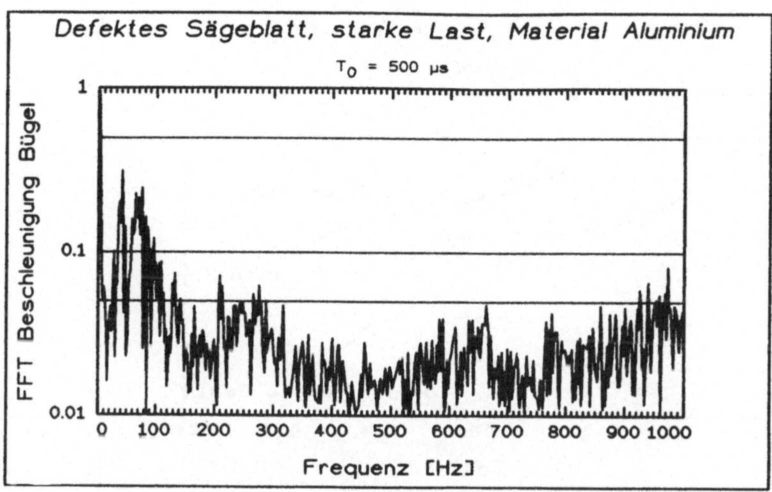

<u>Bild 5b</u>: Frequenzspektrum der Bügelbeschleunigung bei starker Normalkraft und Schnitteingriff, defektes Sägeblatt, grobe Zahnung

Die Drehgeber lieferten direkte digitalisierte Signale für Position φ_1 und Drehgeschwindigkeit $\dot{\varphi}_1$ an der Asynchronmaschine und φ_2 und $\dot{\varphi}_2$ am Kurbeltrieb. Bild 5 zeigt die Spektren der Bügelbeschleunigung für intaktes und defektes Sägeblatt.

Wesentliche Unterschiede in der Form der Spektren erkennt man in den Bereichen bis 100 Hz. Insbesondere eine Komponente bei 40 Hz, die bei leerlaufender Maschine nicht vorhanden ist, zeigt bei verschiedenen Sägeblattzuständen unterschiedliche Amplituden. Es handelt sich um die Zahneingriffsfrequenz, deren Amplitude Auskunft über die Sägeblattbeschaffenheit (Zahnausbrüche, etc.) gibt. Schwingungskomponenten höherer Frequenz stellen Eigenschwingungen der einzelnen Zähne dar und spiegeln sich im Spektrum bei 700 Hz (Bild 5a) bzw. 800 Hz (Bild 5b) wider. Sie sind abhängig vom geometrischen Aufbau des Sägeblattes und ändern sich bei Verschleiß (Zahnkontur).

Zur Beurteilung des Schnittprozesses kann ein dynamisches Schnittkraftmodell verwendet werden. Es zeigt eine PDT2 - Struktur und lautet allgemein im Zeitbereich

$$F_c(t) = -a_1\,\dot{F}_c(t) - a_2\,\ddot{F}_c(t) + b_0\,f(t) + b_1\,\dot{f}(t)$$

für kleine Änderungen um den Betriebspunkt. Hierbei sind $F_c(t)$ die Schnittkraft und $f(t)$ die Bügelposition. Als schnittkraftproportionales Signal wurde die Bügelbeschleunigung gemessen, $f(t)$ kann über eine Kurbeltriebtransformation aus der gemessenen Drehzahl bestimmt werden. Die veränderlichen Ableitungen von $F_c(t)$ und $f(t)$ werden mit Zustandsvariablenfiltern ermittelt /12/. Bei den ZVF handelt es sich um zeitdiskrete Filter, deren Eigenverhalten denen eines Butterworth-Filters entspricht und die das gemessene Ein- und Ausgangssignal des zu identifizierenden Prozesses filtern.

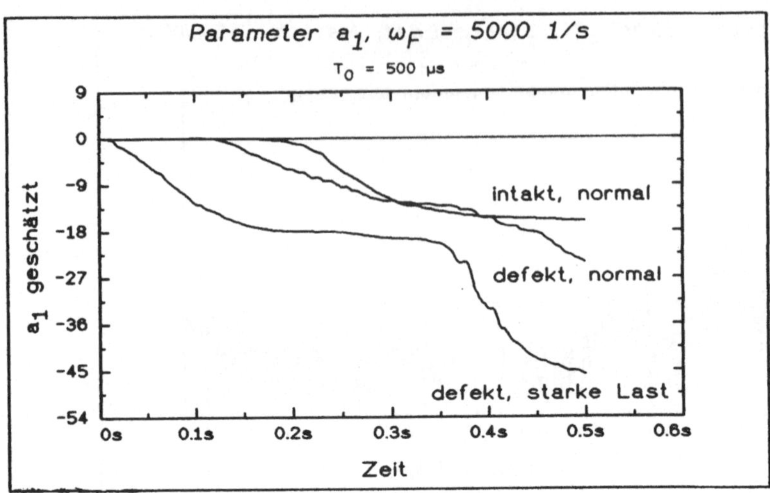

<u>Bild 6:</u> Zeitliche Verläufe des geschätzten Dämpfungsparameters $\hat{a}_1$ des dynamischen Schnittkraftmodelles bei verschiedenen Betriebszuständen

Entsprechend dem dynamischen Schnittkraftmodell (Gl. 2) wurde versucht mit Hilfe von Bügelweg und Bügelbeschleunigung eine Schätzung durchzuführen. Die Ergebnisse sind in Bild 6 beispielhaft für Parameter a_1 wiedergegeben, der verschleißbedingte Änderungen beim Zusammenwirken von Werkzeug und Werkstück zeigt. Es wurden Versuche mit intaktem und defektem Sägeblatt durchgeführt.

Deutlich zeigt sich eine verschleißbedingte Änderung des Dämpfungsparameters a_1. Der steilere Verlauf bei starker Last ist auf die bei größerer Anregung wirkende höhere Signaldynamik in den Ein- und Ausgangssignalen der Schätzung zurückzuführen. Aus diesem Grunde wird hier der Endwert für $\hat{a}_1$ schneller erreicht.

5. SCHLUSSFOLGERUNGEN

Durch eine Kombination von Modellbildung, Spektralanalyse und Parameterschätzverfahren ist es möglich, einen tieferen Einblick in den zu überwachenden Prozeß zu gewinnen. Das vorgestellte System erlaubt es, auftretende Fehler an Werkzeugmaschinen durch Erfassung einfach meßbarer Signale zu erkennen. Eine hardwaremäßige Beschränkung ergibt sich im wesentlichen aus der Rechenleistung des Diagnoserechners, welche aber durch Parallelisierung der Diagnosealgorithmen (Transputersysteme, parallel arbeitende Signalprozessoren) umgangen werden kann. Dadurch wird es möglich, auch komplexere Anlagen mit diesem Diagnosekonzept zu überwachen.

6. LITERATUR

/1/ Harmon, P.; King, D.: *Expertensysteme in der Praxis*, Oldenbourg Verlag München, 1987

/2/ *Wissensbasierte Systeme*, Informatik Fachberichte 155, Springer Verlag Berlin, 1987

/3/ Dal Cin, M.; Phillip, T.: *Expertensysteme für die Fehlerdiagnose*, Informationstechnik, 30. Jahrgang, Heft 4, 1988

/4/ Früchtenicht, H. W.; Wittig, T.: *Ein Ansatz für Echtzeit-Expertensysteme*, Automatisierungstechnische Praxis, 29. Jahrgang, Heft 2, 1987

/5/ Schneider, J.: *Regelbasierte Diagnose zur Verkürzung von Stillstandszeiten bei hochautomatisierten Fertigungseinrichtungen*, Automatisierungstechnik, 38. Jahrgang, Heft 1, 1990

/6/ Isermann, R.: *Process fault detection based on modelling and estimation methods - a survey*, Automatica, Vol. 20, 1984

/7/ Isermann, R.: *Model based fault diagnosis and supervision of machines and drives*, 11th IFAC World Congress on Automatic Control, Tallinn (UdSSR), 1990

/8/ Frank, P. M.: *Evaluation of an analytical redundancy for fault diagnosis in dynamic systems*, IFAC Symposium on Advanced Information Processing in Automatic Control (AIPAC), Nancy (France), 1989

/9/ Isermann, R.: *Wissensbasierte Fehlerdiagnose technischer Prozesse*, Automatisierungstechnik, 36. Jahrgang, Heft 11, 1988

/10/ Neumann, D.; Fuchs, A.: *An on-line knowledge-based fault diagnosis system for machine-tools*, IMEKO Technical Diagnostics '89, Prague (CSFR), 1989

/11/ Goedecke, W.: *Fehlererkennung an einem thermischen Prozeß mit Methoden der Parameterschätzung*, VDI Fortschrittberichte Reihe 8 Nr. 130, VDI Verlag Düsseldorf, 1987

/12/ Young, P.: *Parameter estimation for continuous-time models - a survey*, Automatica, Vol. 17, No. 1, Pergamon Press, 1981

/13/ Freyermuth, B.; Neumann, D.: *Ein Konzept zur Gestaltung wissensbasierter Systeme für die modellgestützte Fehlerfrühdiagnose technischer Prozesse*, VDI/VDE-GMA Automatisierungstechnik, Baden-Baden, 1990

/14/ Neumann, D.; Janik, W.: *Fehlerdiagnose an spanenden Werkzeugmaschinen mit parametrischen Signalmodellen von Spektren*, VDI Schwingungstagung, Mannheim, 1990

PROZESSRECHNERSYSTEM ZUM ADAPTIVEN INNENRUNDSCHLEIFEN

H.K. Tönshoff, A. Walter
Institut für Fertigungstechnik und Spanende Werkzeugmaschinen (IFW)
Universität Hannover

1. Einleitung

Auf Grund ständig steigender Anforderungen an die Qualität technischer Produkte wachsen auch die Ansprüche an die Bearbeitungsgenauigkeit der beim Spanen erzeugten Flächen. Dies trifft besonders auf die Feinbearbeitungsverfahren zu, mit denen die funktionsentscheidenden Oberflächen erzeugt werden. Auch die Herstellung maß- und formgenauer Bohrungen durch Innenrundschleifen bereitet auf Grund der besonderen Merkmale des Verfahrens seit jeher Schwierigkeiten. Zu nennen sind hier die ungünstigen Eingriffsverhältnisse der Schleifscheibe, die Begrenzung des Durchmessers der Schleifscheibe durch den der Bohrung, die geringe Steifigkeit der Maschine sowie die hohen Drehzahlen der Schleifscheibe [1].

Zur Steigerung der Leistungsfähigkeit und der Qualität beim Schleifen wird am Institut für Fertigungstechnik und Spanende Werkzeugmaschinen (IFW) der Universität Hannover die adaptive Prozeßführung eingesetzt, bei der ein Signal aus dem Prozeß (Prozeßkenngröße) abgegriffen und nach einer vorgegebenen Sollwertkurve geregelt wird. Dieses bietet hinsichtlich der Verkürzung der Fertigungszeit, der Verbesserung der Werkstückqualität und der Steigerung der Prozeßsicherheit wesentliche Vorteile gegenüber der konventionellen Prozeßführung mit reiner Vorschubsteuerung [2,3,4].

Obwohl die Vorteile einer adaptiven Prozeßführung schon seit langem bekannt sind [5], existieren nur wenige Steuerungen, die eine solche Prozeßführung realisieren. Aus diesem Grund wurde am IFW ein Prozeßrechnersystem entwickelt, das als Erweiterung herkömmlicher Steuerungen speziell auf die Anforderungen des Innenrundschleifprozesses abgestimmt ist.

Dieses praxisorientierte Prozeßrechnersystem und seine Anwendungsmöglichkeiten sollen in diesem Beitrag vorgestellt werden. Zunächst wird jedoch auf die Möglichkeiten zur Prozeßführung beim Innenrundschleifen eingegangen.

2. Prozeßführung beim Innenrundschleifen

Bislang wird beim Innenrundschleifen fast ausschließlich die konventionelle Prozeßführung eingesetzt. Hierbei erfolgt die radiale Zustellung zwischen Werkzeug und Werkstück mit abschnittsweise konstanter Vorschubgeschwindigkeit. Dementsprechend läßt sich ein Schleifzyklus in die Phasen Eilannäherung, Luftschleifen, Schruppen, Schlichten und Ausfeuern unterteilen.

Während der Eilannäherung wird das Werkzeug mit hoher Geschwindigkeit bis zum größtmöglichen Aufmaß an das Werkstück herangefahren. Hier wird von Eil- auf Schruppzustellung umgeschaltet. Die anschließende Phase des Luftschleifens dauert bis zum tatsächlichen Kontakt zwischen Werkzeug und Werkstück. Diese unproduktive Zeit bis zum Anschnitt hängt vom Aufmaß des Werkstücks ab und ist höchst unerwünscht. Da die Aufmaßschwankungen der Rohteile im Bereich des Aufmaßes selbst liegen können, müssen kostenaufwendige Verfahren wie Vor-sortieren der Werkstücke nach Aufmaß zur Verringerung der Luftschleifzeiten ergriffen werden. Beim anschließenden Schruppen wird der größte Teil des Schleifaufmaßes abgetragen, während die Phasen Schlichten und Ausfeuern der Sicherung der notwendigen Werkstück-qualität dienen.

Mit Hilfe der adaptiven Prozeßführung lassen sich die einzelnen Phasen des Schleifprozesses unter verschiedenen Gesichtspunkten zum Teil erheblich optimieren. Ehe auf die einzelnen Vorteile näher eingegangen wird, soll zunächst der für das adaptive Schleifen erforderliche Regelkreis vorgestellt werden.

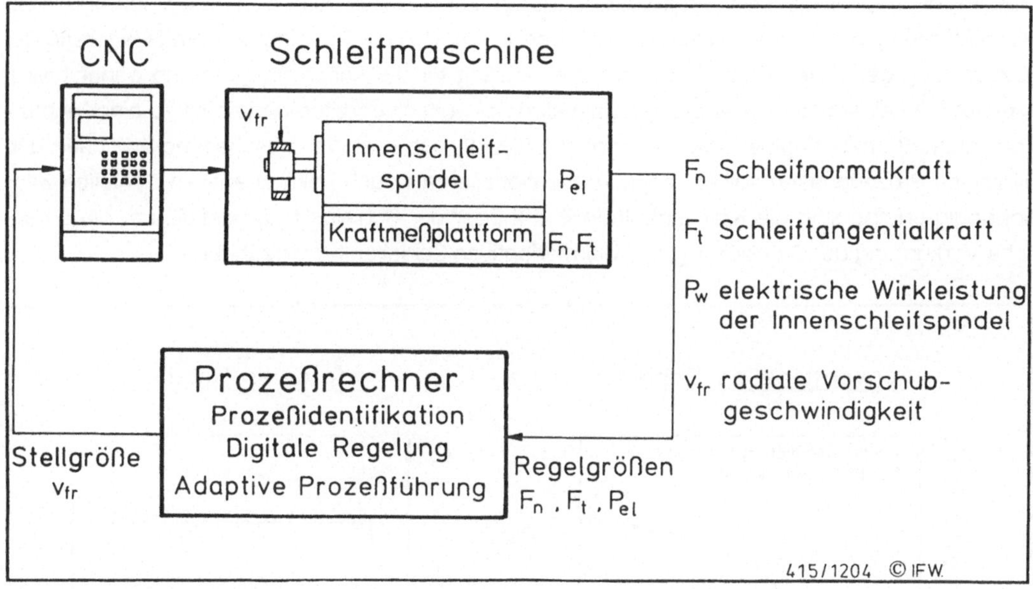

Bild 1: Regelkreis für das adaptive Innenrundschleifen

Während es sich beim konventionellen Innenrundschleifen um eine rein wegabhängige Geschwindigkeitssteuerung handelt, wird bei der adaptiven Prozeßführung eine der Prozeß-kenngrößen (z.B. Schleifnormalkraft, Schleiftangentialkraft, elektrische Wirkleistung der Innen-schleifspindel) als Regelgröße benutzt und nach einer vorgegebenen Sollwertkurve geregelt (Bild 1). Als Stellgröße steht die radiale Geschwindigkeit zwischen Werkstück und Schleifscheibe zur Verfügung. Der Regelkreis wird über den Prozeßrechner geschlossen, der neben der eigentlichen digitalen Regelung des Prozesses und der adaptiven Prozeßführung auch eine Prozeßidentifikation übernimmt. Hierauf wird im nächsten Kapitel näher eingegangen.

Die Meßeinrichtungen zur Erfassung der jeweiligen Regelgröße müssen gute dynamische Übertragungseigenschaften für eine schnelle Anschnitterkennung während der Eilannäherung und gute statische Übertragungseigenschaften während der Regelung beim Schruppen aufweisen [6]. Die geforderten Übertragungseigenschaften sind beim Einsatz von teuren Kraftmeßeinrichtungen (Kraftmeßplattform unterhalb der Schleifspindel, Schleifspindel mit Kraftmeßlager, magnetgelagerte Schleifspindel) erfüllt. Die Schleifleistungsmessung stellt im Vergleich zur Kraftmessung eine kostengünstige Lösung dar, jedoch muß die Eignung dieser Meßeinrichtung noch untersucht werden. Insbesondere weisen einige Frequenzumformer, die für die Steuerung der Drehzahl der Schleifspindel benötigt werden, unzulässige Totzeiten auf. Auf die Realisierung der erforderlichen externen Vorgabe der Zustellgeschwindigkeit an die numerische Steuerung (CNC) wird weiter unten eingegangen.

2.2 Vorteile der adaptiven Prozeßführung

Das Problem unproduktiver Prozeßzeiten durch Luftschleifen läßt sich mit einer adaptiven Prozeßführung in einfacher Weise dadurch beseitigen, daß die Regelgröße bereits während der Eilzustellung gemessen und ihr sprunghafter Anstieg für die Anschnitterkennung genutzt wird. Dies führt zu einer fast vollständigen Elimination der Luftschleifzeiten und somit zu einer nahezu vom Aufmaß unabhängigen Bearbeitungszeit (Bild 2). Obwohl die Umschaltung zwischen den einzelnen Prozeßphasen sowohl beim konventionellen als auch beim adaptiven Schleifen wegabhängig erfolgt, sind die Kraftverläufe in Bild 2 über der Schleifzeit dargestellt, um die erhebliche Verkürzung der Bearbeitungszeit beim adaptiven Schleifen hervorzuheben.

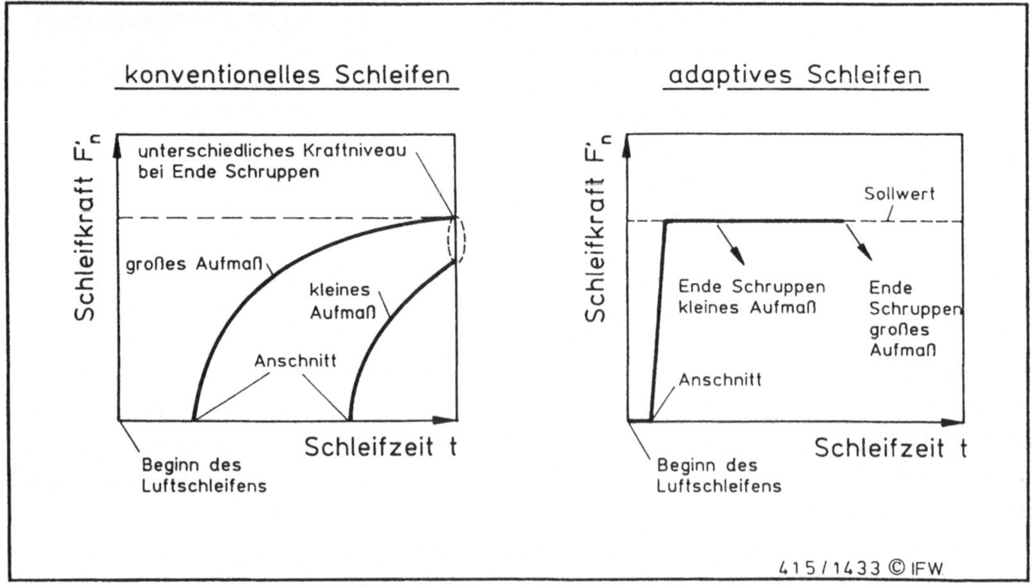

Bild 2: Qualitativer Verlauf der Schleifkraft beim konventionellen und beim adaptiven Schleifen

Während die Elastizitäten im System "Werkzeugmaschine-Schleifprozeß", die im wesentlichen durch die Spindeldurchbiegung verursacht werden, beim konventionellen Schleifen einen langsamen Anstieg der Kraft und damit des Zerspanvolumens zur Folge haben, kann dieser Effekt beim adaptiven Schleifen völlig kompensiert werden. Die Kraft steigt durch die Regelung in kürzester Zeit auf den gewünschten Sollwert an, und der Prozeß befindet sich vollständig in einem stationären Zustand.

Beim konventionellen Schleifen entstehen auf Grund der unterschiedlichen Eingriffszeiten, die durch die Aufmaßschwankungen bedingt sind, auch entsprechend unterschiedliche Auffederungen. Dies hat eine Streuung der Maß- und Formfehler zur Folge, die auch durch die anschließenden Phasen Schlichten und Ausfeuern nicht vollständig abgebaut werden können. Hier bietet das adaptive Schleifen einen grundsätzlichen Vorteil, da am Ende der Bearbeitung (z.B. Schruppen) unabhängig vom Aufmaß des bearbeiteten Werkstückes auf Grund der konstanten Spindeldurchbiegung für alle Werkstücke gleiche Maß- und Formfehler auftreten und so geeignete Maßnahmen zu deren Kompensation ergriffen werden können.

Dieses Konstantkraft- bzw. Konstantleistungschleifen bietet weiterhin den Vorteil einer hohen Prozeßsicherheit, da auf Grund der Regelung keine Überlastung oder gar Beschädigung der Schleifscheibe auftreten kann. Dies ist besonders bei Einsatz kostenintensiver Werkstoffe von entscheidender Bedeutung.

3. Konzept der digitalen Regelung

Bei der Entwicklung des Konzepts der Prozeßregelung stand die Zielsetzung im Vordergrund, eine selbsteinstellende Regelung auf der Basis eines Mikrorechners zu entwickeln. Hiermit läßt sich eine adaptive Prozeßführung realisieren, die für den Bediener keine Kenntnisse der Regelungstechnik erforderlich macht und an einer beliebigen Werkzeugmaschine eingesetzt werden kann.

Eine Voraussetzung für den Aufbau einer digitalen Regelung ist die Kenntnis des Übertragungsverhaltens der Regelstrecke (Bild 3). Das unbekannte Verhalten des Systems "Werkzeugmaschine-Schleifprozeß-Meßeinrichtung" wird mit Hilfe eines parametrischen Prozeßmodells dritter Ordnung beschrieben. Die Schätzung der 6 unbekannten Prozeßparameter a_1-a_3 und b_1-b_3 erfolgt on-line im geschlossenen Regelkreis während des Schleifprozesses nach der rekursiven Methode der "kleinsten Fehlerquadrate" (RLS-Verfahren) und liefert nach jedem Abtasttakt die aktuellen Prozeßparameter. Anhand der so berechneten Parameter wird die optimale Einstellung der Koeffizienten des gewählten Reglers vorgenommen [7].

Als besonders geeigneter Reglertyp hat sich der strukturoptimale Dead-Beat-Regler (Regler mit endlicher Einstellzeit) erwiesen. Die Beschreibung des Dead-Beat-Reglers erfolgt durch eine z-Übertragungsfunktion mit den Koeffizienten p_1-p_3 und q_1-q_3, die sich direkt aus den Prozeßparametern berechnen lassen. Eine diskrete Totzeit der Strecke "d" kann berücksichtigt werden [8].

Mit Hilfe dieser Vorgehensweise kann ein zeitvariables Verhalten des Prozesses, wie es durch den zunehmenden Schleifscheibenverschleiß gegeben ist, berücksichtigt werden. Man spricht hier von einem selbstoptimierenden oder adaptiven Regler. Die Adaption des Reglers an den zu regelnden Prozeß wird vom Prozeßrechner durchgeführt, so daß vom Bediener kein Einfluß auf die Regelgüte ausgeht.

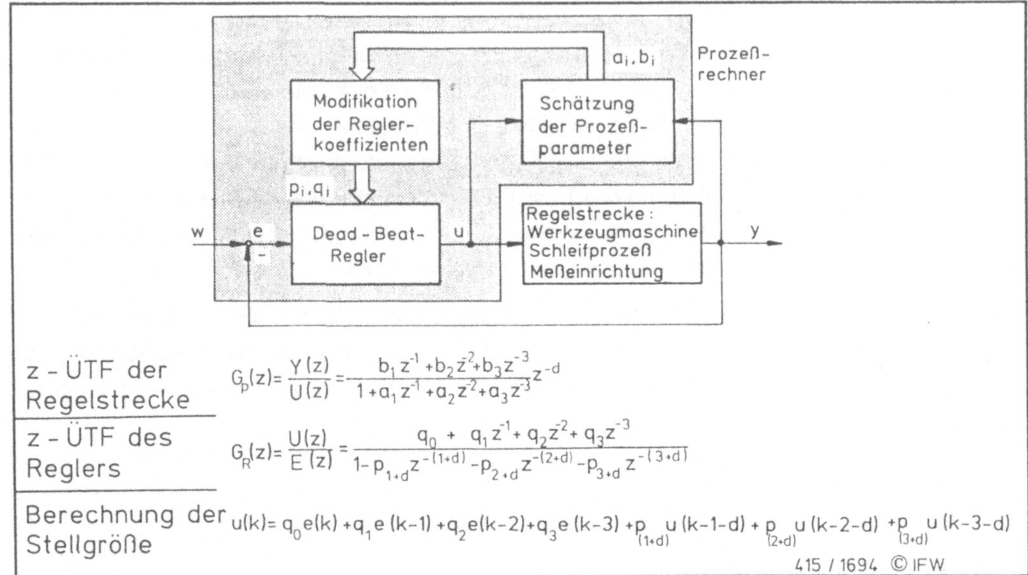

$$z-\ddot{U}TF\ der\ Regelstrecke \qquad G_p(z) = \frac{Y(z)}{U(z)} = \frac{b_1 z^{-1} + b_2 z^{-2} + b_3 z^{-3}}{1 + a_1 z^{-1} + a_2 z^{-2} + a_3 z^{-3}} z^{-d}$$

$$z-\ddot{U}TF\ des\ Reglers \qquad G_R(z) = \frac{U(z)}{E(z)} = \frac{q_0 + q_1 z^{-1} + q_2 z^{-2} + q_3 z^{-3}}{1 - p_{1+d} z^{-(1+d)} - p_{2+d} z^{-(2+d)} - p_{3+d} z^{-(3+d)}}$$

$$Berechnung\ der\ Stellgröße \qquad u(k) = q_0 e(k) + q_1 e(k-1) + q_2 e(k-2) + q_3 e(k-3) + p_{(1+d)} u(k-1-d) + p_{(2+d)} u(k-2-d) + p_{(3+d)} u(k-3-d)$$

415 / 1694 © IFW

Bild 3: Konzept der digitalen Regelung

Die Bohrung der Werkstücke weist in der Regel durch die Vorbearbeitung Unrundheiten auf. Zum schnellen Abbau dieser Unrundheiten ist eine konstante Zustellgeschwindigkeit (Stellgröße) während jeweils einer Werkstückumdrehung erforderlich. Mit Hilfe einer digitalen Regelung kann dies im Gegensatz zu einer analogen Regelung leicht erzielt werden. Dies ist neben der Adaption des Reglers der wesentliche Vorteil beim Einsatz des Prozeßrechnersystems [2].

4. Prozeßrechnersystem

Bei der Entwicklung des Prozeßrechners wurde beachtet, daß er in betrieblicher Umgebung eingesetzt werden mußte. Im Vordergrund standen eine anwenderfreundliche Bedienung, weitgehende Unempfindlichkeit gegen die Umwelteinflüsse in der Nähe der Werkzeugmaschine sowie wirtschaftliche Gesichtspunkte. Die daraus resultierenden, allgemeinen Anforderungen an ein autonomes Prozeßrechnersystem zum adaptiven Innenrundschleifen sowie die zur Erfüllung eingesetzten Hardware-Komponenten sind in Bild 4 aufgeführt. Bei der Realisierung des Prozeßrechnersystems wurde ein zweistufiges Ausbaukonzept gewählt. Es ist dadurch gekennzeichnet, daß ein zur Regelung des Schleifprozesses voll geeignetes, preisgünstiges Grundgerät durch Hinzufügen optionaler Komponenten zu einem komfortablen Entwicklunssystem erweitert werden kann.

Anforderung	wird erfüllt durch
•Integration im Steuerungs- schrank der Maschine	– 19"– Einschubgehäuse, 3HE (Höheneinheiten)
•System ohne Terminal	– Tastatur und LCD-Modul in der Frontplatte
•Grafische Datenausgabe	– grafikfähiges LCD-Modul – hochauflösende Grafik als Karte zusteckbar
•Betrieb als Entwicklungs- system unter RTOS-UH	– Betriebssystem RTOS-UH im EPROM – Speichererweiterung – serielle Schnittstelle V.24 – Floppy-/Winchester-Interface
•Abtastung der Regelgröße	– A/D-Wandler 12 Bit – programmierbares Tiefpaßfilter 2.Ordnung
•Ausgabe der Stellgröße	– analog über 12 Bit D/A-Wandler – digital über Zähler-Baustein AM9513A
•Kommunikation mit der CNC	– 8 digitale Ein-/Ausgänge, interruptfähig

415/1438 © IFW

Bild 4: Anforderungen an ein autonomes Prozeßrechnersystem

Echtzeit – u. Multitasking – Eigenschaft:

- zeitliche Ein – u. Ausplanungen von Tasks (deutsch: Aufgaben)
- gleichzeitige u. prioritätsgesteuerte Bearbeitung mehrerer Tasks
- Reaktion auf externe Ereignisse (Events) in kürzester Zeit
- leichter u. schneller Zugriff auf Hardware von Hochsprache aus

Leistungsmerkmale:

Reaktionszeit auf Prozeßinterrupts in PEARL	< 200 µs
Maximale Anzahl gleichzeitig aktiver Tasks	nicht begrenzt
Taskwechselzeit A – B – A (CPU MC 68000, 8MHz)	< 200 µs
Maximale Anzahl von Prozeßinterrupts	32
Ausführungsgeschwindigkeit des PEARL – Compilers	>500 Zeilen/min

415 / 1693 © IFW

Bild 5: Eigenschaften und Leistungsmerkmale von RTOS-UH [9]

Bei der Auswahl der internen Hardware ist vor allem die Wahl des Betriebssystem entscheidend. Auf Grund der bisher guten Erfahrungen wurde das Betriebssystem RTOS-UH auf dem Prozeß- rechnersystem installiert. Es ist selbst auf kleinen Systemen implementierbar, sofern ein Mikro- prozessor der 68000-Familie benutzt wird. Es besitzt alle für die Regelaufgabe notwendigen Echtzeit- und Multitasking-Eigenschaften (Bild 5). Die Programmentwicklung erfolgt in der Hochsprache PEARL oder in 68000-Assembler [9].

<u>4.1 Hardware-Komponenten des Prozeßrechnersystems</u>

In Bild 6 sind die wesentlichen Hardware-Komponenten des Systems dargestellt. Dabei sind die Systemerweiterungen im rechten unteren Teil des Bildes angeordnet. Kernstück der Hardware ist der Mikroprozessor MC68000 (bzw. MC68010), der mit einer Taktfrequenz von 12 MHz betrieben wird. Über die serielle Schnittstelle (V.24) kann ein Terminal zur Programmentwicklung oder ein Drucker angeschlossen werden. Dieses Terminal wird aber nicht zur Bedienung des Systems benötigt.

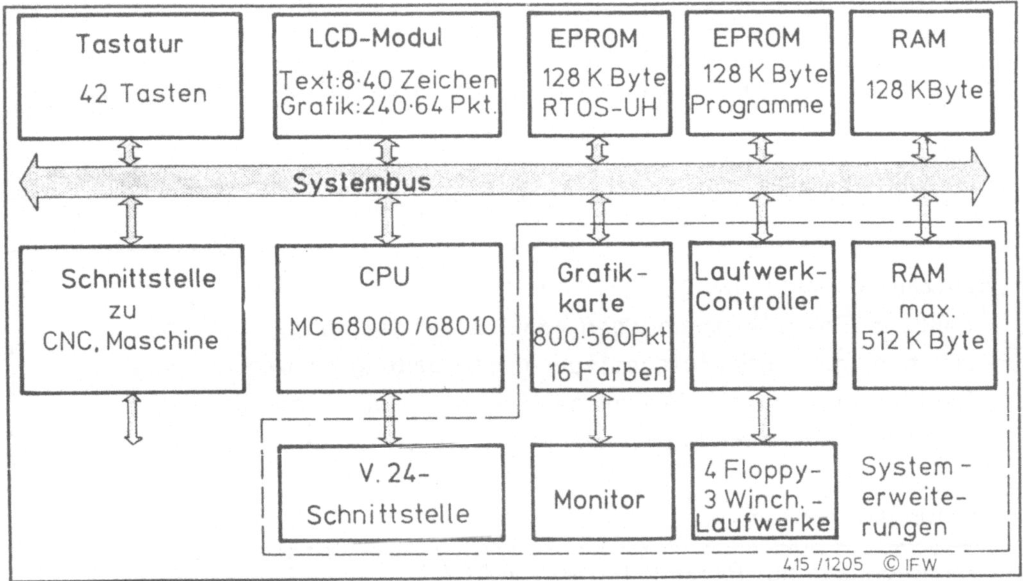

Bild 6: Hardware-Komponenten des Prozeßrechnersystems

Das Betriebssystem und die Anwenderprogramme sind in zwei EPROM-Bänken mit je 128 kByte untergebracht. Da das gesamte Betriebssystem insgesamt nur ca. 100 kByte umfaßt, sind selbst bei dem Grundgerät die Entwicklungsfunktionen wie Editor, Compiler und Assembler integriert. Direkt nach dem Einschalten des Systems wird das Anwenderprogramm geladen und ist unmittelbar ausführbar. Als Arbeitsspeicher stehen 128 kByte batteriegepufferter RAM zur Verfügung. Ein Teil des Arbeitsspeichers ist für die Parameter des letzten Schleifprozesses reserviert, die damit auch bei Ausschalten des Systems nicht verlorengehen.

Die Bedienung des Systems erfolgt ausschließlich über die in der Frontplatte angeordnete Tastatur und das LCD-Modul. Auf dem LCD-Modul können Daten in alphanumerischer Form mit 8 Zeilen zu je 40 Zeichen und/oder Grafiken mit einer Auflösung von 240x64 Punkten dargestellt werden.

Die Schnittstelle zur CNC und Werkzeugmaschine stellt eine wesentliche Komponente des Systems dar. Sie wird im folgenden Abschnitt näher beschrieben.

Das Grundgerät kann durch Erweiterungen zu einem vollständigen und komfortablen Entwicklungssystem aufgerüstet werden. Die Grafikkarte ermöglicht grafische Ausgaben mit hoher Auflösung. Dieses Subsystem stellt einen Bildspeicher von 1024x768 Punkten bei einer Auflösung von 4 Bit/Punkt zur Verfügung. Hiervon lassen sich 800x560 Punkte auf den Monitor ausgeben. Es können daher Monitore mit einer minimalen Auflösung von 800x560 Punkten und 16 Farben bzw. Graustufen angeschlossen werden. Neben einer Erweiterung des Arbeitsspeichers um 512 kByte kann über eine am Grundgerät vorhandene Schnittstelle ein Winchester/Floppy-Laufwerk-Controller angeschlossen werden, der maximal 7 Laufwerke adressiert. Alle Erweiterungen werden von RTOS-UH unterstützt und sind somit sowohl vom Betriebssystem als auch von der Hochsprachenebene aus erreichbar.

4.2 Schnittstelle zu Werkzeugmaschine und CNC

Das grundsätzliche Problem bei der Installation des Prozeßrechnersystems zum adaptiven Innenrundschleifen besteht in der Übergabe der Stellgröße an die Steuerung und betrifft damit die Konfiguration der Schnittstelle zwischen Prozeßrechnersystem und Werkzeugmaschine bzw. CNC. Um das System an verschiedenen Steuerungskonzepten zur externen Vorschubsteuerung einsetzen zu können, kann die Schnittstelle bezüglich der Ausgabe der Stellgröße verschieden genutzt und konfiguriert werden. Die gesamte Hardware-Schnittstelle ist in Bild 7 dargestellt. Sie umfaßt im wesentlichen die A/D-D/A-Wandlerkarte, einen Zählerbaustein (AM9513 mit 5x16 Bit) sowie zwei digitale Ein-/Ausgabe-Ports (MC68901 und MC68230). Die digitalen Signale sind zur galvanischen Trennung über Optokoppler mit der CNC verbunden.

Die A/D-D/A-Wandlerkarte besitzt je zwei analoge Ein- und Ausgänge. Die Regelgröße wird zur Vermeidung von Rückfaltungen zunächst in einem analogen Tiefpaß 2. Ordnung (12dB/Oktave) gefiltert. Die Eckfrequenz des Tiefpaßfilters kann mit Hilfe des Ports MC68230 auf einen Wert zwischen 1 und 1599 Hz programmiert werden. Für die Eilannäherung wird die Eckfrequenz auf ca. 500 Hz eingestellt, um den Anschnitt ohne wesentliche Zeitverzögerung zu erkennen. Nach dem Anschnitt wird die Eckfrequenz in Abhängigkeit der Abtastfrequenz umprogrammiert. Dem Filter sind ein Abtast-Halteglied und ein Analog/Digital-Wandler (12 Bit, 25µs) nachgeschaltet, die das analoge Signal in Zeit und Amplitude quantisieren.

Wie oben erwähnt ist die Ausgabe der Stellgröße an die CNC auf zwei verschiedene Arten möglich. Einerseits kann eine digitale Ausgabe mit Hilfe zweier binärer Signale erfolgen. Hierbei ist eine gegen Störungen unempfindliche Übergabe der Stellgröße gewährleistet. Der Dynamikbereich der Stellgröße beträgt 16 Bit.

Der Betrag wird dabei in Form eines frequenzproportionalen Rechtecksignals übergeben. Die Frequenz bestimmt direkt die Verfahrgeschwindigkeit der Zustellachse, wobei jede positive Flanke zu einem bestimmten Verfahrweg (z.B. 0,5µm) führt. Das Rechtecksignal wird durch zwei kaskadierte 16 Bit-Zähler innerhalb des Bausteins AM9513 erzeugt. Die Richtung der Stellgröße wird über das zweite binäre Signal übergeben. Dies wird in einem zweiten Zähler durch eine Phasenverschiebung des o.g. Signals je nach Verfahrrichtung um +/- 90° erzeugt, wie dies bei inkrementalen Wegmeßsystemen üblich ist. Die Verfahrrichtung wird auch über Bit6 des MC68901 als konstantes Signal ausgegeben. Zusätzlich kann die Stellgröße über einen D/A-Wandler analog (0-10 V) ausgegeben werden. Allerdings ist hierbei nur das Verfahren in einer Richtung möglich.

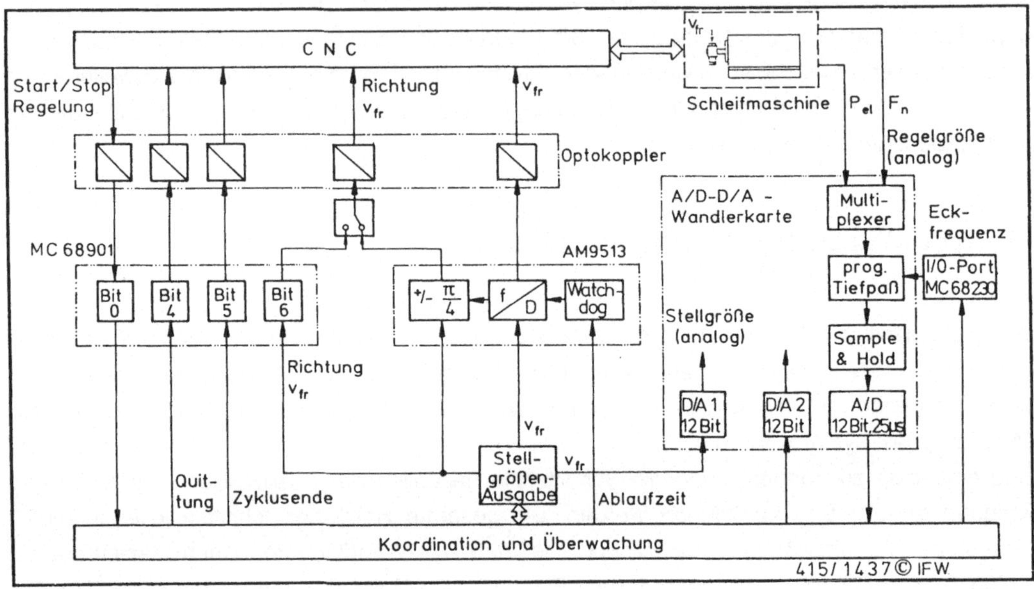

Bild 7: Hardware-Schnittstelle zu Werkzeugmaschine und CNC

Über Bit0, Bit4 und Bit5 des MC68901 wird die digitale Regelung gestartet bzw. gestoppt. Die CNC generiert bei Start bzw. Stop des Schleifzyklus einen Interrupt. Bei Bereitschaft quittiert der Prozeßrechner die Anforderung. Über Bit5 wird die Aufforderung, den Zyklus zu beenden, an die CNC übergeben. Aus Sicherheitsgründen ist den Zählern zur digitalen Stellgrößenausgabe ein weiterer Zähler in Form reiner Hardware überlagert, der als Watchdog fungiert und die Ausgabe der Stellgröße für den Fall einer Fehlfunktion des Prozeßrechnersystems verhindert.

Die gesamte Hardware des Systems ist auf zwei Platinen verteilt, getrennt nach Digital- und Analogteil. Die digitalen Elemente des Mikrorechners (CPU, Speicher, Zählerbaustein, dig. Peripherie) sind auf einer Platine im Doppeleuropaformat untergebracht (Bild 8). Der Analogteil (Tiefpaßfilter, A/D- und D/A-Wandler) befindet sich auf der linken Add-on-Platine. Auf einer weiteren Add-on-Platine ist die Hardware für die Prozeßgrafik untergebracht. Die Kommunikation mit der Peripherie erfolgt über zwei VG-Leisten, die auf diesem Bild nicht zu erkennen sind. Die Add-on-Technik ermöglicht das Hinzustecken weiterer Platinen. Bild 9 zeigt das als komfortables Entwicklungssystem aufgebaute Prozeßrechnersystem mit Tastatur und LCD-Modul in der Frontplatte sowie die Innenrundschleifmaschine mit CNC.

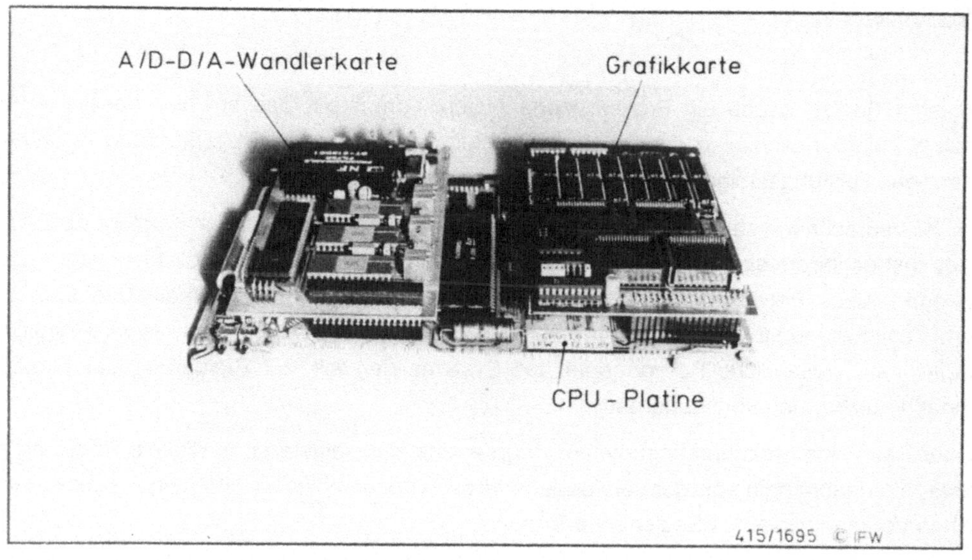

Bild 8: Hardware des Prozeßrechnersystems

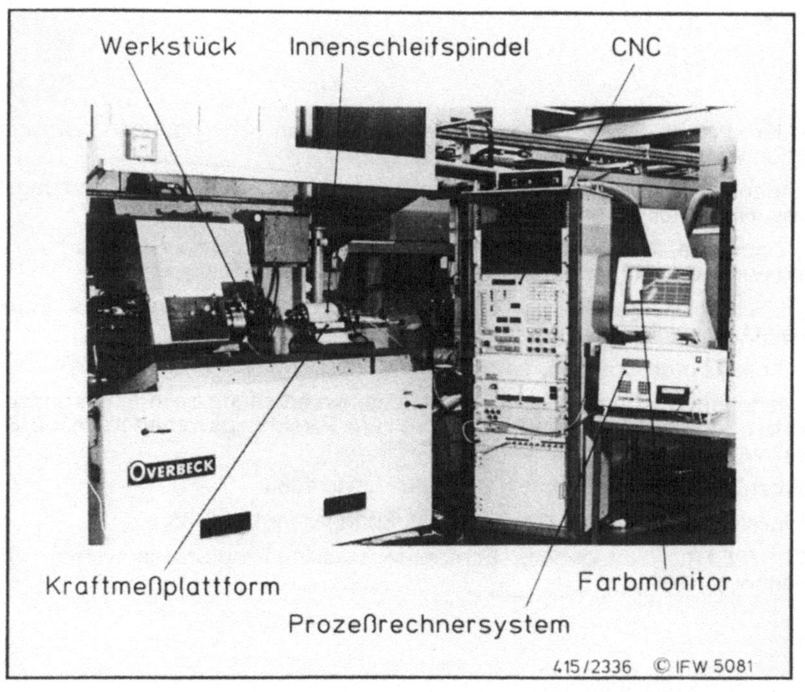

Bild 9: Prozeßrechnersystem und Innenrundschleifmaschine mit CNC

5. Zusammenfassung

In diesem Beitrag wurde ein Prozeßrechnersystem vorgestellt, das am IFW speziell für das adaptive Innenrundschleifen entwickelt wurde. Mit der realisierten Prozeßführung lassen sich wesentliche Leistungssteigerungen beim Schleifen erzielen.

Dem Prozeßrechnersystem wurde insgesamt ein stark praxisorientiertes Konzept zu Grunde gelegt. Insbesondere sind dies der Einsatz als autonomes System, die Möglichkeit zum Ausbau als komfortables Entwicklungssystem sowie die anwenderfreundliche Bedienung über integrierte Tastatur und LCD-Modul. Weiterhin ist die Schnittstelle zu Werkzeugmaschine und CNC, die eine sehr wesentliche Komponente des Systems darstellt, zur Steigerung der Flexibilität mehrfach nutzbar und konfigurierbar.

Auf der Basis des Prozeßrechnersystems wurde eine selbsteinstellende digitale Regelung entwickelt, die unabhängig von der verwendeten Werkzeugmaschine ist und keine regelungstechnischen Vorkenntnisse des Bedieners erfordert.

Abschließend sei angeführt, daß dieses Prozeßrechnersystem im IFW auch zu anderen Aufgaben wie zur Steuerung einer flexiblen Meßzelle für die Qualitätskontrolle beim Schleifen und zur Lageregelung eines Unterwasserroboters eingesetzt wird.

6. Schrifttum

[1] Tönshoff, H.K., Jürgenhake, B.: Ursachen und Wirkungen von Schwingungen beim Innenschleifen, Präzision im Spiegel 1(1979) 1, S. 23-25.

[2] Zinngrebe, M.: Adaptive Prozeßführung beim Innenrundschleifen mit digitalen Grenzregelungen, Dr.-Ing. Diss. Hannover 1990.

[3] Tönshoff, H.K., Zinngrebe, M., Kemmerling, M.: Optimization of Internal Grinding by Microcomputer-Based Force Control, Annals of the CIRP, Vol. 35/1/1986.

[4] Althaus, P.G.: Leistungssteigerung beim Innenrundschleifen durch kubisches Bornitrid (CBN) und neue Maschinenkonzeption, Dr.-Ing. Diss. Hannover 1982.

[5] Hahn, R.S.: Controlled Force Grinding, Trans. of the ASME, Aug. 1964, pp. 287-293.

[6] Tönshoff, H.K., Janocha, H., Zinngrebe, M.: Prozeßdatenverarbeitung beim Innenrundschleifen mit hochharten Schleifstoffen, Abschlußbericht zum Forschungsvorhaben Ja368/3 der Deutschen Forschungsgemeinschaft, Hannover 1988.

[7] Isermann, R.: Prozeßidentifikation, Band 1, Springer-Verlag 1988.

[8] Isermann, R.: Digitale Regelsysteme, Band 1 und 2, Springer-Verlag 1988.

[9] Gerth, W.: RTOS/PEARL - Integriertes Echtzeit-Multitasking-Programmiersystem, Verlag Heinz Heise, Hannover 1989.

Flexibles Automatisierungssystem für eine Versuchsanlage zum Gießwalzen von Stahlband

S. Bernhard, M. Enning, H. Rake
Institut für Regelungstechnik
RWTH Aachen
Steinbachstr. 54
5100 Aachen

1. Einleitung

Die Regelung eines noch vollkcmmen unbekannten Prozesses, etwa im Rahmen der Automatisierung einer Versuchsanlage, stellt den Regelungstechniker häufig vor das folgende Problem: Ein Betrieb ohne Regelung ist nicht möglich, da der Prozeß stark gestört oder möglicherweise sogar instabil ist, die notwendigen Erkenntnisse für eine zufriedenstellende Regelung könnnen aber erst durch Messungen im laufenden Betrieb gewonnen werden.

Ein Ausweg aus dieser Situation besteht nun darin, auf Grundlage einer einfachen Modellbildung eine einfache Regelung zu entwerfen, die zunächst einen sicheren Betrieb gewährleistet, ohne weitergehende Anforderungen an die Regelgüte zu erfüllen. Durch Messungen am laufenden Prozeß und dadurch gestützte theoretische Untersuchungen läßt sich nun ein verfeinertes Prozeßmodell gewinnen, das als Grundlage für eine verbesserte Regelung dienen kann.

Ein in dieser Entwicklungsphase einzusetzendes Automatisierungssystem muß über eine hohe Rechenleistung verfügen, damit es auch für schnelle Prozesse geeignet ist. Unterschiedlichste Strukturen und Versuchsabläufe müssen einfach und mit geringem Programmieraufwand realisierbar sein.

Das Gießwalzen von Stahlband stellt einen im beschriebenen Sinn "schwierigen" Prozeß dar. Im folgenden Beitrag wird die Realisierung eines Automatisierungssystems für eine Versuchsanlage zum Gießwalzen

von Stahlband auf Grundlage eines modularen Bussystems vorgestellt. Im
Anschluß an eine kurze Prozeßbeschreibung werden die Anforderungen an
ein derartiges Automatisierungssystem formuliert. Der nächste Abschnitt
befaßt sich mit der Realisierung und beschreibt in diesem Zusammenhang
die Hardwarekonfigurierung sowie das realisierte Softwaresystem. Abschließend wird über Erfahrungen mit dem Automatisierungssystem berichtet.

2. Prozeßbeschreibung

Die zu automatisierende Versuchsanlage dient zur Untersuchung des sog.
"Zwei-Walzen-Gießverfahrens" zur Herstellung dünner Stahlbänder. Bei
diesem Verfahren wird schmelzflüssiges Material unmittelbar in den
Walzspalt zweier horizontal nebeneinanderliegender, gegensinnig rotie-
render Walzen eingebracht [1].

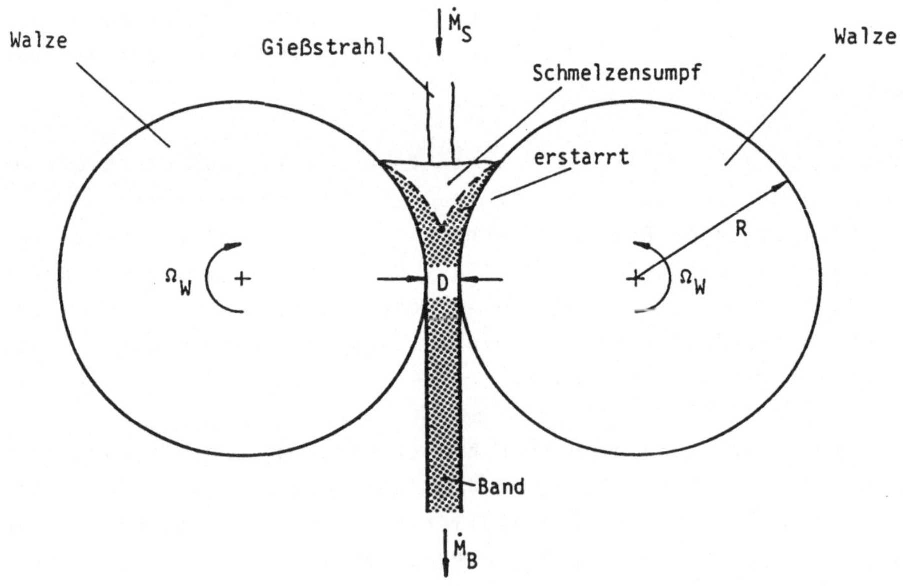

Bild 1: Grundprinzip des Zwei-Walzen-Gießwalzverfahrens

Bild 1 zeigt das Grundprinzip dieses Verfahrens. Entlang der Berührzone
zwischen Schmelze und Walzenoberfläche bildet sich jeweils eine Erstar-
rungsfront aus. Die beiden Fronten wachsen an einem Punkt, der idealer-
weise knapp oberhalb des engsten Walzspaltquerschnitts liegt, zusammen,
so daß nur noch festes Material vorliegt. Dieses wird zum engsten Walzspalt hin mehr oder weniger stark umgeformt und verläßt als Band den
Walzspalt.

3. Anforderungen an ein Automatisierungssystem für eine Versuchsanlage
   zum Gießwalzen von Stahlband

Aus den Zielvorstellungen hinsichtlich der Qualität des zu produzieren-
den Bands resultieren spezielle Anforderungen an einzelne Anlagenkompo-
nenten. Die angestrebte Banddickentoleranz erfordert eine geringe Nach-
giebigkeit von Walzen und Walzenzustellung, die den schnell ablaufenden, aber sonst stabilen Prozeß außerordentlich empfindlich gegenüber
kaum zu vermeidenden Störungen und u.U. sogar instabil werden lassen
[2]. Um dennoch einen sicheren Versuchsbetrieb zu gewährleisten, müssen
neben einer reinen Meßdatenerfassung in erster Linie umfangreiche
Steuerungs-, Regelungs- und Überwachungsfunktionen realisiert werden.

Bei der Planung eines geeigneten Automatisierungssystems steht die Er-
weiterungsfähigkeit und Flexibilität sowohl der Hard- als auch der
Software im Vordergrund, da Umfang und Strukturen der Steuerungs- und
Regelungsaufgaben an die aus dem Versuchsbetrieb resultierenden wech-
selnden Anforderungen angepaßt werden müssen.

Das zu realisierende System von Hard- und Software muß die Verarbeitung
sowohl kontinuierlicher als auch binärer Größen ermöglichen. Dabei muß
es möglich sein, unterschiedlichste Strukturen und Parameter auf einfache Weise durch "Konfigurieren" zu vereinbaren.

Weiterhin ist es u.U. notwendig, eine große Menge von Meßdaten aufzu-
zeichnen, durch deren Auswertung die noch weitgehend unbekannten Zusammenhänge beim Gießwalzen geklärt werden können.Um auch bei lange dau-
ernden Gießwalzexperimenten ein lückenloses Meßprotokoll zu erhalten,
muß die Meßzeit genügend lang sein können. Dies erfordert eine soforti-
ge Ablage der Daten auf ein externes Speichermedium wie Diskette oder

Festplatte. Für die Abtastzeit, mit der die Meßdaten erfaßt werden, ist als minimal möglicher Wert die Reglerabtastzeit von typisch 10 ms vorzusehen.

Aufgrund der hier aufgeführten Anforderungen an ein Automatisierungssystem für die Versuchsanlage konnte unter den am Markt verfügbaren Gerätesystemen eine geeignete Hardwarebasis ausgewählt werden. Insbesondere die Forderung nach möglichst weitgehender Konfigurierbarkeit führte zu dem im weiteren vorgestellten Softwareentwurf.

4. Realisierung

4.1 Hardware

Kleinere Automatisierungsaufgaben insbesondere für Versuchsanlagen werden bereits häufig mit <u>Personal-Computern</u> (PC) erledigt. Die hier gestellten Anforderungen sind jedoch so umfangreich, daß sie mit einem PC nicht bewältigt werden können. Weiterhin sind Echtzeitbetriebssysteme für PC's zur Zeit nur in eingeschränkter Form verfügbar.

<u>Speicherprogrammierbare Steuerungen</u> (SPS) sind zur Lösung von Steuerungsaufgaben prinzipiell gut geeignet. Nachteilig für den hier behandelten Einsatzfall ist allerdings, daß derartige SPS'n nur beschränkt frei programmierbar sind. Nur wenige Typen von Funktionsbausteinen lassen sich zu einer Abarbeitungssequenz zusammenstellen. Sollen zu einem späteren Zeitpunkt z.B. mehrschleifige, gekoppelte Regelungen realisiert werden, reicht der verfügbare Funktionsumfang meist nicht mehr aus.

Gegen den Einsatz eines <u>Prozeßleitsystems</u> spricht dessen eher begrenzte Abtastfrequenz. Ideal geeignet sind derartige Systeme zur Automatisierung großer verfahrenstechnischer Anlagen mit nicht zu kleinen Zeitkonstanten.

Sowohl SPS'n als auch Prozeßleitsysteme können vom Benutzer lediglich konfiguriert werden. Programmierung im eigentlichen Sinne, also das Erstellen von Anwenderprogrammen und deren Übertragen in Maschinensprache, ist weder notwendig noch möglich.

Sollen beliebige Regelungsstrukturen für schnelle Prozesse realisiert werden, führt auch heute noch kein Weg an einem für die jeweilige Aufgabe speziell programmierten Digitalrechner vorbei. Prozeßrechner - die Rechnersysteme, die über ein Echtzeitbetriebssystem und über die notwendigen Hardware-Voraussetzungen hierfür verfügen - werden in jüngster Zeit vielfach auf der Basis von Mikroprozessoren aufgebaut. Dies führt zu einer drastischen Verringerung der Kosten gegenüber "klassischen" Prozeßrechnern mit Großrechnerarchitektur. Die Verwendung multiprozessorfähiger Standardbussysteme bietet darüberhinaus die Möglichkeit, beliebig viele Peripherie-Anschaltbaugruppen und, zur Anpassung der Rechenleistung an den Leistungsbedarf des Prozesses, mehrere Mikroprozessoren einzusetzen.

Zur Automatisierung der Versuchsanlage wurde vor dem Hintergrund der obigen Ausführungen ein frei programmierbares Prozeßrechensystem ausgewählt und dieses mit einer Software ausgestattet, die die im vorigen Abschnitt spezifizierten Anforderungen erfüllt.

Ausgewählt wurde das AMS-Bussystem der Firma SIEMENS. Über den Monoprozessor-Bus SMP kann jede Prozessorbaugruppe weiterhin mit lokalen Ein-/Ausgabebaugruppen aus dem SMP- und/oder AMS- Programm ausgestattet werden; dies ermöglicht die Realisierung eines hierarchischen Buskonzepts, welches die Belastung des AMS-Busses auf ein erträgliches Maß reduziert.

Es wurde ein System mit zwei 80286-Prozessorbaugruppen zusammengestellt; jede Baugruppe ist zusätzlich mit einem 80287-Co-Prozessor ausgestattet. Jede Zentraleinheit verfügt über einen lokalen Speicherbereich von 256 kB, auf den nur der jeweilige Prozessor zugreifen kann. CPU 0 enthält zusätzlich einen als sog. Dual-Ported-RAM ausgeführten globalen Kommunikationsspeicher von 768 kB, auf den beide CPU's zugreifen können. Im Sinne einer effektiven Abarbeitung der Software liegen im globalen Speicher nur Daten, die von beiden Prozessoren manipuliert werden.

Der derzeitige Ausbau des Rechners ist in Bild 2 dargestellt. Deutlich zu erkennen ist das hierarchische Bussystem mit dem globalen AMS-Bus, über den die beiden CPU's miteinander kommunizieren. Dem globalen Bus

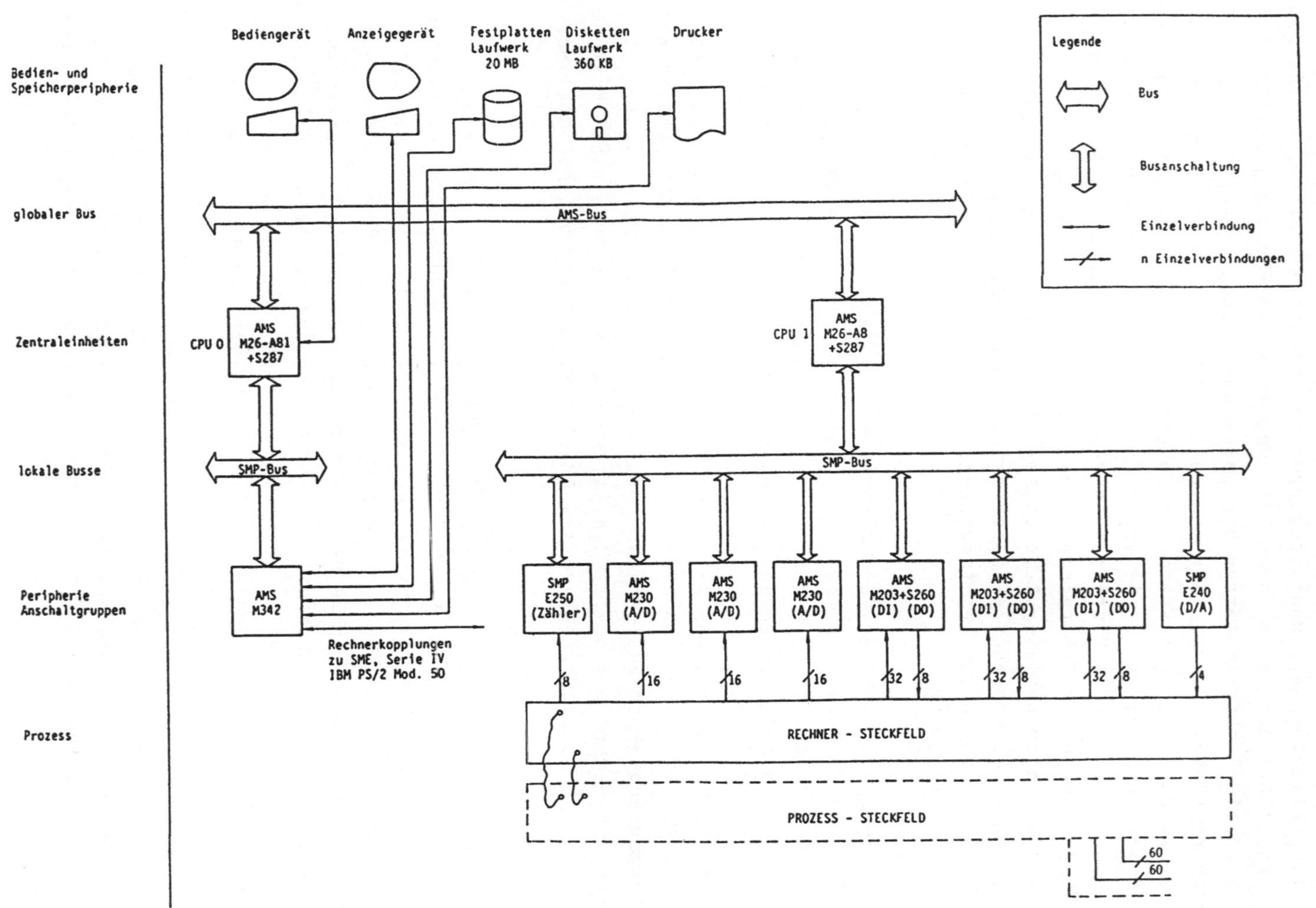

Bild 2: Hardware-Struktur des Prozeßrechners

unterlagert sind zwei SMP-Lokalbusse, über die die jeweilige CPU auf
ihre Peripherie zugreift.

Bei der Aufgabenverteilung auf die beiden Zentraleinheiten muß gewährleistet sein, daß zyklisch ablaufende Programme, z.B. zur Regelung, ein
definiertes zeitliches Verhalten aufweisen. Schwankende Abtastzeiten
sind bei allen digitalen Regelalgorithmen sehr störend. Da sämtlicher
Verkehr mit Peripheriegeräten wie Datensichtgerät, Drucker und Spei-
chergerät interruptgetrieben abläuft, kann die Gesamtbelastung bei mehreren Peripheriegeräten so hoch werden, daß ein zyklisch ablaufendes
Programm kein definiertes zeitliches Verhalten mehr aufweist.

Vor diesem Hintergrund werden die Funktionen Bedienen/Anzeigen/Spei-
chern durch CPU 0 ausgeführt. Sie verfügt über eine Anschaltbaugruppe
für Disketten- und Festplattenlaufwerke sowie über ein Platten- und
Diskettenlaufwerk.

Die Funktionen Regeln/Steuern/Meßdatenerfassung werden von CPU 1 übernommen. Sie verfügt über Ein-/Ausgabebaugruppen mit insgesamt 48 Ana-
log-Eingabekanälen, 4 Analog-Ausgabekanälen, eine Zählerbaugruppe mit 8
Eingängen sowie 96 digitale Ein- /Ausgabekanäle. Die Rechner- Ein-/Aus-
gabekanäle sind auf das sog. Rechner-Steckfeld geführt. Durch Verdrah-
ten mit dem Prozeß-Steckfeld, auf das alle Prozeßsignale bzw. Verbindungen zu den Stellelementen geführt sind, können Verbindungen auf einfache Weise hergestellt und verändert werden.

Die funktionale Aufgabenverteilung auf zwei Zentraleinheiten ermöglicht
die Ausführung von Regelalgorithmen mit kleiner Abtastzeit (typisch
10ms) auch bei umfangreichem Datenverkehr mit dem Benutzer sowie bei
einer simultan ablaufenden Datenspeicherung.

4.2 Software

4.2.1 Betriebssystem und Programmierumgebung

Für das Automatisierungssystem wurde das Echtzeitbetriebssystem RMOS2
von SIEMENS ausgewählt. Es bietet neben der prioritäts- und zeitge-
steuerten Verwaltung mehrerer quasiparallel ablaufender Tasks auf einer
Zentraleinheit die Möglichkeit, mehrere Zentraleinheiten unter der Ver-

waltung eines globalen Betriebssystems (sog. enge Kopplung) zu betrei-
ben. In dieser Betriebsart "enge Kopplung" wurde das Betriebssystem
konfiguriert. Da das Prozeßrechnersystem über eine gut ausgebaute Peripherie verfügt, ist es möglich, mittels eines entsprechenden Software-
adapters die zur Programmentwicklung notwendigen Hilfsmittel wie Editor, Compiler etc. auf dem Anwendungssystem ablaufen zu lassen.

Die Entwicklung der Anwendungssoftware erfolgte ausschließlich auf
Hochsprachenebene in der Programmiersprache PLM-86, die von INTEL speziell auf die Eigenschaften der INTEL-Prozessoren (8086, 80286 ...)
zugeschnitten wurde.

4.2.2   Anwendungssoftware

4.2.2.1 Funktionsumfang und Datenstrukturen

Für das Automatisierungssystem wurde ein Software-Paket erstellt, das
prinzipiell einige der Grundeigenschaften von Prozeßleitsystemen mit
den Möglichkeiten einer SPS und deren Schnelligkeit verbindet [2]. Das
Softwarepaket umfaßt folgende Funktionen und Möglichkeiten:

- Erfassung und Online-Abspeicherung von Meßsignalen, deren Verarbeitung durch beliebige Kombinationen von Standardelementen der Regelungstechnik (z.B. Summierer, verschiedene Regler, digitale Filter,
  Rampengeber), die Weitergabe an analoge Stellausgänge sowie die Ab-
  speicherung zur späteren Auswertung. Zur Verringerung des Datendurch-
  satzes durch das System können für einzelne Signale Vielfache der
  Grundabtastzeit eingestellt werden.

- Spezielle Verarbeitungsbausteine können mit geringem Aufwand zusätzlich programmiert werden.

- Einfache Verarbeitung binärer Größen
  Dies umfaßt die Eingabe binärer Prozeßsignale, deren Verarbeitung mit
  Standardelementen der Boolschen Algebra und die Ausgabe binärer Signale; weiterhin sind Grenzwertüberwachungen und Zeitglieder realisiert.

- Beeinflussung sowohl quasikontinuierlicher Größen als auch binärer
  Funktionen durch manuellen Eingriff in das Prozeßgeschehen über ein
  Bedienterminal.

- Fortlaufende Anzeige einer Auswahl der den Prozeßzustand beschreibenden Prozeßvariablen.

- Realisierung verschiedener Strukturen und Parameter durch menü-
  gestütztes Konfigurieren.

Der aufgelistete Leistungsumfang der Anwendersoftware determiniert bereits die realisierte Ablaufstruktur. Der gesamte Ablauf läßt sich
hierbei in die Editierphase, in der die Konfigurationsdaten menüge-
stützt bearbeitet werden und die Ablaufphase, in der der bei der Konfigurierung festgelegte Versuchsablauf abgearbeitet wird, unterteilen.
Erst in der letzteren werden die Multitaskingeigenschaften des Rechners
benötigt.

In dem Programmsystem ist grundsätzlich zwischen 4 Typen von Daten zu
unterscheiden. <u>Statische Daten</u> beschreiben im wesentlichen unveränder-
liche Eigenschaften des Meßaufbaus und die realisierten Verarbeitungsbausteine. <u>Konfigurationsdaten</u> legen den Ablauf eines Gießwalz-Versu-
ches fest und dienen gleichzeitig als Grundlage für die Dokumentation
des konfigurierten Ablaufs. Der Inhalt der Konfigurierungsdaten wird
vor dem Start in die <u>Ablaufsteuerdaten</u> umkopiert, wobei in erster Linie
symbolische Verknüpfungen durch numerische ersetzt werden. Die <u>Verar-
beitungsdaten</u> schließlich repräsentieren den aktuellen Zustand des
Gießwalzprozesses, der rechnerinternen Signalverarbeitung und des Pro-
grammzustands.

4.2.2.3 Programmstruktur

Die Anwendungssoftware besteht aus einem die Bedienoberfläche enthaltenden Hauptprogramm und mehreren Tasks. Der nach dem Laden auf die
beiden Zentraleinheiten folgende Ablauf des Programmsystems läßt sich
anhand des Bildes 3 verdeutlichen.

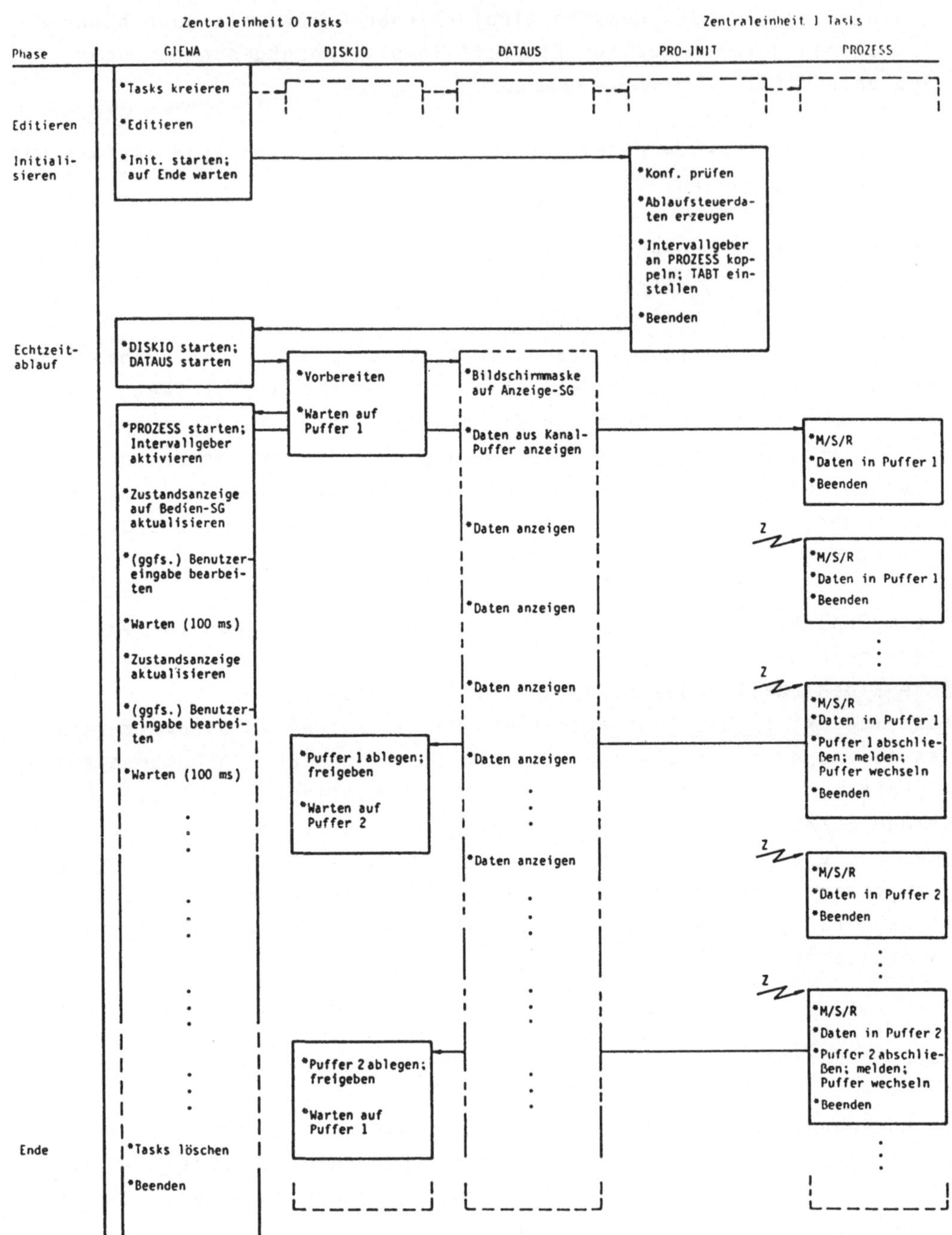

Bild 3: Programmablaufstruktur

Die erste Aktion innerhalb des Hauptprogramms besteht im Kreieren der dynamischen Tasks. In der anschließenden _Editierphase_ erfolgt die menügestützte Bearbeitung der Konfigurierungsdaten; Inkonsistenzen der Konfigurierung werden sofort durch entsprechende Warnungen mitgeteilt.

Nach Beendigung der Editierung werden in der _Initialisierphase_ aus den Konfigurierungsdaten die Ablaufsteuerdaten gewonnen. Da diese aus Gründen des effektiveren Zugriffs im lokalen Speicherbereich der CPU 1 gehalten werden, muß diese Funktion von einer Task (PRO-INIT) ausgeführt werden, die auf der CPU 1 abläuft. Während der Initialisierphase wartet das Bedienprogramm auf die Beendigung der Initialisierungstask. Wenn keine Fehler aufgetreten sind, bindet die Initialisierungstask einen in der Hardware der Zentraleinheitenbaugruppe vorhandenen Echtzeitintervallgeber an die Task "PROZESS" an, welche in der späteren Ablaufphase die Funktionen Messen/Steuern/Regeln wahrnimmt, und stellt die spezifizierte Abtastzeit ein. Dann beendet sie sich selbst und gibt damit den weiteren Ablauf des Bedienprogramms frei.

Die _Echtzeitablaufphase_ beginnt mit dem Start der dynamischen Tasks "DISKIO" und "DATAUS". Während DISKIO für die Ablage der durch PROZESS erfaßten Meßdaten auf das Festplattenlaufwerk zuständig ist, betreibt DATAUS die Ausgabe selektierter Datenwerte auf das Anzeigesichtgerät. Nach den notwendigen Initialisierungen dieser beiden Tasks erfolgt der erstmalige Start der PROZESS-Task durch die Bedientask. Bei diesem Vorgang er folgt gleichzeitig die Aktivierung des Intervallgebers, so daß diese Task von nun an automatisch im Abtasttakt ausgeführt wird.

Der Ablauf der drei Tasks auf CPU 0 (auch das Bedienprogramm ist eine Task, bzw. ein sog. Overlay der Befehlszeileninterpreter-Task) wird durch das Betriebssystem prioritäts- und zeitgesteuert koordiniert und stellt sich wie folgt dar:

Die Task mit der relativ höchsten Priorität ist die Datenablagetask (DISKIO), da diese unmittelbar mit der Echtzeitdatenaufnahme (PROZESS) koordiniert sein muß, damit kein Datenverlust auftritt. Der Datenaustausch zwischen diesen erfolgt über einen doppelt ausgeführten relativ großen Puffer (2 x 4096 Worte), der zwangsläufig im globalen Speicher des Systems liegen muß und im Wechsel durch PROZESS "be-" und DISKIO "entladen" wird. Der Datentransfer zur Platte benötigt vergleichsweise

wenig Zeit, so daß DISKIO in der überwiegenden Zeit auf den Empfang von
Daten wartet. Die Wartezeit kann von der Task mit nächstgeringerer Priorität genutzt werden.

Dies ist die Bedientask, welche in der Echtzeitablaufphase Kommandos
des Benutzers entgegennimmt und einige Anzeigen auf dem Bediensichtge-
rät aktualisiert. Sie führt, sofern sie nicht vorübergehend durch den
Rechenzeitbedarf von DISKIO blockiert ist, dauernd eine Programmschlei-
fe aus, die aus der Prüfung, ob eine Eingabe am Bediengerät erfolgt ist
und deren Bearbeitung, der Aktualisierung der Anzeigen und einer Wartezeit (200 ms) besteht.

Durch die Einfügung der Wartezeit wird Rechenzeit für die Task mit der
nächstniedrigeren Priorität bereitgestellt, ohne daß der Benutzer dies
in der Bedienung der Software merkt. Es handelt sich um die Task DAT-
AUS, die eine Auswahl der den Prozeßzustand beschreibenden Größen auf
das Anzeigesichtgerät numerisch und in halbgraphischer Form (Balkendia-
gramm) ausgibt.

Beispielhaft für die Leistungsfähigkeit und Flexibilität der Anwendungssoftware sei an dieser Stelle der Bausteinkatalog der Funktionsbausteine zur Verarbeitung quasikontinuierlicher Größen aufgeführt
(vgl. Bild 4). Jeder Bausteintyp kann beliebig oft in der Abarbeitungs-
sequenz erscheinen, so daß den Möglichkeiten zur Realisierung z.B. von
Reglern mit Anfahrsteuerungen und Überwachungen kaum Grenzen gesetzt
sind. Der Bausteinkatalog umfaßt die regelungstechnischen Grundfunk-
tionen P-Regler, PID-Regler, die Realisierung digitaler Filter und Regelalgorithmen durch Differenzengleichungen bis max. 3. Ordnung (DIFGL)
sowie das Bilden einer gewichteten Summe von bis zu vier Eingangskanälen (SUMME).

Neben diesen Standardelementen der Regelungstechnik enthält der Bausteinkatalog weitere Funktionsbausteine, die einerseits mit dem Ziel
einer zukünftigen Prozeßidentifikation eingeführt wurden und andererseits zur Realisierung von Anfahrsteuerungen etc. dienen. Hier sind zu
nennen der Umschalter zum Durchschalten eines von zwei Eingangskanälen
auf den Ausgangskanal (UMS), ein Signalgenerator, dessen Ausgangssignal
sich mit vordefinierter Änderungsgeschwindigkeit zwischen zwei Endlagen
ändert (RAMPE), ein BMFS-Testsignalgenerator zur Prozeßidentifikation

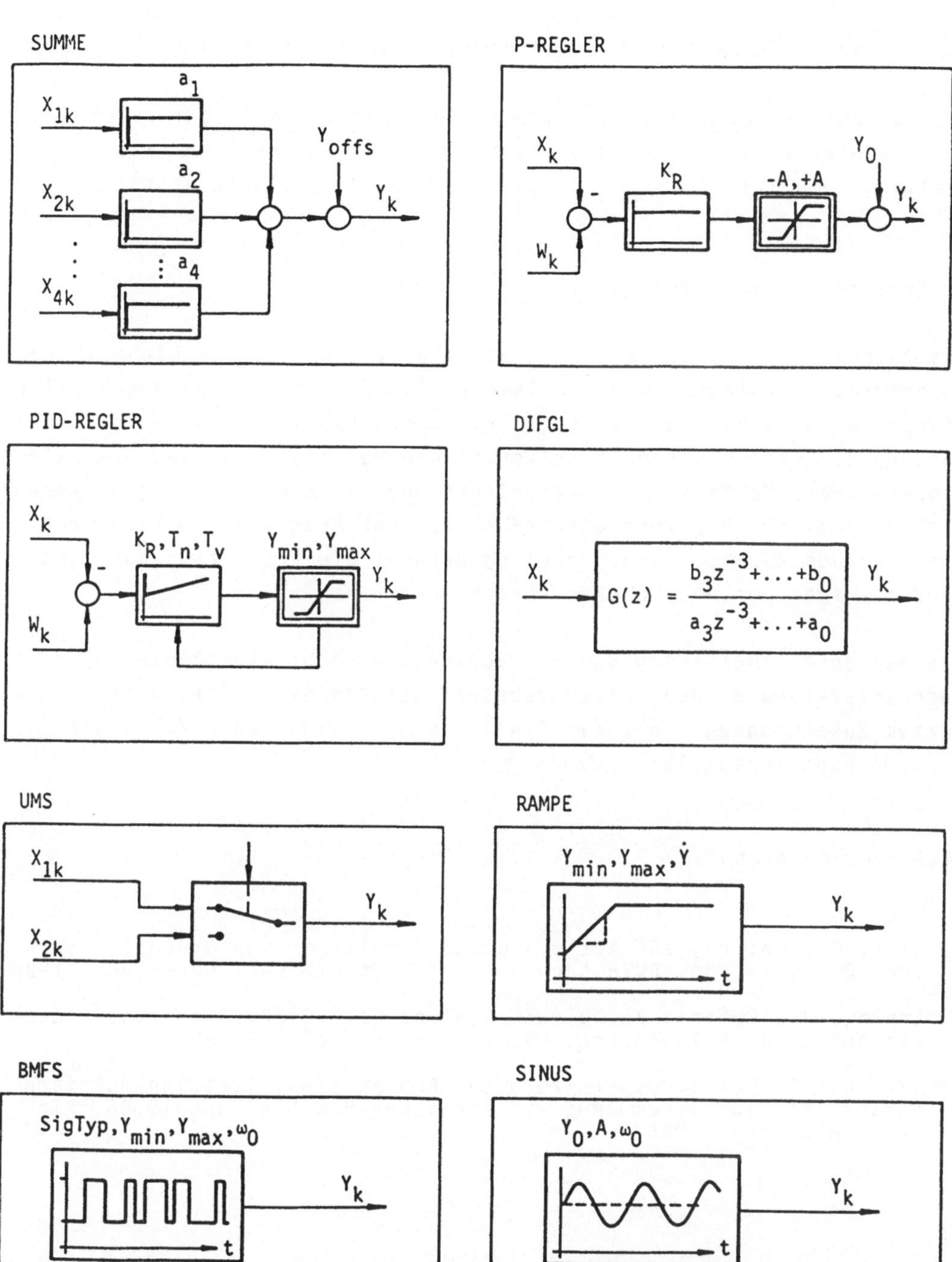

Bild 4: Funktionsbausteine der Anwendungssoftware

(BMFS) sowie ein Generator sinusförmiger Signale (SINUS).

Alle genannten Funktionsbausteine lassen sich manuell oder durch den Zustand binärer Variablen gesteuert ein- oder ausschalten, was die Darstellung komplizierter Steuerungs- und Regelungsabläufe gestattet.

5. Erfahrungen und Ausblick

Das Automatisierungssystem wurde parallel zum Aufbau der Gießwalz-Versuchsanlage entwickelt und realisiert. Es erlaubte von Anfang an eine sichere Beherrschung dieses sehr kritischen Prozesses. Durch die Auswertung der Versuchsergebnisse konnte die Modellbildung des Gießwalzprozesses bis heute so weit verfeinert werden, daß bereits eine wesentlich verbesserte Regelung entworfen und ohne Programmieraufwand realisiert werden konnte. Diese Regelung erlaubt die Produktion von Stahlband mit sehr geringen Dickenschwankungen.

Die geplante Erweiterung der Versuchsanlage macht eine Anpassung der Rechnerleistung an den Leistungsbedarf des Prozesses erforderlich. In diesem Zusammenhang steht die Erweiterung um eine 80386-Zentaleinheitenbaugruppe unmittelbar bevor.

Literaturverzeichnis

1. Hirt, G.: Beitrag zur Auslegung und Simulation von Gießwalzprozessen, Dissertation, RWTH Aachen, Verlag Stahleisen, Düsseldorf 1988.

2. Enning, M.: Beitrag zur Dickenregelung beim Gießwalzen von Bändern, Dissertation, RWTH Aachen, VDI-Verlag, Düsseldorf 1990.

3. Siemens AG: Betriebssystemkern des Realzeit-Multitasking-Multiprozessor-Betriebssystem RMOS 2. Technische Beschreibung 10.86, Siemens, München, 1986.

Automatisierungs- und Steuerungskonzept für ein hochflexibles Handhabungssystem zum Gußputzen

M.Lawo, H.Droll, A.Göbell, K.-H. Häfele, H.Haffner, J.Isele, P.Kohlhepp

Kernforschungszentrum Karlsruhe
Institut für Datenverarbeitung in der Technik
Postfach 3640
7500 Karlsruhe

Zusammenfassung

Im Rahmen des BMFT-Verbundprojektes "Komponenten für fortgeschrittene Handhabungssysteme" wurde das hochflexible System zur Bearbeitung von Gußteilen COMETOS (Coordinate Measuring and Tooling System) entwickelt.
Die Grundprobleme der Automatisierung liegen in den kleinen Losgrößen und den großen Toleranzen der Werkstücke; sie erfordern ein neuartiges Automatisierungskonzept, das auf einer Kombination von bedienergeführten Elementen, Sensorsignalverarbeitung sowie automatischen off-line-Programmkorrekturen beruht und das einen Schwerpunkt des vorliegenden Beitrags bildet.
Die Bearbeitung ist auf mehrere Stationen mit handelsüblichen Industrierobotersteuerungen verteilt. Dabei sind eine Vielzahl nebenläufiger Teilaufgaben durch eine geeignete Zellensteuerung zu koordinieren. Die Hauptmerkmale dieser selbst dezentral auf einem PC-Netz realisierten Steuerung werden beschrieben. Die Echtzeitaspekte werden am Beispiel Interprozeßkommunikation kurz beleuchtet. Ferner wird über einige Erfahrungen bei der ingenieurtechnischen Projektabwicklung berichtet.

1. Problemstellung und Automatisierungsansätze

Die typischerweise zu bearbeitenden Werkstücke sind LKW- und Baumaschinenachsgehäuse in verschiedenen Formen, die aus globularem Grauguß (GGG) in Formkästen der Größe 2800x1200x700 [mm] gefertigt werden (Abb. 1). Beim Gießen entstehen an den Nahtstellen zwischen den Formkastenteilen und in den Zuleitungen des flüssigen Eisens sowie den Kanälen zum Entweichen der Luft teils linien-, teils flächenförmige Materialanhäufungen (Grate), die durch mechanische Bearbeitung zu entfernen sind.

Bisher wird diese Bearbeitung per Hand mit elektrischen und pneumatischen Werkzeugen von max. 2KW Leistung durchgeführt. Die Arbeit ist laut, schmutzig und gefährlich. Sie gehört zu den 7 höchstbelastenden Tätigkeiten. Die Bearbeitungszeit für ein Werkstück wie in Abb. 1 beträgt ca. 45 Minuten. Dabei werden rund 11m Liniengrat entfernt.

Ansätze, um diese Aufgabe zu automatisieren, stehen vor folgenden Herausforderungen:
-   Die genauen Abmessungen sowie die Form, Größe und zum Teil auch Lage der Grate variieren von Werkstück zu Werkstück; die Werkstücke lassen sich auch nicht exakt reproduzierbar aufspannen. Die genauen Geometrieverhältnisse sind daher für jedes Werkstück individuell zu erfassen und in Roboterprogrammkorrekturen umzusetzen (**aktiver Toleranzausgleich**).
-   Wegen der kleinen Losgrößen (ca. 30) und der großen Werkstückvielfalt fällt der Zeitbedarf zum Programmieren (**Teachzeit**) entscheidend ins Gewicht; ein Grenzwert von 5h pro Los (Werkstücktyp) wurde als Projektziel gesetzt.
-   Der Anteil des Automatikbetriebs ist zu maximieren; die dabei erzielten **Bearbeitungsgeschwindigkeiten** sollen mindestens die des Handbetriebs erreichen.
-   Die **Bearbeitungstoleranz**, d.h. die Abweichung der Istkontur des bearbeiteten Werkstücks von der Sollkontur soll bei ca. 0.5mm liegen; es soll aber auch eine gute Schnittqualität erreicht werden.

Nicht zuletzt ist das Kosten-/Nutzen-Verhältnis der zu tätigenden Investitionen zu beachten (**Praktikabilität**). In diesem Zusammenhang stand etwa die Forderung, Standardkomponenten für die Steuerung zu verwenden.

Die bekannten Automatisierungsansätze zum Gußputzen mit Robotern, z.B. [1][2][3], können diesen Forderungen nicht gerecht werden:

Industrieroboter erreichen zwar eine hohe Bearbeitungsgeschwindigkeiten und -qualität, doch sind - bei **konventioneller Programmierung** - die Teachzeiten von teilweise weit über 100h pro Achsbrücke bei kleinen Losgrößen indiskutabel [1]. Konzepte für den Toleranzausgleich fehlen.

Die Teachzeiten lassen sich durch **offline-Programmierung** drastisch senken. Die dazu nötigen detailgetreuen CAD-Modelle der Werkstücke zu erstellen ist aber nur dann wirtschaftlich, wenn diese voll in den eigentlichen Fertigungsprozeß integriert sind, was für Gußteile heute in der Regel (noch) nicht der Fall ist.

Effizient und vor allem bedienerfreundlich programmieren lassen sich diffizile Bearbeitungsaufgaben auch durch **Master-Slave-Manipulatoren**. Wenn aber nicht alle Werkstücke vom Bediener im MSM-Betrieb geputzt werden sollen, stellt sich das Problem, wie die Bearbeitungsbahnen und -parameter sinnvoll auf einen späteren Automatikbetrieb zu übertragen sind, vom Toleranzausgleich ganz zu schweigen.

Ein naheliegender Ansatz für den Toleranzausgleich ist die Anpassung eines groben Automatikpogramms an den genauen Gratverlauf mittels vorausschauender **on-line-Sensorsignalverarbeitung** [2][3]. Diese on-line-Kopplung führt aber zu sehr niedrigen Bearbeitungsgeschwindigkeiten, die neben der Wirtschaftlichkeit auch die Bearbeitungsgüte beeinträchtigen (z.B. Blaufärbung durch Wärmeentwicklung). Sowohl MSM-Steuerungen als auch die on-line-Sensorrückführung erfordern dedizierte Steuerungsentwicklung/-optimierung.

Grundelemente des COMETOS-Automatisierungskonzeptes

Neu ist die Art der Aufgabenteilung zwischen bedienergeführter Programmierung und automatischer Sensorsignalverarbeitung:

- Die Struktur der Bearbeitung, die technologischen Parameter und eine grobe Geometrieinformation stammen aus der Programmierung. Das Wissen des Bedieners über Werkstückform, Grate, geeignete Werkzeuge, Zugänglichkeit, Kollisionsvermeidung etc. wird so ausgenutzt.

  Programmiert wird durch direktes Führen des Roboters im MSM-Betrieb, eingebettet in eine speziell für das Gußputzen entwickelte Bedieneroberfläche, die den Bediener optimal unterstützt.

- Die geometrische Feinstruktur (Toleranzausgleich) stammt aus den Ergebnissen der Vermessung mit Hilfe optischer Sensoren.

  Dies führt jedoch nicht zur on-line-Bahnkorrektur, sondern zur Generierung korrigierter Roboterprogramme. Im Grunde entsteht eine Kette von Programmen, deren Anfangsglieder - aufgrund des Meßverfahrens - grobe Toleranzen erlauben und somit schnell zu teachen sind, deren Ablaufdaten aber nachfolgende Programme mit immer kleineren Toleranzen erzeugen (bootstrapping-Prinzip).

Dadurch wird eine Entkoppelung zwischen Sensorsignalverarbeitungs- und Bearbeitungsgeschwindigkeit erreicht. Ein Zeitvorteil entsteht allerdings nur dann, wenn Messen und Bearbeiten parallel auf getrennten Stationen ablaufen können. Dies wiederum setzt voraus, daß die kinematischen Verhältnisse (Roboter, Werkstückaufspannung) von der Meß- auf die Bearbeitungsstation übertragbar sind.

Der Zellenaufbau und die Teilaufgaben werden in Abschnitt 2 grob beschrieben (in [4] detaillierter). Die Vorgehensweise zur Programmgenerierung folgt in Abschnitt 3.

2. Zellenaufbau und Teilaufgaben

Die Zelle umfaßt derzeit drei Kinematiken: einen kartesischen Meßroboter in Portalbauweise (KUKA IR 461.60), einen kartesischen Bearbeitungsroboter in Portalbauweise (KUKA IR 461.240) und einen im Rahmen des Projektes neu entwickelten Knickarmroboter (NOELL SGR750). Als Robotersteuerung wird die r500 der Fa. AEG mit der Roboterprogrammiersprache DOROB eingesetzt.

Der Meßroboter verfügt über einen nach dem Triangulationsprinzip arbeitenden Laserscanner zur berührungslosen Werkstück- und Gratvermessung. Dieser Sensor nimmt an jedem Meßpunkt ein zweidimensionales Höhenprofil des Werkstücks in einem Fenster von ca. 60x60mm auf. Die Funktionen des Laserscanners und der Bildauswertung (Graterkennung) sind in [5] beschrieben. An der Roboterhand ist ferner eine nachgiebige Sensorkugel (Kraft-Momenten-Sensor) angebracht, um den Meßarm ohne Kraftaufwand zu führen.

Mit dem Knickarmroboter, der mit einer Trennscheibe von 500mm Durchmesser ausgerüstet ist, werden die ziemlich sperrigen Steiger und Angüsse vorab getrennt. Das nachfolgende Abschleifen der Linien- und Flächengrate erledigt der kartesische Bearbeitungsroboter mit mehreren über eine automatische Werkzeugwechseleinrichtung anflanschbaren Werkzeugen (dicke und dünne Schruppscheiben mit 300mm Durchmesser, Hartmetallfräser).

Zwischen den Stationen werden die Werkstücke auf Paletten transportiert. In den Stationen werden die Paletten auf Indexierbolzen abgesetzt und hydraulisch gespannt; die Position desselben Werkstücks zwischen Meß- und Bearbeitungsstation ist dabei auf ca.1/10mm reproduzierbar; unterschiedliche Werkstücke können sich in der Aufspannung um ca. 10mm unterscheiden.

Die Standard-Abfolge der Teilaufgaben für ein Werkstück lautet:

Aufspannen -> Messen(Aufspannung) -> Trennen -> Messen(Gratgeometrie) -> Putzen -> Abspannen.

3. Programmerzeugung - Aufgabenteilung und Informationsflüsse

Das Datenflußmodell in Abb. 2 zeigt, welche Informationen bei der Programmerzeugung in COMETOS wo einfließen.

Die wichtigste Information für die Programmgenerierung stammt aus der **bedienergeführten Programmierung** für das 1. Werkstück jedes Loses, ein Teil aus den Daten des **Automatikbetriebs** der geteachten Programme, und ein Teil sind, zumindest momentan, fest eingerichtete **Anlagenparameter**.

Die Programmierung gliedert sich in das **Teachen eines Meßprogramms**, mit dem die Werkstück- und Gratgeometrie vermessen und zugleich die Grobstruktur und -geometrie des späteren Bearbeitungsprogramms festgelegt wird, und das **Teachen des Bearbeitungsprogramms**, das alle werkzeugabhängigen Festlegungen trifft. Ein Programm ist gegliedert nach **Grattypen** - zuerst werden Flächengrate, dann Liniengrate bearbeitet - und weiter untergliedert nach den einzelnen **Graten** und ggf. Teilgraten.

Für einen einzelnen Grat wird das Meßprogramm durch direktes Führen des Meßroboters programmiert (Abb. 1). Dabei wechseln "Gratbahnelemente", bei denen der Grat ständig im Fenster des Laserscanners sichtbar sein sollte, mit "Luftbahnelementen" ab, die für Umorientierungspunkte, kollisionsfreie Wege zwischen Teilgraten, An-Rückfahrwege von bzw. zur Grundstellung etc. stehen. Diese Programmstruktur ergibt sich aus der Reihenfolge der über pull-down-Menus aktivierten Funktionen der Bedieneroberfläche.

Die beim Führen aufgenommenen Roboterkoordinaten werden zunächst geglättet (reduziert) und die Bahnelemente zu einem ersten Gesamtprogramm, dem sog. **Roh-Meßprogramm** zusammengefügt. Dieses Roh-MP realisiert eine Art 'Stop-and-Go'-Verkehr, in dem abwechselnd der Meßroboter den nächsten Punkt anfährt, kurz anhält und der Laserscanner ein Querprofil des Werkstücks aufnimmt. Dieser Stop-and-go-Verkehr erleichtert die Zuordnung der in sensor-relativen Koordinaten berechneten Gratfußpunkte zu den aktuellen Roboterpositionen in Weltkoordinaten.

Als nächstes muß aus der Folge diskreter Gratfußpunkte ein zusammenhängender Werkstückumriß in Gratlängsrichtung (Soll-Bearbeitungsbahn) rekonstruiert werden. Hier fließt a priori Wissen über das Werkstück ein, z.B. ob der Grat in einer Ebene liegt und wie groß der Minimalabstand der signifikanten Konturänderungen am Werkstück ist. Zunächst einmal müssen genügend viele und genügend dichte Gratfußpunkte vorliegen (Abtastproblem); ggf. wird das Rohmeßprogramm automatisch an den Stellen, wo zuwenig Meßpunkte vorliegen, feiner interpoliert. Führt dies zu keinem Erfolg, so befindet sich das Werkstück/der Grat außerhalb des Scanfensters und der Bahnverlauf des Laserscanners muß an dieser Stelle durch den Bediener nachgebessert (editiert) werden.

Das Verhältnis zwischen Meßaufwand und Abtastfehler, das die Betriebsziele Teachzeit und Bearbeitungstoleranz betrifft, kann verbessert werden, indem eine grobe Polygonzugapproximation des Werkstücks in Gratlängsrichtung durch eine Bahnelementunterteilung beim Führen vorgegeben wird. Durch einen geeigneten Algorithmus werden die gemessenen Umrißelemente - nach Zusammenfassen/Umsortieren - den a priori gegebenen 1:1 zugeordnet und zu einem geschlossenen Polygonzug verbunden, falls möglich. Aus diesem Werkstückumriß wird ein optimiertes **Automatik-Meßprogramm** generiert, das den Sensor senkrecht auf die Werkstückoberfläche mit dem Gratfußpunkt im Zentrum des Sensorfensters stellt, woimmer dies kollisionsfrei möglich ist, und das auf den geraden Werkstückregionen mit wenigen Meßpunkten auskommt. Mit diesem einen Meßprogramm können aufgrund der Toleranz des Sensorfensters alle folgenden Werkstücke dieses Loses vermessen und aus den Scandaten Werkstückumrisse errechnet werden, wobei die Iterations- und Verifikationsschritte des Roh-Meßprogramms nicht mehr durchlaufen werden.

Zugleich ist das Automatik-Meßprogramm Ausgangspunkt für die **Rohversion des Bearbeitungsprogramms**. Nachdem das 1. Werkstück in die Bearbeitungsstation eingelaufen ist, läuft dieses Roh-Bearbeitungsprogramm dort in einem Probelauf verlangsamt, jederzeit unterbrechbar, Bahnelement für Bahnelement ab. Dort bestehen folgende Eingriffsmöglichkeiten:

1.   Die Werkzeugauswahl ist vorzunehmen/zu ändern. Dementsprechend ist die Orientierung des Roboters neu einzustellen; sie ist beim Bearbeiten im allgemeinen eine andere als beim Messen, jenachdem ob eine dicke Schruppscheibe, eine Trennscheibe oder ein Fräser eingesetzt wird. Auch die Zustellung in Richtung der Werkstücknormale wird interaktiv vorgegeben. Die Vorschubgeschwindigkeiten für jedes Werkzeug sind dagegen reine Erfahrungswerte (fest eingerichtete Anlagenparameter).

    Die aus dem Werkstückumriß abgeleitete Bahngeometrie des Meßprogramms wird also angepaßt, wobei kinematische Transformationen zwischen Meß- und Bearbeitungsstation berücksichtigt werden.

2.   Da der Grat meist nicht überall durch jedes Werkzeug kollisionsfrei zugänglich ist, müssen Bahnelemente ggf. aufgetrennt und Werkzeugwechselprogramme eingefügt werden. Um die Zahl der Werkzeugwechsel in Grenzen zu halten, sollte die Bearbeitung an solchen Stellen unterbrochen und an anderer Stelle, wo mit dem alten Werkzeug weiterbearbeitet werden kann, wieder aufgesetzt werden. Dadurch ändert sich die Reihenfolge im Durchlaufen der Bahnelemente. Zusätzlich müssen neue Zwischenwege programmiert werden, um die übersprungenen Werkstückregionen kollisionsfrei zu umfahren.

3.   Schließlich können an besonders kniffligen Stellen Bahnelemente aus dem Meßprogramm ersetzt und im Master-Slave-Betrieb neu geteacht werden.

Während diese Korrekturen nur einmal, beim Programmieren des Bearbeitungsprogramms anfallen, wird der Umriß für jedes Werkstück bestimmt und daraus ein angepaßtes Bearbeitungsprogramm generiert.

4. Zellensteuerung

4.1 Aufgabe der Zellensteuerung

Bei der Bearbeitung von Werkstücken in der Gußputzanlage müssen viele Teilschritte auf verschiedenen Stationen koordiniert ablaufen: die Roboterprogrammierung, das Aufzeichnen der Roboterkoordinaten und Sensordaten, das Generieren neuer Programme aus diesen Daten und das Laden, Ausführen und Überwachen dieser Programme, Werkstücktransport, Werkzeugwechsel und Werkzeugvermessung etc. Die Zellensteuerung, die all diese Aufgaben koordiniert, soll folgendes leisten:

- das *gleichzeitige* Bearbeiten mehrerer Werkstücke, die zu *unvorhersehbaren Zeitpunkten* eintreffen und das System auf *verschiedenen Bearbeitungspfaden* durchlaufen können.
  So fällt die Programmierng der Meß- und Bearbeitungsprogramme nur beim 1. Werkstück jedes Loses an. Bei den folgenden Werkstücken können einzelne Bearbeitungsschritte fehlen (z.B. Trennen) oder übersprungen werden, z.B. wegen Ausfall einer Maschine, und andere hinzukommen, z.B. die Nachbearbeitung unzugänglicher Grate von Hand.
- das korrekte Abwickeln der Abarbeitungspfade jedes einzelnen Werkstücks, d.h. Umsetzen der Schritte in ausführbare elementare Transport-, Rechen- und Bearbeitungsaufträge unter Beachtung ihrer *kausalen Abhängigkeiten* und ihrer *benötigten Betriebsmittel.*
  Z.B. kann eine Achsbrücke erst vermessen werden, nachdem sie in die Meßstation transportiert und das zuständige Automatikprogramm erzeugt und geladen wurde.

Diese Anforderungen sind typisch für flexible Fertigungssteuerungs-Systeme (FFS) [6]. Dezentrale, über lokale Netze verbundene Steuerungssysteme ersetzen heute zunehmend hierarchisch strukturierte, zentral gesteuerte [7]. Die lokalen Stationsrechner sollen möglichst autonom entscheiden können, wann bzw. in welcher Reihenfolge sie die zur Ausführung freigegebenen Aufträge bearbeiten.

4.2 Konzept der Auftragsverwaltung

Die wesentlichen Elemente zur Umsetzung dieser Forderungen werden im folgenden skizziert; das Kanal-Instanz-Netz (Abb.3) gibt einen Überblick über die Datenflüsse.

Ablaufpläne

Durch die Ablaufpläne wird spezifiziert, welche Schritte mit welchen Betriebsmitteln (Stationen) in welcher Reihenfolge und mit welchen Daten auszuführen sind. Bei den Betriebsmitteln wird unterschieden zwischen

- Aufgaben, die auf einer bestimmten Station ablaufen müssen, weil sie dedizierte Prozeßperipherie benötigen wie z.B. den Laserscanner zum Messen

- reinen "Rechenaufgaben" wie Datenreduktion, Graterkennung, Codegenerierung etc., die auf jedem beliebigen freien Prozessor ablaufen können. Solche Rechenaufgaben beanspruchen bei COMETOS einen hohen Anteil der Gesamtzeit.

Ferner wird die Art der Abarbeitung (synchron/asynchron) spezifiziert:

- bei **seriellen** (synchronen) Aufträgen ist deren Ende in jedem Fall abzuwarten, bevor der nächste Auftrag desselben Ablaufplans in Angriff genommen wird

- bei **parallelen** (asynchronen) Aufträgen können weitere Aufträge im selben Ablaufplan nebenläufig gestartet werden; es wird erst später explizit gewartet.

Es gibt **allgemeine** Ablaufpläne, die werkstücktypspezifische Bearbeitungsschemata darstellen, und **individuelle** (werkstück- bzw. palettenspezifische) Ablaufpläne, die erst beim Einlaufen einer Palette in die Anlage durch die Leitrechnerfunktion in Abb. 3 nach einem passenden Schema (allgemeinen Ablaufplan) dynamisch erzeugt und beim Verlassen der Anlage wieder vernichtet werden.

Warteschlangen

Jeder Station ist eine eigene Warteschlange der von ihr zu bearbeitenden Aufträge zugeordnet:

PrivateQueue[i] : Warteschlange der exklusiv auf Station i zu bearbeitenden Aufträge

PublicQueue : Warteschlange der Rechenaufträge, die überall bearbeitet werden können.

Jeder Warteschlangenauftrag ist mit einer Prioritätszahl versehen; die Aufträge sind nach steigenden Prioritäten geordnet. Verschiedene Vorgaben für die Priorität führen zu verschiedenen Strategien, z.B.

- Priorität=Generierungszeit => FIFO-Strategie

- Priorität=verbleibender Spielraum bis zu einer gegebenen Frist => just in time-Strategie.

Stationssteuerung, Scheduling

Die Stationststeuerung ist der Algorithmus, nach dem jede freie Station autonom entscheidet, welchen der freigegebenen Aufträge sie als nächsten bearbeitet.

Im ersten Schritt wird die Auftragswarteschlange ausgewählt:

- mit höchster Priorität wird die zugehörige PrivateQueue bedient, falls nichtleer

- mit zweithöchster Priorität werden Aufträge aus Ablaufplänen bearbeitet, sofern die dort spezifizierten Vorbedingungen erfüllt sind, z.B. die Fertigmeldungen früherer Aufträge vorliegen.

- mit niedrigster Priorität werden Aufträge aus der PublicQueue aufgegriffen.

Die Prioritätenregelung ist dadurch begründet, daß jede Station für die Aufträge aus ihrer PrivateQueue ein unersetzliches Betriebsmittel ist, während bei PublicQueue-Aufträgen auch andere Stationen einspringen und einem Engpaß vorbeugen können.

Im zweiten Schritt wird aus der gewählten Warteschlange der höchstpriore Auftrag entnommen.

Ablaufpläne verfügen über eine Kontrollstruktur (Auftragszeiger) und werden interpretativ abgearbeitet; der Zeiger wird abhängig vom Ergebnis des letzten Schrittes entweder auf den nachfolgenden Auftrag gesetzt, oder der letzte Schritt oder eine ganze Auftragssequenz wiederholt oder (schlimmstenfalls) der gesamte Ablaufplan abgebrochen.

Lokale Auftragsinstanzen

Sie erledigen die eigentliche Arbeit vor Ort. Die einzigen Schnittstellen zur Stationssteuerung sind

-       die **Auftragsanforderung**, die den Auftrag einschließlich Parametern eindeutig spezifiziert, auf einen ausführbaren Algorithmus verweist und dessen Ausführung veranlaßt

-       die **Auftragsfertigmeldung**, die die Auftragsterminierung und ihr Ergebnis signalisiert.

Auftragsinstanzen können bei ihrer Arbeit dynamisch weitere Aufträge für andere Stationen (Private oder PublicQueue) erzeugen und auf ihr Ende warten (Beispiel: Lesen der aktuellen Roboterkoordinaten beim Teachen des Meßprogramms).

Globale Kommunikation, Datenhaltung und Zugriffssynchronisation

Neben der Synchronisation der Abläufe sind auch die Datenzugriffe zu synchronisieren, wenn z.B. mehrere parallele, nicht synchronisierte Aufträge im selben Ablaufplan lesend und schreibend auf gemeinsame Daten zugreifen.

4.3 Realisierung und ingenieurtechnische Abwicklung

Das in 4.2 beschriebene funktionale Modell der Zellensteuerung kann prinzipiell auf mehrere Arten in reale HW/SW "gegossen" werden. Wir diskutieren zunächst den bei der COMETOS-Pilotinstallation eingeschlagenen Weg, und gehen dann noch kurz auf zwei mögliche Alternativen ein.

Steuerungs-Hardware

Jedem Roboter (Meß-,Trenn-,Putzroboter) wird ein eigener Stationsrechner zugeordnet (Abb 4). Diese Stationsrechner (PC_M, PC_T, PC_P) sind IBM-kompatible PC's (386 AT) mit Betriebssystem Dos3.3. Die lokale Kommunikation zwischen Robotersteuerung und PC, d.h. das Laden von DOROB-Programmen, Erfassen von Koordinaten, Übergeben von Sensorkugelsignalen erfolgen über V24 DNC-bzw. Sensorkugelschnittstellen.

Gründe für die Rechnerauswahl waren neben dem Preis v.a. das reichhaltige SW-Angebot (z.B. Bedieneroberflächen). Mit TURBO PASCAL 5.5, einer objektorientierten PASCAL-Erweiterung, steht eine komfortable integrierte SW-Entwicklungsumgebung zur Verfügung. Über Zusatzkarten für Analog/Digital I/O können technische Prozesse gesteuert werden.

Die Koppelung der PC's erfolgt über ein (autark betreibbares) Segment des KfK-LAN auf Ethernet-Basis (IEEE 802.3, XNS-Protokoll). Für die globale Datenhaltung existiert ein PC-Server mit erweitertem Massenspeicherausbau. Alle PC's verfügen über die 3COM-Netz-SW.

4.3.1 COMETOS-Pilotinstallation

Alle lokalen Auftragsinstanzen sind durch DOS-Hauptprogramme realisiert. Die Auftragsanforderung entspricht dem Hauptprogrammstart und erfolgt über Kommandostrings wie in Batchdateien. Auftragsfertigmeldungen bestehen im wesentlichen aus dem DOS-exit-Code bei Verlassen des Hauptprogramms.

Alle Auftragsdaten werden in globalen, vom Server-PC zentral verwalteten Dateien gehalten. Dies gilt auch für die Auftragswarteschlangen und Ablaufpläne. Der Grundmechanismus der Kommunikation ist also der File Transfer bzw. Remote File Access.

Die Stationssteuerung wurde im Auftrag entwickelt [8] und existiert auf jedem PC, ebenfalls als HP, das aber andere HPe startet. Im Grunde handelt es sich um ein anwendungsspezifisches "Betriebssystem".

Da der abgesetzte Filezugriff der DOS-HPe wie der lokale über logische Laufwerknamen erfolgt und jeder Rechner über dieselben Schnittstellendienste der Auftragssteuerung verfügt, ist die physikalische Verteilung der Aufgaben auf die Rechner transparent für die Anwendungsentwicklung. Die Rechnerkonfiguration ist daher sehr leicht änderbar/erweiterbar.

Ein weiterer, bei den engen zeitlichen Randbedingungen des Verbundprojektes sehr wichtiger Vorteil war, daß mit dem vorhandenen Rüstzeug (Arbeitsplatz-PC's, DOS3.3, TURBO 5.5, existierende Anwender-SW, LAN) rasch eine lauffähige, verteilte Pilotinstallation aufgebaut werden konnte. Für andere Alternativen (s.u.) fehlten wesentliche Voraussetzungen, bzw. der Anfangsaufwand, sie zu schaffen, wäre deutlich höher gewesen.

Jeder Warteschlangenauftrag zieht bei dieser Lösung das Laden/Starten eines DOS-HP nach sich. Für den groben Kontrollfluß ist das sicher sinnvoll; für eine feinere Interaktion nebenläufiger Instanzen ist das häufige Laden/Entladen von HPen und Ver-/Entpacken der Daten in globalen Files zu umständlich und ineffizient; echtzeitkritische Kommunikation prozeßnaher Algorithmen erfordert in jedem Fall zusätzliche Mechanismen (vgl. Abschnitt 5). Die Hilfsmittel der 3COM-Netz-SW für die Zugriffssynchronisation auf globale Dateien sind wenig komfortabel. Als ständiges Ärgernis und in gewissem Sinn als Anachronismus erwies sich die 640KB-Beschränkung beim Laden der DOS-HPe, wobei der verfügbare Speicher durch die Bedieneroberfläche, die Stationssteuerung selbst und die interaktive Testumgebung weiter vermindert wurde.

Ein Vorteil der Steuerung ist sicher, daß sie auf Standard-System-SW basiert, um den Preis allerdings, daß wesentliche Funktionen eines komfortableren (Netzwerk-)BS in die Anwendung verlagert oder auf sie verzichtet werden mußte.

4.3.2 Multitasking-(Netzwerk-)Betriebssysteme

Der Einsatz von Multitasking-Betriebssystemen (wie z.B. OS/2 für IBM-PC) liegt nahe, weil sie u.a. standardmäßig Objekte und Operationen für die Warteschlangenverwaltung zur Verfügung stellen.

Auftragsinstanzen sind dann Prozesse, Tasks oder Threads, die wie Hauptprogramme dynamisch geladen/erzeugt werden können, häufiger jedoch speicherresident sind und ständig auf Aufträge warten. Die 640KB-Einschränkung für die rechnerlokale Anwender-SW entfällt. Das Subsystem "Teachen des Meßprogramms" könnte z.B. 4 parallele Prozesse gleichzeitig enthalten: die Bedieneroberfläche, die Bahnelementauswertung im Hintergrund, die Erfassung der Roboterkoordinaten, und die Stationssteuerung.

Der Grundmechanismus zur Interprozeßkommunikation ist der direkte Nachrichtenaustausch über Kommunikationskanäle (mailboxes, ports, bzw. queues und pipes in OS/2-Terminologie), nicht der File-Transfer.

Auftragsanforderungen/-fertigmeldungen werden durch Nachrichten übermittelt, egal ob es sich um "statische", im Ablaufplan stehende oder um dynamisch erzeugte Aufträge handelt. Auch die Auftragsdaten (außer Initialisierungsdaten) können durch Nachrichten übergeben werden. Letztlich erfolgen auch die Regiewechsel zwischen Anwenderprozessen und lokaler Stationssteuerung auf natürliche Weise nachrichtengesteuert.

Der Vorteil dieser Lösung liegt in der größeren Einheitlichkeit und Effizienz der Kommunikations- und Synchronisationsmechanismen und der einfacheren Programmierung.

Verteilungstransparenz wie in 4.3.1 wird jedoch erst durch ein Netzwerk-Betriebssystem (Betriebssystemverbund) erreicht, in dem der Programmierer nicht merkt, auf welchem Rechner die durch Systemdienste manipulierten Prozesse oder Kommunikationskanäle sich befinden. Ein Netzwerk-BS zu OS/2 existiert noch nicht. Die dazu nötigen Dienste mit Hilfe lokaler BS-Dienste und Netbios-Diensten selbst zu implementieren, hätte einen höheren Aufwand als bei der Variante 4.3.1 bedeutet.

4.3.3 MMS-Dienste

MMS ('manufacturing message specification') ist ein Normentwurf der anwendungsorientierten Kommunikationsdienste für die industrielle Fertigung im Rahmen der MAP-Protokollhierarchie [9]. Die Kommunikation basiert auf dem Nachrichtenaustausch nach dem client-server-Modell. Client ist Anforderer eines Dienstes, z.B. die Stationssteuerung; Server der Erbringer des Dienstes, z.B. die Auftragsinstanz eines Stationsrechners. Die konkrete Implementation der Auftragsinstanzen als Prozesse, Hauptprogramme, Unterprogramme etc. wird in MMS nicht vorgeschrieben, sondern durch das abstrakte Objekt *virtual manufacturing device* bzw. seine Unterobjekte (*domain object, program invocation*) subsumiert. Auch Strukturen ähnlich den Ablaufplänen (*transaction object*) stehen standardmäßig zur Verfügung.

MMS bietet also höhere und anwendungsnähere Primitive zur Implementierung der Ablaufsteuerung als Netzwerk-Betriebssysteme und stellt zugleich einen herstellerunabhängigen Standard dar. MAP/MMS erfordert andererseits auch einen erheblichen Aufwand bei der Installation, Erprobung (Tuning) und Einarbeitung. Ein kritischer Punkt ist die Echtzeittauglichkeit der 'Full Map'-Protokollhierarchie; bzw. die Frage, wieweit 'abgemagerte' Versionen noch dem Standard entsprechen. Diese Fragen werden zur Zeit an einer Pilotinstallation auf einem PC-Netz untersucht [10].

5. Echtzeit-Interprozeßkommunikation

Interprozeßkommunikation unter Echtzeitbedingungen tritt beim Teachen des Bearbeitungsprogramms auf, wenn Bahnelemente im MSM-Betrieb programmiert werden. Dabei müssen (vgl. Abb. 4) die Kraftsignale von der Sensorkugel beim Führen des Meßroboters nach Wandlung/Umrechnung in PC_M sowohl an die Robotersteuerung r500_M als auch (über PC_P) an die Steuerung r500_P übertragen werden. Umgekehrt müssen, um eine Kraftreflexion zu realisieren, die Signale des Kraft-Momenten-Sensors am Putzroboter über PC_P an PC_M übertragen werden, um dort auf die Sensorkugel-Schnittstelle der r500_M einzuwirken. Um ausreichende Regelgüte zu erzielen, müssen beide Kommunikationen innerhalb des Bahnplanungs-Zeittaktes der Robotersteuerungen (ca. 38msec) stattfinden.

Dazu wurden im Vorfeld umfangreiche Messungen mit 2 PC's durchgeführt, die Datenpakete konstanter Nutzdatenlänge (128Byte) direkt mit Hilfe der NetBios-Dienste der 3COM-Software austauschen. Unterschiedliche Ethernet-Controller-Karten (3C505,3C503), Softwaretreiber-Versionen (3Open, 3Share), Taktraten der Host-PC's von 8,12 und 16MHz sowie unterschiedliche Interaktionsmuster wurden dabei verwendet [11]. Für den wichtigsten Anwendungsfall "bidirektionaler Datenaustausch mit Warten auf das Übertragungsende" ergaben sich bei der Konfiguration 3C503/3Open *maximale* Antwortzeiten (end-to-end) zwischen 7.3 bei 16MHz und 15.0msec bei 8MHz. Damit konnten die Echtzeitanforderungen der Bahnplanung erfüllt werden.

Überraschenderweise wurden aber bei der Konfiguration mit "intelligenter" Karte 3C505, auf die die SW-Treiber zur Entlastung des Host-PC geladen werden können, durchweg schlechtere Maximalwerte gemessen als mit der "nicht intelligenten" Karte 3C503. Ferner fielen bei der Treiber-Version 3Share einzelne, gegenüber 3Open ca. 10fach höhere Maximalwerte auf, die nur in bestimmten SW-Betriebsarten auftraten, dort aber regelmäßig, und die nicht durch statistische Kollisionsereignisse erklärt werden konnten.

6. Zusammenfassung und Ausblick

Die Anlage wurde Ende 1989 im handhabungstechnischen Labor des Kernforschungszentrums Karlsruhe aufgebaut und in Betrieb genommen. An Musterwerkstücken (LKW-Achsbrücken aus globularem Grauguß GGG50 der Kundengießerei Schubert&Salzer in Ingolstadt) wurden die verschiedenen SW-Bausteine erprobt. Eine zweite Anwendung, die Bearbeitung von Druckmaschinenseitenwänden aus dem Grauguß GG25, ist derzeit in Vorbereitung.

Das Hauptziel, die Programmierzeit zu reduzieren, wurde mit der Bedieneroberfläche und dem Meßverfahren erreicht. Bei der Bearbeitung im Probebetrieb werden die Möglichkeiten, die Bearbeitungszeit durch höhere Bahngeschwindigkeit und größere Werkzeugleistung zu senken, noch nicht ausgeschöpft.

Im Rahmen der Pilotinstallation wurde auch die Zellensteuerung in Betrieb genommen und damit gezeigt, daß mit einfachen Mitteln (z.B. ohne multitasking) ein funktionierendes verteiltes System aufgebaut werden konnte.

Wie mit dem Namen gesagt werden soll, ist mit COMETOS nicht nur Gußputzen möglich. Zukünftige Arbeiten haben u.a. eine Erweiterung des Bearbeitungsspektrums und der Werkstückhandhabung zum Ziel. Ein weiterer Schwerpunkt ist die Entwicklung der Meßprogramme aus CAD-Modellen.

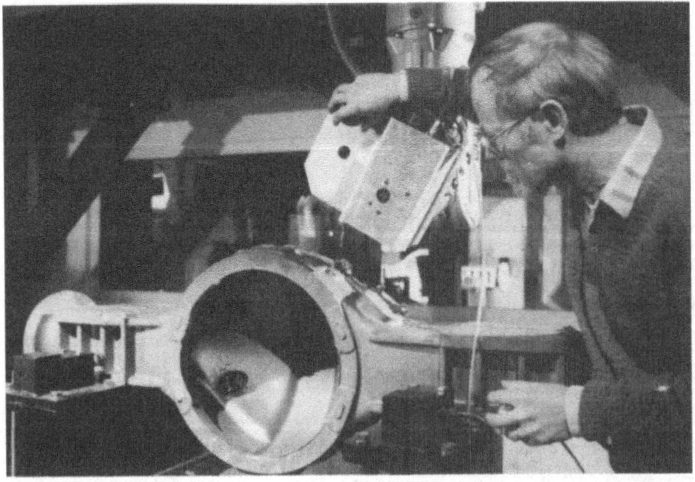

Abb. 1: Führen des Meßroboters über eine LKW-Achsbrücke

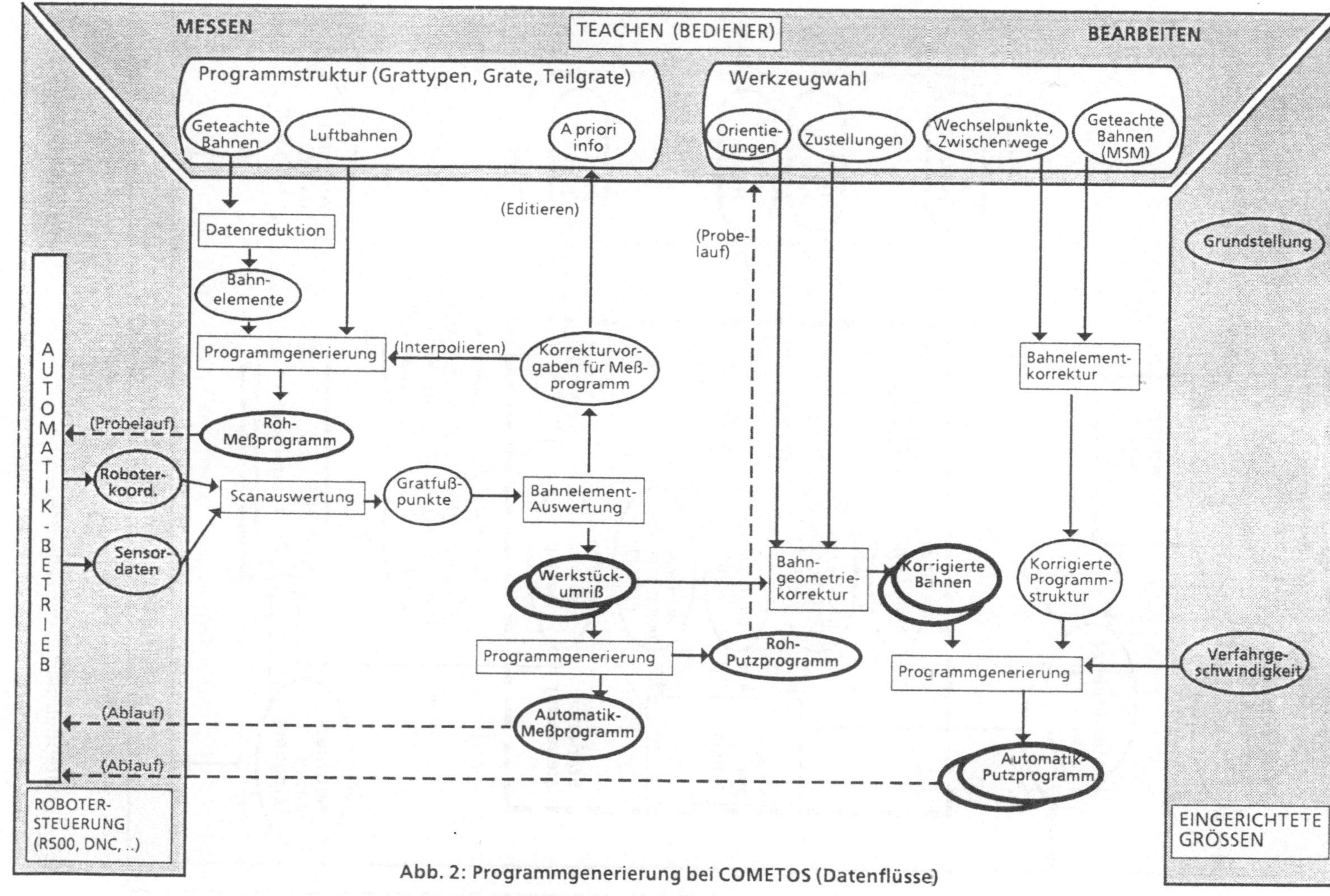

Abb. 2: Programmgenerierung bei COMETOS (Datenflüsse)

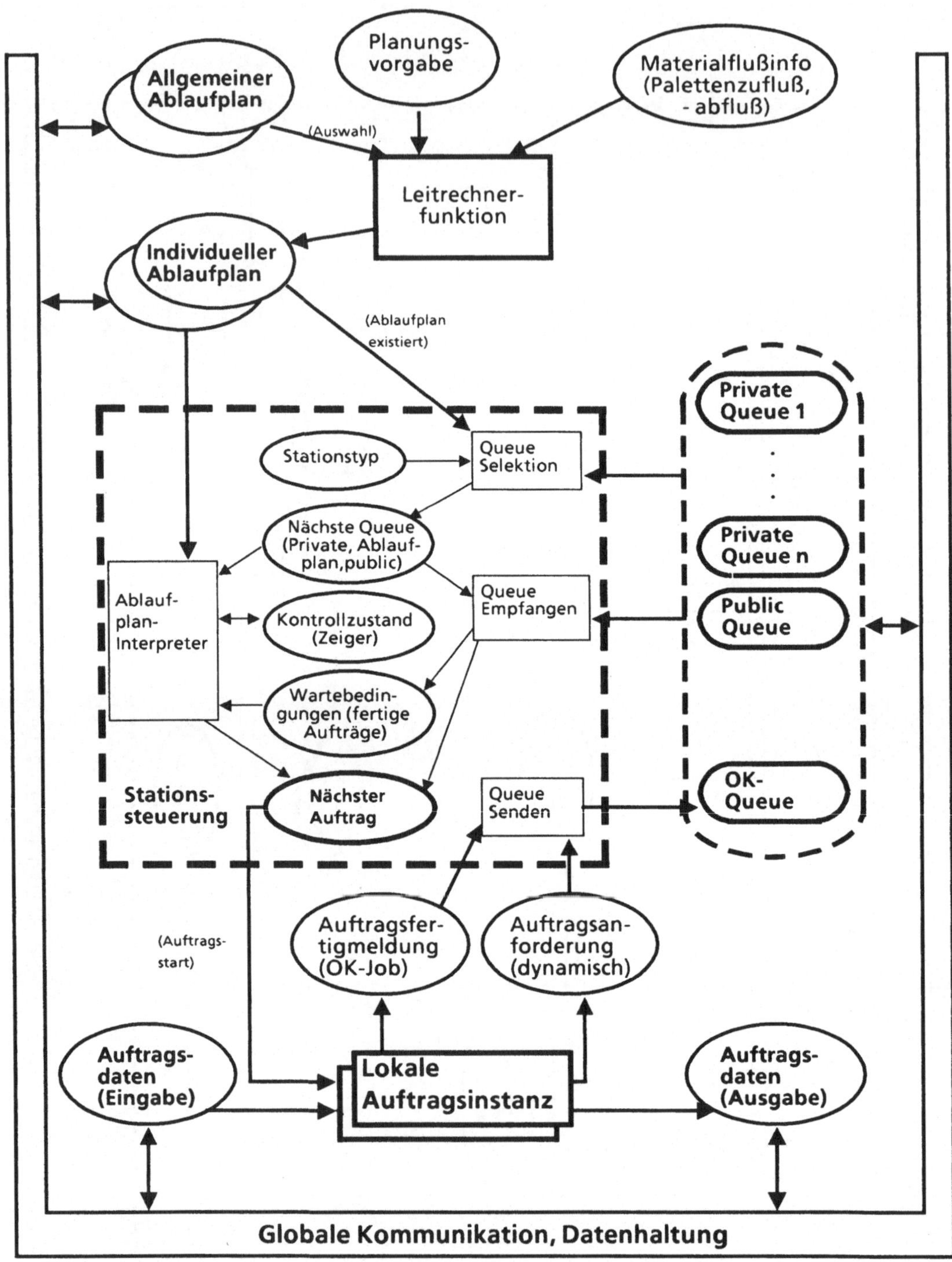

Abb. 3: Datenflußmodell der Zellensteuerung

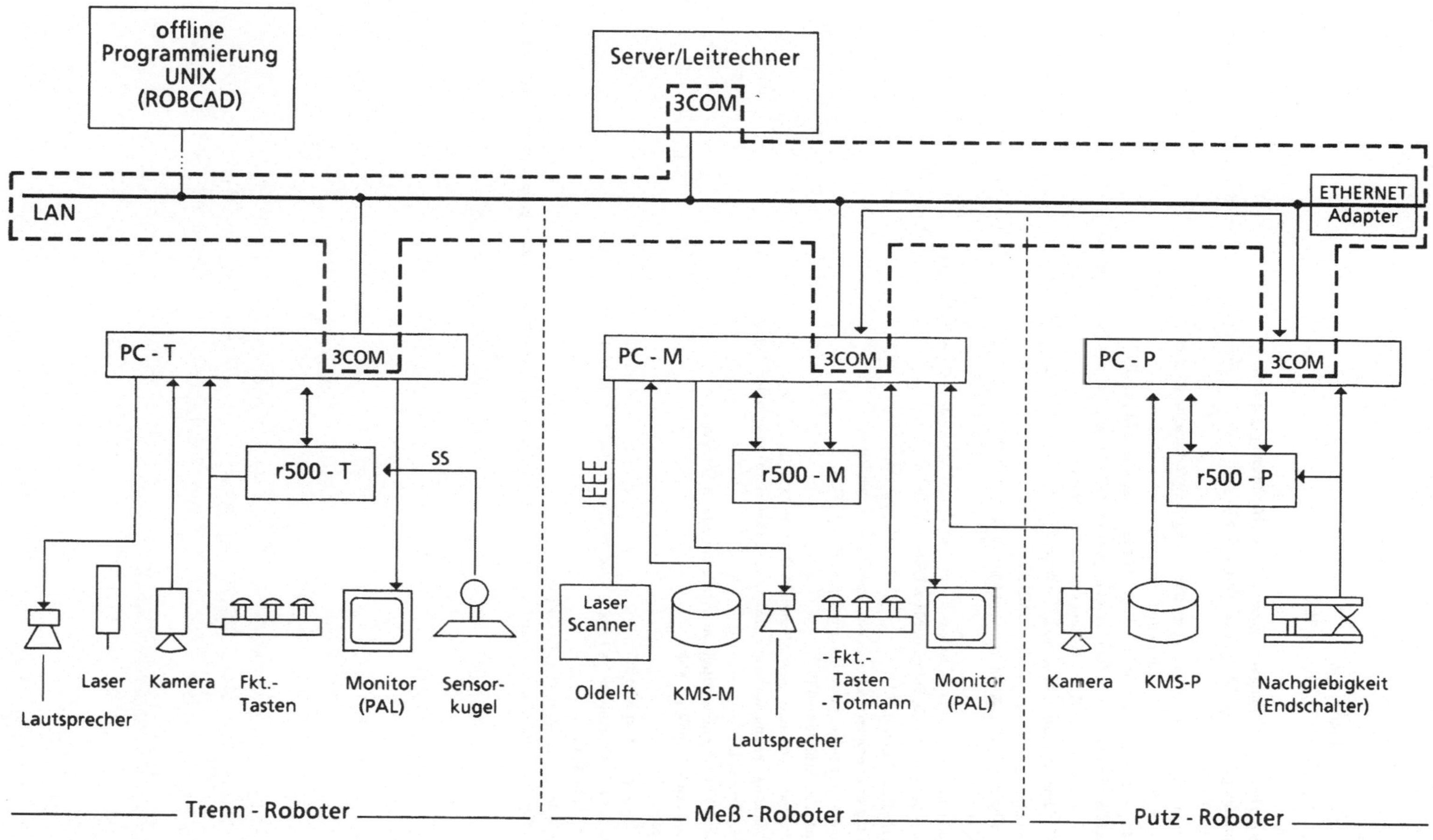

Abb. 4: Zellensteuerung für COMETOS (Hardware-Konfiguration)

7. Literatur

[1]   Klump W., Zauke H.
Entgraten mit Industrierobotern
Werkstattstechnik, Vol.77, 1987, pp.383-387

[2]   Abele E.
Gußputzen mit sensorgeführten programmierbaren Handhabungsgeräten
IPA Forschung und Praxis 72, Springer Verlag, 1983

[3]   Weck M., Tilli T.
Sensorsignalverarbeitung und Lernprogrammierung bei Werkzeugmaschinen
Vortrag HGF-Assistententreffen Berlin, Feb. 1988

[4]   Lawo M.,Droll H.,Göbell A.,Häfele K.-H.,Haffner H.,Isele J.,Kohlhepp P.
COMETOS ein hochflexibles Handhabungssystem zur Bearbeitung mittelgroßer Gußstücke
KfK-Nachrichten Nr.22, Kernforschungszentrum Karlsruhe, 1990

[5]   Droll H.,Göbell A.,Haffner H.,Isele J.,Kohlhepp P.
Programmierung und Sensorik bei COMETOS
KfK-Nachrichten Nr.22, Kernforschungszentrum Karlsruhe, 1990

[6]   Fischer H.
Ein hierarchisch organisiertes, zellenstrukturiertes Steuerungssystem für flexible Fertigungssysteme
Prozeßrechensysteme '88, Stuttgart, März 1988, pp.728-737

[7]   Kowalski H.-J., Thon H.-J.
Steuerung von Fertigungszellen in der rechnerintegrierten Fabrik
atp Sonderheft Fertigungsautomatisierung, 1987, pp.24-31

[8]   Rohnacher P., Schlindwein H.
FM3-Verwaltungssystem. Unveröffentliche Dokumentation,
Kernforschungszentrum Karlsruhe, Institut für Datenverarbeitung in der Technik, 1990

[9]   Brill M., Gramm U.
MMS-Die MAP-Applikationsdienste für die industrielle Fertigung
Elektronik Heft 3, 1989, pp.50-54

[10]   Didic M.
Time Measurements on Map 3.0
EMUG MAP Implementation and Application Meeting, Rüsselsheim, 10.5.1990

[11]   Mellein G., Prestel F.
Messung der Antwortzeiten im 3COM-Netz
Unveröffentlichter Bericht, Kernforschungszentrum Karlsruhe, Institut für Datenverarbeitung in der Technik, Karlsruhe, 1990

Modulares Leitsystem für Fertigung und Montage

Prof. Dr.-Ing. A. Storr
Dipl.-Ing. K. Brantner

Institut für Steuerungstechnik
der Werkzeugmaschinen und Fertigungseinrichtungen
(ISW)
Universität Stuttgart

1 Einleitung

Stetig wachsende Anforderungen im Bereich der Steuerung flexibler Produktionsanlagen führten sowohl im Fertigungs- als auch im Montagebereich zu einer hierarchischen Untergliederung des Gesamtsteuerungssystems /1, 2, 3/. Wesentliche Vorteile ergeben sich dabei hinsichtlich

- Flexibilität,
- Übersichtlichkeit und Transparenz sowie
- autarker Funktionsfähigkeit einzelner Subsysteme über einen bestimmten Zeitraum hinweg.

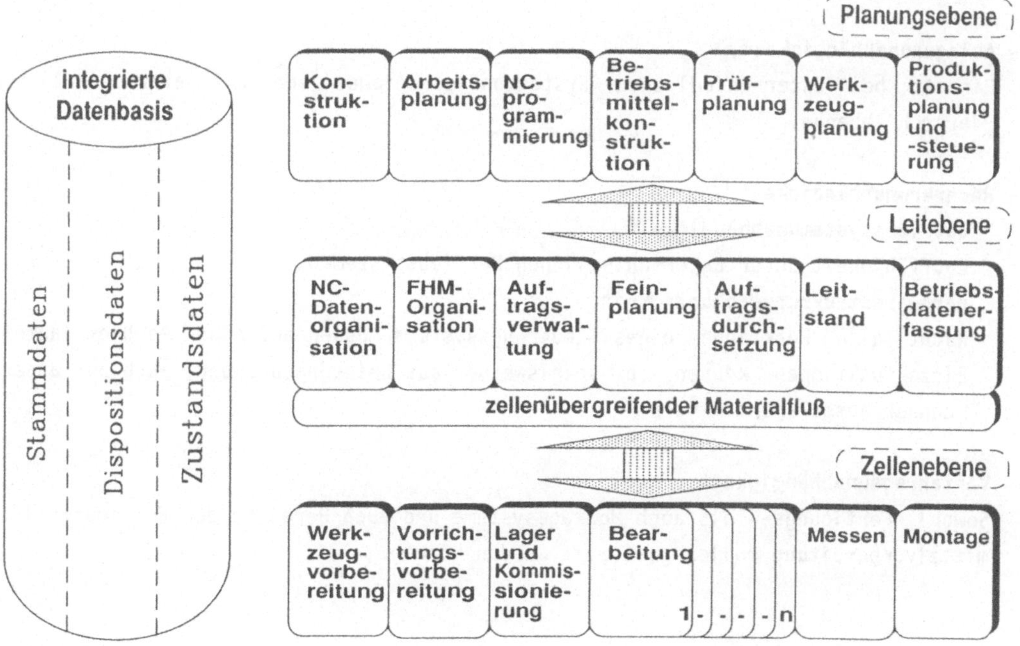

<u>Bild 1:</u> Funktionen und Hierarchieebenen bei der Produktionssteuerung /4/

Bild 1 zeigt eine häufig anzutreffende Untergliederung in Planungs-, Leit- und Zellenebene mit den jeweils typischen Steuerungsfunktionen.

Die Steuerungssoftware selbst stellt bei den Produktionsanlagen zunehmend den entscheidenden Kostenfaktor dar, was wesentlich dadurch bedingt ist, daß sie meist anlagenspezifisch ausgerichtet und somit als Einzellösung anzusehen ist. Diese Tatsache und weitere, zunehmend an Bedeutung gewinnende Anforderungen hinsichtlich

- erhöhter Termintreue auch im Bereich der Softwareentwicklung,
- verkürzter Inbetriebnahmezeiten und
- verringerter Fehlerhäufigkeit trotz wachsender Komplexität

bedingen, daß im Interesse sowohl des Entwicklers als auch des Anwenders verstärkte Anstrengungen hin zu einer **adaptierbaren** Steuerungssoftware unternommen werden. Erste Ansätze hierzu sind bekannt /5, 6, 7/, es fehlt jedoch an einem ganzheitlichen Konzept, das einem **beliebig** hierarchisch strukturierten Steuerungssystem Rechnung trägt und darüber hinaus Anforderungen sowohl des Fertigungs- als auch des Montagebereichs abdeckt.

Aufgrund dieses Defizits wird am Institut für Steuerungstechnik der Werkzeugmaschinen und Fertigungseinrichtungen (ISW) der Universität Stuttgart das adaptierbare Leitsteuerungssystem ALSYS entwickelt, das folgende Charakteristiken aufweist:

- **Anlagenunabhängigkeit,**
  Einsatz bei unterschiedlichen Systemkonfigurationen und auf unterschiedlichen Hierarchieebenen.

- **Rechnerunabhängigkeit,**
  * **Betriebssystemunabhängigkeit,**
    Lauffähigkeit unter unterschiedlichen Betriebssystemen.
  * **Hardwarestrukturunabhängigkeit,**
    Nicht alle Funktionen eines Steuerungssystems müssen auf einem Rechner laufen, Einzelfunktionen können, beispielsweise aus Leistungsgründen, auch auf andere Rechner ausgelagert werden

- **Verfahrensunabhängigkeit,**
  Sowohl Fertigungs- als auch Montagesysteme und auch Bereiche der Fertigunghilfsmittelvorbereitung sollen gesteuert werden können.

2 Anlagenunabhängigkeit von ALSYS

Wesentliche Grundvoraussetzung zur Erreichung der genannten Ziele ist, bei der Entwicklung des Leitsystems auf Erkenntnisse und Prinizipien des zunehmend als eigenständige Ingenieursdisziplin anerkannten Softwareengineerings zurückzugreifen. Die Beachtung allgemeingültiger Prinzipien der SW-Entwicklung, wie Modularisierung, Hierarchisierung, Abstraktion und Lokalität führt im skizzierten Anwendungsfall im Rahmen eines top-down Vorgehens zur Unterteilung der Gesamtsteuerungsaufgabe in möglichst abgeschlossene Funktionseinheiten mit **standardisierten** Schnittstellen. Bild 2 zeigt die oberste Ebene der funktionalen Unterteilung von ALSYS.

<u>Bild 2</u>: Funktionale Gliederung des adaptierbaren Leitsystems ALSYS

Die Funktionseinheiten, im folgenden auch kurz Funktionen genannt, sind weiter zu detaillieren, wobei eine Aufteilung in anlagenunabhängige und anlagenspezifische Funktionen strikt anzustreben ist (siehe Bild 3 am Beispiel des DNC-Systems). Die allgemeingültigen Funktionen stellen mit den standardisierten Schnittstellen gleichsam den Rahmen dar, in den die anlagenspezifischen Komponenten, ähnlich einem Baukastensystem, eingebunden werden.

Zusätzlich zur funktionalen Untergliederung gilt es, eine für Gesamtsystem und Einzelfunktionen gültige Grundstruktur zu erarbeiten, die sowohl den Austausch als auch das Weglassen einzelner Funktionseinheiten unterstützt. Die in Bild 4 am Beispiel der Feinplanung gezeigte Struktur wird diesen Anforderungen gerecht (s.a. /8/, /9/).

Funktion:

Auftraggeber: z.B. Auftragsdurchsetzung od. Bedienung — Festlegung der Auftragsreihenfolge nach Kriterien wie Termin, Kapazität, Priorität, usw.

Anpaßebene 1: z.B. NC-Datenverwaltung — Ermitteln der erforderlichen NC-Daten und Abgleich mit deren Verfügbarkeit in den Stammdaten.

Anpaßebene 2: z.B. NC-Datenübertragung — Aktivierung der NC-Datenübertragung nach Abgleich mit deren Verfügbarkeit an Hand der Zustandsdaten.

Anpaßebene 3: z.B. Maschinenanpassung — Synchronisierung der Datenübertragung mit dem aktuellen Maschinenzustand unter Berücksichtigung der Steuerungseigenschaften.

Anpaßebene 4: z.B. Datenübertragungssystem — Sicherstellung des korrekten Bytetransfers.

Grad der Anlagenabhängigkeit

Bild 3: Anlagenunabhängigkeit durch Untergliederung in Anpaßebenen am Beispiel DNC

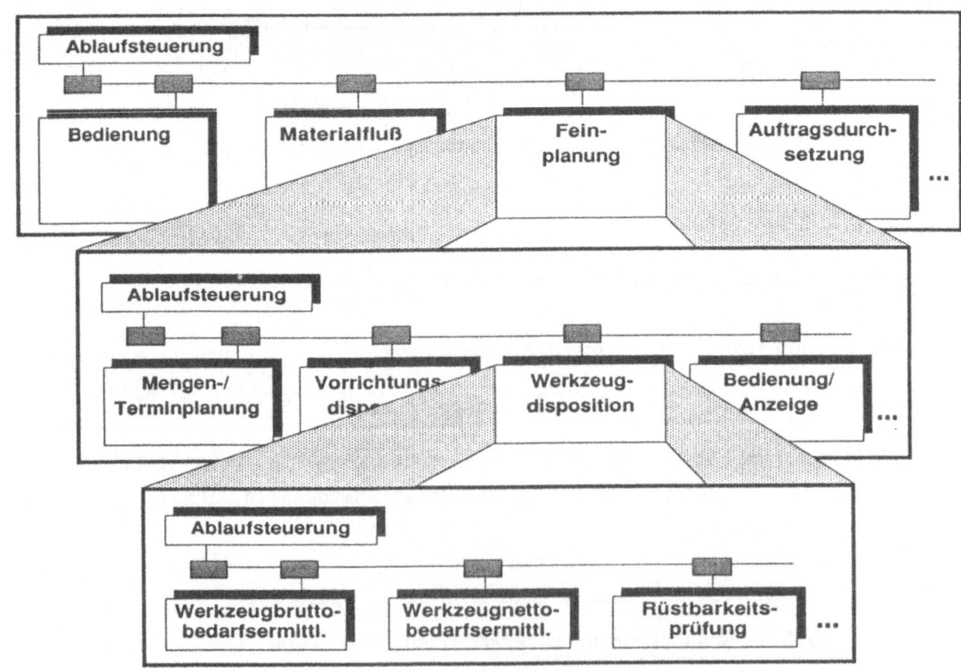

Bild 4: Hierarchische Detaillierung der Funktionseinheit "Feinplanung"

Die gezeigte Strukturierung kann die Anlagenunabhängigkeit der Steuerungssoftware wesentlich unterstützen und erleichtern. Das eigentliche Problem der Adaptierbarkeit ist damit jedoch nur zum Teil gelöst, da jeder einzelne Funktionsbaustein, insbesondere in den unteren Detaillierungsebenen nach Bild 4 zur Funktionserbringung anlagenspezifische Informationen verarbeiten muß. Anlagenspezifische Parameter, Strategien oder Abläufe sind somit weitestgehend **nicht** implizit im Programmcode, sondern explizit in editierbaren, also leicht änderbaren Dateien abzulegen, in Bild 5 als Konfigurationsdatenbasis bezeichnet. Die Mehrheit der ALSYS-Funktionsbausteine ist allgemeingültig programmiert und erfüllt erst durch Interpretation dieser Konfigurationsdaten **zur Laufzeit** ihre anlagenspezifischen Aufgaben. Die Erstellung der Konfigurationsdatenbasis erfolgt je nach Anforderung über grafikunterstützte Dialogmasken oder formale Beschreibungssprachen wie Petri-Netze oder Zustandsgraphen.

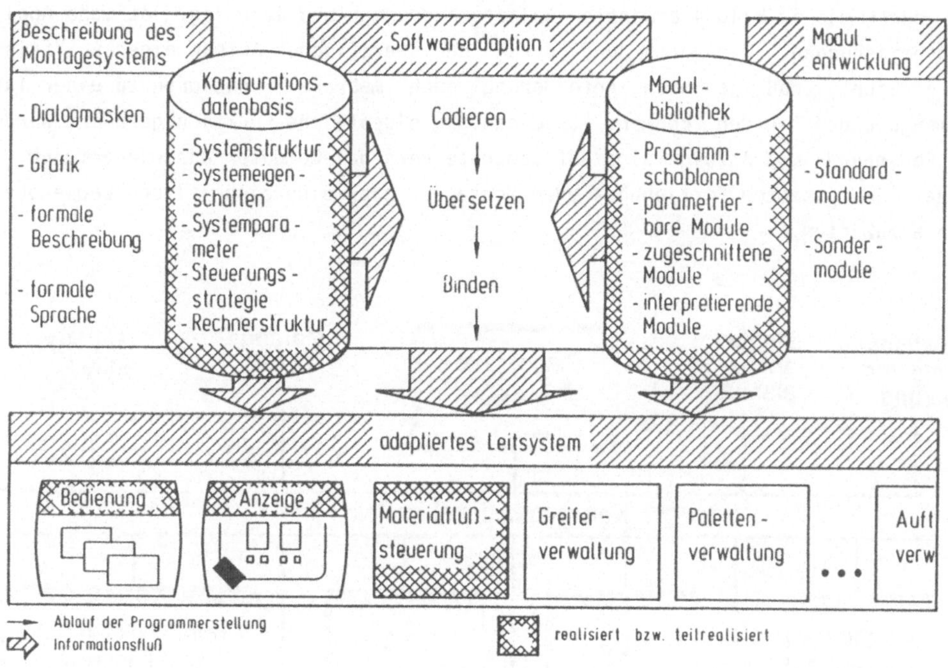

Bild 5: Anpassung von ALSYS an verschiedene Anlagen am Beispiel "Montagesystem"

3 Rechnerunabhängigkeit von ALSYS

3.1 Betriebssystemunabhängigkeit

Die Reduzierung des Anpaßaufwands bei der Portierung auf andere Betriebssysteme wird durch die in Bild 6 dargestellte Softwarestruktur erreicht. Alle ALSYS-**Tasks** sind einheitlich strukturiert, die betriebssystemabhängigen Befehle, insbesondere zur Intertaskkommunikation und -synchronisation, sind in austauschbaren Unterprogrammen zusammengefaßt. Die **Anwendungsfunktion** selbst ist in einer höheren Programmiersprache realisiert. Alle **Tasks** besitzen nur jeweis **eine** Schnittstelle zum sog. Kommunikationsbaustein, mit dem sie gleichberechtigt verbunden sind. Die **logische** Verknüpfung und Synchronisation von **Funktionen** innerhalb eines Steuerungssystems erfolgt durch die in Bild 4 erwähnten Ablaufsteuerungen und über sog. logische Applikationsbeziehungen. Hierdurch und durch den spezifischen Aufbau einer ALSYS-Task wird erreicht, daß je nach Anforderung auch mehrere Funktionen zu einer Task zusammengebunden werden können, ohne daß die eigentliche Funktion geändert werden muß. Bedienung und Anzeige sind auf separate Personal-Computer ausgelagert, um dem Bediener im gesamten Fertigungs- oder Montagebereich eine einheitliche Bedienoberfläche anzubieten.

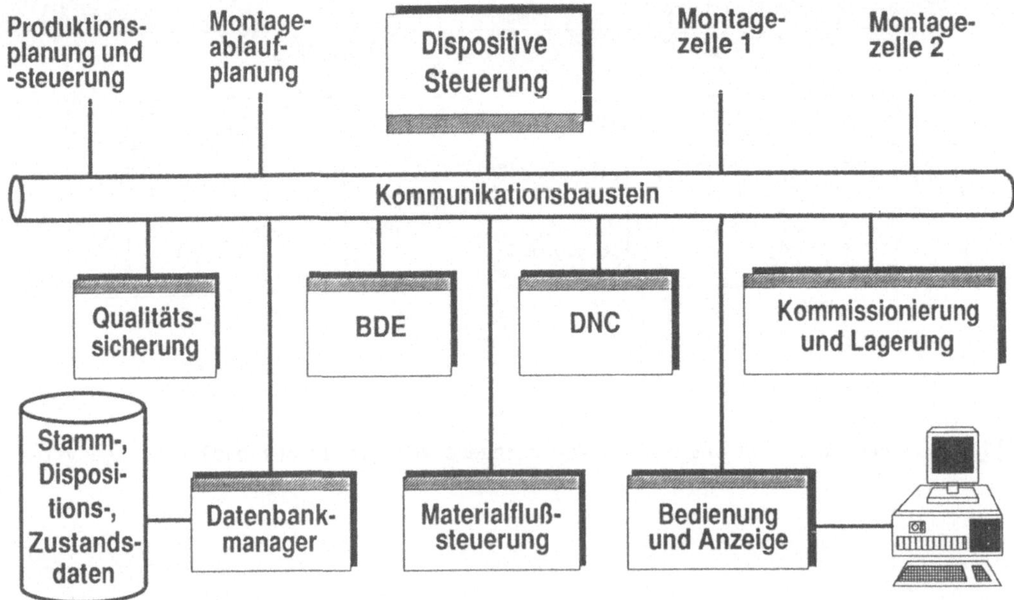

<u>Bild 6</u>: Softwarestruktur von ALSYS

4.2 Hardwarestrukturunabhängigkeit

Zur stufenweisen Erweiterung eines inplementierten Steuerungssystems kann es notwendig werden, aus Leistungsgründen einzelne Funktionen auf andere Rechner auszulagern.

Wesentliche Grundvoraussetung für ein derartig verteiltes Steuerungssystem ist, daß die Kommunikation mit den externen Systemen (Produktionsplanung und -steuerung, Arbeitsvorbereitung, prozeßnahe Steuerungsgeräte ...) **und** die interne Intetaskkommunikation durch **ein** leistungsstarkes Kommunikationssystem übernommen wird. Einzelne **Funktionen** kommunizieren über standardisierte Applikationsbeziehungen miteinander, unabhängig davon, ob die Partnerfunktion in der gleichen Task, auf dem gleichen Rechner oder auf einem anderen Rechner installiert ist.

Gegenüber Standardisierungen und technischen Weiterentwicklungen, beispielsweise im Bereich MAP, ist ALSYS dadurch offen, daß die in Bild 6 schraffiert dargestellten Anpaßmodule nicht nur sämtliche betriebssystemabhängigen Aufrufe beinhalten, sondern auch die Abbildung eines bestimmten Telegramminhaltes auf die Dienste des Kommunikationssystems (z.B. MMS Manufacturing Message Specification) durchführt.

5 Verfahrensunabhängigkeit

Verfahrensunabhängigkeit bedeutet, daß das Steuerungssystem nicht nur im Fertigungssondern auch im Montagebereich (s. Bild 5) oder für die Fertigungshilfsmittelbereitstellung einsetzbar ist. Die vom Steuerungssystem anzustoßenden Aktionen (z.B. fertigen, montieren, Werkzeug einstellen etc.) und die zu steuernden Objekte (Rohteile, Halbzeuge, Fertigteile, Baugruppen, Werkzeuge etc.) sind soweit zu abstrahieren, daß aus der Sicht des Steuerungssystems in allen Bereichen möglichst gleiche Operationen durchzuführen sind.

6 Realisierungsbeispiele

Zur Überprüfung der theoretisch erarbeiteten Ansätze diente zum einen die in Bild 7 gezeigte ISW-Pilotanlage, in der die Funktionen der Leitebene abgedeckt werden. In der in die Pilotanlage integrierten Montagezelle (Bild 8) werden auf Zellenebene angesiedelte Funktionen betont.

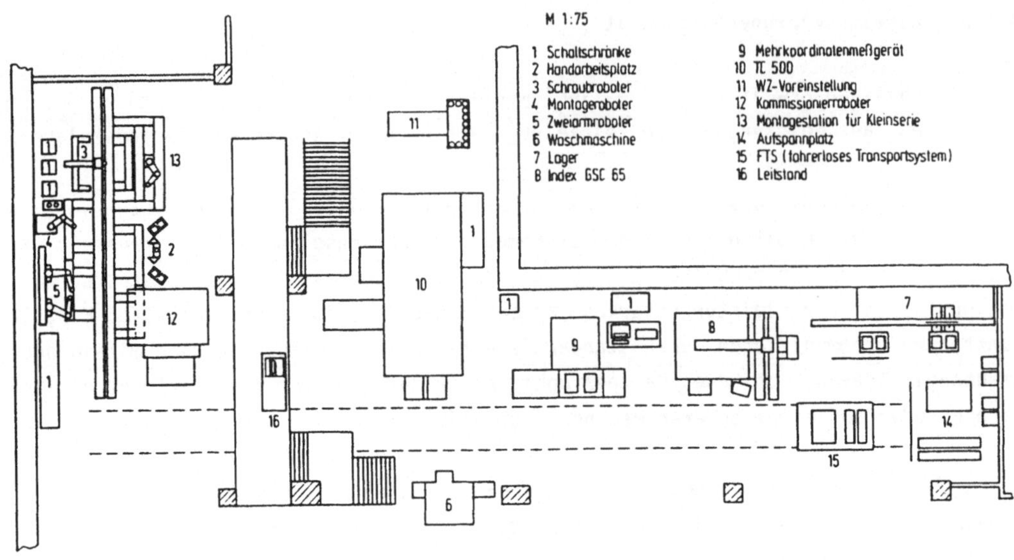

Bild 7: Anlagenlayout der ISW-Pilotanlage

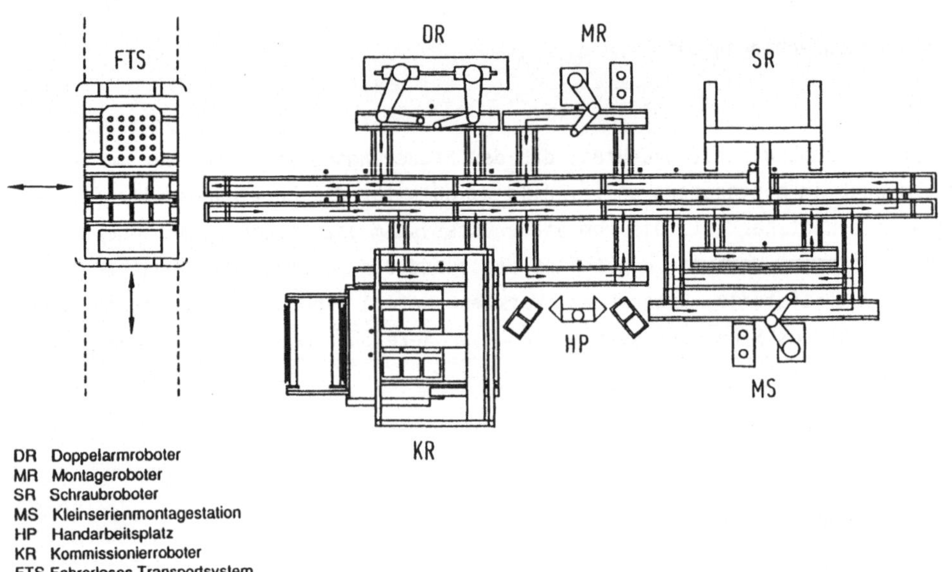

Bild 8: Anlagenlayout der Montagezelle

7 Zusammenfassung und Ausblick

Kürzer werdende Produktlebenszyklen, größere Variantenvielfalt und Forderungen nach höherer Termintreue zwingen die Unternehmen in steigendem Maße zum Aufbau flexibler Produktionsstrukturen und zum Einsatz des Rechners bei der Steuerung der Produktionsanlagen. Die immer komplexer werdende Steuerungs**software** stellt dabei zunehmend den entscheidenden Kostenfaktor dar, was in der Vergangenheit oftmals unterschätzt wurde. Dies ist wesentlich dadurch bedingt, daß sie meist anlagenspezifisch als Einzellösung konzipiert war. Zur Behebung dieses Mangels sind verstärkt wiederverwendbare Steuerungssysteme zu entwickeln, die mit minimalem Anpaßaufwand schnell und einfach an einzelne Anlagen adaptierbar sind. Grundvoraussetzung zur Erreichung dieses Zieles ist zu Beginn eine **detaillierte funktionale Strukturierung** der Gesamtsteuerungsaufgabe und eine **Standardisierung** der Schnittstellen zwischen den Einzelfunktionen. Es ist strikt auf eine Unterteilung in anlagenunabhängige und anlagenabhängige Funktionen zu achten. Der anlagenunabhängige Teil stellt dann für jeden Anwendungsfall ein Grundgerüst dar, welches mit den anlagenabhängigen Funktionen ähnlich einem Baukastensystem ergänzt wird. Am Institut für Steuerungstechnik der Werkzeugmaschinen und Fertigungseinrichtungen der Universität Stuttgart (ISW) sind Arbeiten unter dieser Zielvorgabe im Gange, erste Realisierungen liegen vor. Weitergehende Entwicklungen hinsichtlich des Einsatzes formaler Beschreibungssprachen, neuer Programmiermethoden und von Expertensystemen stehen an.

8 Schrifttum

/1/ Milberg, J., Groha, A.: Der Zellengedanke als Strukturierungsprinzip im Informations- und Materialfluß flexibler Fertigungssysteme. ZwF 81 (1986) H.12, S.682 ... 687.

/2/ Scheller,J., Sommer,E.: Hierarchisches Steuerungskonzept für flexible Montagezellen. Automatisierungstechnische Praxis atp 31 (1989) 4, S.166 ... 173.

/3/ Pritschow, G.: Die flexible Fertigungszelle. wt-Z. ind. Fertig. 75(1985) Nr. 11, S.663 ... 668.

/4/ Walker, M.: Funktionsbausteine für Fertigungsleitsysteme. In: Leittechnik für verkettete Fertigungssysteme. Düsseldorf: VDI-Bildungswerk, 1989.

/5/ Eversheim, W.:   Organisatorische Integration flexibler Fertigungs-
                      systeme in konventionelle Werkstattstrukturen.
                      Vortrag anläßlich des Produktionstechnischen
                      Kolloquiums PTK, Berlin, 1989.

/6/ Groha, A.:   Universelles Zellenrechnerkonzept für flexible
                      Fertigungssysteme. iwb Forschungsbericht 16.
                      Springer-Verlag: Berlin Heidelberg, New York, London,
                      Paris, Tokyo 1988.

/7/ Beier, H.H.:   Fertigungsleitstand und Fertigungsleittechnik,
                      ZwF 84 (1989) 3, S.133 ... 137.

/8/ Pritschow, G.:   Automatisierungstechnik - Eine ganzheitliche steue-
                      rungstechnische Aufgabe. Vortrag anläßlich des Produk-
                      tionstechnischen Kolloquiums PTK, Berlin, 1989.

/9/ Storr. A.,   Das adaptierbare Leitsteuerungssystem ALSYS.
    Brantner, K.   In "Leit- und Steuerungstechnik für flexible Produkti-
                      onsanlagen", München: Hanser Verlag, 1990.

CIM IN DER FERTIGUNGSINDUSTRIE
AM BEISPIEL KUNSTSTOFFVERARBEITUNG
- SCHWERPUNKT QUALITÄTSSICHERUNG -

F. Klose
IBM Deutschland GmbH
AK AIX Marketing

München

ZUSAMMENFASSUNG

Die Bedeutung der Qualitätssicherung hat in den letzten Jahren beständig zugenommen.
Die besondere Stellung der Qualitätssicherung für Spritzgießbetriebe wird einleitend
beschrieben. Darauf aufbauend die Vorgehensweise des Projektes "Qualitätssicherung
für Spritzgießbetriebe" erläutert und auf die Besonderheiten der Verbindung zur Prozeßüberwachung hingewiesen. Besonders wichtig dabei scheint der Punkt zu sein, daß
das Anwendungspaket auf Basis PROQUAM auf einem Offenen System aufbaut und die Vorteile von UNIX für die Einbindung in heterogene Systemlandschaften nutzt.
Der Einsatz von UNIX-basierten Systemen bietet den Anwendern signifikante Vorteile
gegenüber den neuesten herstellerspezifischen Produkten:
- Anwendungsentwicklung weitgehend unabhängig von
  HW-Plattform durchführbar
- Softwarelösungen relativ schnell und einfach auf neue
  HW-Technologien portierbar
- Zunehmende Unabhängigkeit von einzelnen HW-Lieferanten --> Zukunfts-
  sicherheit von Organisation und SW-Investitionen
- Nutzbarkeit von Systemen mit dem besten Preis-Leistungsverhältnis
- Gleiche Systemumgebung über mehrere Rechnerebenen.
Das CAQ-System ist als Kern für ein CIM-Fertigungsleitsystem für die Kunststoffver-
arbeitung anzusehen.

1. INFORMATION UND KUNSTSTOFFVERARBEITUNG

Heute stehen die europäischen Industrieunternehmen unter dem ständig steigen-

den Druck, ihre Wettbewerbsfähigkeit in einem sich schnell verändernden weltwei-

ten Markt zu erhalten. Um die Abwicklung von Aufgaben im Unternehmen zu unter-

stützen - Entwickeln und Herstellen von Produkten und deren Vermarktung -, wurde

in der Vergangenheit erheblich in die Informationstechnologie investiert. Für

die einzelnen Bereiche Entwicklung, Produktion, Verwaltung und für Führungsauf-

gaben steht somit Datenverarbeitung zur Verfügung.

Kürzere Produktlaufzeiten, wachsender nationaler und internationaler Wettbewerb,

steigende Material- und Lohnkosten, technische Neuerungen und die Notwendigkeit

interner Organisationsänderungen sind nur einige Aspekte, mit denen sich ein

Fertigungsunternehmen in der heutigen Zeit beschäftigen muß. Um diesen Anfor-

derungen zu begegnen, müssen ständig neue Wege und Möglichkeiten gesucht werden,

zur

- Senkung der Produktionskosten,

- Verminderung von Rüstzeiten und Lagerbeständen,

- Verbesserung der Produktqualität und
- schnellen Anpassung an sich ändernde Marktgegebenheiten.

Die traditionelle Lösung zur Bewältigung dieser Herausforderungen ist die Automation. Viele Unternehmen nutzen bereits die Vorteile durch computerunterstützte Konstruktion (CAD: Computer Aided Design) bzw. Fertigung (CAM: Computer Aided Manufacturing), aber auch in der Verwaltung durch Unterstützung der Bürotätigkeiten per Computer.

Der Markt für die kunststoffverarbeitende Industrie wird sich weltweit in den nächsten Jahren verdoppeln. Es ist eines der stärkst wachsenden Marktsegmente in Deutschland und der Anteil der Kunststoffe nimmt in allen Bereichen ständig zu. Stellvertretend für die Kunststoffverarbeitung sollen hier einmal die Spritzgießbetriebe betrachtet werden (Bild 1).

Die wirtschaftliche Lage der Spritzgießbetriebe und die an sie gestellten Marktanforderungen haben sich in den letzten Jahren tiefgreifend und nachhaltig geändert. Nicht mehr große Stückzahlen immer gleicher Teile sind gefragt, sondern die wiederholte Lieferung kleiner Lose auf Abruf. Als Zulieferer großer Verbrauchsgüterhersteller haben sie sich in deren logistisches Konzept einer "just-in-time"-Produktion einzufügen. Verschärfter Wettbewerb zwingt sie darüber hinaus, qualitativ höherwertige Artikel mit garantierten Qualitätsmerkmalen anzubieten und dabei nach Möglichkeit steigende Personal-, Material- und Energiekosten in der Produktion aufzufangen. Zusammengenommen erfordern diese Marktanforderungen, nach neuen Methoden und Strukturen der Produktion von Formteilen zu suchen.

Spritzgießbetriebe geraten heute zumeist deswegen unter Kostendruck, weil ihre Kunden in vielen Fällen nicht bereit sind, wegen der teuren Lagerhaltung größere Mengen von Formteilen abzunehmen. Während früher der Kunde bei seinem Lieferanten meist Aufträge mit großen Stückzahlen abrief, bestellt er heute bei gleicher Gesamtmenge sehr viel kleinere Stückzahlen, die jedoch zu exakt vorgegebenen Terminen angeliefert werden müssen. Speziell die Automobilindustrie - einer der größten Abnehmer der kunststoff- und kautschukverarbeitenden Industrie - verlangt eine weitgehende Anpassung der Liefermengen an die Bedürfnisse ihrer eigenen Fertigungsstraßen, d. h. die zur Montage benötigten Teilemengen sollen erst unmittelbar vor ihrem Einbau angeliefert werden. Somit ist das Lagerproblem vom Abnehmer zum Lieferanten verlagert worden.

Diese Tatsache zwingt dem Spritzgießbetrieb entweder eine umfangreiche und damit kostspielige Lagerhaltung auf oder erfordert eine "flexible Fertigung". Diese ermöglicht, bei kleinstmöglicher Lagerhaltung, geringen Zeit- und Materialverlusten und somit geringen Herstellkosten kurzfristig kleinere Losgrößen unterschiedlicher Formteile herzustellen, um somit umlaufendes Kapital und Lagerkosten zu reduzieren.

Um den Herausforderungen der veränderten Marktbedingungen gerecht zu werden, genügt es nicht, Einzelmaßnahmen zu ergreifen, etwa den Prozeß höher zu automatisieren. Vielmehr muß das Produktionssystem als Ganzes in allen seinen Teilaspekten neu gestaltet werden. Insbesondere müssen die Planung der technischen Einrichtungen zur flexiblen Automatisierung, eine den Flexibiltätsansprüchen genügende, qualifikationsorientierte Arbeitsorganisation geringer Arbeitsteilung und die Personalentwicklung aufeinander bezogen und gemeinsam konzipiert werden. Diese Umgestaltung betrifft die ganze Fabrik. Sie kann allein auf der Grundlage eines solchen Gesamtkonzeptes der künftigen Produktion schrittweise vollzogen werden.

Bei dem erforderlichen Investitionsvolumen ist es jedoch kaum denkbar, daß solche Lösungen von heute auf morgen realisiert werden können. Vielmehr muß die Umgestaltung der gesamten Produktion zumeist Schritt für Schritt durch gezielte Investitionen durchgeführt werden. Gerade hierin aber liegt die große Chance der Spritzgießverarbeiter, da oftmals auch vorhandene Maschinen durch entsprechende Nachrüstung den Ansprüchen einer flexiblen Fertigung Rechnung tragen können.

Diese Produktionsstrategie kann aber nur zum wirtschaftlichen Erfolg führen, wenn sie die Gestaltung einer angemessenen Arbeitsorganisation und eine umfassende und systematische Personalentwicklung einschließt. Dies lehren die Einsatzerfahrungen mit flexiblen Fertigungssystemen in der Metallbearbeitung. Für den Betrieb eines flexiblen Fertigungssystems ist qualifizierte Gruppenarbeit anzustreben, welche die Arbeitsaufgaben des Rüstens, Programmierens, Disponierens von Aufträgen, das Sichern der Qualität und die routinemäßige Wartung der Anlage umfaßt. Das Personal auf ein derartiges Aufgabenspektrum frühzeitig und schrittweise vorzubereiten, muß als ebenso wichtige Investition begriffen werden wie die Beschaffung der neuen Produktmittel.

2. QUALITÄTSSICHERUNG

Die Qualitätssicherung (CAQ: Computer Aided Quality) ist in den Unternehmen in letzter Zeit sehr in den Vordergrund getreten. Die hohe Abhängigkeit vom Exporterfolg und der immer stärker werdende Wettbewerb bringen es mit sich, daß die Unternehmen unter anderem durch hohe Qualität und Flexibilität vorhandene Stand- oder Währungsnachteile ausgleichen müssen. Die Qualitätssicherung wird zum Bestandteil der strategischen Unternehmensführung, mit deren Hilfe der Absatz verbessert und wirtschaftlicher produziert werden kann.
Die Bedeutung der Qualitätssicherung hat in den letzten Jahren beständig zugenommen. Begleitend haben sich auch die Maßnahmen zur Sicherstellung der geforderten Eigenschaften eines Produkts verändert. Sie bestehen nicht mehr nur aus der Prüfung einiger ausgesuchter Merkmale am fertigen Werkstück, sondern aus der gesamtheitlichen Optimierung des Prozesses der Produktherstellung. Das bedeutet, daß heute durch eine mehrfach kontrollierte und regelkreisorientierte Prozeßsteuerung qualitative Abweichungen von einer Sollvorgabe von vornherein

vermieden werden sollen. Die Prozeßbeherrschung umfaßt dabei alle Maßnahmen,
die das Zusammenwirken von Personal, Maschinen, Einrichtungen, Rohmaterial, Methoden und Arbeitsumwelt beschreiben.

Neben der Kontrolle der Produktqualität steht heute vermehrt die Qualität des Prozesses selbst im Vordergrund. Speziell bei automatisierten Fertigungsabläufen kommt der Prozeßoptimierung besondere Bedeutung zu. Diese kann allerdings nur dann sinnvoll durchgeführt werden, wenn eine Rückführung von Qaualitätsdaten an die Verarbeitungsmaschine erfolgt. Problematisch hierbei ist die Komplexität der Wirkungsmechanismen im Prozeß, d. h. der Zusammenhänge zwischen Maschineneinstellung und Qualitätsmerkmalen. Entsprechende Hilfsmittel auf der Grundlage statistischer Prozeßmodelle ermöglichen dem Einrichter eine optimale Maschineneinstellung.

Die Qualität eines Produktes ist zentraler Aspekt der Fertigung. Zur Erfassung und Bewertung von Qualtätsmerkmalen müssen in einer automatisierten Fertigung Systeme eingesetzt werden, die objektiv und reproduzierbar arbeiten.

Die Problematik der Spritzgießprozeßdokumentation mittels an der Spritzgießmaschine erfaßten Fertigungsparametern (Druck, Temparatur etc.) besteht darin, daß quantifizierbare Beziehungen zwischen den Qualitätsmerkmalen des Formteils und den gemessenen Fertigungsparametern bekannt sein müssen.

Mit Hilfe eines statistischen Versuchsplanes werden systematisch Variationen des Betriebspunkteinstelldatensatzes der Spritzgießmaschine durchgeführt. Dabei werden die Fertigungsparameter während des Fertigungszyklusses und die Qualitätsmerkmale der produzierten Formteile erfaßt.

Aus den aufgenommenen Verläufen der Fertigungsparameter werden Kennzahlen, wie z. B. die Einspritzarbeit, berechnet, die mit den aufgenommenen Qualitätsmerkmalen des Formteiles korreliert werden können.

Die als signifikant erkannten Kennzahlen werden dazu benutzt, um statistisch abgesicherte Prozeßmodelle zu ermitteln, welche die Qualitätsmerkmale des Formteiles als Funktion der Kennzahlen beschreiben.

Mit Hilfe dieser Prozeßmodelle ist es möglich, Sollgrößen und Toleranzen der Kennzahlen zu bestimmen, die für die geforderten Qualitätsmerkmale des Formteiles einzuhalten sind.

Das Prozeßüberwachungssystem bewertet und dokumentiert mit den ermittelten Prozeßmodellen den aktuellen Fertigungszustand. Dies geschieht über die Führung von Regelkarten bzw. durch Meldung von Ausschußteilen z. B. an einen Leitrechner.

Somit ist es möglich geworden, die statistische Prozeßkontrolle mit Hilfe der Fertigungsparameter der Spritzgießmaschine durchzuführen und die Formteilendkontrolle auf ein Minimum zu reduzieren.

3. ZIELSETZUNG DES PROJEKTES

Um die regionale Wettbewerbsfähigkeit der heimischen, mittelständischen Industrie in bezug auf den bevorstehenden, europäischen Markt 1993 zu fördern, hat
das Land Nordrhein-Westfalen zu Beginn 1990 ein Förderprogramm zur Qualitätssicherung (CAQ) für Spritzgießbetriebe aufgesetzt. Ein Team aus dem Projektfüh-
rer (AGIT), Systemintegrator (GRP), Hochschulinstitut (IKV) und Hard- und System-
softwarehersteller (IBM) erarbeiten gemeinsam eine Lösung auf Basis des Anwendungspaketes PROQUAM unter dem Betriebssystem UNIX für mittelständische Betriebe
(Bild 2). Die Laufzeit des Projektes beträgt zwei Jahre, und es handelt sich
dabei um eine beispielhafte Lösung für einen Innovations- und Technologietransfer. Ziel des Projektes ist es, den Qualitätsstandard der Fertigprodukte zu
steigern, die Flexibilität der Fertigung, die Produktivität und Lieferbereit-
schaft zu erhöhen und durch Verwendung des offenen Betriebssystems UNIX einen
damit verbundenen Investitionsschutz herzustellen.

Die Aufgaben verteilen sich auf die einzelnen Beteiligten wie folgt:

- Die AGIT schließt Verträge mit den entsprechenden
  12 - 15 mittelständischen Spritzgießunternehmen ab, die vom
  Land Nordrhein-Westfalen mit 50 % der Kosten bezuschußt werden.

- Das IKV analysiert den Bedarf der einzelnen Anforderungen,
  erstellt das Pflichtenheft, berät bei der Einführung von CAQ
  in den Betrieben und schult deren Mitarbeiter.

- GRP installiert die Anwendung auf Basis von PROQUAM und
  modifiziert das System nach Angaben des IKV.

- IBM liefert die Systeme für die mittelständischen
  Unternehmen und stellt GRP zu günstigen Bedingungen
  Entwicklungs- und Vorführsysteme zur Verfügung. Ebenso
  wird mit dem IKV ein Studienvertrag zur Erstellung eines
  Lehr- und Lernprogrammes abgeschlossen. Bei der Por-
  tierung und Vermarktung des Komplett-Paketes wird aktive
  Unterstützung geleistet. Zum Einsatz kommt der zur Zeit
  modernste Arbeitsplatzrechner IBM RISC System/6000
  (Reduced Instruction Set Cycles) mit dem Betriebssystem
  AIX V.3 (Advanced Interactive Executive).

Um möglichst schnell, zu praxisnahen Erfahrungen zu kommen, wurde der Weg
gewählt, zu einem möglichst frühen Zeitpunkt (Anfang 1991) ein CAQ-Grundpaket
auf Basis von PROQUAM (es gibt mehrere Referenzinstallationen unter UNIX) bei
den ausgewählten Firmen zu installieren. Damit können die Anwender frühzeitig
Erfahrung sammeln, und das IKV kann mit ihnen in der echten Produktion die Analysen durchführen, um danach die Modifikationen und Erweiterungen an PROQUAM
festzulegen. Aus dem Anforderungsprofil wird dann durch das IKV ein Pflichtenheft aus Sicht des Anwenders entwickelt, welches detailliert den Leistungsum-

fang der zu erstellenden Softwareergänzungen festlegt; daraus wird dann ein umfassendes Anwendungspaket zur Qualitätssicherung für Spritzgießbetriebe entwickelt. Durch die strenge Verbindung zwischen Praxis im Betrieb und Forschung im Hochschulinstitut, die sich gegenseitig befruchten, dürfte ein sehr realitätsnahes und auch kostengünstiges Anwendungspaket entstehen.

4. <u>FUNKTIONSUMFANG DES ANWENDUNGSPAKETES ZUR QUALITÄTSSICHERUNG</u>
<u>IN SPRITZGIESSBETRIEBEN</u>

Die Qualitätssicherung stellt im allgemeinen in den Betrieben eine eigenständige Einheit mit den Funktionen:
- Qualitätsplanung
- Qualitätsprüfung
- Qualitätslenkung

dar. Auf die Aufgaben der o. g. Einzelmodule, die einen recht großen Umfang erlangen können, z. B mit der Verwaltung von Stichprobenplänen, Dynamisierung der Prüfung, Prüfmittelverwaltung sowie Berichtswesen oder Lieferanten- und Eigenbewertungen, soll hier nicht näher eingegangen werden. Vielmehr soll der besondere Aspekt der Einbeziehung der Produktionssteuerung in den täglichen Ablauf der Qualitätssicherung aufgezeigt werden.

Bei einem integrierten Informationssystem ist sicherzustellen, daß die Auslösung der Prüfaufträge automatisch mit Start des Produktionsauftrages erfolgt. Dadurch wird erreicht, daß auch bei kurzfristigen Dispositionsänderungen der Arbeitsvorbereitung der richtige Prüfauftrag gleichzeitig mit dem Produktionsbeginn ausgelöst wird.

Bei der Auslösung der Prüfaufträge werden alle bereits im Produktionsauftrag bekannten organisatorischen Daten ohne manuelles Zutun in den Prüfauftrag übernommen. Die von der Arbeitsvorbereitung festgelegten Daten wie
- verwendetes Werkzeug
- Produktions-Start- und Endtermin
- Sollwerte für Zykluszeit und Toleranz etc.

werden direkt in den Prüfauftrag eingestellt und dort archiviert (Bild 3).
Anhand der im Produktionsauftrag bekannten Artikelnummer wird der zugehörige Prüfplan selektiert und die entsprechenden Prüfanweisungen in den Prüfauftrag übernommen.

Ein weiterer wichtiger Aspekt bei der Verknüpfung zwischen Produktionssteuerung und Qualitätssicherung ist die Dynamisierung der Stichprobenprüfung aufgrund des betreffenden Maschinenstatus. So ist die Stichprobenfrequenz und der Stichprobenumfang bei Produktionsanlauf oder Produktionsstörungen dem betreffenden Maschinenstatus automatisch anzupassen.

Die Erfassung der einzelnen Prüfergebnisse erfolgt entweder durch manuelle Eingabe oder über Schnittstellen des Leitrechners. An diesen Schnittstellen können unterschiedliche Meßgeräte wie Meßschieber, Waagen oder Koordinatenmeßtische angeschlossen werden.

Neben den reinen Prüfdaten werden alle wichtigen Informationen aus dem Produktionsbereich mit im Prüfauftrag dokumentiert.

Alle Veränderungen von

- Zykluszeiten,
- Werkzeugdaten,
- Formnestzahlen,
- Personal- oder
- Materialänderungen etc.

werden im Prüfauftrag dokumentiert, ohne daß dies seitens der Qualitätssicherung ausgelöst werden müßte.

Eine derartige Dokumentation von Produktionsereignissen ermöglicht eine schnelle und sichere Entscheidung der Qualitätssicherung. Dies sind wichtige Erfordernisse zur Dokumentation im Rahmen der Produzentenhaftung.

Neben der klassischen Qualitätsprüfung tritt die statistische Prozeßkontrolle SPC immer stärker in den Vordergrund.

Die statistische Prozeßkontrolle (nach Ford Q 101) ermöglicht die Überprüfung der Maschinen- und Prozeßfähigkeit anhand von einzelnen Qualitäts- oder Prozeßmerkmalen.

Sie ist produktionsbegleitend und beginnt mit der Urwerterfassung, bei welcher Einzelwerte eines Qualitätsmerkmales erfaßt und analysiert werden. Aus dieser Urwertanalyse ergibt sich unter Berücksichtigung der Verteilungsfunktion die Maschinenfähigkeit, nach deren Überprüfung die laufende Produktion mit den Regeln der statistischen Prozeßkontrolle auf die Einhaltung der Prozeßfähigkeit hin überprüft wird. Bei jeder Eingabe einer neuen Stichprobe wird diese auf ihre Lage innerhalb der Regelkarte geprüft. Alle Überschreitungen der Eingriffs- oder Spezifikationsgrenzen werden sofort gemeldet, damit ein Eingriff in die Produktion erfolgen kann.

Meldungen der SPC-Analyse werden zudem in einer speziellen Meldedatei abgespeichert und können dort z. B. tageweise entnommen werden. Neben den erfaßten Überschreitungen von Grenzwerten werden auch Trendmeldungen der Regelkarte erfaßt und ausgewertet.

Die im Qualitätssicherungssystem archivierten Daten können nach den unterschiedlichsten Kriterien analysiert und ausgewertet werden.

In die kundenspezifisch anpaßbaren Berichte fließen

- Histogramme,
- Wahrscheinlichkeitsnetz und
- Regelkarten

ein. Aufgrund der Verknüpfung des Qualitätssicherungssystems mit der Produktionsdatenerfassung können wichtige kundenspezifische Produktionsdaten mit in die Berichte und Auswertungen aufgenommen werden.

Aufgabe der Prozeßdatenerfassung ist es, bereits im Vorfeld der Qualitätssicherung die Einhaltung der gewünschten Produkteigenschaften sicherzustellen. Die Überwachung von qualitätsrelevanten Prozeßgrößen wie Drücke, Temperaturen, Wege, Zeiten und Geschwindigkeiten ermöglicht es, bereits während der Formteilbildungsphase Aussagen über die Produktqualität zu ermöglichen und somit Ausschußteile direkt nach der Herstellung auszusondern.

Die Bedeutung der Qualitätskontrolle durch Prozeßüberwachung wird erst richtig in einer Fertigung deutlich, bei der die Materialentsorgung automatisch durch FTS-Systeme erfolgt oder bei der ohne eine 3-schichtige Besetzung der Qualitätskontrolle (z. B. personalreduzierte Nacht- oder Wochenendschichten) gefertigt wird (Bild 4). Hier übernimmt die Prozeßüberwachung die 100 % Kontrolle, separiert Ausschußteile und schaltet gegebenenfalls automatisch die Maschinen ab.

Die Verknüpfung der Prozeßüberwachung mit der Produktionssteuerung setzt bereits bei Auftragsbeginn ein. Anhand der im organisatorischen Teil der Folgeaufträge festgelegten artikel-, werkzeug-, maschinen- und materialspezifischen Datenfelder werden einmal als gut erkannte Grenzwerte der Prozeßgrößen sowie die maschinenspezifischen Eingriffsgrenzen auftragsbezogen vom BDE-Leitrechner zu den Gruppenterminals in die Fertigung übertragen. Die Prozeßsollwerte werden dabei in einer hierarchisch strukturierten Datenbank auf dem UNIX-Rechner hinterlegt. Die Dokumentation der verdichteten Prozeßdaten erfolgt dabei ebenfalls auftrags- bzw. chargenbezogen auf dem BDE-Leitrechner.

Bei modernen NC-gesteuerten Produktionsmaschinen können Prozeßdaten über eine serielle Schnittstelle ausgelesen werden, so daß keine zusätzlichen Meßwertaufnehmer zu installieren sind. Zudem kann die Datenaufbereitung in der Maschinensteuerung erfolgen.

Eine universelle von den Maschinenherstellern sowie den unterschiedlichen Steuerungstypen weitgehend unabhängige Lösung ist die Verarbeitung von Analogwerten im Gruppenterminal. Die Analogsignale der in jeder Maschinensteuerung bereits integrierten Meßverstärker sind herauszuführen und an das Gruppenterminal anzuschließen. Die Aufbereitung und Verarbeitung der Meßsignale erfolgt dann einheitlich im Gruppenterminal. Ein Auswechseln der Verarbeitungsmaschine bedingt keinerlei Änderung an der bestehenden Maschinensoftware.

Auch Maschinen älterer Bauart sind problemlos in dieses Konzept zu integrieren, sofern die notwendigen Meßverstärker nachgerüstet werden.

Die im Zuge der sich ändernden Marktverhältnisse entstehenden erhöhten Ansprüche an die Qualität erfordern ein durchgängiges Konzept zur Qualitätssicherung. Es muß u. a. folgende Komponenten enthalten:

- Materialeingangsprüfung
- Überwachung der Fertigung (sowohl Einzelteilfertigung als auch Montage)
- Prüfungen am Fertigteil (Geometrie, Gewicht usw.)
- Zusammenführung der Qualitätsinformation, Auswertung.

Diese Module müssen bei einer flexiblen Fertigung mit eingebaut werden.
Die Durchführung von Materialeingangsprüfungen stellt derzeit, vorwiegend in
der Elastomerverarbeitung, immer noch ein Problem dar.
Mit zunehmender Compoundierung von Thermoplasten ist jedoch auch hier eine
Wareneingangskontrolle notwendig, um Störquellen für die nachfolgenden Prozesse
auszuschalten.
Die Überprüfung bei Gummi-Mischungen ist insofern notwendig, daß die Aufberei-
tung beim Verarbeiter selbst durchgeführt wird. Untersucht wird insbesondere
das Vulkanisationsverhalten und das Fließverhalten. Die Materialprüfung im
Vulkameter liefert Daten, die mittels Berechnungsprogrammen eine Vorausbestim-
stimmung artikelspezifischer Vulkanisationszeiten erlauben.
Zur Beschreibung des Fließverhaltens gibt es noch keine genormten Prüfmethoden
oder solche, die sich in der Praxis bereits durchgesetzt haben.
Qualitätssicherung durch Meßdatenerfassung und Auswertung während
der Produktion:
-    Überwachung qualitätsrelevanter Produktionsparameter durch die
     Spritzgießmaschine
-    Erfassung frei wählbarer Produktionsparameter mit teilweiser
     Auswertung durch die Spritzgießmaschine
-    Statistische Auswertung der ausgewählten Qualitätsparameter
-    Anschlußmöglichkeit für teilebezogene Kontrolle
     (Meßschieber, Waage, Taster)
-    Integrationsmöglichkeit in ein übergeordnetes CAQ-System
-    Produktionsdatenprotokoll.
Qualitätssicherung durch Trennung von Gut- und Schlechtteilen:
-    Selektiereinheit im Ausfallbereich der Spritzlinge
-    Toleranzüberwachung entscheidet über Stellung der Sortierklappe
-    In Anfahrzyklen gefertigte Teile werden nach Werkzeugwechsel
     automatisch aussortiert
-    Trennung von Anguß und Spritzlingen ist möglich.
Punkte, die die Qualitätssprüfung und die Qualitätsplanung sowie eventuell Ver-
bindung zur Prozeßüberwachung betreffen, werden vom IKV in einer Analyse heraus-
ausgearbeitet und als Ergänzungen bereitgestellt, dabei soll auch die FMEA-Methode zum Einsatz kommen.
Neue gesetzliche Auflagen und gesteigerte Erwartungen der Anwender erfordern
immer größere Anstrengungen, zuverlässige und sichere Produkte und Produktsysteme
herzustellen. In diesem Zusammenhang findet die FMEA-Methode, das ist die Abkür-
zung für Failure Mode Effects Analysis, immer größere Beachtung. Dieses Verfahren
zur Risikovermeidung und Kostenminderung wird im deutschen Sprachgebrauch "Feh-
lermöglichkeits- und Einfluß-Analyse" genannt. FMEA geht davon aus, daß es weitaus wirtschaftlicher ist, potentielle Fehler zu vermeiden, als aufgetretene
Fehler zu beseitigen, und findet als Methode zur frühzeitigen Schwachstellen-

aufdeckung und -vermeidung in Entwicklung, Konstruktion, Planung und Ferti-
gungsvorbereitung Anwendung.

5. OFFENE SYSTEME UND OFFENE KOMMUNIKATION

Die Unternehmen sind heute gefordert, eine einheitliche Lösung für die Informationsverarbeitung in allen Bereichen zu installieren. Neben den zahlreichen Hard-
ware- und Softwarelösungen spielen daher zunehmend die Integrationsfähigkeiten
von Rechnersystemen in heterogenen Systemlandschaften eine wichtige Rolle. Ein
Hineinwachsen in bestehende Umgebungen ist begleitet von Forderungen noch kon-
tinuierlich steigender Funktionalität und steigendem Leistungsvermögen.
Die wohl wichtigsten Anforderungen des Marktes an UNIX-basierende Systeme lassen
sich durch die folgenden 5 Punkte beschreiben:
-    Breites Leistungsangebot
-    Hohe Funktionalität
-    Vielfältige Ausbau- und Wachstumsmöglichkeiten
-    Integration und Vernetzung
-    Konformität mit Standards.
Der Ruf der DV-Anwender nach offenen Systemen ist unüberhörbar. Der Kunde möchte
seine Investitionen in Soft- und Hardware geschützt sehen. Die Portabilität
seiner Anwendungen wird durch die Forderung nach Standards erreicht.
Heutige DV-Anwender verlangen Arbeitsplatzsysteme mit hoher Leistung. Funktiona-
lität und definiertem Antwortzeitverhalten. Möglichkeiten zum Wachstum mit
breitem Einsatzspektrum für Ein- und Mehrplatzsysteme ist ebenso eine wichtige
Forderung wie die insbesondere gerade dabei wichtige Möglichkeit, in gewohnten
Umgebungen mit bekannten und benutzerfreundlichen Hilfsmitteln zu arbeiten.
Unternehmen wachsen mit der zur Verfügung stehenden Datenverarbeitung zunehmend
in heterogene Systemlandschaften. Die Idee der Unternehmenskommunikation gene-
riert einen hohen Bedarf an Vernetzung und Integration.
Betriebssysteme mit Zukunft sind solche, die die Forderungen für Offene Systeme
erfüllen.
Eine Antwort auf diese Anforderungen im Markt der Offenen Systeme ist z. B. das
IBM RISC System/6000 und der standard-konformen UNIX-basierenden Betriebssystem-
Software AIX Version 3 für RISC System/6000.
Dabei wird nicht nur auf Standards, wie sie von offiziellen Standardisierungs-
gremien (IEEE, ANSI, ISO) erstellt wurden, sondern auch auf Standards von Or-
ganisationen wie zum Beispiel der X/Open und der OSF Rücksicht genommen.
IBM AIX Version 3 erfüllt folgende Standards:
-    IEEE POSIX 1003.1 (Portable Operating System for Computer Environments)
-    X/Open Common Application Environment (CAE); ab dem 3ten Quartal 1990
     werden IBM AIX Systeme ausgeliefert, die dem X/Open Portability Guide,
     Ausgabe 3, entsprechen.

- OSF Application Environment Specifikationen (AES) Level 0
- Konformität zu AT&T System V.2
- Konformität zu BSD 4.3.

Das Anwendungspaket zur Qualitätssicherung für Spritzgießbetriebe wird auf dem IBM RISC System/6000 unter AIX V.3 verfügbar sein. Bei der Erstellung wird objektorientierte Programmierung mit C++ angewandt. Die Echtzeitfähigkeit des AIX wird entweder direkt oder über Vormultiprozessor für die Prozeßüberwachung eingesetzt. Das neue Dateisystem bietet erhöhte Integrität und durch Spiegeldateien erhöhte Datensicherheit. Die Stärke der hohen Rechengeschwindigkeit kann für Varianten- und Modellberechnungen herangezogen werden. Ein noch sehr wesentlicher Punkt ist die Einbindung der Anwendung in heterogene Systemlandschaften und gerade dazu werden eine große Anzahl von Kommunikationsmöglichkeiten, wie sie AIX V.3 bietet, notwendig. Die Anschlüsse zu allen möglichen Industriestandards, wie Ethernet, V.24, Tokenring und SDLC müssen unterstützt werden. Protokolle wie TCP/IP und ein Netzwerkmanagement SNMP müssen vorhanden sein.

Das Rückgrat jeglicher Automatisierung in unterschiedlichster Systemumgebung ist die Kommunikationstechnik, die von UNIX hervorragend unterstützt wird.

Die anwenderseitig immer stärker vertretene Forderung nach Offenheit der Rechnersysteme bzgl. Hard- und Software und nach ihrer Intregration in heterogenen Netzen hat nachhaltig den UNIX Markt geprägt und ist Richtschnur für Weiterentwicklungen. Daß gerade UNIX als das Betriebssystem gesehen wird, dieser Forderung am besten nachkommen zu können, dokumentiert sich weltweit in den konzertierten Aktivitäten namhafter HW/SW-Hersteller und Anwender und die Zusammenarbeit mit nationalen und internationalen Normungsgremien.

Eine Zielsetzung dieser Bestrebungen ist die Vernetzung von Systemen in einem Verbund, die dem Anwender, unabhängig von der eingesetzten Hardware, den Zugriff auf verteilte Dateien, Programme, Rechnerleistung und Peripherie ermöglichen soll. Diese Funktionen werden mit dem umfangreichen Kommunikationspaket der IBM RISC System/6000 Familie unter Berücksichtigung offizieller Normen als auch etablierter Industriestandards in hohem Maße erfüllt.

In UNIX-Umgebungen steht das TCP/IP Communication Subsystem zur Verfügung. TCP/IP ist ein Bündel von Protokollen, die in technischen und kommerziellen Umgebungen weit verbreitet sind. Sie bieten elementare Funktionen wie Dateitransfer, Remote Login und elektronische Post zwischen unterschiedlichen Rechnern in lokalen und Weitverkehrsnetzen.

Vor dem Hintergrund des Gesamt DV-Marktes verlangt die Anwenderforderung nach Offenheit und Integration die Fähigkeit von UNIX Systemen, sich auch im Rahmen der außerhalb UNIX gegebenen Standards an andere DV-Welten, z. B. SAA anschließen zu können.

Verteilten Datenzugriff bietet das Network File System NFS von SUN Microsystems Inc.. Es ermöglicht dem Benutzer, einen transparenten Zugriff auf entfernte Dateien, d. h. er sieht die Daten des entfernten Systems, als wären sie lokal

gespeichert. NFS läuft auf LAN Netzen (Ethernet, Token Ring) und setzt das
Netzwerkprotokoll TCP/IP voraus.
Mit Network Computing System NCS von Apollo Domain, das wie NFS auf TCP/IP ba-
siert, kann der Zugriff auf entfernte Rechnerleistung und Peripherie realisiert
werden. So können unterschiedliche Rechner innerhalb eines LAN im Rahmen einer
Arbeits- und Lastverteilung speziell nach ihren Funktionen, z. B. Vektorrechner,
Datenbanken, eingesetzt werden.
Unter TCP/IP verbirgt sich ein Bündel von Protokollen und Funktionen für die
Kommunikation zwischen heterogenen Systemen in lokalen und Weitverkehrsnetzen.
Als heterogen werden Systeme bezeichnet, wenn sie unterschiedliche Hersteller,
Architekturen oder Betriebssysteme haben.
TCP/IP ermöglicht elementare Funktionen zwischen unterschiedlichen Rechnern
und zwischen Netzen wie z. B. Filetransfer, Remote Login, Electronic Mail.
TCP/IP ist damit Basis für eine offene Systemkommunikation (Bild 5).
In der Verbundverarbeitung innerhalb lokaler Netze mit dem Protokoll TCP/IP
haben sich für verteilte Dateisysteme NFS, für verteilte Rechnerleistung NCS
und für Netzwerkmanagement SNMP (Simple Network Management Protocol) als Indu-
striestandard durchgesetzt und sind im AIX V.3 enthalten.

6.   ERFAHRUNGEN UND ERWEITERUNGEN

Sowohl bei der Firma Burger als auch bei Geberit ist das Basispaket PROQUAM
seit mehreren Jahren in Verbindung mit Betriebsdatenerfassung (BDE) im Einsatz.
Mittels CAQ konnten beide Firmen ihr Image als Qualitätslieferant weiter verbessern und die Durchlaufzeit sowie die Losgrößen drastisch verringern.
Sieht man das Leitsystem zur Qualitätssicherung für Spritzgießbetriebe als Kern-
baustein für ein Fertigungsleitsystem an, welches Module für BDE, MDE, DNC, CAM,
Prozeßregelung, Materialver- und -entsorgung, Instandhaltung sowie Schnittstellen zu vorgelagerten CAD- und PPS-Systemen enthält, so ist damit ein integriertes
Werkstattsystem vorhanden, welches alle Vorkommnisse in der Produktion überwacht
und steuert (Bild 6). Als weiterer zukunftsträchtiger Zusatz des CAQ-Systems
wäre eine Bildauswertung zur Erkennung von Fehlern und Schwachstellen anzusehen.
Dieses Qualitätssicherungssystem versetzt die Spritzgießbetriebe gegenüber dem
Wettbewerb in eine gute Ausgangsposition, um kontrollierte Qualität zu produzieren und nicht im Nachhinein zu erprüfen und damit den Ausschuß drastisch
zu senken. Ebenso ist das CAQ-System als Kernzelle für CIM bei Spritzgießbe-
trieben zu sehen. Durch Verwendung von UNIX ist die Offenheit des Systems ge-
wahrt, die Schnittstellen sind bekannt und die Einbindung in eine vorhandene
heterogene Systemlandschaft ist ohne große Schwierigkeiten möglich, um ein integriertes System aufzubauen. CAQ berührt alle Bereiche eines Unternehmens, wie
z. B. Wareneingang, Fertigung, Warenausgang, Konstruktion usw. und CAQ ist daher
hervorragend für den Einstieg in die integrierte Fertigung geeignet; denn mit
CAQ kann man am raschesten einen Nutzen feststellen.

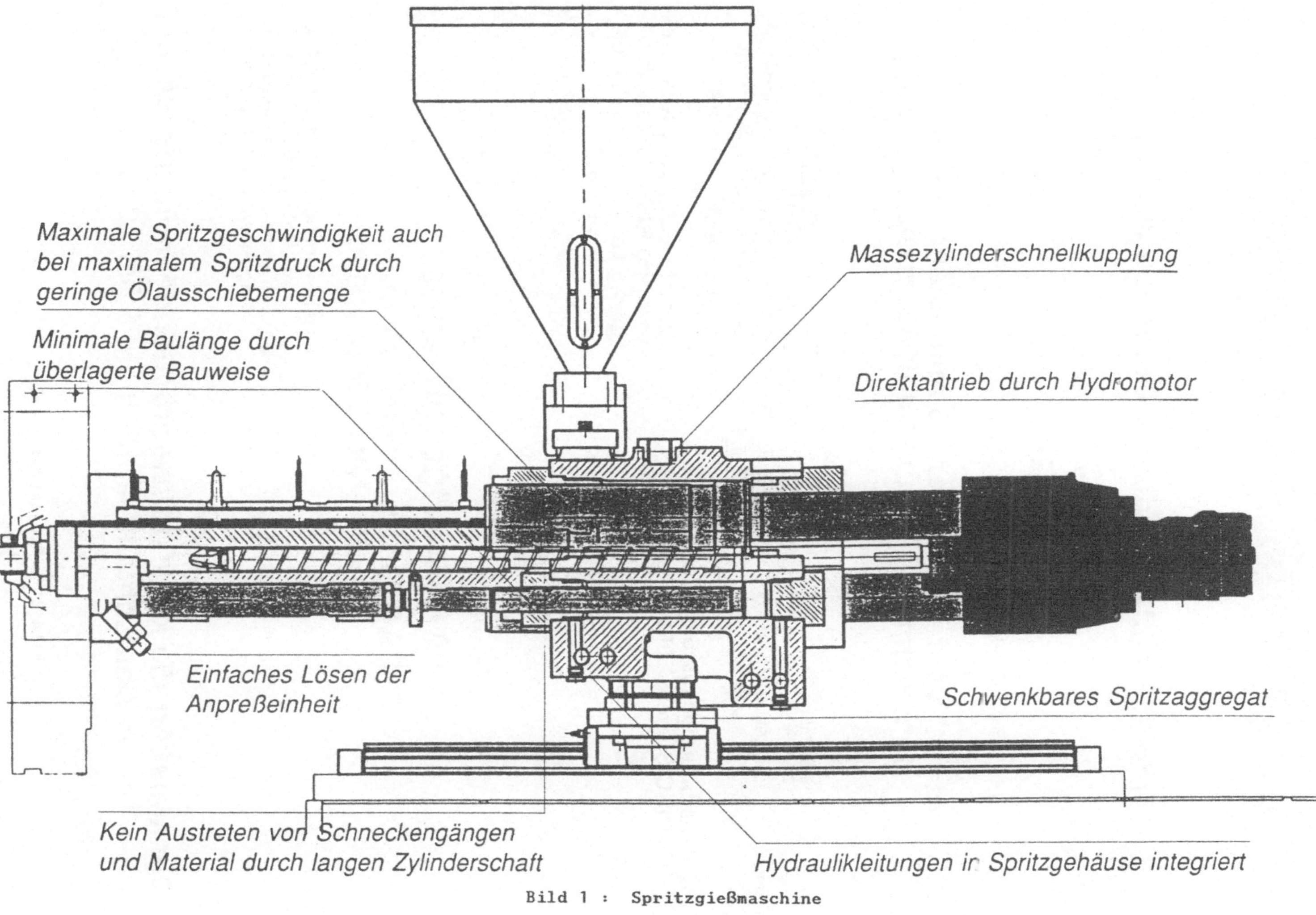

Bild 1 : Spritzgießmaschine

Land Nordrhein–Westfalen (NRW)

fördert 15 mittelständische Unternehmen über

(Aachener Gesellschaft für Innovation und Technologietransfer mbH)

macht Verträge mit den 15 mittelständischen Unternehmen für Kunststoffverarbeitung

(Institut für Kunststoffverarbeitung der RWTH Aachen)

(Entwicklungs– und Vertriebsgesellschaft für Produktionsdatenerfassungsanlagen mbH)

1 x Ausbildungssystem IBM RISC System/6000

1 x Entwicklungssystem und 15 x IBM RISC System/6000 für die ausgewählten Unternehmen:

Anwendungspaket
CAQ für Spritzgießbetriebe
Basis PROQUAM

Referenzinstallationen unter AIX:
Burger, Geberit, GF, Kautex,
BASF, Kabelwerke–Rheinshagen,
King Plastik, WOCO, usw.

Der Markt für die kunststoffverarbeitende Industrie wächst in Deutschland jährlich um 8 bis 10 %

Bild 2 : Projektübersicht

dfkg99

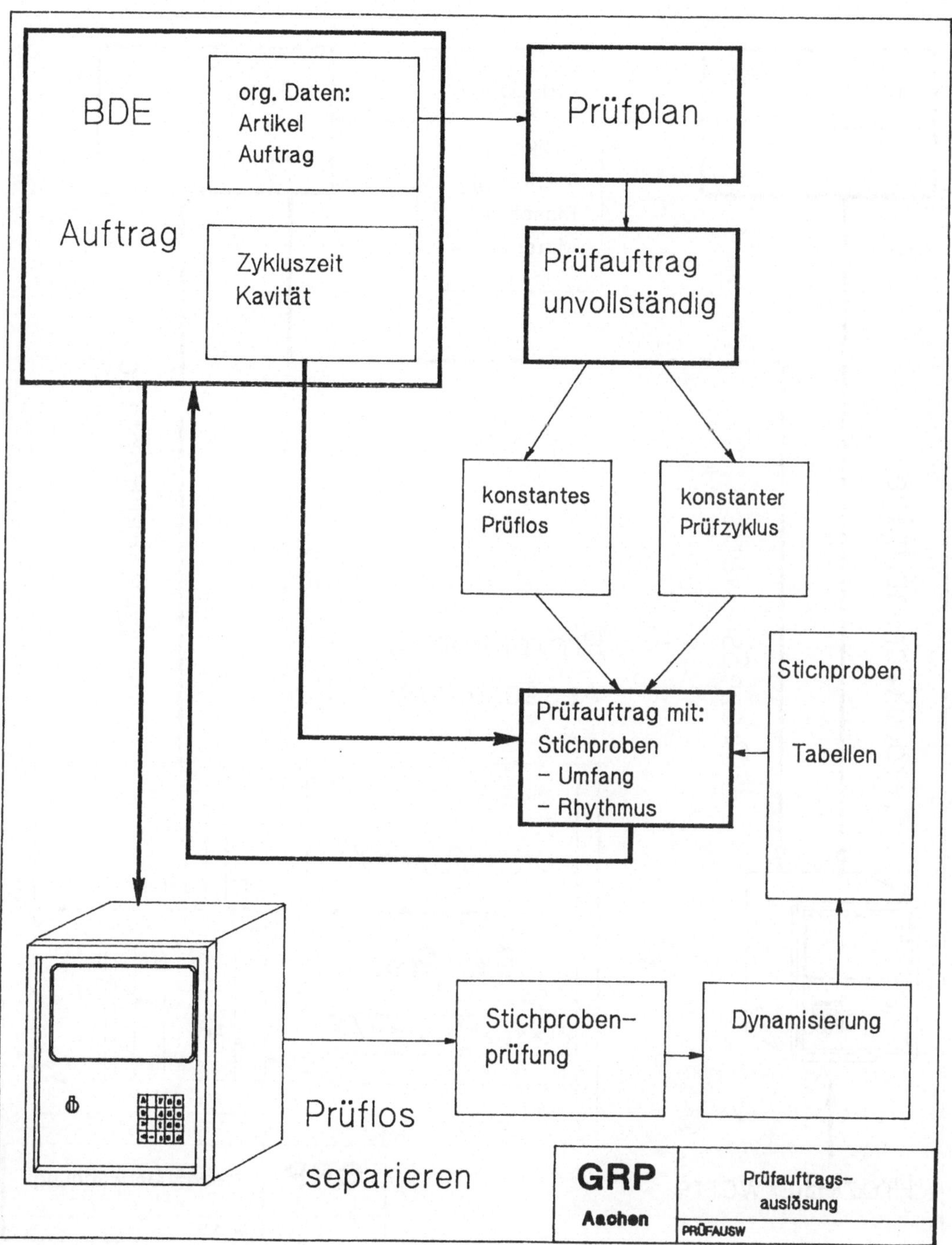

Bild 3 : Prüfauftragsauslösung

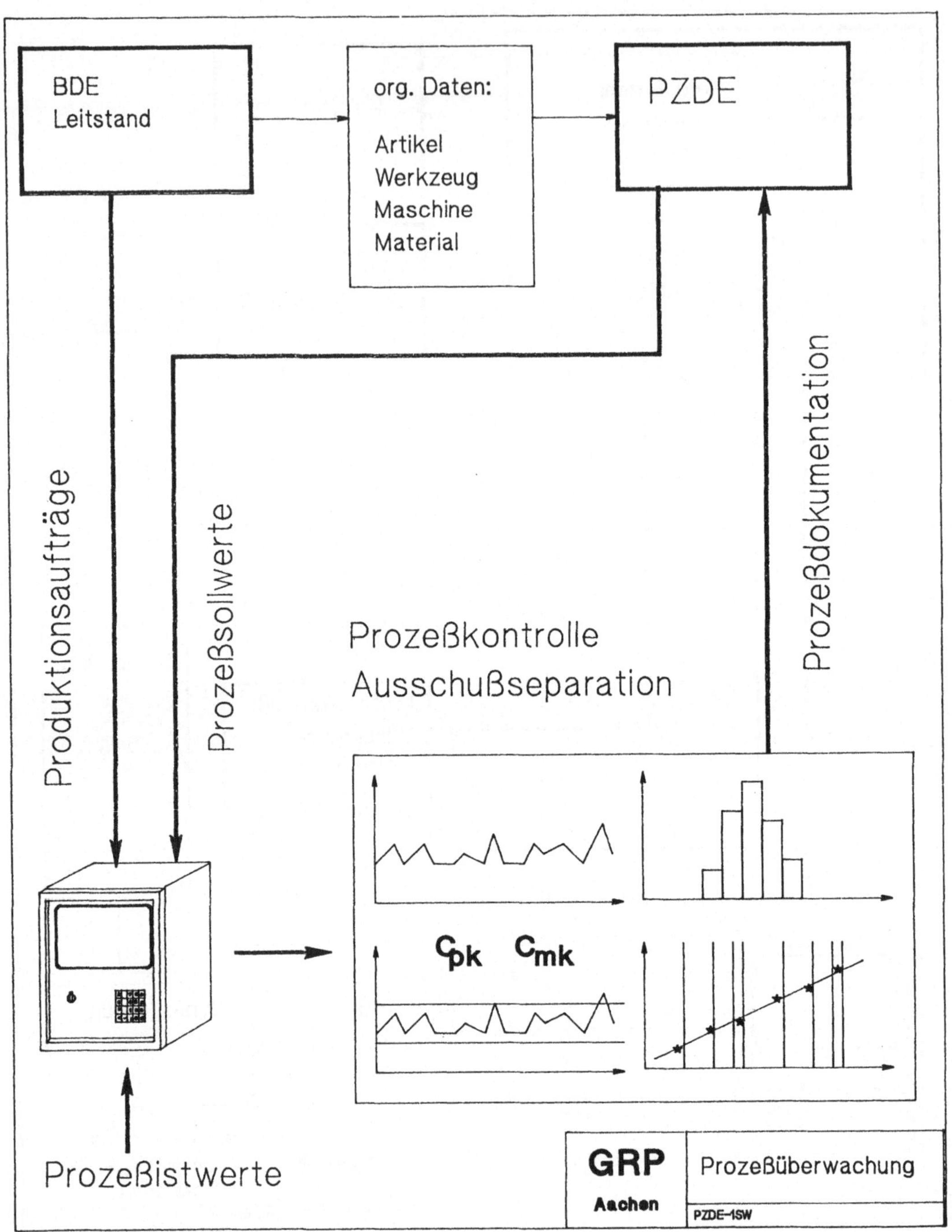

Bild 4 : Prozeßüberwachung

IBM AIX Version 3 für RISC System/6000

IBM

TCP/IP Funktionen

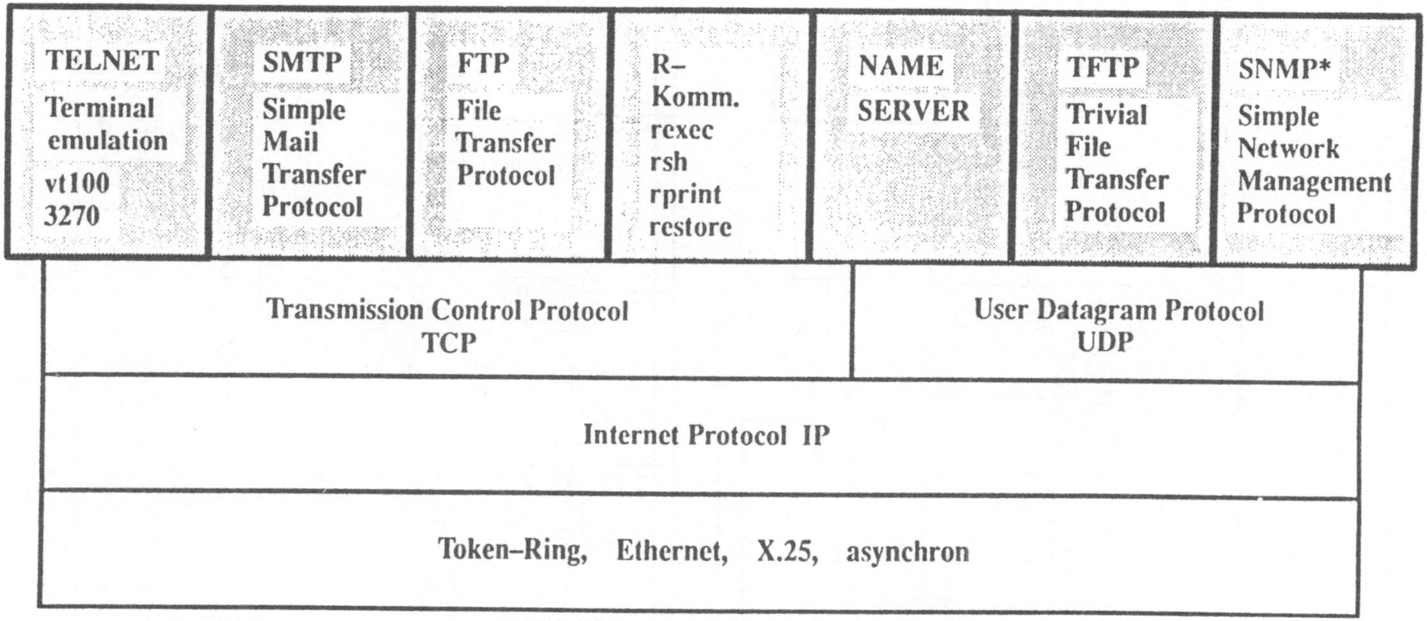

TELNET	SMTP	FTP	R–	NAME	TFTP	SNMP*
Terminal emulation vt100 3270	Simple Mail Transfer Protocol	File Transfer Protocol	Komm. rexec rsh rprint restore	SERVER	Trivial File Transfer Protocol	Simple Network Management Protocol

Transmission Control Protocol TCP	User Datagram Protocol UDP

Internet Protocol IP

Token–Ring, Ethernet, X.25, asynchron

▶ *Gebrauchsfertige Funktionen auf Anwenderebene*

* in IBM AIX Network Management/6000

Bild 5 : TCP/IP Funktionen

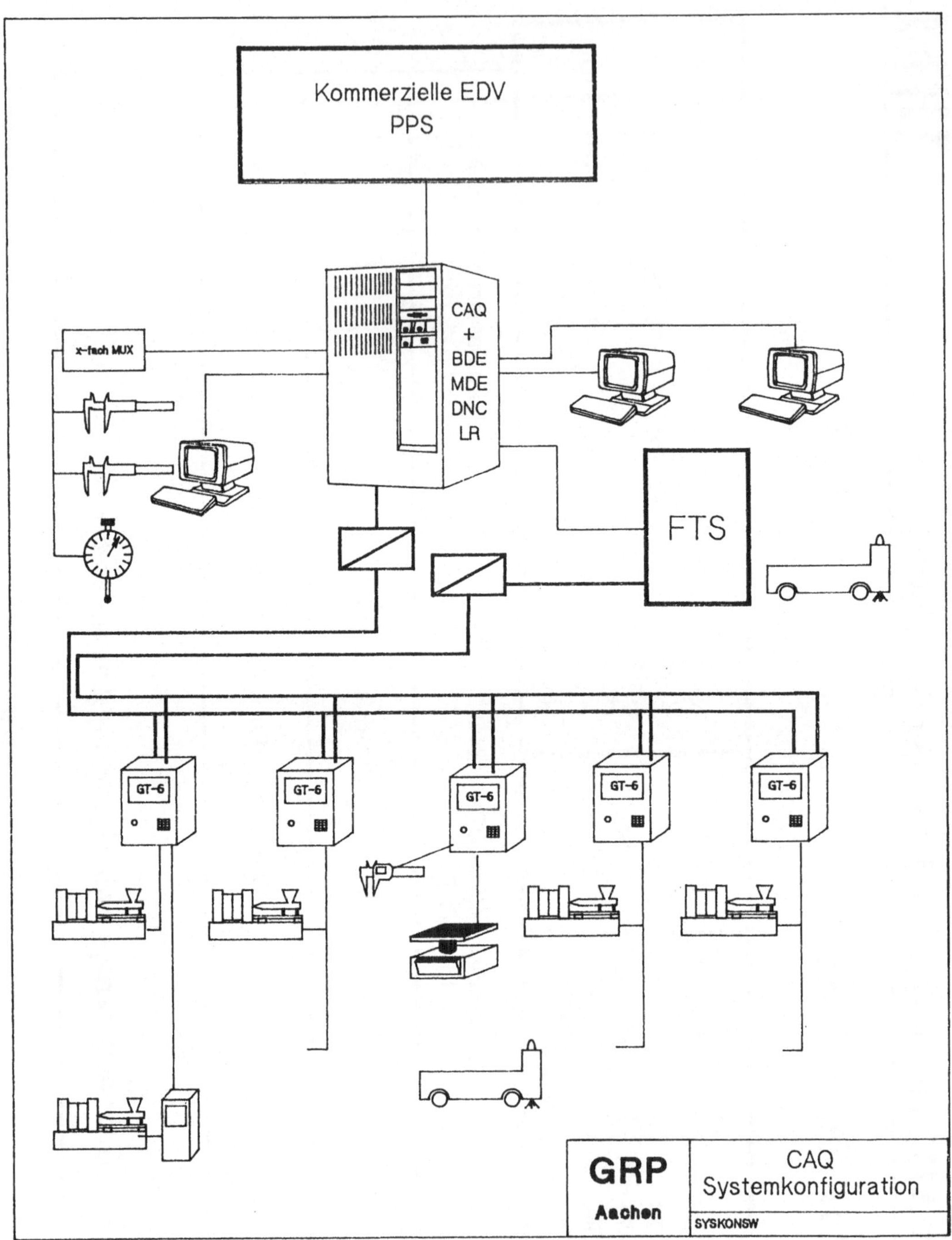

Bild 6 : CAQ-Leitsystem

Erfahrungen bei der Anwendung von Echtzeitexpertensystemen für die Diagnose und Steuerung

Balzer, D.; Kirbach, V.; May,V.

Technische Hochschule Leipzig
Fachbereich Automatisierung
Institut für Prozeßführung

Abstracts

This paper presents a knowledge based system of automation to be applied to diagnosis and process safety tasks in industrial applications. It shows how an efficient low cost real-time expert system for supervision/failure detection can be built. This expert system, which runs on a PC, ist embedded in a conventionally programmed process control system. The requirement and features of the real-time expert system and the resulting hybrid software system as well as the performance of the integrated system are discussed and critically evaluated.

1. Einleitung

Künstliche Intelligenz (KI) bedeutet Nachbildung des menschlichen Denkprozesses. Auf dem Gebiet der Expertensysteme, einem Teilgebiet der KI, wird an der Technischen Hochschule Leipzig seit einigen Jahren theoretisch und praktisch gearbeitet, vor allem bezogen auf Anwendungen in der Automatisierungstechnik. Dabei wurde darauf orientiert, Expertensysteme zu entwickeln, die die Lösung zweier Kategorien von Ingenieuraufgaben unterstützen (Tabelle 1).

Tabelle 1:   Zwei Kategorien von Ingenieuraufgaben

	Technische Vorbereitung	Betreiben
Aufgaben des Ingenieurs	Planung (der Investition), Entwurf und Konstruktion, Projektierung, technologische Vorbereitung	Planung (der Produktion), Steuerung, Regelung, Lenkung, Leitung, Wartung, Instandhaltung
Aufgaben für Expertensysteme	Bestimmung der optimalen Lösung ohne Zeitbeschränkung (Erzeugung von Software- und Hardware-Projekten)	Bestimmung einer zulässigen Lösung mit Zeitbeschränkung (Erzeugung von Steuerungen)

Für den praktischen Einsatz der Expertensystem-Technologie sind folgende Gesichtspunkte entscheidend:

1. Das Wissen und die Erfahrung von Fachleuten müssen umfassend genutzt werden,
2. Die Mensch-Maschine-Kommunikation (Mensch-Rechner-Kommunikation) sollte weitgehend in natürlicher Sprache erfolgen,
3. Die Möglichkeiten zur Verbesserung Software-Technologie sind konsequent zu nutzen.

Von besonderer Wichtigkeit ist der Punkt 3. Die Vorteile beim Einsatz von Expertensystemen liegen in dieser Hinsicht vor allem darin, daß eine bedeutende Erhöhung der Qualität der Software in Hinsicht auf folgende Qualitätsmerkmale erreicht werden kann: Portabilität, Zuverlässigkeit, Änderbarkeit, Nutzerfreundlichkeit, Wiederverwendbarkeit. Unsere Erfahrungen zeigen z.B., daß eine Erhöhung der Zuverlässigkeit der Software um den Faktor 10 erreicht werden kann und daß die Produktivität der Software-Technologie durch den Einsatz von Expertensystemen um den Faktor 2-3 verbessert werden kann.

2. Anforderungen an ein Expertensystem zur Steuerung von Produktionsprozessen

Expertensysteme können eingesetzt werden sowohl im geschlossenen Wirkungskreis (sogenannten Expertenregler) als auch zur Unterstützung des Anlagenpersonals bei der Steuerung komplexer Systeme (Prozeßsteuerung), d.h. zur Prozeßoptimierung, -stabilisierung und -sicherung (einschließlich Qualitätssicherung) [1]-[3]. Bei der Prozeßsteuerung gewinnen folgende Probleme an Bedeutung:
1. Lösung von Automatisierungs- und Steuerungsaufgaben höherer Dimensionen (über 5 frei wählbare Steuergrößen).
2. Dekomposition und Koordinierung bei der Steuerung großer Systeme.
3. Einhaltung komplizierter Nebenbedingungen (komplizierte mathematische Modelle, eine Vielzahl logischer Bedingungen, usw.).

Mit den bekannten mathematischen Methoden (z.B. Mehrebenenoptimierung, Pontrjaginprinzip, Stabilitätstheorie) sind diese Probleme im Echtzeitregime nicht oder nur schwer lösbar. Eine kombinierte Anwendung der Expertensystemtechnologie und der "klassischen" mathematischen Methoden ist zu empfehlen.
Unter Berücksichtigung der bisherigen Erfahrungen auf den Gebieten der Prozeßsteuerung und der Expertensysteme muß ein Expertensystem-Shell für die Lösung von Aufgaben der Prozeßautomatisierung folgende Besonderheiten besitzen [4], [5]:

1. Durch eine Trennung von Diagnose und Therapie (Synthese der Steuerung) sowie durch die Anwendung einer speziellen Hypothese- und Teststrategie kann der Inferenzmechanismus extrem beschleunigt werden. Dadurch kann die Echtzeitfähigkeit gewährleistet werden.
2. Die Diagnose des Prozeßzustandes erfolgt durch eine Verarbeitung von Meßdaten, die durch automatische Meßwerterfassung gewonnen werden (On-line-Kopplung zwischen Expertensystem und technologischem Prozeß).
3. Algebraische Algorithmen (z.B. zur numerischen Simulation die Optimierung) müssen über spezielle Schnittstellen problemlos in das Expertensystem integriert werden können.

3. Architektur des Expertensystem-Shells und Repräsentationskonzept

Bei Beachtung der oben formulierten Besonderheiten bzw. Anforderungen an das Echtzeitexpertensystem-Shell wurde PROCON als ein universelles, in sich geschlossenes Softwarewerkzeug zur Implementation, Verifikation und Nutzung wissensbasierter Systeme entwickelt. Es wurde in LISP implementiert und unterstützt

- ein modernes, objektorientiertes Konzept zur expliziten Strukturierung von Wissen,
- verschiedene, kooperative Problemlösungsprozesse für Diagnose, Expertise und Prozeßkommunikation,
- eine komfortable Akquisition zur problemnahen Wissenseingabe und Übersetzung in eine effiziente Repräsentationsform,
- eine Konsultation mit einer Dialog- sowie separaten Trace- und Erklärungskomponente zur Verifikation und Konsultation implementierter Wissensbasen (vgl. auch [6] und [7]).

Alle Komponenten ordnen sich in ein streng modulares Schalenkonzept mit definierten Software-schnittstellen ein (siehe Bild 1).

Dem Repräsentations-konzept liegt das Konzept zur Wissens-modellierung zugrunde (siehe Bild 2 und Bild 3), wobei die Knoten in einem Situations- bzw. Ereignisbaum als Schema bezeichnet werden. Der Kernrepräsenta-tion von Schemata liegt ein allgemeiner, flexibler, objektorientierter Ansatz zugrunde, der in sich deklarative Aspekte (Prototypbe-

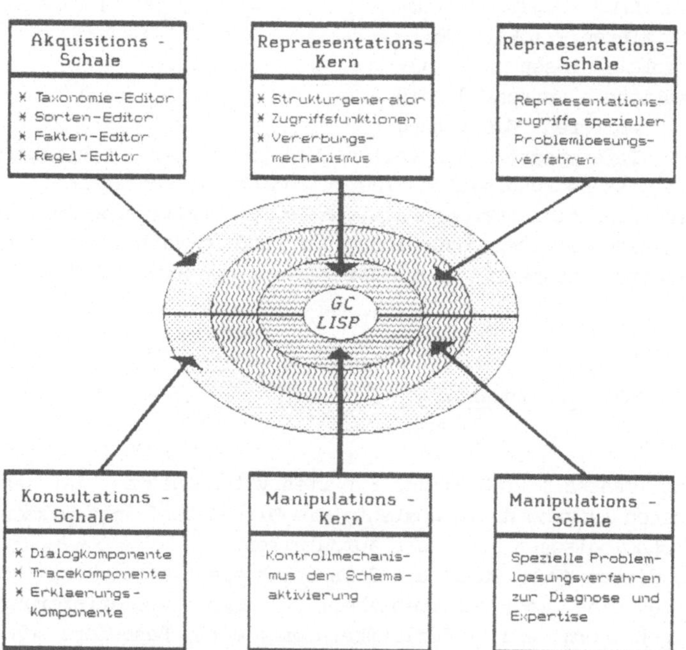

Bild 1   Schalenkonzept des Shells PROCON

schreibung, Instantiierung, Vererbung) framebasierter Repräsentationssprachen mit funktionalen Aspekten (Einkapselung, Nachrichtenaustausch) objektorientierter Programmiersprachen vereint und folgende drei Grundprinzipien verkörpert:

- Jedes Schema ist Träger sämtlichen deklarativen (Fakten), funktionalen (Funktionen) und prozeduralen (Regeln) Wissens einer prototypischen Problemklasse (Steuer-, Zustands- oder Vorhersageklasse).
- Schemata bilden eine Problemhierarchie, in der ein spezieller Vererbungsmechanismus die Weitergabe deklarativen, funktionalen und prozeduralen Wissens innerhalb ihres Gültigkeitsbereiches, d.h. von höheren zu niederen Abstraktionsebenen entlang der Taxonomiezweige sicherstellt.

- Jedes Schema ist Träger lokaler Problemlösungsprozesse, die über Nachrichten aktivierbar sind und auf das gesamte Wissen der Vererbungshierarchie zugreifen können.

Schemata stellen damit aktive, in sich geschlossene Datenobjekte zur Repräsentation und Manipulation von Wissen dar, die über Nachrichten mit ihrer Umgebung kommunizieren, lokale Teilprobleme lösen und maßgebliche Entscheidungen zur weiteren Gestaltung des Inferenzprozesses fällen. Von besonderer Wichtigkeit für den Anwender ist die Wissensorganisation (Auffüllen der Wissensbasis).

Die Akquisitionskomponente von PROCON stellt eine in sich geschlossene, problemnahe Entwicklungsumgebung dar und umfaßt interaktive Werkzeuge zur Wissenseingabe, -inspektion, -transformation und -verifikation.

Inferenz-abschnitt	Ziel des Abschnittes	Modell der Schemata	Beispieltaxonomie einer Wissensbasis
SITUATION	Diagnose und Expertise des aktuellen Prozeßzustandes	prototypische Zustandsklassen	Wurzel-Schema / Schema 1
AKTION	Diagnose und Expertise der resultierenden Therapiesteuerung	prototypische Steuerklassen	Schema 2
PREDIKTION	Diagnose und Expertise zu erwartender Auswirkungen der Therapiesteuerung	prototypische Vorhersageklassen	Schema 3

Bild 2: Wissensmodellierung

4. Praktische Anwendungen

Das zentrale Problem des Expertensystemeinsatzes ist das Wissensingenieurwesen, dessen Aufgabe darin besteht, die Wissensbasis mit Wissen über die Steuerung des technologischen Prozesses aufzufüllen.

Zu diesem Zweck wurde ein 3-Phasen-Konzept entwickelt (Bild 4), dessen Realisierung eine enge Zusammenarbeit zwischen Automatisierungstechnikern, Verfahrenstechnikern und Informatikern erfordert. Besonders schwierig sind die ersten beiden Phasen, wo es auf die Formulierung der Steuerungsaufgabe sowie auf die Gewinnung und die Strukturierung des Wissens ankommt. An dieser Stelle sollen allgemeine Gesichtspunkte für die Arbeit des Wissensingenieurs genannt werden, die teilweise in den vorangegangenen Abschnitten bereits angedeutet wurden:

- Trennung von Diagnose und Steuerung,

- kombinierte Anwendung von Oberflächen- und Tiefenwissen,

- Orientierung auf hierarchische Steuerungsstrukturen mit mehreren Koordinierungsebenen (Koordinierung von Teilsystemen durch heuristische Algorithmen, Steuerung der einzelnen Teilsysteme durch algebraische Algorithmen).

Die Einbindung von Expertensystemen in Real-Time-Umgebungen ist mit einer Reihe gravierender Probleme verbunden. So wird z.B. im System PROCON die Diagnose auf einem sog. "eingefrorenen" Prozeßzustand durchgeführt, der durch aktuelle vom Prozeßleitsystem übernommene Prozeßinformationen charakterisiert ist.

Im Zusammenhang mit der Echtzeitfähigkeit ist auch die dynamische Speicherverwaltung des Expertensystems von Bedeutung (Garbage Collection/GC). Eine vollständige Eliminierung des GC bzgl. des Laufzeitverhaltens ist generell nicht möglich. In PROCON wurde sein Einfluß jedoch minimiert, indem an zeitkri-

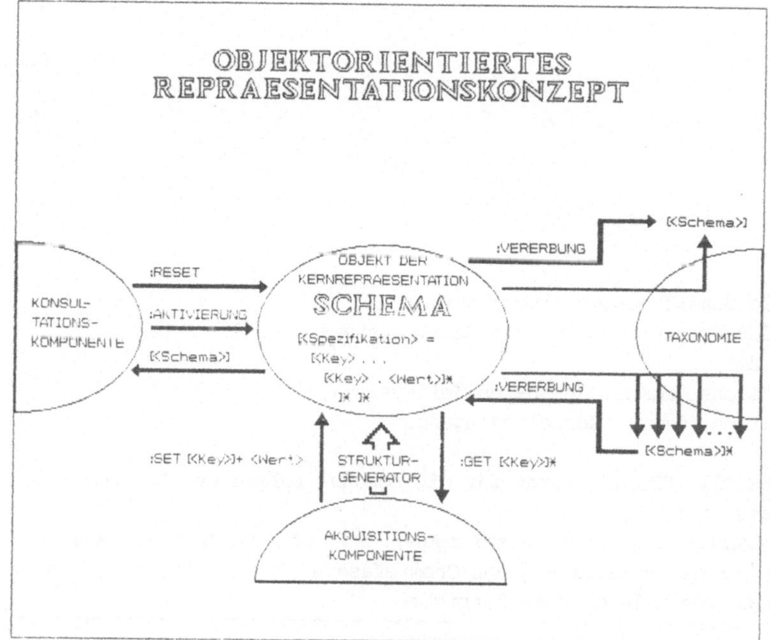

Bild 3: Konzept der Wissensrepraesentation

tischen Stellen (z.B. Zeitpunkt unmittelbar nach Ausgabe der 1. Lösungsspezifikation) softwaremäßig ausgelöst wird. Damit steht der nächsten Konsultation wieder der gesamte Speicherplatz zur Verfügung.

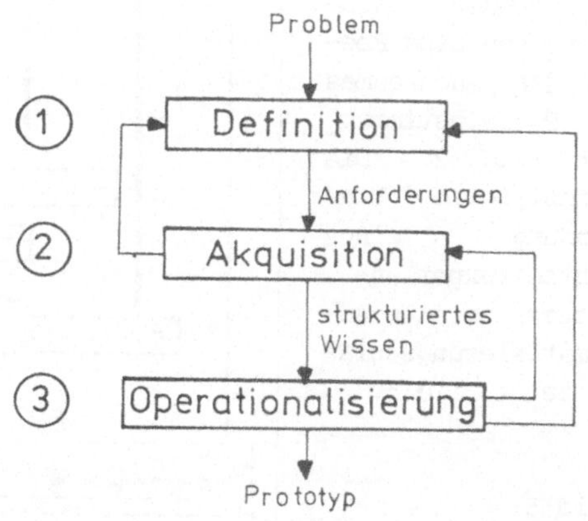

Bild 4: 3-Phasen-Konzept

Bei der Benutzung von Tiefenwissen gehen wir davon aus, daß das Gesamtsystem in N Teilsysteme zerlegt wurde. Für Aufgaben der Prozeßautomatisierung liegt das Tiefenwissen in Form von Lösungsalgorithmen für folgende Aufgabe vor:

$$Q(X) = \sum_{n=1}^{N} Q_n(X_n) \longrightarrow \min_{x} \tag{1}$$

$$f_n(X_n) = 0 \quad , \quad n = 1,2,\ldots,N \tag{2}$$

$$G(X) = \sum_{n=1}^{N} g_n(X_n) = 0 \tag{3}$$

Dabei wurden folgende Bezeichnungen eingeführt:

x, x_n – Vektor der Prozeßparameter des Gesamtsystems bzw. des n-ten Teilsystems,

$f_n, (X_n){=}0$ – mathematisches Modell des n-ten Teilsystems,

$Q(X)$ – Zielfunktion, $G(X)$ – Koppelbedingungen.

Das Expertensystem-Shell PROCON wurde für die Lösung folgender Probleme der Prozeßsteuerung eingesetzt:
- operative Fehlererkennung und -beseitigung bei Off-set-Druckprozessen
- Prozeßsicherung bei der Herstellung von Chemiefasern
- Prozeßführung eines metallurgischen Betriebes

Auf zwei der genannten Applikationen soll im weiteren näher eingegangen werden. Die Ebenen der entsprechenden Wissensbasen sind Ebenen des Suchbaumes in Übereinstimmung mit Bild 2. Die prinzipielle Einbindung eines Echtzeitexpertensystems in eine Automatisierungsanlage zeigt Bild 5.

Bild 5:
Integration von PROCON in eine Automatisierungsanlage

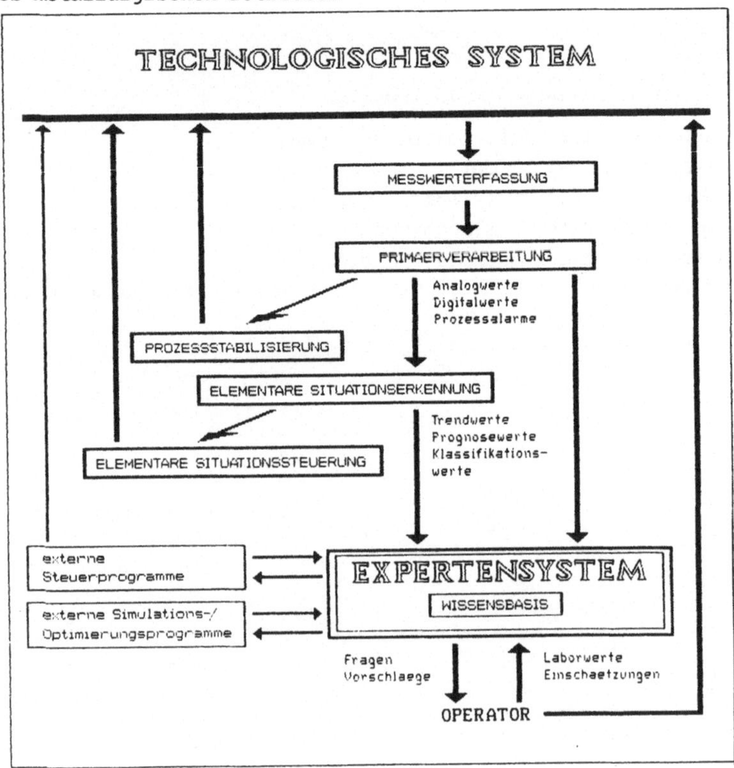

4.1. Fehlererkennung und -beseitigung bei Off-set-Durckprozessen

Die Zielstellung des intelligenten Diagnosesystems besteht in
- der Erkennung von Druckfehlern auf jedem Druckbogen,
- der Benennung von Druckfehlern durch Interpretation der Fehlerbilder in Abhängigkeit vom technologischen Hintergrund und
- der Ableitung von Steuerempfehlungen,

um eine gleichbleibend hohe Qualität der Druckerzeugnisse zu sichern und den Makulaturanteil zu senken.

Auf Grund bestehender Echtzeitforderungen wurde das Gesamtsystem in eine schnelle (konventionelle) Vorverarbeitungseinheit zur Bildaufnahme, Fehlerdedektion und Merkmalsextraktion sowie ein Expertensystem für die

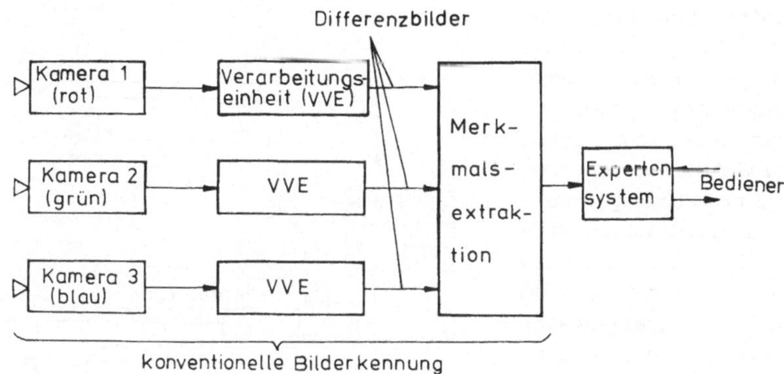

Bild 6: Expertensystem zur Diagnose von Druckfehlern

Fehleridentifikation und Ableitung einer Therapiesteuerung aufgespalten (vgl Bild 6). Die Vorverarbeitungseinheit ist als Spezialhardware ausgeführt.

Für die Realisierung der Prototypen wurden 16 Druckfehler, die am häufigsten auftretenden, determinierten und die stochastischen mit der gravierendsten Wirkung, ausgewählt. Die Fehlerdiagnose erfolgt ausschließlich anhand der extrahierten Bildmerkmale (ca. 25), während meßbare Umgebungsbedingungen (Temperatur, Luftfeuchtigkeit) und Ausgleichsvorgänge derzeit noch nicht berücksichtigt werden. Die prototypische Wissensbasis umfaßt ca. 25 Objekte in 3 Ebenen:
- Ebene 1: strategische Ebene
- Ebene 2: Identifikationsebene
- Ebene 3: Entscheidungsebene

4.2. Prozeßsicherung bei der Herstellung von Chemiefasern

Die zu steuernde Anlage ist der chemische Teil einer Polyesterfaser-Anlage. Hauptrohstoffe sind Terephtalsäure (PTA) und Ethylen-Glykol (EG). Weiterhin werden noch Antimontrioxid (SB_2O_3) als Katalysator und Titandioxid (TiO_2) als Mattierungsmittel eingesetzt (Bild 7).

Diese Komponenten werden in der ersten Prozeßstufe (Pastemischer) dosiert und homogenisiert. Es schließt sich eine zweistufige Veresterung (VE) an, gefolgt von einer dreistufigen Vorkondensation (VK). In der letzten Stufe, dem Ringscheiben-Reaktor (RSR), findet die Polykondensation des Produktes statt, das nun spinnfähig ist und, durch Lochdüsen zu Fäden gedrückt, versponnen wird.

Die zur Aufrechterhaltung des Prozesses, der endotherm verläuft, notwendige Wärmeenergie wird einem Wärmeträger (Diphyldampf) entnommen, der in einer Wärmeträger-Heizanlage (WTH) aufgeheizt wird. Außerdem benötigt man noch Vakuum für die Abführung von Kondensat sowie für die Einstellung der richtigen Reaktionsbedingungen. Dieser kontinuierliche Prozeß ist weitestgehend vollautomatisiert und wird über ein Prozeßleitsystem (PLS) TDC 3000 gesteuert.

Hauptziel der technologischen Prozeßführung ist es, unter gleichbleibenden Produktionsbedingungen ein spinnfähiges Produkt hoher Ausbeute mit gleichbleibend hohen Qualitätsparametern zu erzeugen.

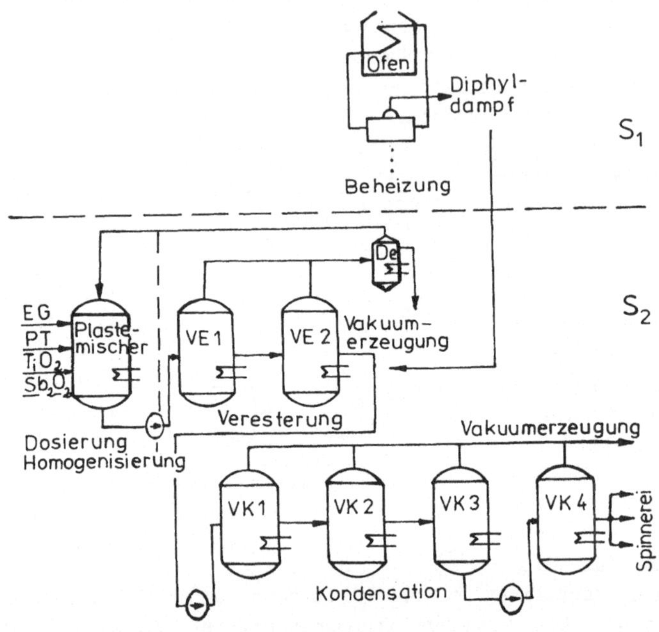

Bild 7: Prozeßstufen der zu steuernden Polyesterfaser-Anlage

Der Inhalt der Wissensbasis für den Fall, daß die WTH ausfällt, ist in Tabelle 2 dargestellt.

Tabelle 2: Struktur der Wissensbasis zu Abschn. 4.2.

Ebene	Inhalt der Wissensbasis	Ziel der Prozeßführung
0	Standardsituation (z.B. Ausfall WTH)	Diagnose
1	Wiederherstellung der Funktionstüchtigkeit des WTH	Steuerung
2	Zeitdauer des Ausfalls des WTH	
3	Verbrauch und Temperatur des Wärmeträgers, Temperatur des Reaktionsgemisches	Diagnose

5. Zusammenfassung

Die Anwendungen zur Therapiesteuerung der Polyesterfaser-Anlage, des fertigungstechnischen Prozesses einer Gießereitaktstraße und zur Maschinendiagnose bei Druckprozessen in Auswertung einer vorgeschalteten Bildverarbeitung haben bereits das Stadium einer Prototyplösung erreicht. Positiv hervorgehoben wird von allen Anwendern insbesondere die komfortable, problemnahe Entwicklungsumgebung sowie die hohe Laufzeiteffizienz des Problemlösungsprozesses (Konsultationszeiten von 1 bis 3 s), die PROCON gegenüber anderen ExpertensystemShells für die Lösung von Echtzeitaufgaben favorisiert.

Ausgehend von den bisherigen Erfahrungen mit dem Expertensystem-Shell PROCON sind folgende Weiterentwicklungen vorgesehen:

a) Schaffung von Möglichkeiten zur Erzeugung von expliziten zeitabhängigen Steuerfunktionen (implizit bereits heute möglich)

b) Kopplung mit Simulationssystemen ("Tiefenwissen")

c) Einbinden einer automatisierten Lernkomponente in das Expertensystem. Damit soll ein automatisiertes Lernen an Beispielen, durch Analogien oder an Erfahrungen möglich werden.

d) Weiterführung der methodischen Arbeit zum Wissensingenieurwesen durch Verallgemeinerung weiterer Applikationen.

Literatur

[1] May, V. (1989): Grundlagen und Entwicklung von Rahmenexpertensystemen für Aufgaben der Prozeßführung – Das Expertensystem-Shell PROCON, Dissertation A, TH Leipzig, 1989

[2] Balzer, D.; Kirbach, V.; May, V. (1989): A Real-Time Expert System for Process Control, AIRTC '89 2nd IFAC Workshop on Artificial Intelligence in Real-Time Control, Shenyang China 1989, Proceedings

[3] Voss, H.: Architectural Issues for Expert Systems in Real-Time Control. IFAC Workshop on Artificial Intelligence in Real-Time Control, Swansea, Sept. 1988, Preprints pp. 1–5

[4] Arzen, K.-E. (1989): Knowledge-Based Control Systems: Aspects on the Unification of Conventional Control Systems and Knowledge-Based Systems. Proc. 1989 American Control Conference, Pittsburgh, Vol. 3, pp 2233–2238

[5] Rosenhof, H. (1989): Real-Time Expert Systems in Process Control. EPRI Conference on Expert Systems Applications for the Electric Power Industry, Orlands

[6] Merris, D. (1989): Gensym Real-Time Expert Systems move into applications arena. Flexible Automation, Feb. 1989,4

[7] Früchtenicht, H.W.; Wittig, T.: Ein Ansatz für Echtzeit-Expertensysteme. Automatisierungstechnik 29 (1987), pp 78–82